INTRODUCTION
TO THE THEORY
OF STATISTICS

McGRAW-HILL SERIES IN PROBABILITY AND STATISTICS

DAVID BLACKWELL and HERBERT SOLOMON, *Consulting Editors*

BHARUCHA-REID: Elements of the Theory of Markov Processes and Their Applications
DRAKE: Fundamentals of Applied Probability Theory
EHRENFELD AND LITTAUER: Introduction to Statistical Methods
GIBBONS: Nonparametric Statistical Inference
GRAYBILL: Introduction to Linear Statistical Models, Volume I
JEFFREY: The Logic of Decision
LI: Introduction to Experimental Statistics
MILLER: Simultaneous Statistical Inference
MOOD, GRAYBILL, AND BOES: Introduction to the Theory of Statistics
MORRISON: Multivariate Statistical Methods
PEARCE: Biological Statistics: An Introduction
PFEIFFER: Concepts of Probability Theory
RAJ: The Design of Sample Surveys
RAJ: Sampling Theory
THOMASIAN: The Structure of Probability Theory with Applications
WADSWORTH AND BRYAN: Application of Probability and Random Variables
WASAN: Parametric Estimation
WEISS: Statistical Decision Theory
WOLF: Elements of Probability and Statistics

McGRAW-HILL
BOOK COMPANY
New York
St. Louis
San Francisco
Düsseldorf
Johannesburg
Kuala Lumpur
London
Mexico
Montreal
New Delhi
Panama
Rio de Janeiro
Singapore
Sydney
Toronto

ALEXANDER M. MOOD
Professor of Administration and
Director of Public Policy Research Organization
University of California, Irvine

FRANKLIN A. GRAYBILL
Department of Statistics
Colorado State University
Fort Collins, Colorado

DUANE C. BOES
Department of Statistics
Colorado State University
Fort Collins, Colorado

Introduction to the Theory of Statistics

THIRD EDITION

To HARRIET A.M.M.
To my GRANDCHILDREN F.A.G.
To JOAN, LISA, and KARIN D.C.B.

Library of Congress Cataloging in Publication Data

Mood, Alexander McFarlane, 1913-
Introduction to the theory of statistics.

(McGraw-Hill series in probability and statistics)
Bibliography: p.
1. Mathematical statistics. I. Graybill, Franklin A., joint author. II. Boes, Duane C., joint author. III. Title.
QA276.M67 1974 519.5 73-292
ISBN 0-07-042864-6

INTRODUCTION
TO THE THEORY
OF STATISTICS

234567890 KPKP 7987654

This book was set in Times Roman.
The editors were Brete C. Harrison and Madelaine Eichberg;
the cover was designed by Nicholas Krenitsky;
and the production supervisor was Ted Agrillo.
The drawings were done by Oxford Illustrators Limited.
The printer and binder was Kinsport Press, Inc.

CONTENTS

PREFACE TO THE THIRD EDITION

The purpose of the third edition of this book is to give a sound and self-contained (in the sense that the necessary probability theory is included) introduction to classical or mainstream statistical theory. It is not a statistical-methods-cookbook, nor a compendium of statistical theories, nor is it a mathematics book. The book is intended to be a textbook, aimed for use in the traditional full-year upper-division undergraduate course in probability and statistics, or for use as a text in a course designed for first-year graduate students. The latter course is often a "service course," offered to a variety of disciplines.

No previous course in probability or statistics is needed in order to study the book. The mathematical preparation required is the conventional full-year calculus course which includes series expansion, multiple integration, and partial differentiation. Linear algebra is not required. An attempt has been made to talk to the reader. Also, we have retained the approach of presenting the theory with some connection to practical problems. The book is not mathematically rigorous. Proofs, and even exact statements of results, are often not given. Instead, we have tried to impart a "feel" for the theory.

The book is designed to be used in either the quarter system or the semester system. In a quarter system, Chaps. I through V could be covered in the first quarter, Chaps. VI through part of VIII the second quarter, and the rest of the book the third quarter. In a semester system, Chaps. I through VI could be covered the first semester and the remaining chapters the second semester. Chapter VI is a "bridging" chapter; it can be considered to be a part of "probability" or a part of "statistics." Several sections or subsections can be omitted without disrupting the continuity of presentation. For example, any of the following could be omitted: Subsec. 4.5 of Chap. II; Subsecs., 2.6, 3.5, 4.2, and 4.3 of Chap. III; Subsec. 5.3 of Chap. VI; Subsecs. 2.3, 3.4, 4.3 and Secs. 6 through 9 of Chap. VII; Secs. 5 and 6 of Chap. VIII; Secs. 6 and 7 of Chap. IX; and all or part of Chaps. X and XI. Subsection 5.3 of Chap VI on extreme-value theory is somewhat more difficult than the rest of that chapter. In Chap. VII, Subsec. 7.1 on Bayes estimation can be taught without Subsec. 3.4 on loss and risk functions but Subsec. 7.2 cannot. Parts of Sec. 8 of Chap. VII utilize matrix notation. The many problems are intended to be essential for learning the material in the book. Some of the more difficult problems have been starred.

ALEXANDER M. MOOD
FRANKLIN A. GRAYBILL
DUANE C. BOES

EXCERPTS FROM THE FIRST AND SECOND EDITION PREFACES

This book developed from a set of notes which I prepared in 1945. At that time there was no modern text available specifically designed for beginning students of mathematical statistics. Since then the situation has been relieved considerably, and had I known in advance what books were in the making it is likely that I should not have embarked on this volume. However, it seemed sufficiently different from other presentations to give prospective teachers and students a useful alternative choice.

The aforementioned notes were used as text material for three years at Iowa State College in a course offered to senior and first-year graduate students. The only prerequisite for the course was one year of calculus, and this requirement indicates the level of the book. (The calculus class at Iowa State met four hours per week and included good coverage of Taylor series, partial differentiation, and multiple integration.) No previous knowledge of statistics is assumed.

This is a statistics book, not a mathematics book, as any mathematician will readily see. Little mathematical rigor is to be found in the derivations simply because it would be boring and largely a waste of time at this level. Of course rigorous thinking is quite essential to good statistics, and I have been at some pains to make a show of rigor and to instill an appreciation for rigor by pointing out various pitfalls of loose arguments.

While this text is primarily concerned with the theory of statistics, full cognizance has been taken of those students who fear that a moment may be wasted in mathematical frivolity. All new subjects are supplied with a little scenery from practical affairs, and, more important, a serious effort has been made in the problems to illustrate the variety of ways in which the theory may be applied.

The problems are an essential part of the book. They range from simple numerical examples to theorems needed in subsequent chapters. They include important subjects which could easily take precedence over material in the text; the relegation of subjects to problems was based rather on the feasibility of such a procedure than on the priority of the subject. For example, the matter of correlation is dealt with almost entirely in the problems. It seemed to me inefficient to cover multivariate situations twice in detail, i.e., with the regression model and with the correlation model. The emphasis in the text proper is on the more general regression model.

The author of a textbook is indebted to practically everyone who has touched the field, and I here bow to all statisticians. However, in giving credit to contributors one must draw the line somewhere, and I have simplified matters by drawing it very high; only the most eminent contributors are mentioned in the book.

I am indebted to Catherine Thompson and Maxine Merrington, and to E. S. Pearson, editor of *Biometrika*, for permission to include Tables III and V, which are abridged versions of tables published in *Biometrika*. I am also indebted to Professors R. A. Fisher and Frank Yates, and to Messrs. Oliver and Boyd, Ltd., Edinburgh, for permission to reprint Table IV from their book "Statistical Tables for Use in Biological, Agricultural and Medical Research."

Since the first edition of this book was published in 1950 many new statistical techniques have been made available and many techniques that were only in the domain of the mathematical statistician are now useful and demanded by the applied statistician. To include some of this material we have had to eliminate other material, else the book would have come to resemble a compendium. The general approach of presenting the theory with some connection to practical problems apparently contributed significantly to the success of the first edition and we have tried to maintain that feature in the present edition.

I

PROBABILITY

1 INTRODUCTION AND SUMMARY

The purpose of this chapter is to define *probability* and discuss some of its properties. Section 2 is a brief essay on some of the different meanings that have been attached to probability and may be omitted by those who are interested only in mathematical (axiomatic) probability, which is defined in Sec. 3 and used throughout the remainder of the text. Section 3 is subdivided into six subsections. The first, Subsec. 3.1, discusses the concept of probability models. It provides a real-world setting for the eventual mathematical definition of probability. A review of some of the set theoretical concepts that are relevant to probability is given in Subsec. 3.2. Sample space and event space are defined in Subsec. 3.3. Subsection 3.4 commences with a recall of the definition of a function. Such a definition is useful since many of the words to be defined in this and coming chapters (e.g., probability, random variable, distribution, etc.) are defined as particular functions. The indicator function, to be used extensively in later chapters, is defined here. The probability axioms are presented, and the probability function is defined. Several properties of this probability function are stated. The culmination of this subsection is the definition of a probability space. Subsection 3.5 is devoted to examples of probabilities defined on finite sample spaces. The related concepts of independence of events and conditional probability are discussed in the sixth and final subsection. Bayes' theorem, the multiplication rule, and the theorem of total probabilities are proved or derived, and examples of each are given.

Of the three main sections included in this chapter, only Sec. 3, which is by far the longest, is vital. The definitions of probability, probability space, conditional probability, and independence, along with familiarity with the properties of probability, conditional and unconditional and related formulas, are the essence of this chapter. This chapter is a background chapter; it introduces the language of probability to be used in developing *distribution theory*, which is the backbone of the theory of statistics.

2 KINDS OF PROBABILITY

2.1 Introduction

One of the fundamental tools of statistics is probability, which had its formal beginnings with games of chance in the seventeenth century.

Games of chance, as the name implies, include such actions as spinning a roulette wheel, throwing dice, tossing a coin, drawing a card, etc., in which the outcome of a trial is uncertain. However, it is recognized that even though the outcome of any particular trial may be uncertain, there is a *predictable* long-term outcome. It is known, for example, that in many throws of an ideal (balanced, symmetrical) coin about one-half of the trials will result in heads. It is this long-term, predictable regularity that enables gaming houses to engage in the business.

A similar type of uncertainty and long-term regularity often occurs in experimental science. For example, in the science of genetics it is uncertain whether an offspring will be male or female, but in the long run it is known approximately what percent of offspring will be male and what percent will be female. A life insurance company cannot predict which persons in the United States will die at age 50, but it can predict quite satisfactorily *how many* people in the United States will die at that age.

First we shall discuss the classical, or a priori, theory of probability; then we shall discuss the frequency theory. Development of the axiomatic approach will be deferred until Sec. 3.

2.2 Classical or A Priori Probability

As we stated in the previous subsection, the theory of probability in its early stages was closely associated with games of chance. This association prompted the classical definition. For example, suppose that we want the probability of the event that an ideal coin will turn up heads. We argue in this manner: Since there are only two ways that the coin can fall, heads or tails, and since the coin is well balanced, one would expect that the coin is just as likely to fall heads as tails; hence, the probability of the event of a head will be given the value $\frac{1}{2}$. This kind of reasoning prompted the following classical definition of probability.

> **Definition 1 Classical probability** If a random experiment can result in n mutually exclusive and equally likely outcomes and if n_A of these outcomes have an attribute A, then the *probability* of A is the fraction n_A/n. ////

We shall apply this definition to a few examples in order to illustrate its meaning.

If an ordinary die (one of a pair of dice) is tossed—there are six possible outcomes—any one of the six numbered faces may turn up. These six outcomes are *mutually exclusive* since two or more faces cannot turn up simultaneously. And if the die is fair, or true, the six outcomes are *equally likely;* i.e., it is expected that each face will appear with about equal relative frequency in the long run. Now suppose that we want the probability that the result of a toss be an even number. Three of the six possible outcomes have this attribute. The probability that an even number will appear when a die is tossed is therefore $\frac{3}{6}$, or $\frac{1}{2}$. Similarly, the probability that a 5 will appear when a die is tossed is $\frac{1}{6}$. The probability that the result of a toss will be greater than 2 is $\frac{2}{3}$.

To consider another example, suppose that a card is drawn at random from an ordinary deck of playing cards. The probability of drawing a spade is readily seen to be $\frac{13}{52}$, or $\frac{1}{4}$. The probability of drawing a number between 5 and 10, inclusive, is $\frac{24}{52}$, or $\frac{6}{13}$.

The application of the definition is straightforward enough in these simple cases, but it is not always so obvious. Careful attention must be paid to the qualifications "mutually exclusive," "equally likely," and "random." Suppose that one wishes to compute the probability of getting two heads if a coin is tossed twice. He might reason that there are three possible outcomes for the two tosses: two heads, two tails, or one head and one tail. One of these three outcomes has the desired attribute, i.e., two heads; therefore the probability is $\frac{1}{3}$. This reasoning is faulty because the three given outcomes are not equally likely. The third outcome, one head and one tail, can occur in two ways since the head may appear on the first toss and the tail on the second or the head may appear on the second toss and the tail on the first. Thus there are four equally likely outcomes: HH, HT, TH, and TT. The first of these has the desired attribute, while the others do not. The correct probability is therefore $\frac{1}{4}$. The result would be the same if two ideal coins were tossed simultaneously.

Again, suppose that one wished to compute the probability that a card drawn from an ordinary well-shuffled deck will be an ace or a spade. In enumerating the favorable outcomes, one might count 4 aces and 13 spades and reason that there are 17 outcomes with the desired attribute. This is clearly incorrect because these 17 outcomes are not mutually exclusive since the ace of spades is both an ace and a spade. There are 16 outcomes that are favorable to an ace or a spade, and so the correct probability is $\frac{16}{52}$, or $\frac{4}{13}$.

We note that by the classical definition the probability of event A is a number between 0 and 1 inclusive. The ratio n_A/n must be less than or equal to 1 since the total number of possible outcomes cannot be smaller than the number of outcomes with a specified attribute. If an event is certain to happen, its probability is 1; if it is certain not to happen, its probability is 0. Thus, the probability of obtaining an 8 in tossing a die is 0. The probability that the number showing when a die is tossed is less than 10 is equal to 1.

The probabilities determined by the classical definition are called *a priori* probabilities. When one states that the probability of obtaining a head in tossing a coin is $\frac{1}{2}$, he has arrived at this result purely by deductive reasoning. The result does not require that any coin be tossed or even be at hand. We say that if the coin is true, the probability of a head is $\frac{1}{2}$, but this is little more than saying the same thing in two different ways. Nothing is said about how one can determine whether or not a particular coin is true.

The fact that we shall deal with ideal objects in developing a theory of probability will not trouble us because that is a common requirement of mathematical systems. Geometry, for example, deals with conceptually perfect circles, lines with zero width, and so forth, but it is a useful branch of knowledge, which can be applied to diverse practical problems.

There are some rather troublesome limitations in the classical, or a priori, approach. It is obvious, for example, that the definition of probability must be modified somehow when the total number of possible outcomes is infinite. One might seek, for example, the probability that an integer drawn *at random* from the positive integers be even. The intuitive answer to this question is $\frac{1}{2}$.

If one were pressed to justify this result on the basis of the definition, he might reason as follows: Suppose that we limit ourselves to the first 20 integers; 10 of these are even so that the ratio of favorable outcomes to the total number is $\frac{10}{20}$, or $\frac{1}{2}$. Again, if the first 200 integers are considered, 100 of these are even, and the ratio is also $\frac{1}{2}$. In general, the first $2N$ integers contain N even integers; if we form the ratio $N/2N$ and let N become infinite so as to encompass the whole set of positive integers, the ratio remains $\frac{1}{2}$. The above argument is plausible, and the answer is plausible, but it is no simple matter to make the argument stand up. It depends, for example, on the natural ordering of the positive integers, and a different ordering could produce a different result. Thus, one could just as well order the integers in this way: 1, 3, 2; 5, 7, 4; 9, 11, 6; ..., taking the first pair of odd integers then the first even integer, the second pair of odd integers then the second even integer, and so forth. With this ordering, one could argue that the probability of drawing an even integer is $\frac{1}{3}$. The integers can also be ordered so that the ratio will oscillate and never approach any definite value as N increases.

There is another difficulty with the classical approach to the theory of probability which is deeper even than that arising in the case of an infinite number of outcomes. Suppose that we toss a coin known to be biased in favor of heads (it is bent so that a head is more likely to appear than a tail). The two possible outcomes of tossing the coin are not equally likely. What is the probability of a head? The classical definition leaves us completely helpless here.

Still another difficulty with the classical approach is encountered when we try to answer questions such as the following: What is the probability that a child born in Chicago will be a boy? Or what is the probability that a male will die before age 50? Or what is the probability that a cookie bought at a certain bakery will have less than three peanuts in it? All these are legitimate questions which we want to bring into the realm of probability theory. However, notions of "symmetry," "equally likely," etc., cannot be utilized as they could be in games of chance. Thus we shall have to alter or extend our definition to bring problems similar to the above into the framework of the theory. This more widely applicable probability is called *a posteriori* probability, or *frequency*, and will be discussed in the next subsection.

2.3 A Posteriori or Frequency Probability

A coin which seemed to be well balanced and symmetrical was tossed 100 times, and the outcomes recorded in Table 1. The important thing to notice is that the relative frequency of heads is close to $\frac{1}{2}$. This is not unexpected since the coin was symmetrical, and it was anticipated that in the long run heads would occur about one-half of the time. For another example, a single die was thrown 300 times, and the outcomes recorded in Table 2. Notice how close the relative frequency of a face with a 1 showing is to $\frac{1}{6}$; similarly for a 2, 3, 4, 5, and 6. These results are not unexpected since the die which was used was quite symmetrical and balanced; it was expected that each face would occur with about equal frequency in the long run. This suggests that we might be willing to use this relative frequency in Table 1 as an approximation for the probability that the particular coin used will come up heads or we might be willing to use the relative frequencies in Table 2 as approximations for the probabilities that various numbers on this die will appear. Note that although the relative frequencies of the different outcomes are predictable, the actual outcome of an individual throw is unpredictable.

In fact, it seems reasonable to assume for the coin experiment that there exists a number, label it p, which is the probability of a head. Now if the coin appears well balanced, symmetrical, and true, we might use Definition 1 and state that p is approximately equal to $\frac{1}{2}$. It is only an approximation to set p equal to $\frac{1}{2}$ since for this particular coin we cannot be certain that the two cases, heads and tails, are exactly equally likely. But by examining the balance and symmetry of the coin it may seem quite reasonable to assume that they are. Alternatively, the coin could be tossed a large number of times, the results recorded as in Table 1, and the relative frequency of a head used as an approximation for p. In the experiment with a die, the probability p_2 of a 2 showing could be approximated by using Definition 1 or by using the relative frequency in Table 2. The important thing is that we postulate that there is a number p which is defined as the probability of a head with the coin or a number p_2 which is the probability of a 2 showing in the throw of the die. Whether we use Definition 1 or the relative frequency for the probability seems unimportant in the examples cited.

Table 1 RESULTS OF TOSSING A COIN 100 TIMES

Outcome	Observed Frequency	Observed relative frequency	Long-run expected relative frequency of a balanced coin
H	56	.56	.50
T	44	.44	.50
Total	100	1.00	1.00

Suppose, as described above, that the coin is unbalanced so that we are quite certain from an examination that the two cases, heads and tails, are *not* equally likely to happen. In these cases a number p can still be postulated as the probability that a head shows, but the classical definition will not help us to find the value of p. We must use the frequency approach or possibly some physical analysis of the unbalanced coin.

In many scientific investigations, observations are taken which have an element of uncertainty or unpredictability in them. As a very simple example, suppose that we want to predict whether the next baby born in a certain locality will be a male or a female. This is individually an uncertain event, but the results of groups of births can be dealt with satisfactorily. We find that a certain long-run regularity exists which is similar to the long-run regularity of the frequency ratio of a head when a coin is thrown. If, for example, we find upon examination of records that about 51 percent of the births are male, it might be reasonable to postulate that the probability of a male birth in this locality is equal to a number p and take .51 as its approximation.

To make this idea more concrete, we shall assume that a series of observations (or experiments) can be made under quite uniform conditions. That is, an observation of a random experiment is made; then the experiment is repeated under similar conditions, and another observation taken. This is repeated many times, and while the conditions are similar each time, there is an uncontrollable variation which is haphazard or random so that the observations are individually unpredictable. In many of these cases the observations fall into certain classes wherein the relative frequencies are quite stable. This suggests that we postulate a number p, called the probability of the event, and approximate p by the relative frequency with which the repeated observations satisfy the

Table 2 RESULTS OF TOSSING A DIE 300 TIMES

Outcome	Observed Frequency	Observed relative frequency	Long-run expected relative frequency of a balanced die
1	51	.170	.1667
2	54	.180	.1667
3	48	.160	.1667
4	51	.170	.1667
5	49	.163	.1667
6	47	.157	.1667
Total	300	1.000	1.000

event. For instance, suppose that the experiment consists of sampling the population of a large city to see how many voters favor a certain proposal. The outcomes are "favor" or "do not favor," and each voter's response is unpredictable, but it is reasonable to postulate a number p as the probability that a given response will be "favor." The relative frequency of "favor" responses can be used as an approximate value for p.

As another example, suppose that the experiment consists of sampling transistors from a large collection of transistors. We shall postulate that the probability of a given transistor being defective is p. We can approximate p by selecting several transistors at random from the collection and computing the relative frequency of the number defective.

The important thing is that we can *conceive* of a series of observations or experiments under rather uniform conditions. Then a number p can be postulated as the probability of the event A happening, and p can be approximated by the *relative frequency* of the event A in a series of experiments.

3 PROBABILITY—AXIOMATIC

3.1 Probability Models

One of the aims of science is to predict and describe events in the world in which we live. One way in which this is done is to construct mathematical models which adequately describe the real world. For example, the equation $s = \frac{1}{2}gt^2$ expresses a certain relationship between the symbols s, g, and t. It is a mathematical model. To use the equation $s = \frac{1}{2}gt^2$ to predict s, the distance a body falls, as a function of time t, the gravitational constant g must be known. The latter is a physical constant which must be measured by experimentation if the equation $s = \frac{1}{2}gt^2$ is to be useful. The reason for mentioning this equation is that we do a similar thing in probability theory; we construct a probability model which can be used to describe events in the real world. For example, it might be desirable to find an equation which could be used to predict the sex of each birth in a certain locality. Such an equation would be very complex, and none has been found. However, a probability model can be constructed which, while not very helpful in dealing with an individual birth, is quite useful in dealing with groups of births. Therefore, we can postulate a number p which represents the probability that a birth will be a male. From this fundamental probability we can answer questions such as: What is the probability that in ten births at least three will be males? Or what is the probability that there will be three consecutive male births in the next five? To answer questions such as these and many similar ones, we shall develop an *idealized probability model*.

The two general types of probability (a priori and a posteriori) defined above have one important thing in common: They both require a conceptual experiment in which the various outcomes can occur under somewhat uniform conditions. For example, repeated tossing of a coin for the a priori case, and repeated birth for the a posteriori case. However, we might like to bring into the realm of probability theory situations which cannot conceivably fit into the framework of repeated outcomes under somewhat similar conditions. For example, we might like to answer questions such as: What is the probability my wife loves me? Or what is the probability that World War III will start before January 1, 1985? These types of problems are certainly a legitimate part of general probability theory and are included in what is referred to as *subjective probability*. We shall not discuss subjective probability to any great extent in this book, but we remark that the axioms of probability from which we develop probability theory are rich enough to include a priori probability, a posteriori probability, and subjective probability.

To start, we require that every possible outcome of the experiment under study can be enumerated. For example, in the coin-tossing experiment there are two possible outcomes: heads and tails. We shall associate probabilities only with these outcomes or with collections of these outcomes. We add, however, that even if a particular outcome is impossible, it can be included (its probability is 0). The main thing to remember is that every outcome which *can occur* must be included.

Each conceivable outcome of the conceptual experiment under study will be defined as a sample point, and the totality of conceivable outcomes (or sample points) will be defined as the sample space.

Our object, of course, is to assess the probability of certain outcomes or collections of outcomes of the experiment. Discussion of such probabilities is conveniently couched in the language of set theory, an outline of which appears in the next subsection. We shall return to formal definitions and examples of sample space, event, and probability.

3.2 An Aside—Set Theory

We begin with a collection of objects. Each object in our collection will be called a *point* or *element*. We assume that our collection of objects is large enough to include all the points under consideration in a given discussion. The totality of all these points is called the *space, universe,* or *universal set*. We will call it the space (anticipating that it will become the sample space when we speak of probability) and denote it by Ω. Let ω denote an element or point in Ω. Although a *set* can be defined as any collection of objects, we shall assume, unless otherwise stated, that all the sets mentioned in a given discussion consist of points in the space Ω.

EXAMPLE 1 $\Omega = R_2$, where R_2 is the collection of points ω in the plane and $\omega = (x, y)$ is any pair of real numbers x and y. ////

EXAMPLE 2 $\Omega = \{$all United States citizens$\}$. ////

We shall usually use capital Latin letters from the beginning of the alphabet, with or without subscripts, to denote sets. If ω is a point or element belonging to the set A, we shall write $\omega \in A$; if ω is not an element of A, we shall write $\omega \notin A$.

Definition 2 Subset If every element of a set A is also an element of a set B, then A is defined to be a *subset* of B, and we shall write $A \subset B$ or $B \supset A$; read "A is contained in B" or "B contains A." ////

Definition 3 Equivalent sets Two sets A and B are defined to be *equivalent*, or *equal*, if $A \subset B$ and $B \subset A$. This will be indicated by writing $A = B$. ////

Definition 4 Empty set If a set A contains no points, it will be called the *null set*, or *empty set*, and denoted by ϕ. ////

Definition 5 Complement The *complement* of a set A with respect to the space Ω, denoted by $\bar{A}$, A^c, or $\Omega - A$, is the set of all points that are in Ω but not in A. ////

Definition 6 Union Let A and B be any two subsets of Ω; then the set that consists of all points that are in A or B or both is defined to be the *union* of A and B and written $A \cup B$. ////

Definition 7 Intersection Let A and B be any two subsets of Ω; then the set that consists of all points that are in both A and B is defined to be the *intersection* of A and B and is written $A \cap B$ or AB. ////

Definition 8 Set difference Let A and B be any two subsets of Ω. The set of all points in A that are not in B will be denoted by $A - B$ and is defined as *set difference*. ////

EXAMPLE 3 Let $\Omega = \{(x, y): 0 \le x \le 1 \text{ and } 0 \le y \le 1\}$, which is read the collection of all points (x, y) for which $0 \le x \le 1$ and $0 \le y \le 1$. Define the following sets:

$$A_1 = \{(x, y): 0 \le x \le 1; 0 \le y \le \tfrac{1}{2}\},$$
$$A_2 = \{(x, y): 0 \le x \le \tfrac{1}{2}; 0 \le y \le 1\},$$
$$A_3 = \{(x, y): 0 \le x \le y \le 1\},$$
$$A_4 = \{(x, y): 0 \le x \le \tfrac{1}{2}; 0 \le y \le \tfrac{1}{2}\}.$$

(We shall adhere to the practice initiated here of using braces to embrace the points of a set.)
The set relations below follow.

$$A_4 \subset A_1; \qquad A_4 \subset A_2; \qquad A_1 \cap A_2 = A_1 A_2 = A_4;$$
$$A_2 \cup A_3 = A_4 \cup A_3; \qquad \bar{A}_1 = \{(x, y): 0 \le x \le 1; \tfrac{1}{2} < y \le 1\};$$
$$A_1 - A_4 = \{(x, y): \tfrac{1}{2} < x \le 1; 0 \le y \le \tfrac{1}{2}\}.$$ ////

EXAMPLE 4 Let Ω, A_1, A_2, and A_3 be as indicated in the diagrams in Fig. 1 which are called *Venn diagrams.* ////

The *set operations* of complement, union, and intersection have been defined in Definitions 5 to 7, respectively. These set operations satisfy quite a number of laws, some of which follow, stated as theorems. Proofs are omitted.

Theorem 1 Commutative laws $A \cup B = B \cup A$ and $A \cap B = B \cap A$. ////

Theorem 2 Associative laws $A \cup (B \cup C) = (A \cup B) \cup C$, and $A \cap (B \cap C) = (A \cap B) \cap C$. ////

Theorem 3 Distributive laws $A \cap (B \cup C) = (A \cap B) \cup (A \cap C)$, and $A \cup (B \cap C) = (A \cup B) \cap (A \cup C)$. ////

Theorem 4 $(A^C)^C = \overline{(\bar{A})} = A$; in words, the complement of A complement equals A. ////

Theorem 5 $A\Omega = A$; $A \cup \Omega = \Omega$; $A\phi = \phi$; and $A \cup \phi = A$. ////

Theorem 6 $A\bar{A} = \phi$; $A \cup \bar{A} = \Omega$; $A \cap A = A$; and $A \cup A = A$. ////

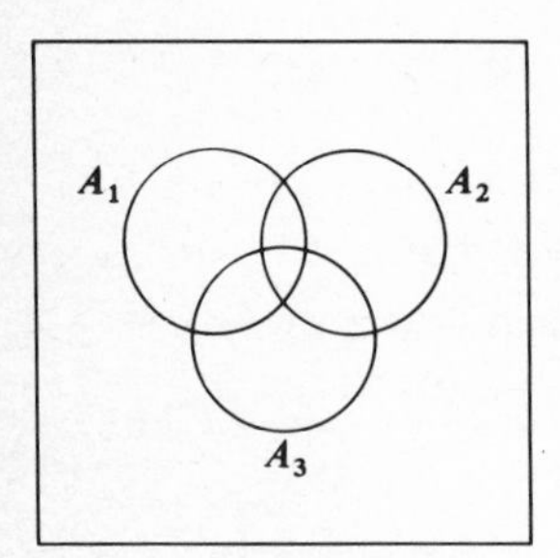

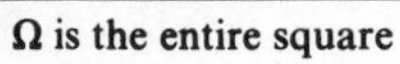
Ω is the entire square

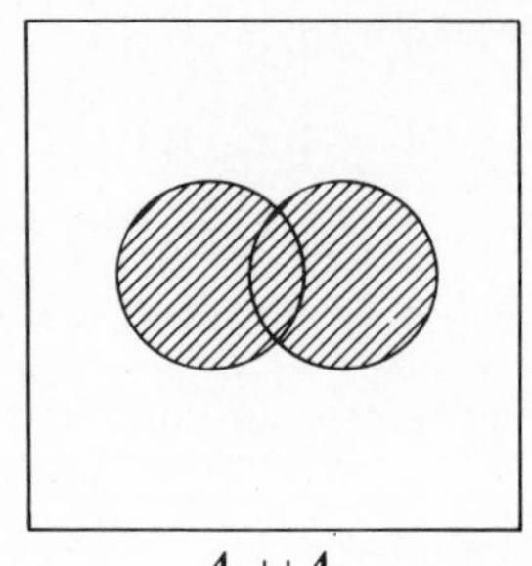
$A_1 \cup A_2$

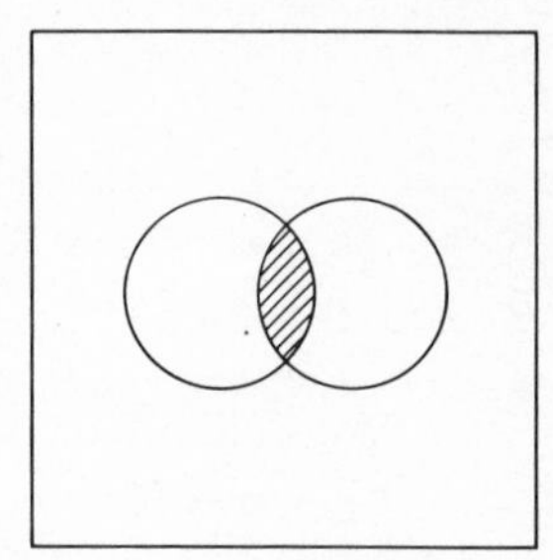
$A_1 \cap A_2$

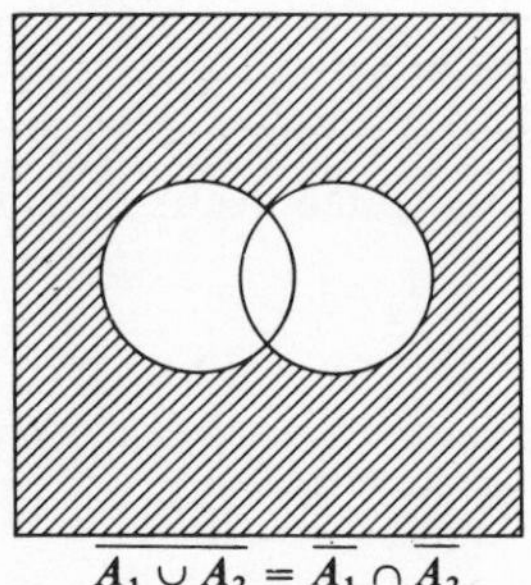
$\overline{A_1 \cup A_2} = \overline{A_1} \cap \overline{A_2}$

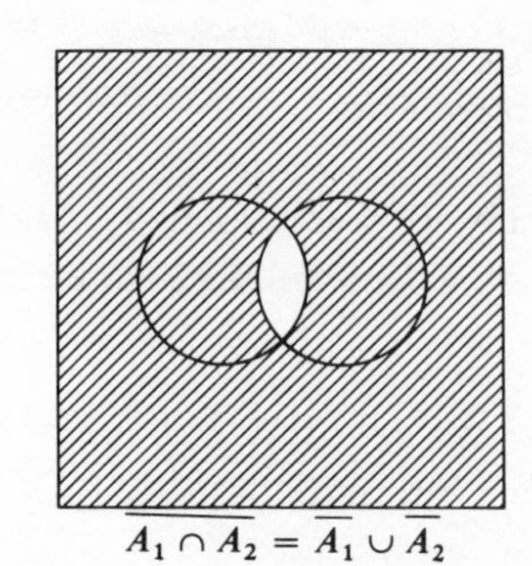
$\overline{A_1 \cap A_2} = \overline{A_1} \cup \overline{A_2}$

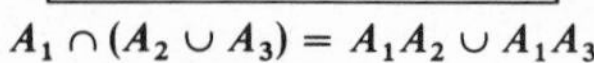
$A_1 \cap (A_2 \cup A_3) = A_1A_2 \cup A_1A_3$

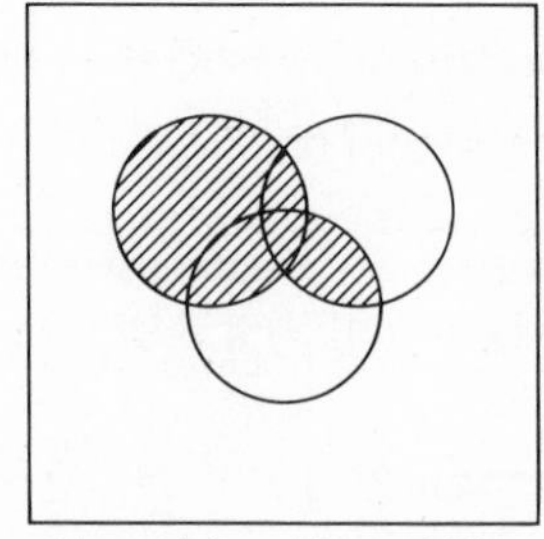
$A_1 \cup (A_2 \cap A_3) = (A_1 \cup A_2)(A_1 \cup A_3)$

FIGURE 1

Theorem 7 $\overline{(A \cup B)} = \bar{A} \cap \bar{B}$, and $\overline{(A \cap B)} = \bar{A} \cup \bar{B}$. These are known as *De Morgan's laws.* ////

Theorem 8 $A - B = A\bar{B}$. ////

Several of the above laws are illustrated in the *Venn diagrams* in Fig. 1. Although we will feel free to use any of the above laws, it might be instructive to give a proof of one of them just to illustrate the technique. For example, let us show that $\overline{(A \cup B)} = \bar{A} \cap \bar{B}$. By definition, two sets are equal if each is contained in the other. We first show that $\overline{(A \cup B)} \subset \bar{A} \cap \bar{B}$ by proving that if $\omega \in \overline{A \cup B}$, then $\omega \in \bar{A} \cap \bar{B}$. Now $\omega \in \overline{(A \cup B)}$ implies $\omega \notin A \cup B$, which implies that $\omega \notin A$ and $\omega \notin B$, which in turn implies that $\omega \in \bar{A}$ and $\omega \in \bar{B}$; that is, $\omega \in \bar{A} \cap \bar{B}$. We next show that $\bar{A} \cap \bar{B} \subset \overline{(A \cup B)}$. Let $\omega \in \bar{A} \cap \bar{B}$, which means ω belongs to both $\bar{A}$ and $\bar{B}$. Then $\omega \notin A \cup B$ for if it did, ω must belong to at least one of A or B, contradicting that ω belongs to both $\bar{A}$ and $\bar{B}$; however, $\omega \notin A \cup B$ means $\omega \in \overline{(A \cup B)}$, completing the proof.

We defined union and intersection of two sets; these definitions extend immediately to more than two sets, in fact to an arbitrary number of sets. It is customary to distinguish between the sets in a collection of subsets of Ω by assigning names to them in the form of subscripts. Let Λ (Greek letter capital lambda) denote a catalog of names, or *indices.* Λ is also called an *index set.* For example, if we are concerned with only two sets, then our index set Λ includes only two indices, say 1 and 2; so $\Lambda = \{1, 2\}$.

Definition 9 Union and intersection of sets Let Λ be an index set and $\{A_\lambda : \lambda \in \Lambda\} = \{A_\lambda\}$, a collection of subsets of Ω indexed by Λ. The set of points that consists of all points that belong to A_λ for at least one λ is called the *union* of the sets $\{A_\lambda\}$ and is denoted by $\bigcup_{\lambda \in \Lambda} A_\lambda$. The set of points that consists of all points that belong to A_λ for every λ is called the *intersection* of the sets $\{A_\lambda\}$ and is denoted by $\bigcap_{\lambda \in \Lambda} A_\lambda$. If Λ is empty, then define $\bigcup_{\lambda \in \Lambda} A_\lambda = \phi$ and $\bigcap_{\lambda \in \Lambda} A_\lambda = \Omega$. ////

EXAMPLE 5 If $\Lambda = \{1, 2, \ldots, N\}$, i.e., Λ is the index set consisting of the first N integers, then $\bigcup_{\lambda \in \Lambda} A_\lambda$ is also written as

$$\bigcup_{n=1}^{N} A_n = A_1 \cup A_2 \cup \cdots \cup A_N. \qquad ////$$

One of the most fundamental theorems relating unions, intersections, and complements for an arbitrary collection of sets is due to *De Morgan*.

Theorem 9 De Morgan's theorem Let Λ be an index set and $\{A_\lambda\}$ a collection of subsets of Ω indexed by Λ. Then,

(i) $\overline{\bigcup_{\lambda \in \Lambda} A_\lambda} = \bigcap_{\lambda \in \Lambda} \bar{A}_\lambda$.

(ii) $\overline{\bigcap_{\lambda \in \Lambda} A_\lambda} = \bigcup_{\lambda \in \Lambda} \bar{A}_\lambda$. ////

We will not give a proof of this theorem. Note, however, that the special case when the index set Λ consists of only two names or indices is Theorem 7 above, and a proof of part of Theorem 7 was given in the paragraph after Theorem 8.

Definition 10 Disjoint or mutually exclusive Subsets A and B of Ω are defined to be *mutually exclusive* or *disjoint* if $A \cap B = \phi$. Subsets $A_1, A_2, \ldots$ are defined to be *mutually exclusive* if $A_i A_j = \phi$ for every $i \neq j$. ////

Theorem 10 If A and B are subsets of Ω, then (i) $A = AB \cup A\bar{B}$, and (ii) $AB \cap A\bar{B} = \phi$.

PROOF (i) $A = A \cap \Omega = A \cap (B \cup \bar{B}) = AB \cup A\bar{B}$. (ii) $AB \cap A\bar{B} = AAB\bar{B} = A\phi = \phi$. ////

Theorem 11 If $A \subset B$, then $AB = A$, and $A \cup B = B$.

PROOF Left as an exercise. ////

3.3 Definitions of Sample Space and Event

In Subsec. 3.1 we described what might be meant by a probability model. There we said that we had in mind some conceptual experiment whose possible outcomes we would like to study by assessing the probability of certain outcomes or collection of outcomes. In this subsection, we will give two important definitions, along with some examples, that will be used in assessing these probabilities.

Definition 11 Sample space The *sample space*, denoted by Ω, is the collection or totality of all possible outcomes of a conceptual experiment. ////

One might try to understand the definition by looking at the individual words. Use of the word "space" can be justified since the sample space is the total collection of objects or elements which are the outcomes of the experiment. This is in keeping with our use of the word "space" in set theory as the collection of all objects of interest in a given discussion. The word "sample" is harder to justify; our experiment is random, meaning that its outcome is uncertain so that a given outcome is just one sample of many possible outcomes.

Some other symbols that are used in other texts to denote a sample space, in addition to Ω, are S, Z, R, E, $\mathfrak{X}$, and A.

Definition 12 Event and event space An *event* is a subset of the sample space. The class of all events associated with a given experiment is defined to be the *event space*. ////

The above does not precisely define what an event is. An event will always be a subset of the sample space, but for sufficiently large sample spaces not all subsets will be events. Thus the class of all subsets of the sample space will not necessarily correspond to the event space. However, we shall see that the class of all events can always be selected to be large enough so as to include all those subsets (events) whose probability we may want to talk about. If the sample space consists of only a finite number of points, then the corresponding event space will be the class of all subsets of the sample space.

Our primary interest will not be in events per se but will be in the probability that an event does or does not occur or happen. An event A is said to occur if the experiment at hand results in an outcome (a point in our sample space) that belongs to A. Since a point, say ω, in the sample space is a subset (that subset consisting of the point ω) of the sample space Ω, it is a candidate to be an event. Thus ω can be viewed as a point in Ω or as a subset of Ω. To distinguish, let us write $\{\omega\}$, rather than just ω, whenever ω is to be viewed as a subset of Ω. Such a *one-point* subset will always be an event and will be called an *elementary event*. Also ϕ and Ω are both subsets of Ω, and both will always be events. Ω is sometimes called the *sure* event.

We shall attempt to use only capital Latin letters (usually from the beginning of the alphabet), with or without affixes, to denote events, with the exception that ϕ will be used to denote the empty set and Ω the sure event. The event space will always be denoted by a script Latin letter, and usually $\mathscr{A}$. $\mathscr{B}$ and $\mathscr{F}$, as well as other symbols, are used in some texts to denote the class of all events.

The sample space is basic and generally easy to define for a given experiment. Yet, as we shall see, it is the event space that is really essential in defining probability. Some examples follows.

EXAMPLE 6 The experiment is the tossing of a single die (a regular six-sided polyhedron or cube marked on each face with one to six spots) and noting which face is up. Now the die can land with any one of the six faces up; so there are six possible outcomes of the experiment;

$$\Omega = \{ ⚀, ⚁, ⚂, ⚃, ⚄, ⚅ \}.$$

Let $A = \{\text{even number of spots up}\}$. A is an event; it is a subset of Ω. $A = \{⚁, ⚃, ⚅\}$. Let $A_i = \{i \text{ spots up}\}$; $i = 1, 2, \ldots, 6$. Each A_i is an elementary event. For this experiment the sample space is finite; hence the event space is all subsets of Ω. There are $2^6 = 64$ events, of which only 6 are elementary, in $\mathscr{A}$ (including both ϕ and Ω). See Example 19 of Subsec. 3.5, where a technique for counting the number of events in a finite sample space is presented. ////

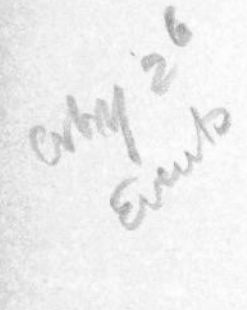

EXAMPLE 7 Toss a penny, nickel, and dime simultaneously, and note which side is up on each. There are eight possible outcomes of this experiment. $\Omega = \{$(H, H, H), (H, H, T), (H, T, H), (T, H, H), (H, T, T), (T, H, T), (T, T, H), (T, T, T)$\}$. We are using the first position of $(\cdot, \cdot, \cdot)$, called a 3-*tuple*, to record the outcome of the penny, the second position to record the outcome of the nickel, and the third position to record the outcome of the dime. Let $A_i = \{\text{exactly } i \text{ heads}\}$; $i = 0, 1, 2, 3$. For each i, A_i is an event. Note that A_0 and A_3 are each elementary events. Again all subsets of Ω are events; there are $2^8 = 256$ of them. ////

EXAMPLE 8 The experiment is to record the number of traffic deaths in the state of Colorado next year. Any nonnegative integer is a conceivable outcome of this experiment; so $\Omega = \{0, 1, 2, \ldots\}$. $A = \{\text{fewer than } 500 \text{ deaths}\} = \{0, 1, \ldots, 499\}$ is an event. $A_i = \{\text{exactly } i \text{ deaths}\}$, $i = 0, 1, \ldots$, is an elementary event. There is an infinite number of points in the sample space, and each point is itself an (elementary) event; so there is an infinite number of events. Each subset of Ω is an event. ////

EXAMPLE 9 Select a light bulb, and record the time in hours that it burns before burning out. Any nonnegative number is a conceivable outcome of this experiment; so $\Omega = \{x : x \geq 0\}$. For this sample space not all subsets of Ω are events; however, any subset that can be exhibited will be an event. For example, let

$$A = \{\text{bulb burns for at least } k \text{ hours but burns out before } m \text{ hours}\} = \{x: k \le x < m\};$$

then A is an event for any $0 \le k < m$. ////

EXAMPLE 10 Consider a random experiment which consists of counting the number of times that it rains and recording in inches the total rainfall next July in Fort Collins, Colorado. The sample space could then be represented by

$$\Omega = \{(i, x): i = 0, 1, 2, \dots \quad \text{and} \quad 0 \le x\},$$

where in the 2-tuple $(\cdot, \cdot)$ the first position indicates the number of times that it rains and the second position indicates the total rainfall. For example, $\omega = (7, 2.251)$ is a point in Ω corresponding to there being seven different times that it rained with a total rainfall of 2.251 inches. $A = \{(i, x): i = 5, \dots, 10 \text{ and } x \ge 3\}$ is an example of an event. ////

EXAMPLE 11 In an agricultural experiment, the yield of five varieties of wheat is examined. The five varieties are all grown under rather uniform conditions. The outcome is a collection of five numbers $(y_1, y_2, y_3, y_4, y_5)$, where y_i represents the yield of the ith variety in bushels per acre. Each y_i can conceivably be any real number greater than or equal to 0. In this example let the event A be defined by the conditions that y_2, y_3, y_4, and y_5 are each 10 or more bushels per acre larger than y_1, the standard variety. In our notation we write

$$A = \{(y_1, y_2, y_3, y_4, y_5): y_j \ge y_1 + 10; j = 2, 3, 4, 5; 0 \le y_1\}. \qquad ////$$

Our definition of sample space is precise and satisfactory, whereas our definitions of event and event space are not entirely satisfactory. We said that if the sample space was "sufficiently large" (as in Examples 9 to 11 above), not all subsets of the sample space would be events; however, we did not say exactly which subsets would be events and which would not. Rather than developing the necessary mathematics to precisely define which subsets of Ω constitute our event space $\mathscr{A}$, let us state some properties of $\mathscr{A}$ that it seems reasonable to require:

(i) $\Omega \in \mathscr{A}$.

(ii) If $A \in \mathscr{A}$, then $\bar{A} \in \mathscr{A}$.

(iii) If A_1 and $A_2 \in \mathscr{A}$, then $A_1 \cup A_2 \in \mathscr{A}$.

We said earlier that we were interested in events mainly because we would be interested in the probability that an event happens. Surely, then, we would want $\mathscr{A}$ to include Ω, the sure event. Also, if A is an event, meaning we can talk about the probability that A occurs, then $\bar{A}$ should also be an event so that we can talk about the probability that A does not occur. Similarly, if A_1 and A_2 are events, so should $A_1 \cup A_2$ be an event.

Any collection of events with properties (i) to (iii) is called a *Boolean algebra*, or just *algebra*, of events. We might note that the collection of all subsets of Ω necessarily satisfies the above properties. Several results follow from the above assumed properties of $\mathscr{A}$.

Theorem 12 $\phi \in \mathscr{A}$.

PROOF By property (i) $\Omega \in \mathscr{A}$; by (ii) $\bar{\Omega} \in \mathscr{A}$; but $\bar{\Omega} = \phi$; so $\phi \in \mathscr{A}$. ////

Theorem 13 If A_1 and $A_2 \in \mathscr{A}$, then $A_1 \cap A_2 \in \mathscr{A}$.

PROOF $\bar{A}_1$ and $\bar{A}_2 \in \mathscr{A}$; hence $\bar{A}_1 \cup \bar{A}_2$, and $\overline{(\bar{A}_1 \cup \bar{A}_2)} \in \mathscr{A}$, but $\overline{(\bar{A}_1 \cup \bar{A}_2)} = \bar{\bar{A}}_1 \cap \bar{\bar{A}}_2 = A_1 \cap A_2$ by De Morgan's law. ////

Theorem 14 If $A_1, A_2, \ldots, A_n \in \mathscr{A}$, then $\bigcup_{i=1}^{n} A_i$ and $\bigcap_{i=1}^{n} A_i \in \mathscr{A}$.

PROOF Follows by induction. ////

We will always assume that our collection of events $\mathscr{A}$ is an algebra—which partially justifies our use of $\mathscr{A}$ as our notation for it. In practice, one might take that collection of events of interest in a given consideration and enlarge the collection, if necessary, to include (i) the sure event, (ii) all complements of events already included, and (iii) all finite unions and intersections of events already included, and this will be an algebra $\mathscr{A}$. Thus far, we have not explained why $\mathscr{A}$ cannot always be taken to be the collection of all subsets of Ω. Such explanation will be given when we define probability in the next subsection.

3.4 Definition of Probability

In this section we give the axiomatic definition of probability. Although this formal definition of probability will not in itself allow us to achieve our goal of assigning actual probabilities to events consisting of certain outcomes of random experiments, it is another in a series of definitions that will ultimately lead to that goal. Since probability, as well as forthcoming concepts, is defined as a particular function, we begin this subsection with a review of the notion of a function.

The definition of a function The following terminology is frequently used to describe a function: A *function*, say $f(\cdot)$, is a rule (law, formula, recipe) that associates each point in one set of points with one and only one point in another set of points. The first collection of points, say A, is called the *domain*, and the second collection, say B, the *counterdomain*.

> **Definition 13 Function** A *function*, say $f(\cdot)$, with domain A and counterdomain B, is a collection of ordered pairs, say (a, b), satisfying (i) $a \in A$ and $b \in B$; (ii) each $a \in A$ occurs as the first element of some ordered pair in the collection (each $b \in B$ is not necessarily the second element of some ordered pair); and (iii) no two (distinct) ordered pairs in the collection have the same first element. ////

If $(a, b) \in f(\cdot)$, we write $b = f(a)$ (read "b equals f of a") and call $f(a)$ the *value* of $f(\cdot)$ at a. For any $a \in A$, $f(a)$ is an element of B; whereas $f(\cdot)$ is a set of ordered pairs. The set of all values of $f(\cdot)$ is called the *range* of $f(\cdot)$; i.e., the range of $f(\cdot) = \{b \in B: b = f(a) \text{ for some } a \in A\}$ and is always a subset of the counterdomain B but is not necessarily equal to it. $f(a)$ is also called the *image* of a under $f(\cdot)$, and a is called the *preimage* of $f(a)$.

EXAMPLE 12 Let $f_1(\cdot)$ and $f_2(\cdot)$ be the two functions, having the real line for their domain and counterdomain, defined by

$$f_1(\cdot) = \{(x, y): y = x^3 + x + 1, \ -\infty < x < \infty\}$$

and

$$f_2(\cdot) = \{(x, y): y = x^2, \ -\infty < x < \infty\}.$$

The range of $f_1(\cdot)$ is the counterdomain, the whole real line, but the range of $f_2(\cdot)$ is all nonnegative real numbers, not the same as the counterdomain. ////

Of particular interest to us will be a class of functions that are called *indicator* functions.

Definition 14 Indicator function Let Ω be any space with points ω and A any subset of Ω. The *indicator function* of A, denoted by $I_A(\cdot)$, is the function with domain Ω and counterdomain equal to the set consisting of the two real numbers 0 and 1 defined by

$$I_A(\omega) = \begin{cases} 1 & \text{if} \quad \omega \in A \\ 0 & \text{if} \quad \omega \notin A. \end{cases}$$

$I_A(\cdot)$ clearly "indicates" the set A. ////

Properties of Indicator Functions Let Ω be any space and $\mathscr{A}$ any collection of subsets of Ω:

(i) $I_A(\omega) = 1 - I_{\bar{A}}(\omega)$ for every $A \in \mathscr{A}$.

(ii) $I_{A_1 A_2 \cdots A_n}(\omega) = I_{A_1}(\omega) \cdot I_{A_2}(\omega) \cdots I_{A_n}(\omega)$ for $A_1, \ldots, A_n \in \mathscr{A}$.

(iii) $I_{A_1 \cup A_2 \cup \cdots \cup A_n}(\omega) = \max [I_{A_1}(\omega), I_{A_2}(\omega), \ldots, I_{A_n}(\omega)]$ for $A_1, \ldots, A_n \in \mathscr{A}$.

(iv) $I_A^2(\omega) = I_A(\omega)$ for every $A \in \mathscr{A}$.

Proofs of the above properties are left as an exercise.

The indicator function will be used to "indicate" subsets of the real line; e.g.,

$$I_{\{[0,1)\}}(x) = I_{[0,1)}(x) = \begin{cases} 1 & \text{if} \quad 0 \le x < 1 \\ 0 & \text{otherwise,} \end{cases}$$

and if I^+ denotes the set of positive integers,

$$I_{I^+}(x) = \begin{cases} 1 & \text{if } x \text{ is some positive integer} \\ 0 & \text{otherwise.} \end{cases}$$

Frequent use of indicator functions will be made throughout the remainder of this book. Often the utility of the indicator function is just notational efficiency as the following example shows.

EXAMPLE 13 Let the function $f(\cdot)$ be defined by

$$f(x) = \begin{cases} 0 & \text{for} \quad x \le 0 \\ x & \text{for} \quad 0 < x \le 1 \\ 2 - x & \text{for} \quad 1 < x \le 2 \\ 0 & \text{for} \quad 2 < x. \end{cases}$$

By using the indicator function, $f(x)$ can be written as

$$f(x) = xI_{(0,1]}(x) + (2 - x)I_{(1,2]}(x),$$

or also by using the absolute value symbol as

$$f(x) = (1 - |1 - x|)I_{(0,2]}(x). \qquad ////$$

Another type of function that we will have occasion to discuss is the *set function* defined as any functon which has as its domain a collection of sets and as its counterdomain the real line including, possibly, infinity. Examples of set functions follow.

EXAMPLE 14 Let Ω be the sample space corresponding to the experiment of tossing two dice, and let $\mathscr{A}$ be the collection of all subsets of Ω. For any $A \in \mathscr{A}$ define $N(A)$ = number of outcomes, or points in Ω, that are in A. Then $N(\phi) = 0$, $N(\Omega) = 36$, and $N(A) = 6$ if A is the event containing those outcomes having a total of seven spots up. ////

The *size-of-set* function alluded to in the above example can be defined, in general, for any set A as the number of points in A, where A is a member of an arbitrary collection of sets $\mathscr{A}$.

EXAMPLE 15 Let Ω be the plane or two-dimensional euclidean space and $\mathscr{A}$ any collection of subsets of Ω for which area is meaningful. Then for any $A \in \mathscr{A}$ define $Q(A)$ = area of A. For example, if $A = \{(x, y): 0 \le x \le 1, 0 \le y \le 1\}$, then $Q(A) = 1$; if $A = \{(x, y): x^2 + y^2 = r^2\}$, then $Q(A) = \pi r^2$; and if $A = \{(0, 0), (1, 1)\}$ then $Q(A) = 0$. ////

The probability function to be defined will be a particular set function.

Probability function Let Ω denote the sample space and $\mathscr{A}$ denote a collection of events assumed to be an algebra of events (see Subsec. 3.3) that we shall consider for some random experiment.

Definition 15 Probability function A *probability function* $P[\cdot]$ is a set function with domain $\mathscr{A}$ (an algebra of events)* and counterdomain the interval [0, 1] which satisfies the following axioms:

(i) $P[A] \geq 0$ for every $A \in \mathscr{A}$.

(ii) $P[\Omega] = 1$.

(iii) If $A_1, A_2, \ldots$ is a sequence of mutually exclusive events in $\mathscr{A}$ (that is, $A_i \cap A_j = \phi$ for $i \neq j$; $i, j = 1, 2, \ldots$) and if $A_1 \cup A_2 \cup \cdots = \bigcup_{i=1}^{\infty} A_i \in \mathscr{A}$, then $P\left[\bigcup_{i=1}^{\infty} A_i\right] = \sum_{i=1}^{\infty} P[A_i]$. ////

These axioms are certainly motivated by the definitions of classical and frequency probability. This definition of probability is a mathematical definition; it tells us which set functions can be called probability functions; it does not tell us what value the probability function $P[\cdot]$ assigns to a given event A. We will have to model our random experiment in some way in order to obtain values for the probability of events.

$P[A]$ is read "the probability of event A" or "the probability that event A occurs," which means the probability that any outcome in A occurs.

We have used brackets rather than parentheses in our notation for a probability function, and we shall continue to do the same throughout the remainder of this book.

*In defining a probability function, many authors assume that the domain of the set function is a sigma-algebra rather than just an algebra. For an algebra $\mathscr{A}$, we had the property

$$\text{if } A_1 \text{ and } A_2 \in \mathscr{A}, \text{ then } A_1 \cup A_2 \in \mathscr{A}.$$

A sigma-algebra differs from an algebra in that the above property is replaced by

$$\text{if } A_1, A_2, \ldots, A_n, \ldots \in \mathscr{A}, \text{ then } \bigcup_{n=1}^{\infty} A_n \in \mathscr{A}.$$

It can be shown that a sigma-algebra is an algebra, but not necessarily conversely. If the domain of the probability function is taken to be a sigma-algebra then axiom (iii) can be simplified to

$$\text{if } A_1, A_2, \ldots \text{ is a sequence of mutually exclusive events in } \mathscr{A},\ P\left[\bigcup_{i=1}^{\infty} A_i\right] = \sum_{i=1}^{\infty} P[A_i].$$

A fundamental theorem of probability theory, called the *extension theorem*, states that if a probability function is defined on an algebra (as we have done) then it can be extended to a sigma-algebra. Since the probability function can be extended from an algebra to a sigma-algebra, it is reasonable to begin by assuming that the probability function is defined on a sigma-algebra.

EXAMPLE 16 Consider the experiment of tossing two coins, say a penny and a nickel. Let $\Omega = \{(H, H), (H, T), (T, H), (T, T)\}$ where the first component of $(\cdot, \cdot)$ represents the outcome for the penny. Let us model this random experiment by assuming that the four points in Ω are equally likely; that is, assume $P[\{(H, H)\}] = P[\{(H, T)\}] = P[\{(T, H)\}] = P[\{(T, T)\}]$. The following question arises: Is the $P[\cdot]$ function that is implicitly defined by the above really a probability function; that is, does it satisfy the three axioms? It can be shown that it does, and so it is a probability function.

In our definitions of event and $\mathscr{A}$, a collection of events, we stated that $\mathscr{A}$ cannot always be taken to be the collection of all subsets of Ω. The reason for this is that for "sufficiently large" Ω the collection of all subsets of Ω is so large that it is impossible to define a probability function consistent with the above axioms.

We are able to deduce a number of properties of our function $P[\cdot]$ from its definition and three axioms. We list these as theorems.

It is in the statements and proofs of these properties that we will see the convenience provided by assuming $\mathscr{A}$ is an algebra of events. $\mathscr{A}$ is the domain of $P[\cdot]$; hence only members of $\mathscr{A}$ can be placed in the dot position of the notation $P[\cdot]$. Since $\mathscr{A}$ is an algebra, if we assume that A and $B \in \mathscr{A}$, we know that $\bar{A}$, $A \cup B$, AB, $\bar{A}B$, etc., are also members of $\mathscr{A}$, and so it makes sense to talk about $P[\bar{A}]$, $P[A \cup B]$, $P[AB]$, $P[\bar{A}B]$, etc.

Properties of $P[\cdot]$ For each of the following theorems, assume that Ω and $\mathscr{A}$ (an algebra of events) are given and $P[\cdot]$ is a probability function having domain $\mathscr{A}$.

Theorem 15 $P[\phi] = 0$.

PROOF Take $A_1 = \phi$, $A_2 = \phi$, $A_3 = \phi$, ...; then by axiom (iii)

$$P[\phi] = P\left[\bigcup_{i=1}^{\infty} A_i\right] = \sum_{i=1}^{\infty} P[A_i] = \sum_{i=1}^{\infty} P[\phi],$$

which can hold only if $P[\phi] = 0$. ////

Theorem 16 If $A_1, \ldots, A_n$ are mutually exclusive events in $\mathscr{A}$, then

$$P[A_1 \cup \cdots \cup A_n] = \sum_{i=1}^{n} P[A_i].$$

PROOF Let $A_{n+1} = \phi$, $A_{n+2} = \phi$, ..., ; then $\bigcup_{i=1}^{n} A_i = \bigcup_{i=1}^{\infty} A_i \in \mathscr{A}$, and

$$P\left[\bigcup_{i=1}^{n} A_i\right] = P\left[\bigcup_{i=1}^{\infty} A_i\right] = \sum_{i=1}^{\infty} P[A_i] = \sum_{i=1}^{n} P[A_i]. \qquad ////$$

Theorem 17 If A is an event in $\mathscr{A}$, then

$$P[\bar{A}] = 1 - P[A].$$

PROOF $A \cup \bar{A} = \Omega$, and $A \cap \bar{A} = \phi$; so

$$P[\Omega] = P[A \cup \bar{A}] = P[A] + P[\bar{A}].$$

But $P[\Omega] = 1$ by axiom (ii); the result follows. ////

Theorem 18 If A and $B \in \mathscr{A}$, then $P[A] = P[AB] + P[A\bar{B}]$, and $P[A - B] = P[A\bar{B}] = P[A] - P[AB]$.

PROOF $A = AB \cup A\bar{B}$, and $AB \cap A\bar{B} = \phi$; so $P[A] = P[AB] + P[A\bar{B}]$. ////

Theorem 19 For every two events A and $B \in \mathscr{A}$, $P[A \cup B] = P[A] + P[B] - P[AB]$. More generally, for events $A_1, A_2, \ldots, A_n \in \mathscr{A}$

$$P[A_1 \cup A_2 \cup \cdots \cup A_n] = \sum_{j=1}^{n} P[A_j] - \sum\sum_{i<j} P[A_i A_j]$$
$$+ \sum\sum\sum_{i<j<k} P[A_i A_j A_k] - \cdots + (-1)^{n+1} P[A_1 A_2 \ldots A_n].$$

PROOF $A \cup B = A \cup \bar{A}B$, and $A \cap \bar{A}B = \phi$; so

$$P[A \cup B] = P[A] + P[\bar{A}B]$$
$$= P[A] + P[B] - P[AB].$$

The more general statement is proved by mathematical induction. (See Problem 16.) ////

Theorem 20 If A and $B \in \mathscr{A}$ and $A \subset B$, then $P[A] \le P[B]$.

PROOF $B = BA \cup B\bar{A}$, and $BA = A$; so $B = A \cup B\bar{A}$, and $A \cap B\bar{A} = \phi$; hence $P[B] = P[A] + P[B\bar{A}]$. The conclusion follows by noting that $P[B\bar{A}] \ge 0$. ////

Theorem 21 Boole's inequality If $A_1, A_2, \ldots, A_n \in \mathscr{A}$, then

$$P[A_1 \cup A_2 \cup \cdots \cup A_n] \leq P[A_1] + P[A_2] + \cdots + P[A_n].$$

PROOF $P[A_1 \cup A_2] = P[A_1] + P[A_2] - P[A_1 A_2] \leq P[A_1] + P[A_2]$. The proof is completed using mathematical induction. ////

We conclude this subsection with one final definition.

Definition 16 Probability space A *probability space* is the triplet $(\Omega, \mathscr{A}, P[\cdot])$, where Ω is a sample space, $\mathscr{A}$ is a collection (assumed to be an algebra) of events (each a subset of Ω), and $P[\cdot]$ is a probability function with domain $\mathscr{A}$. ////

Probability space is a single term that gives us an expedient way to assume the existence of all three components in its notation. The three components are related; $\mathscr{A}$ is a collection of subsets of Ω, and $P[\cdot]$ is a function that has $\mathscr{A}$ as its domain. The probability space's main use is in providing a convenient method of stating background assumptions for future definitions, theorems, etc. It also ties together the main definitions that we have covered so far, namely, definitions of sample space, event space, and probability.

3.5 Finite Sample Spaces

In previous subsections we formally defined sample space, event, and probability, culminating in the definition of probability space. We remarked there that these formal definitions did not in themselves enable us to compute the value of the probability for an event A, which is our goal. We said that we had to appropriately model the experiment. In this section we show how this can be done for finite sample spaces, that is, sample spaces with only a finite number of elements or points in them.

In certain kinds of problems, of which games of chance are notable examples, the sample space contains a finite number of points, say $N = N(\Omega)$. [Recall that $N(A)$ is the size of A, that is, the number of sample points in A.] Some of these problems can be modeled by assuming that points in the sample space are equally likely. Such problems are the subject to be discussed next.

Finite sample space with equally likely points For certain random experiments there is a finite number of outcomes, say N, and it is often realistic to assume that the probability of each outcome is $1/N$. The classical definition of probability is generally adequate for these problems, but we shall show how the axiomatic definition is applicable as well. Let $\omega_1, \omega_2, \ldots, \omega_N$ be the N sample points in a finite space Ω. Suppose that the set function $P[\cdot]$ with domain the collection of all subsets of Ω satisfies the following conditions:

(i) $P[\{\omega_1\}] = P[\{\omega_2\}] = \cdots = P[\{\omega_N\}]$.

(ii) If A is any subset of Ω which contains $N(A)$ sample points [has size $N(A)$], then $P[A] = N(A)/N$.

Then it is readily checked that the set function $P[\cdot]$ satisfies the three axioms and hence is a probability function.

Definition 17 Equally likely probability function The probability function $P[\cdot]$ satisfying conditions (i) and (ii) above is defined to be an *equally likely probability function.* ////

Given that a random experiment can be realistically modeled by assuming equally likely sample points, the only problem left in determining the value of the probability of event A is to find $N(\Omega) = N$ and $N(A)$. Strictly speaking this is just a problem of counting—count the number of points in A and the number of points in Ω.

EXAMPLE 17 Consider the experiment of tossing two dice (or of tossing one die twice). Let $\Omega = \{(i_1, i_2): i_1 = 1, 2, \ldots, 6; i_2 = 1, 2, \ldots, 6\}$. Here i_1 = number of spots up on the first die, and i_2 = number of spots up on the second die. There are $6 \cdot 6 = 36$ sample points. It seems reasonable to attach the probability of $\frac{1}{36}$ to each sample point. Ω can be displayed as a lattice as in Fig. 2. Let A_7 = event that the total is 7; then $A_7 = \{(1, 6), (2, 5), (3, 4), (4, 3), (5, 2), (6, 1)\}$; so $N(A_7) = 6$, and $P[A_7] = N(A_7)/N(\Omega) = \frac{6}{36} = \frac{1}{6}$. Similarly $P[A_j]$ can be calculated for A_j = total of j; $j = 2, \ldots, 12$. In this example the number of points in any event A can be easily counted, and so $P[A]$ can be evaluated for any event A. ////

If $N(A)$ and $N(\Omega)$ are large for a given random experiment with a finite number of equally likely outcomes, the counting itself can become a difficult problem. Such counting can often be facilitated by use of certain combinatorial formulas, some of which will be developed now.

Assume now that the experiment is of such a nature that each outcome can be represented by an n-tuple. The above example is such an experiment; each outcome was represented by a 2-tuple. As another example, if the experiment is one of drawing a sample of size n, then n-tuples are particularly

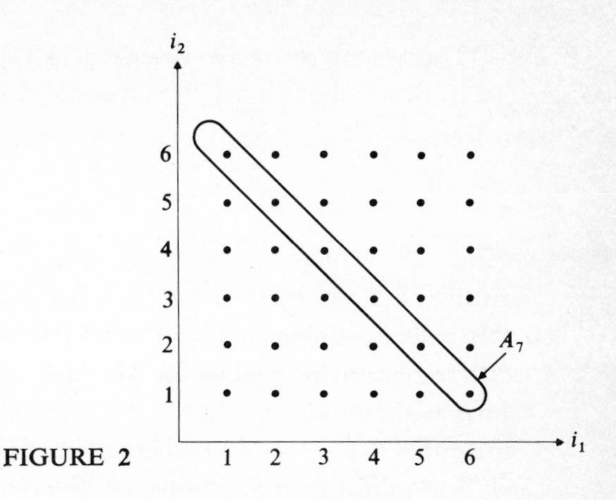

FIGURE 2

useful in recording the results. The terminology that is often used to describe a basic random experiment known generally by *sampling* is that of balls and urns. It is assumed that we have an urn containing, say, M balls, which are numbered 1 to M. The experiment is to select or draw balls from the urn one at a time until n balls have been drawn. We say we have drawn a *sample* of size n. The drawing is done in such a way that at the time of a particular draw each of the balls in the urn at that time has an equal chance of selection. We say that a ball has been selected *at random*. Two basic ways of drawing a sample are *with replacement* and *without replacement*, meaning just what the words say. A sample is said to be drawn with replacement, if after each draw the ball drawn is itself returned to the urn, and the sample is said to be drawn without replacement if the ball drawn is not returned to the urn. Of course, in sampling without replacement the size of the sample n must be less than or equal to M, the original number of balls in the urn, whereas in sampling with replacement the size of sample may be any positive integer. In reporting the results of drawing a sample of size n, an n-tuple can be used; denote the n-tuple by $(z_1, \ldots, z_n)$, where z_i represents the number of the ball drawn on the ith draw.

In general, we are interested in the size of an event that is composed of points that are n-tuples satisfying certain conditions. The size of such a set can be computed as follows: First determine the number of objects, say N_1, that may be used as the first component. Next determine the number of objects, say N_2, that may be used as the second component of an n-tuple given that the first component is known. (We are assuming that N_2 does not depend on which object has occurred as the first component.) And then determine the number of objects, say N_3, that may be used as the third component given that the first and second components are known. (Again we are assuming N_3 does not depend on which objects have occurred as the first and second components.) Continue in this manner until N_n is determined. The size $N(A)$ of the set A of n-tuples then equals $N_1 \cdot N_2 \cdots N_n$.

EXAMPLE 18 The total number of different ordered samples of n balls that can be obtained by drawing balls from an urn containing M distinguishable balls (distinguished by numbers 1 to M) is M^n if the sampling is done with replacement and is $M(M-1) \cdots (M-n+1)$ if the sampling is done without replacement. An ordered sample can be represented by an n-tuple, say $(z_1, \ldots, z_n)$, where z_j is the number of the ball obtained on the jth draw and the total number of different ordered samples is the same as the total number of n-tuples. In sampling with replacement, there are M choices of numbers for the first component, M choices of numbers for the second component, and finally M choices for the nth component. Thus there are M^n such n-tuples. In sampling without replacement, there are M choices of numbers for the first component, $M-1$ choices for the second, $M-2$ choices for the third, and finally $M-n+1$ choices for the nth component. In total, then, there are $M(M-1)(M-2) \cdots (M-n+1)$ such n-tuples. $M(M-1) \cdots (M-n+1)$ is abbreviated $(M)_n$ (see Appendix A). ////

EXAMPLE 19 Let S be any set containing M elements. How many subsets does S have? First let us determine the number of subsets of size n that S has. Let x_n denote this number, that is, the number of subsets of S of size n. A subset of size n is a collection of n objects, the objects not arranged in any particular order. For example the subset $\{s_1, s_5, s_7\}$ is the same as the subset $\{s_5, s_1, s_7\}$ since they contain the same three objects. If we take a given subset of S which contains n elements, $n!$ different ordered samples can be obtained by sampling from the given subset without replacement. If for each of the x_n different subsets there are $n!$ different ordered samples of size n, then there are $(n!)x_n$ different ordered samples of size n in sampling without replacement from the set S of M elements. But we know from the previous example that this number is $(M)_n$; hence $(n!)x_n = (M)_n$, or

$$x_n = \frac{(M)_n}{n!} = \binom{M}{n} = \text{number of subsets of size } n \text{ that may be formed from the elements of a set of size } M. \tag{1}$$

The total number of subsets of S, where S is a set of size M, is $\sum_{n=0}^{M}\binom{M}{n}$. This includes the empty set (set with no elements in it) and the whole set, both of which are subsets. Using the binomial theorem (see Appendix A)

$$(a+b)^M = \sum_{n=0}^{M}\binom{M}{n}a^n b^{M-n}, \qquad \text{with} \quad a = b = 1,$$

we see that

$$2^M = \sum_{n=0}^{M}\binom{M}{n}; \tag{2}$$

thus a *set of size M has* 2^M *subsets.* ////

EXAMPLE 20 Suppose an urn contains M balls numbered 1 to M, where the first K balls are defective and the remaining $M - K$ are nondefective. The experiment is to draw n balls from the urn. Define A_k to be the event that the sample of n balls contains exactly k defectives. There are two ways to draw the sample: (i) with replacement and (ii) without replacement. We are interested in $P[A_k]$ under each method of sampling. Let our sample space $\Omega = \{(z_1, \ldots, z_n): z_j = \text{number of the ball drawn on the } j\text{th draw}\}$. Now

$$P[A_k] = \frac{N(A_k)}{N(\Omega)}.$$

From Example 18 above, we know $N(\Omega) = M^n$ under (i) and $N(\Omega) = (M)_n$ under (ii). A_k is that subset of Ω for which exactly k of the z_j's are ball numbers 1 to K inclusive. These k ball numbers must fall in some subset of k positions from the total number of n available positions. There are $\binom{n}{k}$ ways of selecting the k positions for the ball numbers 1 to K inclusive to fall in. For each of the $\binom{n}{k}$ different positions, there are $K^k(M-K)^{n-k}$ different n-tuples for case (i) and $(K)_k(M-K)_{n-k}$ different n-tuples for case (ii). Thus A_k has size $\binom{n}{k}K^k(M-K)^{n-k}$ for case (i) and size $\binom{n}{k}(K)_k(M-K)_{n-k}$ for case (ii); so,

$$P[A_k] = \frac{\binom{n}{k}K^k(M-K)^{n-k}}{M^n} \tag{3}$$

in sampling with replacement, and

$$P[A_k] = \frac{\binom{n}{k}(K)_k(M-K)_{n-k}}{(M)_n} \tag{4}$$

in sampling without replacement. This latter formula can be rewritten as

$$P[A_k] = \frac{\binom{K}{k}\binom{M-K}{n-k}}{\binom{M}{n}}. \tag{5}$$

It might be instructive to derive Eq. (5) in another way. Suppose that our sample space, denoted now by Ω', is made up of *subsets* of size n, rather than n-tuples; that is, $\Omega' = \{\{z_1, \ldots, z_n\}: z_1, \ldots, z_n$ are the numbers on the n balls drawn$\}$. There are $\binom{M}{n}$ subsets of size n of the M balls; so $N(\Omega') = \binom{M}{n}$. If it is assumed that each of these subsets of size n is just as likely as any other subset of size n (one can think of selecting all n balls at once rather than one at a time), then $P[A_k'] = N(A_k')/N(\Omega')$. Now $N(A_k')$ is the size of the event consisting of those subsets of size n which contain exactly k balls from the balls that are numbered 1 to K inclusive. The k balls from the balls that are numbered 1 to K can be selected in $\binom{K}{k}$ ways, and the remaining $n-k$ balls from the balls that are numbered $K+1$ to M can be selected in $\binom{M-K}{n-k}$ ways; hence $N(A_k') = \binom{K}{k}\binom{M-K}{n-k}$, and finally

$$P[A_k'] = N(A_k')/N(\Omega') = \frac{\binom{K}{k}\binom{M-K}{n-k}}{\binom{M}{n}}.$$

We have derived the probability of exactly k defectives in sampling without replacement by considering two different sample spaces; one sample space consisted of n-tuples, the other consisted of subsets of size n.

To aid in remembering the formula given in Eq. (5), note that $K + M - K = M$ and $k + n - k = n$; i.e., the sum of the "upper" terms in the numerator equals the "upper" term in the denominator, and the sum of the "lower" terms in the numerator equals the "lower" term in the denominator. ////

EXAMPLE 21 The formula given in Eq. (5) is particularly useful to calculate certain probabilities having to do with card games. For example, we might ask the probability that a certain 13-card hand contains exactly 6 spades. There are $M = 52$ total cards, and one can model the card shuffling and dealing process by assuming that the 13-card hand represents a sample of size 13 drawn without replacement from the 52 cards. Let A_6 denote the event of exactly 6 spades. There are a total of 13 spades (defective balls in sampling terminology); so

$$P[A_6] = \frac{\binom{13}{6}\binom{52-13}{13-6}}{\binom{52}{13}}$$

by Eq. (5). ////

Many other formulas for probabilities of specified events defined on finite sample spaces with equally likely sample points can be derived using methods of combinatorial analysis, but we will not undertake such derivations here. The interested reader is referred to Refs. 10 and 8.

Finite sample space without equally likely points We saw for finite sample spaces with equally likely sample points that $P[A] = N(A)/N(\Omega)$ for any event A. For finite sample spaces without equally likely sample points, things are not quite as simple, but we can completely define the values of $P[A]$ for each of the $2^{N(\Omega)}$ events A by specifying the value of $P[\cdot]$ for each of the $N = N(\Omega)$ elementary events. Let $\Omega = \{\omega_1, \ldots, \omega_N\}$, and assume $p_j = P[\{\omega_j\}]$ for $j = 1, \ldots, N$. Since

$$1 = P[\Omega] = P\left[\bigcup_{j=1}^{N}\{\omega_j\}\right] = \sum_{j=1}^{N} P[\{\omega_j\}],$$

$$\sum_{j=1}^{N} p_j = 1.$$

For any event A, define $P[A] = \Sigma p_j$, where the summation is over those ω_j belonging to A. It can be shown that $P[\cdot]$ so defined satisfies the three axioms and hence is a probability function.

EXAMPLE 22 Consider an experiment that has N outcomes, say $\omega_1, \omega_2, \ldots, \omega_N$, where it is known that outcome ω_{j+1} is twice as likely as outcome ω_j, where $j = 1, \ldots, N-1$; that is, $p_{j+1} = 2p_j$, where $p_i = P[\{\omega_i\}]$. Find $P[A_k]$, where $A_k = \{\omega_1, \omega_2, \ldots, \omega_k\}$. Since

$$\sum_{j=1}^{N} p_j = \sum_{j=1}^{N} 2^{j-1} p_1 = p_1(1 + 2 + 2^2 + \cdots + 2^{N-1}) = p_1(2^N - 1) = 1,$$

$$p_1 = \frac{1}{2^N - 1}$$

and

$$p_j = 2^{j-1}/(2^N - 1);$$

hence

$$P[A_k] = \sum_{j=1}^{k} p_j = \sum_{j=1}^{k} 2^{j-1}/(2^N - 1) = \frac{2^k - 1}{2^N - 1}. \qquad ////$$

3.6 Conditional Probability and Independence

In the application of probability theory to practical problems it is not infrequent that the experimenter is confronted with the following situation: Such and such has happened; now what is the probability that something else will happen? For example, in an experiment of recording the life of a light bulb, one might be interested in the probability that the bulb will last 100 hours given that it has already lasted for 24 hours. Or in an experiment of sampling from a box containing 100 resistors of which 5 are defective, what is the probability that the third draw results in a defective given that the first two draws resulted in defectives? Probability questions of this sort are considered in the framework of conditional probability, the subject that we study next.

Conditional probability We begin by assuming that we have a probability space, say $(\Omega, \mathscr{A}, P[\cdot])$; that is, we have at hand some random experiment for which a sample space Ω, collection of events $\mathscr{A}$, and probability function $P[\cdot]$ have all been defined.

Given two events A and B, we want to define the conditional probability of event A given that event B has occurred.

Definition 18 Conditional probability Let A and B be two events in $\mathscr{A}$ of the given probability space $(\Omega, \mathscr{A}, P[\cdot])$. The *conditional probability* of event A given event B, denoted by $P[A \mid B]$, is defined by

$$P[A \mid B] = \frac{P[AB]}{P[B]} \qquad \text{if} \qquad P[B] > 0, \tag{6}$$

and is left undefined if $P[B] = 0$. ////

Remark A formula that is evident from the definition is $P[AB] = P[A \mid B]P[B] = P[B \mid A]P[A]$ if both $P[A]$ and $P[B]$ are nonzero. This formula relates $P[A \mid B]$ to $P[B \mid A]$ in terms of the unconditional probabilities $P[A]$ and $P[B]$. ////

We might note that the above definition is compatible with the frequency approach to probability, for if one observes a large number, say N, of occurrences of a random experiment for which events A and B are defined, then $P[A \mid B]$ represents the proportion of occurrences in which B occurred that A also occurred, that is,

$$P[A \mid B] = \frac{N_{AB}}{N_B},$$

where N_B denotes the number of occurrences of the event B in the N occurrences of the random experiment and N_{AB} denotes the number of occurrences of the event $A \cap B$ in the N occurrences. Now $P[AB] = N_{AB}/N$, and $P[B] = N_B/N$; so

$$\frac{P[AB]}{P[B]} = \frac{N_{AB}/N}{N_B/N} = \frac{N_{AB}}{N_B} = P[A \mid B],$$

consistent with our definition.

EXAMPLE 23 Let Ω be any finite sample space, $\mathscr{A}$ the collection of all subsets of Ω, and $P[\cdot]$ the equally likely probability function. Write $N = N(\Omega)$. For events A and B,

$$P[A \mid B] = \frac{P[AB]}{P[B]} = \frac{N(AB)/N}{N(B)/N},$$

where, as usual, $N(B)$ is the size of set B. So for any finite sample space with equally likely sample points, the values of $P[A \mid B]$ are defined for any two events A and B provided $P[B] > 0$. ////

EXAMPLE 24 Consider the experiment of tossing two coins. Let $\Omega = \{(\text{H}, \text{H}), (\text{H}, \text{T}), (\text{T}, \text{H}), (\text{T}, \text{T})\}$, and assume that each point is equally likely. Find (i) the probability of two heads given a head on the first coin and (ii) the probability of two heads given at least one head. Let $A_1 = \{\text{head on first coin}\}$ and $A_2 = \{\text{head on second coin}\}$; then the probability of two heads given a head on the first coin is

$$P[A_1 A_2 \mid A_1] = \frac{P[A_1 A_2 A_1]}{P[A_1]} = \frac{P[A_1 A_2]}{P[A_1]} = \frac{\frac{1}{4}}{\frac{1}{2}} = \frac{1}{2}.$$

The probability of two heads given at least one head is

$$P[A_1A_2 | A_1 \cup A_2] = \frac{P[A_1A_2 \cap (A_1 \cup A_2)]}{P[A_1 \cup A_2]} = \frac{P[A_1A_2]}{P[A_1 \cup A_2]} = \frac{\frac{1}{4}}{\frac{3}{4}} = \frac{1}{3}.$$

We obtained numerical answers to these two questions, but to do so we had to model the experiment; we assumed that the four sample points were equally likely.

When speaking of conditional probabilities we are conditioning on some given event B; that is, we are assuming that the experiment has resulted in some outcome in B. B, in effect, then becomes our "new" sample space. One question that might be raised is: For given event B for which $P[B] > 0$, is $P[\cdot | B]$ a probability function having $\mathscr{A}$ as its domain? In other words, does $P[\cdot | B]$ satisfy the three axioms? Note that:

(i) $P[A | B] = P[AB]/P[B] \geq 0$ for every $A \in \mathscr{A}$.

(ii) $P[\Omega | B] = P[\Omega B]/P[B] = P[B]/P[B] = 1$.

(iii) If $A_1, A_2, \ldots$ is a sequence of mutually exclusive events in $\mathscr{A}$ and $\bigcup_{i=1}^{\infty} A_i \in \mathscr{A}$, then

$$P\left[\bigcup_{i=1}^{\infty} A_i | B\right] = \frac{P\left[\left(\bigcup_{i=1}^{\infty} A_i\right) B\right]}{P[B]} = \frac{P\left[\bigcup_{i=1}^{\infty} (A_i B)\right]}{P[B]} = \frac{\sum_{i=1}^{\infty} P[A_i B]}{P[B]} = \sum_{i=1}^{\infty} P[A_i | B].$$

Hence, $P[\cdot | B]$ for given B satisfying $P[B] > 0$ is a probability function, which justifies our calling it a conditional probability. $P[\cdot | B]$ also enjoys the same properties as the unconditional probability. The theorems listed below are patterned after those in Subsec. 3.4.

Properties of $P[\cdot | B]$ Assume that the probability space $(\Omega, \mathscr{A}, P[\cdot])$ is given, and let $B \in \mathscr{A}$ satisfy $P[B] > 0$.

Theorem 22 $P[\phi | B] = 0$. ////

Theorem 23 If $A_1, \ldots, A_n$ are mutually exclusive events in $\mathscr{A}$, then

$$P[A_1 \cup \cdots \cup A_n | B] = \sum_{i=1}^{n} P[A_i | B]. \quad ////$$

Theorem 24 If A is an event in $\mathscr{A}$, then

$$P[\bar{A} | B] = 1 - P[A | B]. \quad ////$$

Theorem 25 If A_1 and $A_2 \in \mathscr{A}$, then

$$P[A_1 \mid B] = P[A_1 A_2 \mid B] + P[A_1 \bar{A}_2 \mid B]. \qquad ////$$

Theorem 26 For every two events A_1 and $A_2 \in \mathscr{A}$,

$$P[A_1 \cup A_2 \mid B] = P[A_1 \mid B] + P[A_2 \mid B] - P[A_1 A_2 \mid B]. \qquad ////$$

Theorem 27 If A_1 and $A_2 \in \mathscr{A}$ and $A_1 \subset A_2$, then

$$P[A_1 \mid B] \le P[A_2 \mid B]. \qquad ////$$

Theorem 28 If $A_1, A_2, \ldots, A_n \in \mathscr{A}$, then

$$P[A_1 \cup A_2 \cup \cdots \cup A_n \mid B] \le \sum_{j=1}^{n} P[A_i \mid B]. \qquad ////$$

Proofs of the above theorems follow from known properties of $P[\cdot]$ and are left as exercises.

There are a number of other useful formulas involving conditional probabilities that we will state as theorems. These will be followed by examples.

Theorem 29 Theorem of total probabilities For a given probability space $(\Omega, \mathscr{A}, P[\cdot])$, if $B_1, B_2, \ldots, B_n$ is a collection of mutually disjoint events in $\mathscr{A}$ satisfying $\Omega = \bigcup_{j=1}^{n} B_j$ and $P[B_j] > 0$ for $j = 1, \ldots, n$, then for every $A \in \mathscr{A}$, $P[A] = \sum_{j=1}^{n} P[A \mid B_j]P[B_j]$.

PROOF Note that $A = \bigcup_{j=1}^{n} AB_j$ and the AB_j's are mutually disjoint; hence

$$P[A] = P\left[\bigcup_{j=1}^{n} AB_j\right] = \sum_{j=1}^{n} P[AB_j] = \sum_{j=1}^{n} P[A \mid B_j]P[B_j]. \qquad ////$$

Corollary For a given probability space $(\Omega, \mathscr{A}, P[\cdot])$ let $B \in \mathscr{A}$ satisfy $0 < P[B] < 1$; then for every $A \in \mathscr{A}$

$$P[A] = P[A \mid B]P[B] + P[A \mid \bar{B}]P[\bar{B}]. \qquad ////$$

Remark Theorem 29 remains true if $n = \infty$. ////

Theorem 29 (and its corollary) is particularly useful for those experiments that have stages; that is, the experiment consists of performing first one thing (first stage) and then another (second stage). Example 25 provides an example of such an experiment; there, one first selects an urn and then selects a ball from the selected urn. For such experiments, if B_j is an event defined only in terms of the first stage and A is an event defined in terms of the second stage, then it may be easy to find $P[B_j]$; also, it may be easy to find $P[A \mid B_j]$, and then Theorem 29 evaluates $P[A]$ in terms of $P[B_j]$ and $P[A \mid B_j]$ for $j = 1, \ldots, n$. In an experiment consisting of stages it is natural to condition on results of a first stage.

Theorem 30 Bayes' formula For a given probability space $(\Omega, \mathscr{A}, P[\cdot])$, if $B_1, B_2, \ldots, B_n$ is a collection of mutually disjoint events in $\mathscr{A}$ satisfying $\Omega = \bigcup_{j=1}^{n} B_j$ and $P[B_j] > 0$ for $j = 1, \ldots, n$, then for every $A \in \mathscr{A}$ for which $P[A] > 0$

$$P[B_k \mid A] = \frac{P[A \mid B_k]P[B_k]}{\sum_{j=1}^{n} P[A \mid B_j]P[B_j]}.$$

PROOF

$$P[B_k \mid A] = \frac{P[B_k A]}{P[A]} = \frac{P[A \mid B_k]P[B_k]}{\sum_{j=1}^{n} P[A \mid B_j]P[B_j]}$$

by using both the definition of conditional probability and the theorem of total probabilities. ////

Corollary For a given probability space $(\Omega, \mathscr{A}, P[\cdot])$ let A and $B \in \mathscr{A}$ satisfy $P[A] > 0$ and $0 < P[B] < 1$; then

$$P[B \mid A] = \frac{P[A \mid B]P[B]}{P[A \mid B]P[B] + P[A \mid \bar{B}]P[\bar{B}]}. \qquad ////$$

Remark Theorem 30 remains true if $n = \infty$. ////

As was the case with the theorem of total probabilities, Bayes' formula is also particularly useful for those experiments consisting of stages. If B_j, $j = 1, \ldots, n$, is an event defined in terms of a first stage and A is an event defined in terms of the whole experiment including a second stage, then asking for $P[B_k \mid A]$ is in a sense backward; one is asking for the probability of an event defined in terms of a first stage of the experiment conditioned on what happens in a later stage of the experiment. The natural conditioning would be to condition on what happens in the first stage of the experiment, and this is precisely what Bayes' formula does; it expresses $P[B_k|A]$ in terms of the natural conditioning given by $P[A|B_j]$ and $P[B_j]$, $j = 1, \ldots, n$.

Theorem 31 Multiplication rule For a given probability space $(\Omega, \mathscr{A}, P[\cdot])$, let $A_1, \ldots, A_n$ be events belonging to $\mathscr{A}$ for which $P[A_1 \cdots A_{n-1}] > 0$; then

$$P[A_1 A_2 \cdots A_n] = P[A_1]P[A_2|A_1]P[A_3|A_1A_2] \cdots P[A_n|A_1 \cdots A_{n-1}].$$

PROOF The proof can be attained by employing mathematical induction and is left as an exercise. ////

As with the two previous theorems, the multiplication rule is primarily useful for experiments defined in terms of stages. Suppose the experiment has n stages and A_j is an event defined in terms of stage j of the experiment; then $P[A_j|A_1A_2 \cdots A_{j-1}]$ is the conditional probability of an event described in terms of what happens on stage j conditioned on what happens on stages $1, 2, \ldots, j-1$. The multiplication rule gives $P[A_1 A_2 \cdots A_n]$ in terms of the natural conditional probabilities $P[A_j|A_1A_2 \cdots A_{j-1}]$ for $j = 2, \ldots, n$.

EXAMPLE 25 There are five urns, and they are numbered 1 to 5. Each urn contains 10 balls. Urn i has i defective balls and $10 - i$ nondefective balls, $i = 1, 2, \ldots, 5$. For instance, urn 3 has three defective balls and seven nondefective balls. Consider the following random experiment: First an urn is selected at random, and then a ball is selected at random from the selected urn. (The experimenter does not know which urn was selected.) Let us ask two questions: (i) What is the probability that a defective ball will be selected? (ii) If we have already selected the ball and noted that it is defective, what is the probability that it came from urn 5?

SOLUTION Let A denote the event that a defective ball is selected and B_i the event that urn i is selected, $i = 1, \ldots, 5$. Note that $P[B_i] = \frac{1}{5}$, $i = 1, \ldots, 5$, and $P[A|B_i] = i/10$, $i = 1, \ldots, 5$. Question (i) asks, What is $P[A]$? Using the theorem of total probabilities, we have

$$P[A] = \sum_{i=1}^{5} P[A|B_i]P[B_i] = \sum_{i=1}^{5} \frac{i}{10} \cdot \frac{1}{5} = \frac{1}{50} \sum_{i=1}^{5} i = \frac{1}{50} \frac{5 \cdot 6}{2} = \frac{3}{10}.$$

Note that there is a total of 50 balls of which 15 are defective! Question (ii) asks, What is $P[B_5 | A]$? Since urn 5 has more defective balls than any of the other urns and we selected a defective ball, we suspect that $P[B_5 | A] > P[B_i | A]$ for $i = 1$, 2, 3, or 4. In fact, we suspect $P[B_5 | A] > P[B_4 | A] > \cdots > P[B_1 | A]$. Employing Bayes' formula, we find

$$P[B_5 | A] = \frac{P[A | B_5]P[B_5]}{\sum_{i=1}^{5} P[A | B_i]P[B_i]} = \frac{\frac{1}{2} \cdot \frac{1}{5}}{\frac{3}{10}} = \frac{1}{3}.$$

Similarly,

$$P[B_k | A] = \frac{(k/10) \cdot \frac{1}{5}}{\frac{3}{10}} = \frac{k}{15}, \qquad k = 1, \ldots, 5,$$

substantiating our suspicion. Note that unconditionally all the B_i's were equally likely whereas, conditionally (conditioned on occurrence of event A), they were not. Also, note that

$$\sum_{k=1}^{5} P[B_k | A] = \sum_{k=1}^{5} \frac{k}{15} = \frac{1}{15} \sum_{k=1}^{5} k = \frac{1}{15} \frac{5 \cdot 6}{2} = 1. \qquad ////$$

EXAMPLE 26 Assume that a student is taking a multiple-choice test. On a given question, the student either knows the answer, in which case he answers it correctly, or he does not know the answer, in which case he guesses hoping to guess the right answer. Assume that there are five multiple-choice alternatives, as is often the case. The instructor is confronted with this problem: Having observed that the student got the correct answer, he wishes to know what is the probability that the student knew the answer. Let p be the probability that the student will know the answer and $1 - p$ the probability that the student guesses. Let us assume that the probability that the student gets the right answer given that he guesses is $\frac{1}{5}$. (This may not be a realistic assumption since even though the student does not know the right answer, he often would know that certain alternatives are wrong, in which case his probability of guessing correctly should be better than $\frac{1}{5}$.) Let A denote the event that the student got the right answer and B denote the event that the student knew the right answer. We are seeking $P[B | A]$. Using Bayes' formula, we have

$$P[B | A] = \frac{P[A | B]P[B]}{P[A | B]P[B] + P[A | \bar{B}]P[\bar{B}]} = \frac{1 \cdot p}{1 \cdot p + \frac{1}{5}(1 - p)}.$$

Note that

$$\frac{p}{p + \frac{1}{5}(1 - p)} \geq p. \qquad ////$$

EXAMPLE 27 An urn contains ten balls of which three are black and seven are white. The following game is played: At each trial a ball is selected at random, its color is noted, and it is replaced along with two additional balls of the same color. What is the probability that a black ball is selected in each of the first three trials? Let B_i denote the event that a black ball is selected on the ith trial. We are seeking $P[B_1 B_2 B_3]$. By the multiplication rule,

$$P[B_1 B_2 B_3] = P[B_1]P[B_2 | B_1]P[B_3 | B_1 B_2] = \tfrac{3}{10} \cdot \tfrac{5}{12} \cdot \tfrac{7}{14} = \tfrac{1}{16}. \qquad ////$$

EXAMPLE 28 Suppose an urn contains M balls of which K are black and $M - K$ are white. A sample of size n is drawn. Find the probability that the jth ball drawn is black given that the sample contains k black balls. (We intuitively expect the answer to be k/n.) We have to consider sampling (i) with replacement and (ii) without replacement.

SOLUTION Let A_k denote the event that the sample contains exactly k black balls and B_j denote the event that the jth ball drawn is black. We seek $P[B_j | A_k]$. Consider (i) first.

$$P[A_k] = \binom{n}{k} \frac{K^k(M-K)^{n-k}}{M^n} \quad \text{and} \quad P[A_k | B_j] = \binom{n-1}{k-1} \frac{K^{k-1}(M-K)^{n-k}}{M^{n-1}}$$

by Eq. (3) of Subsec. 3.5. Since the balls are replaced, $P[B_j] = K/M$ for any j. Hence,

$$P[B_j | A_k] = \frac{P[A_k | B_j]P[B_j]}{P[A_k]} = \frac{\binom{n-1}{k-1}[K^{k-1}(M-K)^{n-k}/M^{n-1}] \cdot \dfrac{K}{M}}{\binom{n}{k} K^k(M-K)^{n-k}/M^n} = \frac{k}{n}.$$

For case (ii),

$$P[A_k] = \frac{\binom{K}{k}\binom{M-K}{n-k}}{\binom{M}{n}} \quad \text{and} \quad P[A_k | B_j] = \frac{\binom{K-1}{k-1}\binom{M-K}{n-k}}{\binom{M-1}{n-1}}$$

by Eq. (5) of Subsec. 3.5. $P[B_j] = \sum_{i=0}^{j-1} P[B_j | C_i]P[C_i]$, where C_i denotes the event of exactly i black balls in the first $j - 1$ draws. Note that

$$P[C_i] = \frac{\binom{K}{i}\binom{M-K}{j-1-i}}{\binom{M}{j-1}}$$

and

$$P[B_j | C_i] = \frac{K - i}{M - j + 1},$$

and so

$$P[B_j] = \sum_{i=0}^{j-1} \frac{K - i}{M - j + 1} \frac{\binom{K}{i}\binom{M - K}{j - 1 - i}}{\binom{M}{j - 1}} = \frac{K}{M}.$$

Finally,

$$P[B_j | A_k] = \frac{P[A_k | B_j]P[B_j]}{P[A_k]} = \frac{\left[\binom{K - 1}{k - 1}\binom{M - K}{n - k}\Big/\binom{M - 1}{n - 1}\right]\frac{K}{M}}{\binom{K}{k}\binom{M - K}{n - k}\Big/\binom{M}{n}} = \frac{k}{n}.$$

Thus we obtain the same answer under either method of sampling. ////

Independence of events If $P[A | B]$ does not depend on event B, that is, $P[A | B] = P[A]$, then it would seem natural to say that event A is independent of event B. This is given in the following definition.

Definition 19 Independent events For a given probability space $(\Omega, \mathscr{A}, P[\cdot])$, let A and B be two events in $\mathscr{A}$. Events A and B are defined to be *independent* if and only if any one of the following conditions is satisfied:

(i) $P[AB] = P[A]P[B]$.

(ii) $P[A | B] = P[A]$ if $P[B] > 0$.

(iii) $P[B | A] = P[B]$ if $P[A] > 0$. ////

Remark Some authors use "statistically independent," or "stochastically independent," instead of "independent." ////

To argue the equivalence of the above three conditions, it suffices to show that (i) implies (ii), (ii) implies (iii), and (iii) implies (i). If $P[AB] = P[A]P[B]$, then $P[A | B] = P[AB]/P[B] = P[A]P[B]/P[B] = P[A]$ for $P[B] > 0$; so (i) implies (ii). If $P[A | B] = P[A]$, then $P[B | A] = P[A | B]P[B]/P[A] = P[A]P[B]/P[A] = P[B]$ for $P[A] > 0$ and $P[B] > 0$; so (ii) implies (iii). And if $P[B | A] = P[B]$, then $P[AB] = P[B | A]P[A] = P[B]P[A]$ for $P[A] > 0$. Clearly $P[AB] = P[A]P[B]$ if $P[A] = 0$ or $P[B] = 0$.

EXAMPLE 29 Consider the experiment of tossing two dice. Let A denote the event of an odd total, B the event of an ace on the first die, and C the event of a total of seven. We pose three problems:

(i) Are A and B independent?
(ii) Are A and C independent?
(iii) Are B and C independent?

We obtain $P[A|B] = \frac{1}{2} = P[A]$, $P[A|C] = 1 \neq P[A] = \frac{1}{2}$, and $P[C|B] = \frac{1}{6} = P[C] = \frac{1}{6}$; so A and B are independent, A is not independent of C, and B and C are independent. ////

The property of independence of two events A and B and the property that A and B are mutually exclusive are distinct, though related, properties. For example, two mutually exclusive events A and B are independent if and only if $P[A]P[B] = 0$, which is true if and only if either A or B has zero probability. Or if $P[A] \neq 0$ and $P[B] \neq 0$, then A and B independent implies that they are not mutually exclusive, and A and B mutually exclusive implies that they are not independent. Independence of A and B implies independence of other events as well.

Theorem 32 If A and B are two independent events defined on a given probability space $(\Omega, \mathscr{A}, P[\cdot])$, then A and $\bar{B}$ are independent, $\bar{A}$ and B are independent, and $\bar{A}$ and $\bar{B}$ are independent.

PROOF

$$P[A\bar{B}] = P[A] - P[AB] = P[A] - P[A]P[B] = P[A](1 - P[B]) = P[A]P[\bar{B}].$$

Similarly for the others. ////

The notion of independent events may be extended to more than two events.

Definition 20 Independence of several events For a given probability space $(\Omega, \mathscr{A}, P[\cdot])$, let $A_1, A_2, \ldots, A_n$ be n events in $\mathscr{A}$. Events $A_1, A_2, \ldots, A_n$ are defined to be *independent* if and only if

$$P[A_i A_j] = P[A_i]P[A_j] \qquad \text{for } i \neq j$$

$$P[A_i A_j A_k] = P[A_i]P[A_j]P[A_k] \qquad \text{for } i \neq j, j \neq k, i \neq k$$

$$\vdots$$

$$P\left[\bigcap_{i=1}^{n} A_i\right] = \prod_{i=1}^{n} P[A_i].$$

////

One might inquire whether all the above conditions are required in the definition. For instance, does $P[A_1A_2A_3] = P[A_1]P[A_2]P[A_3]$ imply $P[A_1A_2] = P[A_1]P[A_2]$? Obviously not, since $P[A_1A_2A_3] = P[A_1]P[A_2]P[A_3]$ if $P[A_3] = 0$, but $P[A_1A_2] \neq P[A_1]P[A_2]$ if A_1 and A_2 are not independent. Or does pairwise independence imply independence? Again the answer is negative, as the following example shows.

EXAMPLE 30 Pairwise independence does not imply independence. Let A_1 denote the event of an odd face on the first die, A_2 the event of an odd face on the second die, and A_3 the event of an odd total in the random experiment that consists of tossing two dice. $P[A_1]P[A_2] = \frac{1}{2} \cdot \frac{1}{2} = P[A_1A_2]$, $P[A_1]P[A_3] = \frac{1}{2} \cdot \frac{1}{2} = P[A_3 | A_1]P[A_1] = P[A_1A_3]$, and $P[A_2A_3] = \frac{1}{4} = P[A_2]P[A_3]$; so A_1, A_2, and A_3 are pairwise independent. However $P[A_1A_2A_3] = 0 \neq \frac{1}{8} = P[A_1]P[A_2]P[A_3]$; so A_1, A_2, and A_3 are not independent. ////

In one sense, independence and conditional probability are each used to find the same thing, namely, $P[AB]$, for $P[AB] = P[A]P[B]$ under independence and $P[AB] = P[A | B]P[B]$ under nonindependence. The nature of the events A and B may make calculations of $P[A]$, $P[B]$, and possibly $P[A | B]$ easy, but direct calculation of $P[AB]$ difficult, in which case our formulas for independence or conditional probability would allow us to avoid the difficult direct calculation of $P[AB]$. We might note that $P[AB] = P[A | B]P[B]$ is valid whether or not A is independent of B provided that $P[A | B]$ is defined.

The definition of independence is used not only to check if two given events are independent but also to model experiments. For instance, for a given experiment the nature of the events A and B might be such that we are willing to assume that A and B are independent; then the definition of independence gives the probability of the event $A \cap B$ in terms of $P[A]$ and $P[B]$. Similarly for more than two events.

EXAMPLE 31 Consider the experiment of sampling with replacement from an urn containing M balls of which K are black and $M - K$ white. Since balls are being replaced after each draw, it seems reasonable to assume that the outcome of the second draw is independent of the outcome of the first. Then P[two blacks in first two draws] = P[black on first draw]P[black on second draw] $= (K/M)^2$. ////

PROBLEMS

To solve some of these problems it may be necessary to make certain assumptions, such as sample points are equally likely, or trials are independent, etc., when such assumptions are not explicitly stated. Some of the more difficult problems, or those that require special knowledge, are marked with an *.

1 One urn contains one black ball and one gold ball. A second urn contains one white and one gold ball. One ball is selected at random from each urn.

(*a*) Exhibit a sample space for this experiment.

(*b*) Exhibit the event space.

(*c*) What is the probability that both balls will be of the same color?

(*d*) What is the probability that one ball will be green?

2 One urn contains three red balls, two white balls, and one blue ball. A second urn contains one red ball, two white balls, and three blue balls.

(*a*) One ball is selected at random from each urn.

(i) Describe a sample space for this experiment.

(ii) Find the probability that both balls will be of the same color.

(iii) Is the probability that both balls will be red greater than the probability that both will be white?

(*b*) The balls in the two urns are mixed together in a single urn, and then a sample of three is drawn. Find the probability that all three colors are represented, when (i) sampling with replacement and (ii) without replacement.

3 If A and B are disjoint events, $P[A] = .5$, and $P[A \cup B] = .6$, what is $P[B]$?

4 An urn contains five balls numbered 1 to 5 of which the first three are black and the last two are gold. A sample of size 2 is drawn with replacement. Let B_1 denote the event that the first ball drawn is black and B_2 denote the event that the second ball drawn is black.

(*a*) Describe a sample space for the experiment, and exhibit the events B_1, B_2, and B_1B_2.

(*b*) Find $P[B_1]$, $P[B_2]$, and $P[B_1B_2]$.

(*c*) Repeat parts (*a*) and (*b*) for sampling without replacement.

5 A car with six spark plugs is known to have two malfunctioning spark plugs. If two plugs are pulled at random, what is the probability of getting both of the malfunctioning plugs?

6 In an assembly-line operation, $\frac{1}{3}$ of the items being produced are defective. If three items are picked at random and tested, what is the probability:

(*a*) That exactly one of them will be defective?

(*b*) That at least one of them will be defective?

7 In a certain game a participant is allowed three attempts at scoring a hit. In the three attempts he must alternate which hand is used; thus he has two possible strategies: right hand, left hand, right hand; or left hand, right hand, left hand. His chance of scoring a hit with his right hand is .8, while it is only .5 with his left hand. If he is successful at the game provided that he scores at least two hits in a row, what strategy gives the better chance of success? Answer the same question if .8 is replaced by p_1 and .5 by p_2. Does your answer depend on p_1 and p_2?

(*a*) Suppose that A and B are two equally strong teams. Is it more probable that A will beat B in three games out of four or in five games out of seven?

(*b*) Suppose now that the probability that A beats B in an individual game is p. Answer part (*a*). Does your answer depend on p?

9 If $P[A] = \frac{1}{3}$ and $P[\bar{B}] = \frac{1}{4}$, can A and B be disjoint? Explain.

10 Prove or disprove: If $P[A] = P[B] = p$, then $P[AB] \leq p^2$.

11 Prove or disprove: If $P[A] = P[\bar{B}]$ then $\bar{A} = B$.

12 Prove or disprove: If $P[A] = 0$, then $A = \phi$.

13 Prove or disprove: If $P[A] = 0$, then $P[AB] = 0$.

14 Prove: If $P[\bar{A}] = \alpha$ and $P[\bar{B}] = \beta$, then $P[AB] \geq 1 - \alpha - \beta$.

15 Prove properties (i) to (iv) of indicator functions.

16 Prove the more general statement in Theorem 19.

17 Exhibit (if such exists) a probability space, denoted by $(\Omega, \mathscr{A}, P[\cdot])$, which satisfies the following. For A_1 and A_2 members of $\mathscr{A}$, if $P[A_1] = P[A_2]$, then $A_1 = A_2$.

18 Four drinkers (say I, II, III, and IV) are to rank three different brands of beer (say A, B, and C) in a blindfold test. Each drinker ranks the three beers as 1 (for the beer he likes best), 2, and 3, and then the assigned ranks of each brand of beer are summed. Assume that the drinkers really cannot discriminate between beers so that each is assigning his rankings at random.

(*a*) What is the probability that beer A will receive a total score of 4?

(*b*) What is the probability that some beer will receive a total score of 4?

(*c*) What is the probability that some beer will receive a total score of 5 or less?

19 The following are three of the classical problems in probability.

(*a*) Compare the probability of a total of 9 with a total of 10 when three fair dice are tossed once (Galileo and Duke of Tuscany).

(*b*) Compare the probability of at least one 6 in 4 tosses of a fair die with the probability of at least one double-6 in 24 tosses of two fair dice (Chevalier de Méré).

(*c*) Compare the probability of at least one 6 when six dice are rolled with the probability of at least two 6s when twelve dice are rolled (Pepys to Newton).

20 A seller has a dozen small electric motors, two of which are faulty. A customer is interested in the dozen motors. The seller can crate the motors with all twelve in one box or with six in each of two boxes; he knows that the customer will inspect two of the twelve motors if they are all crated in one box and one motor from each of the two smaller boxes if they are crated six each to two smaller boxes. He has three strategies in his attempt to sell the faulty motors: (i) crate all twelve in one box; (ii) put one faulty motor in each of the two smaller boxes; or (iii) put both of the faulty motors in one of the smaller boxes and no faulty motors in the other. What is the probability that the customer will not inspect a faulty motor under each of the three strategies?

21 A sample of five objects is drawn from a larger population of N objects ($N \geq 5$). Let N_w or N_{wo} denote the number of different samples that could be drawn depending, respectively, on whether sampling is done with or without replacement. Give the values for N_w and N_{wo}. Show that when N is very large, these two values are approximately equal in the sense that their ratio is close to 1 but not in the sense that their difference is close to 0.

22 Out of a group of 25 persons, what is the probability that all 25 will have different birthdays? (Assume a 365-day year and that all days are equally likely.)

23 A bridge player knows that his two opponents have exactly five hearts between the two of them. Each opponent has thirteen cards. What is the probability that there is a three-two split on the hearts (that is, one player has three hearts and the other two)?

24 (*a*) If r balls are randomly placed into n urns (each ball having probability $1/n$ of going into the first urn), what is the probability that the first urn will contain exactly k balls?

(*b*) Let $n \to \infty$ and $r \to \infty$ while $r/n = m$ remains constant. Show that the probability you calculated approaches $e^{-m}m^k/k!$.

25 A biased coin has probability p of landing heads. Ace, Bones, and Clod toss the coin successively, Ace tossing first, until a head occurs. The person who tosses the first head wins. Find the probability of winning for each.

**26* It is told that in certain rural areas of Russia marital fortunes were once told in the following way: A girl would hold six strings in her hand with the ends protruding above and below; a friend would tie together the six upper ends in pairs and then tie together the six lower ends in pairs. If it turned out that the friend had tied the six strings into at least one ring, this was supposed to indicate that the girl would get married within a year. What is the probability that a single ring will be formed when the strings are tied at random? What is the probability that at least one ring will be formed? Generalize the problem to $2n$ strings.

27 Mr. Bandit, a well-known rancher and not so well-known part-time cattle rustler, has twenty head of cattle ready for market. Sixteen of these cattle are his own and consequently bear his own brand. The other four bear foreign brands. Mr. Bandit knows that the brand inspector at the market place checks the brands of 20 percent of the cattle in any shipment. He has two trucks, one which will haul all twenty cattle at once and the other that will haul ten at a time. Mr. Bandit feels that he has four different strategies to follow in his attempt to market the cattle without getting caught. The first is to sell all twenty head at once; the others are to sell ten head on two different occasions, putting all four stolen cattle in one set of ten, or three head in one shipment and one in the other, or two head in each of the shipments of ten. Which strategy will minimize Mr. Bandit's probability of getting caught, and what is his probability of getting caught under each strategy?

28 Show that the formula of Eq. (4) is the same as the formula of Eq. (5).

29 Prove Theorem 31.

30 Either prove or disprove each of the following (you may assume that none of the events has zero probability):

(*a*) If $P[A|B] > P[A]$, then $P[B|A] > P[B]$.

(*b*) If $P[A] > P[B]$, then $P[A|C] > P[B|C]$.

31 A certain computer program will operate using either of two subroutines, say A and B, depending on the problem; experience has shown that subroutine A will be used 40 percent of the time and B will be used 60 percent of the time. If A is used, then there is a 75 percent probability that the program will run before its time limit is exceeded; and if B is used, there is a 50 percent chance that it will do so. What is the probability that the program will run without exceeding the time limit?

32 Suppose that it is known that a fraction .001 of the people in a town have tuberculosis (TB). A tuberculosis test is given with the following properties: If the person does have TB, the test will indicate it with a probability .999. If he does not have TB, then there is a probability .002 that the test will erroneously indicate that he does. For one randomly selected person, the test shows that he has TB. What is the probability that he really does?

**33* Consider the experiment of tossing two fair regular tetrahedra (a polyhedron with four faces numbered 1 to 4) and noting the numbers on the downturned faces.

(*a*) Give three proper events (an event A is proper if $0 < P[A] < 1$) which are independent (if such exist).

(*b*) Give three proper events which are pairwise independent but not independent (if such exist).

(*c*) Give four proper events which are independent (if such exist).

34 Prove or disprove:

(*a*) If A and B are independent events, then $P[AB|C] = P[A|C]P[B|C]$.

(*b*) If $P[A|B] = P[B]$, then A and B are independent.

35 Prove or disprove:

(*a*) If $P[A|B] \geq P[A]$, then $P[B|A] \geq P[B]$.

(*b*) If $P[B|\bar{A}] = P[B|A]$, then A and B are independent.

(*c*) If $a = P[A]$ and $b = P[B]$, then $P[A|B] \geq (a + b - 1)/b$.

36 Consider an urn containing 10 balls of which 5 are black. Choose an integer n at random from the set 1, 2, 3, 4, 5, 6, and then choose a sample of size n without replacement from the urn. Find the probability that all the balls in the sample will be black.

37 A die is thrown as long as necessary for a 6 to turn up. Given that the 6 does not turn up at the first throw, what is the probability that more than four throws will be necessary?

38 Die A has four red and two blue faces, and die B has two red and four blue faces. The following game is played: First a coin is tossed once. If it falls heads, the game continues by repeatedly throwing die A; if it falls tails, die B is repeatedly tossed.

(a) Show that the probability of red at any throw is $\frac{1}{2}$.

(b) If the first two throws of the die resulted in red, what is the probability of red at the third throw?

(c) If red turns up at the first n throws, what is the probability that die A is being used?

39 Urn A contains two white and two black balls; urn B contains three white and two black balls. One ball is transferred from A to B; one ball is then drawn from B and turns out to be white. What is the probability that the transferred ball was white?

**40* It is known that each of four people A, B, C, and D tells the truth in a given instance with probability $\frac{1}{3}$. Suppose that A makes a statement, and then D says that C says that B says that A was telling the truth. What is the probability that A was actually telling the truth?

41 In a T maze, a laboratory animal is given a choice of going to the left and getting food or going to the right and receiving a mild electric shock. Assume that before any conditioning (in trial number 1) animals are equally likely to go to the left or to the right. After having received food on a particular trial, the probabilities of going to the left and right become .6 and .4, respectively, on the following trial. However, after receiving a shock on a particular trial, the probabilities of going to the left and right on the next trial are .8 and .2, respectively. What is the probability that the animal will turn left on trial number 2? On trial number 3?

**42* In a breeding experiment, the male parent is known to have either two dominant genes (symbolized by AA) or one dominant and one recessive (Aa). These two cases are equally likely. The female parent is known to have two recessive genes (aa). Since the offspring gets one gene from each parent, it will be either Aa or aa, and it will be possible to say with certainty which one.

(a) If we suppose one offspring is Aa, what is the probability that the male parent is AA?

(b) If we suppose two offspring are both Aa, what is the probability that the male parent is AA?

(c) If one offspring is aa, what is the probability that the male parent is Aa?

43 The constitution of two urns is

I	three black two white	II	four black six white

A draw is made by selecting an urn by a process which assigns probability p to the selection of urn I and probability $1 - p$ to the selection of urn II. The selection of a ball from either urn is by a process which assigns equal probability to all balls in the urn. What value of p makes the probability of obtaining a black ball the same as if a single draw were made from an urn with seven black and eight white balls (all balls equally probable of being drawn)?

44 Given $P[A] = .5$ and $P[A \cup B] = .6$, find $P[B]$ if:
(*a*) A and B are mutually exclusive.
(*b*) A and B are independent.
(*c*) $P[A \mid B] = .4$.

45 Three fair dice are thrown once. Given that no two show the same face:
(*a*) What is the probability that the sum of the faces is 7?
(*b*) What is the probability that one is an ace?

46 Given that $P[A] > 0$ and $P[B] > 0$, prove or disprove:
(*a*) If $P[A] = P[B]$, then $P[A \mid B] = P[B \mid A]$.
(*b*) If $P[A \mid B] = P[B \mid A]$, then $P[A] = P[B]$.

47 Five percent of the people have high blood pressure. Of the people with high blood pressure, 75 percent drink alcohol; whereas, only 50 percent of the people without high blood pressure drink alcohol. What percent of the drinkers have high blood pressure?

48 A distributor of watermelon seeds determined from extensive tests that 4 percent of a large batch of seeds will not germinate. He sells the seeds in packages of 50 seeds and guarantees at least 90 percent germination. What is the probability that a given package will violate the guarantee?

49 If A and B are independent, $P[A] = \frac{1}{3}$, and $P[\bar{B}] = \frac{1}{4}$, find $P[A \cup B]$.

50 Mr. Stoneguy, a wealthy diamond dealer, decides to reward his son by allowing him to select one of two boxes. Each box contains three stones. In one box two of the stones are real diamonds, and the other is a worthless imitation; and in the other box one is a real diamond, and the other two worthless imitations. If the son were to choose randomly between the two boxes, his chance of getting two real diamonds would be $\frac{1}{2}$. Mr. Stoneguy, being a sporting type, allows his son to draw one stone from one of the boxes and to examine it to see if it is a real diamond. The son decides to take the box that the stone he tested came from if the tested stone is real and to take the other box otherwise. Now what is the probability that the son will get two real diamonds?

51 If $P[A] = P[B] = P[B \mid A] = \frac{1}{2}$, are A and B independent?

52 If A and B are independent and $P[A] = P[B] = \frac{1}{2}$, what is $P[A\bar{B} \cup \bar{A}B]$?

53 If $P[B] = P[A \mid B] = P[C \mid AB] = \frac{1}{2}$, what is $P[ABC]$?

54 If A and B are independent and $P[A] = P[B \mid A] = \frac{1}{2}$, what is $P[A \cup B]$?

55 Suppose B_1, B_2, and B_3 are mutually exclusive. If $P[B_j] = \frac{1}{3}$ and $P[A \mid B_j] = j/6$ for $j = 1, 2, 3$, what is $P[A]$?

**56* The game of craps is played by letting the thrower toss two dice until he either wins or loses. The thrower wins on the first toss if he gets a total of 7 or 11; he loses on the first toss if he gets a total of 2, 3, or 12. If he gets any other total on his first toss, that total is called his *point*. He then tosses the dice repeatedly until he obtains a total of 7 or his point. He wins if he gets his point and loses if he gets a total of 7. What is the thrower's probability of winning?

57 In a dice game a player casts a pair of dice twice. He wins if the two totals thrown do not differ by more than 2 with the following exceptions: If he gets a

3 on the first throw, he must produce a 4 on the second throw; if he gets an 11 on the first throw, he must produce a 10 on the second throw. What is his probability of winning?

58 Assume that the conditional probability that a child born to a couple will be male is $\frac{1}{2} + m\varepsilon_1 - f\varepsilon_2$, where ε_1 and ε_2 are certain small constants, m is the number of male children already born to the couple, and f is the number of female children already born to the couple.

(*a*) What is the probability that the third child will be a boy given that the first two are girls?

(*b*) Find the probability that the first three children will be all boys.

(*c*) Find the probability of at least one boy in the first three children.

(Your answers will be expressed in terms of ε_1 and ε_2.)

**59* A network of switches a, b, c, and d is connected across the power lines A and B as shown in the sketch. Assume that the switches operate electrically and have independent operating mechanisms. All are controlled simultaneously by the same impulses; that is, it is intended that on an impulse all switches shall close simultaneously. But each switch has a probability p of failure (it will not close when it should).

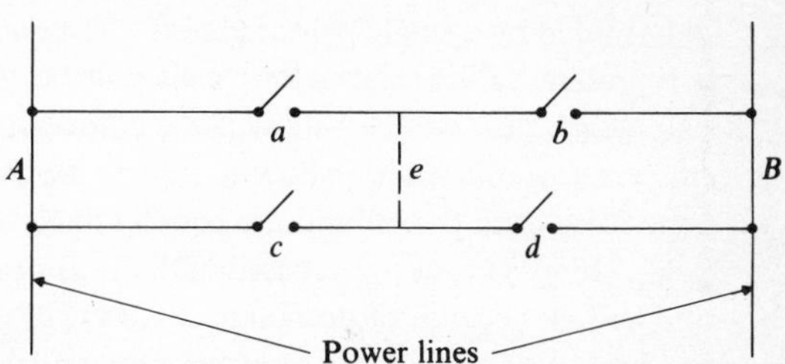

(*a*) What is the probability that the circuit from A to B will fail to close?

(*b*) If a line is added on at e, as indicated in the sketch, what is the probability that the circuit from A to B will fail to close?

(*c*) If a line and switch are added at e, what is the probability that the circuit from A to B will fail to close?

60 Let $B_1, B_2, \ldots, B_n$ be mutually disjoint, and let $B = \bigcup_{j=1}^{n} B_j$. Suppose $P[B_j] > 0$ and $P[A \mid B_j] = p$ for $j = 1, \ldots, n$. Show that $P[A \mid B] = p$.

61 In a laboratory experiment, an attempt is made to teach an animal to turn right in a maze. To aid in the teaching, the animal is rewarded if it turns right on a given trial and punished if it turns left. On the first trial the animal is just as likely to turn right as left. If on a particular trial the animal was rewarded, his probability of turning right on the next trial is $p_1 > \frac{1}{2}$, and if on a given trial the animal was punished, his probability of turning right on the next trial is $p_2 > p_1$.

(*a*) What is the probability that the animal will turn right on the third trial?

(*b*) What is the probability that the animal will turn right on the third trial, given that he turned right on the first trial?

*62 You are to play ticktacktoe with an opponent who on his turn makes his mark by selecting a space at random from the unfilled spaces. You get to mark first. Where should you mark to maximize your chance of winning, and what is your probability of winning? (Note that your opponent cannot win, he can only tie.)

63 Urns I and II each contain two white and two black balls. One ball is selected from urn I and transferred to urn II; then one ball is drawn from urn II and turns out to be white. What is the probability that the transferred ball was white?

64 Two regular tetrahedra with faces numbered 1 to 4 are tossed repeatedly until a total of 5 appears on the down faces. What is the probability that more than two tosses are required?

65 Given $P[A] = .5$ and $P[A \cup B] = .7$:

(a) Find $P[B]$ if A and B are independent.

(b) Find $P[B]$ if A and B are mutually exclusive.

(c) Find $P[B]$ if $P[A \mid B] = .5$.

66 A single die is tossed; then n coins are tossed, where n is the number shown on the die. What is the probability of exactly two heads?

*67 In simple Mendelian inheritance, a physical characteristic of a plant or animal is determined by a single pair of genes. The color of peas is an example. Let y and g represent yellow and green; peas will be green if the plant has the color-gene pair (g, g); they will be yellow if the color-gene pair is (y, y) or (y, g). In view of this last combination, yellow is said to be *dominant* to green. Progeny get one gene from each parent and are equally likely to get either gene from each parent's pair. If (y, y) peas are crossed with (g, g) peas, all the resulting peas will be (y, g) and yellow because of dominance. If (y, g) peas are crossed with (g, g) peas, the probability is .5 that the resulting peas will be yellow and is .5 that they will be green. In a large number of such crosses one would expect about half the resulting peas to be yellow, the remainder to be green. In crosses between (y, g) and (y, g) peas, what proportion would be expected to be yellow? What proportion of the yellow peas would be expected to be (y, y)?

*68 Peas may be smooth or wrinkled, and this is a simple Mendelian character. Smooth is dominant to wrinkled so that (s, s) and (s, w) peas are smooth while (w, w) peas are wrinkled. If (y, g) (s, w) peas are crossed with (g, g) (w, w) peas, what are the possible outcomes, and what are their associated probabilities? For the (y, g) (s, w) by (g, g) (s, w) cross? For the (y, g) (s, w) by (y, g) (s, w) cross?

69 Prove the two unproven parts of Theorem 32.

70 A supplier of a certain testing device claims that his device has high reliability inasmuch as $P[A \mid B] = P[\bar{A} \mid \bar{B}] = .95$, where $A = \{$device indicates component is faulty$\}$ and $B = \{$component is faulty$\}$. You hope to use the device to locate the faulty components in a large batch of components of which 5 percent are faulty.

(a) What is $P[B \mid A]$?

(b) Suppose you want $P[B \mid A] = .9$. Let $p = P[A \mid B] = P[\bar{A} \mid \bar{B}]$. How large does p have to be?

II

RANDOM VARIABLES, DISTRIBUTION FUNCTIONS, AND EXPECTATION

1 INTRODUCTION AND SUMMARY

The purpose of this chapter is to introduce the concepts of *random variable*, *distribution and density functions*, and *expectation*. It is primarily a "definitions-and-their-understanding" chapter; although some other results are given as well. The definitions of random variable and cumulative distribution function are given in Sec. 2, and the definitions of density functions are given in Sec. 3. These definitions are easily stated since each is just a particular function. The cumulative distribution function exists and is defined for each random variable; whereas, a density function is defined only for particular random variables. Expectations of functions of random variables are the underlying concept of all of Sec. 4. This concept is introduced by considering two particular, yet extremely important, expectations. These two are the mean and variance, defined in Subsecs. 4.1 and 4.2, respectively. Subsection 4.3 is devoted to the definition and properties of expectation of a function of a random variable. A very important result in the chapter appears in Subsec. 4.4 as the Chebyshev inequality and a generalization thereof. It is nice to be able to attain so famous a result so soon and with so little weaponry. The Jensen inequality is given in

Subsec. 4.5. Moments and moment generating functions, which are expectations of particular functions, are considered in the final subsection. One major unproven result, that of the uniqueness of the moment generating function, is given there. Also included is a brief discussion of some measures of some characteristics, such as location and dispersion, of distribution or density functions.

This chapter provides an introduction to the language of *distribution theory*. Only the univariate case is considered; the bivariate and multivariate cases will be considered in Chap. IV. It serves as a preface to, or even as a companion to, Chap. III, where a number of parametric families of distribution functions is presented. Chapter III gives many examples of the concepts defined in Chap. II.

2 RANDOM VARIABLE AND CUMULATIVE DISTRIBUTION FUNCTION

2.1 Introduction

In Chap. I we defined what we meant by a probability space, which we denoted by the triplet $(\Omega, \mathscr{A}, P[\cdot])$. We started with a conceptual random experiment; we called the totality of possible outcomes of this experiment the sample space and denoted it by Ω. $\mathscr{A}$ was used to denote a collection of subsets, called events, of the sample space. Finally our probability function $P[\cdot]$ was a set function having domain $\mathscr{A}$ and counterdomain the interval $[0, 1]$. Our object was, and still is, to assess probabilities of events. In other words, we want to model our random experiment so as to be able to give values to the probabilities of events. The notion of *random variable*, to be defined presently, will be used to *describe* events, and a *cumulative distribution function* will be used to give the probabilities of certain events defined in terms of random variables; so both concepts will assist us in defining probabilities of events, our goal. One advantage that a cumulative distribution function will have over its counterpart, the probability function (they both give probabilities of events), is that it is a function with domain the real line and counterdomain the interval $[0, 1]$. Thus we will be able to graph it. It will become a convenient tool in modeling random experiments. In fact, we will often model a random experiment by assuming certain things about a random variable and its distribution function and in so doing completely bypass describing the probability space.

2.2 Definitions

We commence by defining a random variable.

Definition 1 Random Variable For a given probability space $(\Omega, \mathscr{A}, P[\cdot])$, a *random variable*, denoted by X or $X(\cdot)$, is a function with domain Ω and counterdomain the real line. The function $X(\cdot)$ must be such that the set A_r, defined by $A_r = \{\omega: X(\omega) \leq r\}$, belongs to $\mathscr{A}$ for every real number r. ////

If one thinks in terms of a random experiment, Ω is the totality of outcomes of that random experiment, and the function, or random variable, $X(\cdot)$ with domain Ω makes some real number correspond to each outcome of the experiment. That is the important part of our definition. The fact that we also require the collection of ω's for which $X(\omega) \leq r$ to be an event (i.e., an element of $\mathscr{A}$) for each real number r is not much of a restriction for our purposes since our intention is to use the notion of random variable only in describing events. We will seldom be interested in a random variable per se; rather we will be interested in events defined in terms of random variables. One might note that the $P[\cdot]$ of our probability space $(\Omega, \mathscr{A}, P[\cdot])$ is not used in our definition.

The use of words "random" and "variable" in the above definition is unfortunate since their use cannot be convincingly justified. The expression "random variable" is a misnomer that has gained such widespread use that it would be foolish for us to try to rename it.

In our definition we denoted a random variable by either $X(\cdot)$ or X. Although $X(\cdot)$ is a more complete notation, one that emphasizes that a random variable is a function, we will usually use the shorter notation of X. For many experiments, there is a need to define more than one random variable; hence further notations are necessary. We will try to use capital Latin letters with or without affixes from near the end of the alphabet to denote random variables. Also, we use the corresponding small letter to denote a value of the random variable.

EXAMPLE 1 Consider the experiment of tossing a single coin. Let the random variable X denote the number of heads. $\Omega = \{\text{head, tail}\}$, and $X(\omega) = 1$ if ω = head, and $X(\omega) = 0$ if ω = tail; so, the random variable X associates a real number with each outcome of the experiment. We called X a random variable so mathematically speaking we should show

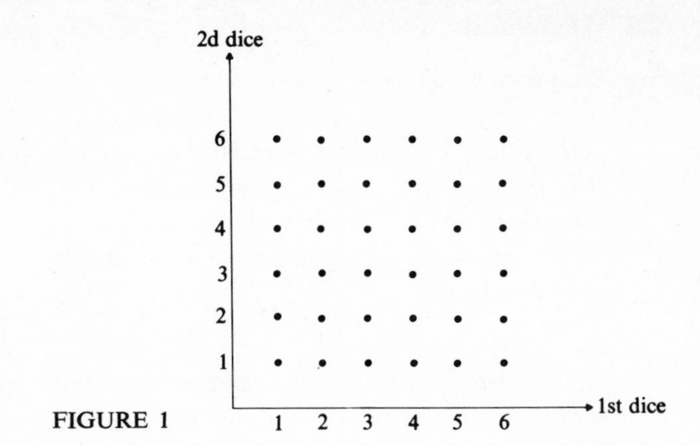

FIGURE 1

that it satisfies the definition; that is, we should show that $\{\omega\colon X(\omega) \le r\}$ belongs to $\mathscr{A}$ for every real number r. $\mathscr{A}$ consists of the four subsets: ϕ, {head}, {tail}, and Ω. Now, if $r < 0$, $\{\omega\colon X(\omega) \le r\} = \phi$; and if $0 \le r < 1$, $\{\omega\colon X(\omega) \le r\} = \{\text{tail}\}$; and if $r \ge 1$, $\{\omega\colon X(\omega) \le r\} = \Omega = \{\text{head, tail}\}$. Hence, for each r the set $\{\omega\colon X(\omega) \le r\}$ belongs to $\mathscr{A}$; so $X(\cdot)$ is a random variable. ////

EXAMPLE 2 Consider the experiment of tossing two dice. Ω can be described by the 36 points displayed in Fig. 1. $\Omega = \{(i, j)\colon i = 1, \ldots, 6$ and $j = 1, \ldots, 6\}$. Several random variables can be defined; for instance, let X denote the sum of the upturned faces; so $X(\omega) = i + j$ if $\omega = (i, j)$. Also, let Y denote the absolute difference between the upturned faces; then $Y(\omega) = |i - j|$ if $\omega = (i, j)$. It can be shown that both X and Y are random variables. We see that X can take on the values 2, 3, ..., 12 and Y can take on the values 0, 1, ..., 5. ////

In both of the above examples we described the random variables in terms of the random experiment rather than in specifying their functional form; such will usually be the case.

Definition 2 Cumulative distribution function The *cumulative distribution function* of a random variable X, denoted by $F_X(\cdot)$, is defined to be that function with domain the real line and counterdomain the interval

[0, 1] which satisfies $F_X(x) = P[X \leq x] = P[\{\omega\colon X(\omega) \leq x\}]$ for every real number x. ////

A cumulative distribution function is uniquely defined for each random variable. If it is known, it can be used to find probabilities of events defined in terms of its corresponding random variable. (One might note that it is in this definition that we use the requirement that $\{\omega\colon X(\omega) \leq r\}$ belong to $\mathscr{A}$ for every real r which appears in our definition of random variable X.) Note that different random variables can have the same cumulative distribution function. See Example 4 below.

The use of each of the three words in the expression "cumulative distribution function" is justifiable. A cumulative distribution function is first of all a *function*; it is a *distribution* function inasmuch as it tells us how the values of the random variable are distributed, and it is a *cumulative* distribution function since it gives the distribution of values in cumulative form. Many writers omit the word "cumulative" in this definition. Examples and properties of cumulative distribution functions follow.

EXAMPLE 3 Consider again the experiment of tossing a single coin. Assume that the coin is fair. Let X denote the number of heads. Then,

$$F_X(x) = \begin{cases} 0 & \text{if } x < 0 \\ \frac{1}{2} & \text{if } 0 \leq x < 1 \\ 1 & \text{if } 1 \leq x. \end{cases}$$

Or $F_X(x) = \frac{1}{2}I_{[0,1)}(x) + I_{[1,\infty)}(x)$ in our indicator function notation. ////

EXAMPLE 4 In the experiment of tossing two fair dice, let Y denote the absolute difference. The cumulative distribution of Y, $F_Y(\cdot)$, is sketched in Fig. 2. Also, let X_k denote the value on the upturned face of the kth die for $k = 1, 2$. X_1 and X_2 are different random variables, yet both have the same cumulative distribution function, which is $F_{X_k}(x) = \sum_{i=1}^{5} \frac{i}{6} I_{[i,i+1)}(x) + I_{[6,\infty)}(x)$ and is sketched in Fig. 3. ////

Careful scrutiny of the definition and above examples might indicate the following properties of any cumulative distribution function $F_X(\cdot)$.

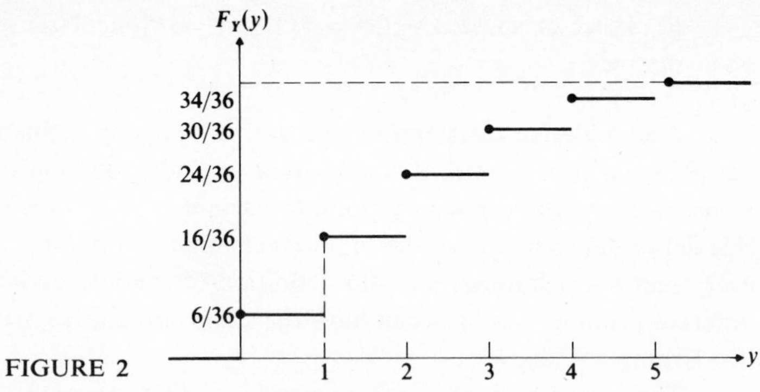

FIGURE 2

Properties of a Cumulative Distribution Function $F_X(\cdot)$

(i) $F_X(-\infty) \equiv \lim_{x \to -\infty} F_X(x) = 0$, and $F_X(+\infty) \equiv \lim_{x \to +\infty} F_X(x) = 1$.

(ii) $F_X(\cdot)$ is a monotone, nondecreasing function; that is, $F_X(a) \leq F_X(b)$ for $a < b$.

(iii) $F_X(\cdot)$ is continuous from the right; that is,

$$\lim_{0<h \to 0} F_X(x + h) = F_X(x).$$

Except for (ii), we will not prove these properties. Note that the event $\{\omega\colon X(\omega) \leq b\} = \{X \leq b\} = \{X \leq a\} \cup \{a < X \leq b\}$ and $\{X \leq a\} \cap \{a < X \leq b\} = \phi$; hence, $F_X(b) = P[X \leq b] = P[X \leq a] + P[a < X \leq b] \geq P[X \leq a] = F_X(a)$ which proves (ii). Property (iii), the continuity of $F_X(\cdot)$ from the right, results from our defining $F_X(x)$ to be $P[X \leq x]$. If we had defined, as some authors do, $F_X(x)$ to be $P[X < x]$, then $F_X(\cdot)$ would have been continuous from the left.

Definition 3 Cumulative distribution function Any function $F(\cdot)$ with domain the real line and counterdomain the interval [0, 1] satisfying the above three properties is defined to be a *cumulative distribution function.* ////

This definition allows us to use the term "cumulative distribution function" without mentioning random variable.

After defining what is meant by continuous and discrete random variables in the first two subsections of the next section, we will give another property that cumulative distribution functions possess, the property of decomposition into three parts.

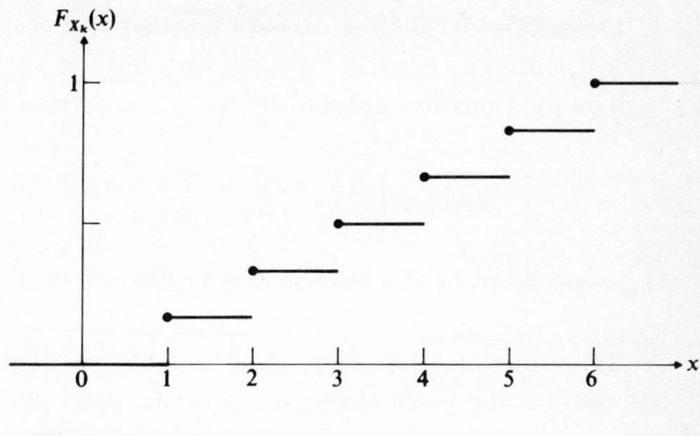

FIGURE 3

The cumulative distribution functions defined here are univariate; the introduction of bivariate and multivariate cumulative distribution functions will be deferred until Chap. IV.

3 DENSITY FUNCTIONS

Random variable and the cumulative distribution function of a random variable have been defined. The cumulative distribution function described the distribution of values of the random variable. For two distinct classes of random variables, the distribution of values can be described more simply by using *density functions.* These two classes, distinguished by the words "discrete" and "continuous," are considered in the next two subsections.

3.1 Discrete Random Variables

Definition 4 Discrete random variable A random variable X will be defined to be *discrete* if the range of X is countable. If a random variable X is discrete, then its corresponding cumulative distribution function $F_X(\cdot)$ will be defined to be *discrete.* ////

By the range of X being countable we mean that there exists a finite or denumerable set of real numbers, say $x_1, x_2, x_3, \ldots$, such that X takes on values only in that set. If X is discrete with distinct values $x_1, x_2, \ldots, x_n, \ldots$, then $\Omega = \bigcup_n \{\omega\colon X(\omega) = x_n\} = \bigcup_n \{X = x_n\}$, and $\{X = x_i\} \cap \{X = x_j\} = \phi$ for $i \neq j$; hence $1 = P[\Omega] = \sum_n P[X = x_n]$ by the third axiom of probability.

Definition 5 Discrete density function of a discrete random variable If X is a discrete random variable with distinct values $x_1, x_2, \ldots, x_n, \ldots$, then the function, denoted by $f_X(\cdot)$ and defined by

$$f_X(x) = \begin{cases} P[X = x_j] & \text{if } x = x_j, j = 1, 2, \ldots, n, \ldots \\ 0 & \text{if } x \neq x_j \end{cases} \tag{1}$$

is defined to be the *discrete density function* of X. ////

The values of a discrete random variable are often called *mass points*; and, $f_X(x_j)$ denotes the *mass* associated with the *mass point* x_j. *Probability mass function*, *discrete frequency function*, and *probability function* are other terms used in place of *discrete density function*. Also, the notation $p_X(\cdot)$ is sometimes used intead of $f_X(\cdot)$ for discrete density functions. $f_X(\cdot)$ is a function with domain the real line and counterdomain the interval [0, 1]. If we use the indicator function,

$$f_X(x) = \sum_{n=1}^{\infty} P[X = x_n] I_{\{x_n\}}(x), \tag{2}$$

where $I_{\{x_n\}}(x) = 1$ if $x = x_n$ and $I_{\{x_n\}}(x) = 0$ if $x \neq x_n$.

Theorem 1 Let X be a discrete random variable. $F_X(\cdot)$ can be obtained from $f_X(\cdot)$, and vice versa.

PROOF Denote the mass points of X by $x_1, x_2, \ldots$. Suppose $f_X(\cdot)$ is given; then $F_X(x) = \sum_{\{j: x_j \leq x\}} f_X(x_j)$. Conversely, suppose $F_X(\cdot)$ is given; then $f_X(x_j) = F_X(x_j) - \lim_{0<h\to 0} F_X(x_j - h)$; hence $f_X(x_j)$ can be found for each mass point x_j; however, $f_X(x) = 0$ for $x \neq x_j, j = 1, 2, \ldots$, so $f_X(x)$ is determined for all real numbers. ////

EXAMPLE 5 To illustrate what is meant in Theorem 1, consider the experiment of tossing a single die. Let X denote the number of spots on the upper face:

$$f_X(x) = \left(\frac{1}{6}\right) I_{\{1, 2, \ldots, 6\}}(x),$$

and

$$F_X(x) = \sum_{i=1}^{5} (i/6) I_{[i, i+1)}(x) + I_{[6, \infty)}(x).$$

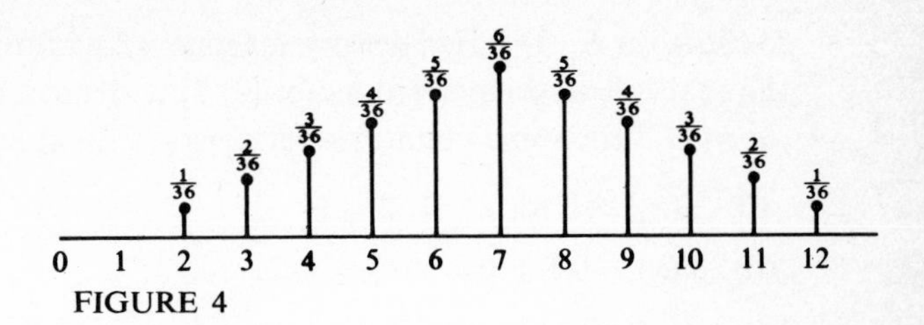

FIGURE 4

According to Theorem 1, for given $f_X(\cdot)$, $F_X(x)$ can be found for any x; for instance, if $x = 2.5$,

$$F_X(2.5) = \sum_{\{j:\, x_j \leq 2.5\}} f_X(x_j) = f_X(1) + f_X(2) = \frac{2}{6}.$$

And, if $F_X(\cdot)$ is given, $f_X(x)$ can be found for any x. For example, for $x = 3$,

$$f_X(3) = F_X(3) - \lim_{0<h\to 0} F_X(3-h) = \left(\frac{3}{6}\right) - \left(\frac{2}{6}\right) = \frac{1}{6}. \qquad ////$$

The cumulative distribution function of a discrete random variable has steps at the mass points; that is, at the mass point x_j, $F_X(\cdot)$ has a step of size $f_X(x_j)$, and $F_X(\cdot)$ is flat between mass points.

EXAMPLE 6 Consider the experiment of tossing two dice. Let X denote the total of the upturned faces. The mass points of X are 2, 3, ..., 12. $f_X(\cdot)$ is sketched in Fig. 4. Let Y denote the absolute difference of the upturned faces; then $f_Y(\cdot)$ is given in tabular form by

y	0	1	2	3	4	5
$f_Y(y)$	$\frac{6}{36}$	$\frac{10}{36}$	$\frac{8}{36}$	$\frac{6}{36}$	$\frac{4}{36}$	$\frac{2}{36}$

////

The discrete density function tells us how likely or probable each of the values of a discrete random variable is. It also enables one to calculate the probability of events described in terms of the discrete random variable X. For example, let X have mass points $x_1, x_2, \ldots, x_n, \ldots$; then $P[a < X \leq b] = \sum_{j:\{a<x_j\leq b\}} f_X(x_j)$ for $a < b$.

Definition 6 Discrete density function Any function $f(\cdot)$ with domain the real line and counterdomain [0, 1] is defined to be a *discrete density function* if for some countable set $x_1, x_2, \ldots, x_n, \ldots$,

(i) $f(x_j) > 0$ for $j = 1, 2, \ldots$.

(ii) $f(x) = 0$ for $x \neq x_j; j = 1, 2, \ldots$.

(iii) $\sum f(x_j) = 1$, where the summation is over the points $x_1, x_2, \ldots, x_n, \ldots$. ////

This definition allows us to speak of discrete density functions without reference to some random variable. Hence we can talk about properties that a density function might have without referring to a random variable.

3.2 Continuous Random Variables

Definition 7 Continuous random variable A random variable X is called *continuous* if there exists a function $f_X(\cdot)$ such that $F_X(x) = \int_{-\infty}^{x} f_X(u)du$ for every real number x. The cumulative distribution function $F_X(\cdot)$ of a continuous random variable X is called *absolutely continuous.* ////

Definition 8 Probability density function of a continuous random variable If X is a continuous random variable, the function $f_X(\cdot)$ in $F_X(x) = \int_{-\infty}^{x} f_X(u)\,du$ is called the *probability density function* of X. ////

Other names that are used instead of probability density function include *density function*, *continuous density function*, and *integrating density function.*

Note that strictly speaking *the* probability density function $f_X(\cdot)$ of a random variable X is not uniquely defined. All that the definition requires is that the integral of $f_X(\cdot)$ gives $F_X(x)$ for every x, and more than one function $f_X(\cdot)$ may satisfy such requirement. For example, suppose $F_X(x) = xI_{[0,1)}(x) + I_{[1,\infty)}(x)$; then $f_X(u) = I_{(0,1)}(u)$ satisfies $F_X(x) = \int_{-\infty}^{x} f_X(u)\,du$ for every x, and so $f_X(\cdot)$ is a probability density function of X. However $f_X(u) = I_{(0,\frac{1}{2})}(u) + 69I_{\{\frac{1}{2}\}}(u) + I_{(\frac{1}{2},1)}(u)$ also satisfies $F_X(x) = \int_{-\infty}^{x} f_X(u)\,du$. (The idea is that if the value of a function is changed at only a "few" points, then its integral is unchanged.) In practice a unique choice of $f_X(\cdot)$ is often dictated by continuity considerations and for this reason we will usually allow ourselves the liberty of speaking of *the* probability density when in fact *a* probability density is more correct.

One should point out that the word "continuous" in "continuous random variable" is not used in its usual sense. Although a random variable is a function and the notion of a continuous function is fairly well established in mathematics, "continuous" here is not used in that usual mathematical sense. In fact it is not clear in what sense it is used. Two possible justifications do come to mind. In contrasting discrete random variables with continuous random variables, one notes that a discrete random variable takes on a finite or denumerable set of values whereas a continuous random variable takes on a nondenumerable set of values. Possibly it is the connection between "nondenumerable" and "continuum" that justifies use of the word "continuous." All the continuous random variables that we shall encounter will take on a continuum of values. The second justification arises when one notes that the absolute continuity of the cumulative distribution function is the regular mathematical definition of an absolutely continuous function (in words, a function is called absolutely continuous if it can be written as the integral of its derivative); the "continuous," then, in a corresponding continuous random variable could be considered just an abbreviation of "absolutely continuous."

Theorem 2 Let X be a continuous random variable. Then $F_X(\cdot)$ can be obtained from an $f_X(\cdot)$, and vice versa.

PROOF If X is a continuous random variable and an $f_X(\cdot)$ is given, then $F_X(x)$ is obtained by integrating $f_X(\cdot)$; that is, $F_X(x) = \int_{-\infty}^{x} f_X(u)\,du$. On the other hand, if $F_X(\cdot)$ is given, then an $f_X(x)$ can be obtained by differentiation; that is, $f_X(x) = dF_X(x)/dx$ for those points x for which $F_X(x)$ is differentiable. ////

The notations for discrete density function and probability density function are the same, yet they have quite different interpretations. For discrete random variables $f_X(x) = P[X = x]$, which is not true for continuous random variables. For continuous random variables,

$$f_X(x) = \frac{dF_X(x)}{dx} = \lim_{\Delta x \to 0} \frac{F_X(x + \Delta x) - F_X(x - \Delta x)}{2\Delta x};$$

hence $f_X(x)2\Delta x \approx F_X(x + \Delta x) - F_X(x - \Delta x) = P[x - \Delta x < X \leq x + \Delta x]$; that is, the probability that X is in a *small* interval containing the value x is approximately equal to $f_X(x)$ times the width of the interval. For discrete random variables $f_X(\cdot)$ is a function with domain the real line and counterdomain the interval [0, 1]; whereas, for continuous random variables $f_X(\cdot)$ is a function with domain the real line and counterdomain the infinite interval $[0, \infty)$.

Remark We will use the term "density function" without the modifier of "discrete" or "probability" to represent either kind of density. ////

EXAMPLE 7 Let X be the random variable representing the length of a telephone conversation. One could model this experiment by assuming that the distribution of X is given by $F_X(x) = (1 - e^{-\lambda x})I_{[0, \infty)}(x)$, where λ is some positive number. The corresponding probability density function would be given by $f_X(x) = \lambda e^{-\lambda x} I_{[0, \infty)}(x)$. If we assume that telephone conversations are measured in minutes, $P[5 < X \leq 10] = \int_5^{10} \lambda e^{-x\lambda}\, dx = e^{-5\lambda} - e^{-10\lambda} = e^{-1} - e^{-2} \approx .23$ for $\lambda = \frac{1}{5}$, or $P[5 < X \leq 10] = P[X \leq 10] - P[X \leq 5] = (1 - e^{-\lambda 10}) - (1 - e^{-\lambda 5}) = e^{-1} - e^{-2}$ for $\lambda = \frac{1}{5}$. ////

The probability density function is used to calculate the probability of events defined in terms of the corresponding continuous random variable X. For example, $P[a < X \leq b] = \int_a^b f_X(x)\, dx$ for $a < b$.

Definition 9 **Probability density function** Any function $f(\cdot)$ with domain the real line and counterdomain $[0, \infty)$ is defined to be a *probability density function* if and only if

(i) $f(x) \geq 0$ for all x.

(ii) $\int_{-\infty}^{\infty} f(x)\, dx = 1$. ////

With this definition we can speak of probability density functions without reference to random variables. We might note that a probability density function of a continuous random variable as defined in Definition 8 does indeed possess the two properties in the above definition.

3.3 Other Random Variables

Not all random variables are either continuous or discrete, or not all cumulative distribution functions are either absolutely continuous or discrete.

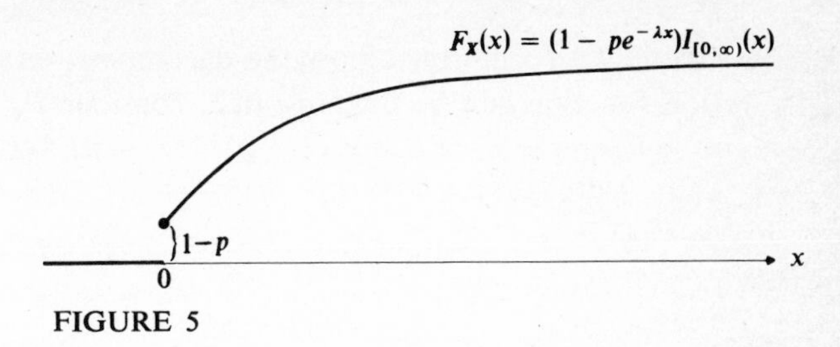

FIGURE 5

EXAMPLE 8 Consider the experiment of recording the delay that a motorist encounters at a one-way traffic stop sign. Let X be the random variable that represents the delay that the motorist experiences after making the required stop. There is a certain probability that there will be no opposing traffic so that the motorist will be able to proceed with no delay. On the other hand, if the motorist has to wait, he may have to wait for any of a continuum of possible times. This experiment could be modeled by assuming that X has a cumulative distribution function given by $F_X(x) = (1 - pe^{-\lambda x})I_{[0,\infty)}(x)$. This $F_X(x)$ has a jump of $1 - p$ at $x = 0$ but is continuous for $x > 0$. See Fig. 5. ////

Many practical examples of cumulative distribution functions that are partly discrete and partly absolutely continuous can be given. Yet there are still other types of cumulative distribution functions. There are continuous cumulative distribution functions, called *singular continuous*, whose derivative is 0 at almost all points. We will not consider such distribution functions other than to note the following result.

Decomposition of a cumulative distribution function Any cumulative distribution function $F(x)$ may be represented in the form

$$F(x) = p_1 F^{d}(x) + p_2 F^{ac}(x) + p_3 F^{sc}(x), \qquad \text{where } p_i \geq 0,\ i = 1, 2, 3. \tag{3}$$

$\sum_{i=1}^{3} p_i = 1$, and $F^{d}(\cdot)$, $F^{ac}(\cdot)$, and $F^{sc}(\cdot)$ are each cumulative distribution functions with $F^{d}(\cdot)$ discrete, $F^{ac}(\cdot)$ absolutely continuous, and $F^{sc}(\cdot)$ singular continuous.

Cumulative distributions studied in this book will have at most a discrete part and an absolutely continuous part; that is, the p_3 in Eq. (3) will always be 0 for the $F(\cdot)$ that we will study.

EXAMPLE 9 To illustrate how the decomposition of a cumulative distribution function can be implemented, consider $F_X(x) = (1 - pe^{-\lambda x})I_{[0, \infty)}(x)$ as in Example 8. $F_X(x) = (1 - p)F^{\text{d}}(x) + pF^{\text{ac}}(x)$, where $F^{\text{d}}(x) = I_{[0, \infty)}(x)$ and $F^{\text{ac}}(x) = (1 - e^{-\lambda x})I_{[0, \infty)}(x)$. Note that $F_X(x) = (1 - p)F^{\text{d}}(x) + pF^{\text{ac}}(x) = (1 - p)I_{[0, \infty)}(x) + p(1 - e^{-\lambda x})I_{[0, \infty)}(x) = (1 - pe^{-\lambda x})I_{[0, \infty)}(x)$. ////

A density function corresponding to a cumulative distribution that is partly discrete and partly absolutely continuous could be defined as follows: If $F(x) = (1 - p)F^{\text{d}}(x) + pF^{\text{ac}}(x)$, where $0 < p < 1$ and $F^{\text{d}}(\cdot)$ and $F^{\text{ac}}(\cdot)$ are, respectively, discrete and absolutely continuous cumulative distribution functions, let the density function $f(x)$ corresponding to $F(x)$ be defined by $f(x) = (1 - p)f^{\text{d}}(x) + pf^{\text{ac}}(x)$, where $f^{\text{d}}(\cdot)$ is the discrete density function corresponding to $F^{\text{d}}(\cdot)$ and $f^{\text{ac}}(\cdot)$ is the probability density function corresponding to $F^{\text{ac}}(\cdot)$. Such a density function would require careful interpretation; so when considering cumulative distribution functions that are partly discrete and partly continuous, we will tend to work with the cumulative distribution function itself rather than with a density function.

Remark In future chapters we will frequently have to state that a random variable has a certain distribution. We will make such a statement by giving either the cumulative distribution function or the density function of the random variable of interest. ////

4 EXPECTATIONS AND MOMENTS

An extremely useful concept in problems involving random variables or distributions is that of *expectation*. The subsections of this section give definitions and results regarding expectations.

4.1 Mean

Definition 10 Mean Let X be a random variable. The *mean* of X, denoted by μ_X or $\mathscr{E}[X]$, is defined by:

(i)
$$\mathscr{E}[X] = \sum x_j f_X(x_j) \tag{4}$$

if X is discrete with mass points $x_1, x_2, \ldots, x_j, \ldots$.

(ii)
$$\mathscr{E}[X] = \int_{-\infty}^{\infty} x f_X(x)dx \tag{5}$$

if X is continuous with probability density function $f_X(x)$.

(iii)
$$\mathscr{E}[X] = \int_0^{\infty} [1 - F_X(x)]\, dx - \int_{-\infty}^{0} F_X(x)\, dx \tag{6}$$

for an arbitrary random variable X. ////

In (i), $\mathscr{E}[X]$ is defined to be the indicated series provided that the series is absolutely convergent; otherwise, we say that the mean does not exist. And in (ii), $\mathscr{E}[X]$ is defined to be the indicated integral if the integral exists; otherwise, we say that the mean does not exist. Finally, in (iii), we require that both integrals be finite for the existence of $\mathscr{E}[X]$.

Note what the definition says: In $\sum_j x_j f_X(x_j)$, the summand is the jth value of the random variable X multiplied by the probability that X equals that jth value, and then the summation is over all values. So $\mathscr{E}[X]$ is an "average" of the values that the random variable takes on, where each value is weighted by the probability that the random variable is equal to that value. Values that are more probable receive more weight. The same is true in integral form in (ii). There the value x is multiplied by the approximate probability that X equals the value x, namely $f_X(x)\, dx$, and then integrated over all values.

Several remarks are in order.

Remark In the definition of a mean of a random variable, only density functions [in (i) and (ii)] or distribution functions [in (iii)] were used; hence we have really defined the mean for these functions without reference to random variables. We then call the defined mean the mean of the cumulative distribution function or of the appropriate density function. Hence, we can and will speak of the mean of a distribution or density function as well as the mean of a random variable. ////

Remark $\mathscr{E}[X]$ is the center of gravity (or *centroid*) of the unit mass that is determined by the density function of X. So the mean of X is a measure of where the values of the random variable X are "centered." Other measures of "location" or "center" of a random variable or its corresponding density are given in Subsec. 4.6. ////

Remark (iii) of the definition is for all random variables; whereas, (i) is for discrete random variables, and (ii) is for continuous random variables. Of course, $\mathscr{E}[X]$ could have been defined by just giving (iii). The reason for including (i) and (ii) is that they are more intuitive for their respective cases. It can be proved, although we will not do it, that (i) follows from (iii) in the case of discrete random variables and (ii) follows from (iii) in the case of continuous random variables. Our main use of (iii) will be in finding the mean of a random variable X that is neither discrete nor continuous. See Example 12 below. ////

EXAMPLE 10 Consider the experiment of tossing two dice. Let X denote the total of the two dice and Y their absolute difference. The discrete density functions for X and Y are given in Example 6.

$$\mathscr{E}[Y] = \sum y_j f_Y(y_j) = \sum_{i=0}^{5} i f_Y(i) = 0 \cdot \tfrac{6}{36} + 1 \cdot \tfrac{10}{36}$$

$$+ 2 \cdot \tfrac{8}{36} + 3 \cdot \tfrac{6}{36} + 4 \cdot \tfrac{4}{36} + 5 \cdot \tfrac{2}{36} = \tfrac{70}{36}.$$

$$\mathscr{E}[X] = \sum_{i=2}^{12} i f_X(i) = 7.$$

Note that $\mathscr{E}[Y]$ is not one of the possible values of Y. ////

EXAMPLE 11 Let X be a continuous random variable with probability density function $f_X(x) = \lambda e^{-\lambda x} I_{[0, \infty)}(x)$.

$$\mathscr{E}[X] = \int_{-\infty}^{\infty} x f_X(x)\, dx = \int_0^{\infty} x \lambda e^{-\lambda x}\, dx = \frac{1}{\lambda}.$$

The corresponding cumulative distribution function is

$$F_X(x) = (1 - e^{-\lambda x}) I_{[0, \infty)}(x); \text{ so } \mathscr{E}[X] = \int_0^{\infty} [1 - F_X(x)]\, dx$$

$$- \int_{-\infty}^{0} F_X(x)\, dx = \int_0^{\infty} (1 - 1 + e^{-\lambda x})\, dx = 1/\lambda. \qquad ////$$

EXAMPLE 12 Let X be a random variable with cumulative distribution function given by $F_X(x) = (1 - pe^{-\lambda x}) I_{[0, \infty)}(x)$; then

$$\mathscr{E}[X] = \int_0^{\infty} [1 - F_X(x)]\, dx - \int_{-\infty}^{0} F_X(x)\, dx = \int_0^{\infty} pe^{-\lambda x}\, dx = \frac{p}{\lambda}.$$

Here, we have used Eq. (6) to find the mean of a random variable that is partly discrete and partly continuous. ////

EXAMPLE 13 Let X be a random variable with probability density function given by $f_X(x) = x^{-2}I_{[1,\infty)}(x)$; then

$$\mathscr{E}[X] = \int_1^\infty x \frac{dx}{x^2} = \lim_{b\to\infty} \log_e b = \infty,$$

so we say that $\mathscr{E}[X]$ does not exist. We might also say that the mean of X is infinite since it is clear here that the integral that defines the mean is infinite. ////

4.2 Variance

The mean of a random variable X, defined in the previous subsection, was a measure of *central location* of the density of X. The *variance* of a random variable X will be a measure of the *spread* or *dispersion* of the density of X.

Definition 11 Variance Let X be a random variable, and let μ_X be $\mathscr{E}[X]$. The *variance of* X, denoted by σ_X^2 or var $[X]$, is defined by

(i)
$$\text{var}\,[X] = \sum_j (x_j - \mu_X)^2 f_X(x_j) \tag{7}$$

if X is discrete with mass points $x_1, x_2, \ldots, x_j, \ldots$.

(ii)
$$\text{var}\,[X] = \int_{-\infty}^{\infty} (x - \mu_X)^2 f_X(x)\, dx \tag{8}$$

if X is continuous with probability density function $f_X(x)$.

(iii)
$$\text{var}\,[X] = \int_0^\infty 2x[1 - F_X(x) + F_X(-x)]\, dx - \mu_X^2 \tag{9}$$

for an arbitrary random variable X. ////

The variances are defined only if the series in (i) is convergent or if the integrals in (ii) and (iii) exist. Again, the variance of a random variable is defined in terms of the density function or cumulative distribution function of the random variable; hence variance could be defined in terms of these functions without reference to a random variable.

Note what the definition says: In (i), the square of the difference between the jth value of the random variable X and the mean of X is multiplied by the probability that X equals the jth value, and then these terms are summed. More weight is assigned to the more probable squared differences. A similar comment applies for (ii). Variance is a *measure of spread* since if the values of a random variable X tend to be far from their mean, the variance of X will be larger than the variance of a comparable random variable Y whose values tend to be near their mean. It is clear from (i) and (ii) and true for (iii) that variance is nonnegative. We saw that a mean was the center of gravity of a density; similarly (for those readers familiar with elementary physics or mechanics), variance represents the moment of inertia of the same density with respect to a perpendicular axis through the center of gravity.

Definition 12 Standard deviation If X is a random variable, the *standard deviation* of X, denoted by σ_X, is defined as $+\sqrt{\text{var}[X]}$. ////

The standard deviation of a random variable, like the variance, is a measure of the spread or dispersion of the values of the random variable. In many applications it is preferable to the variance as such a measure since it will have the same measurement units as the random variable itself.

EXAMPLE 14 Let X be the total of the two dice in the experiment of tossing two dice.

$$\begin{aligned}\text{var}[X] &= \sum (x_j - \mu_X)^2 f_X(x_j)\\ &= (2-7)^2\tfrac{1}{36} + (3-7)^2\tfrac{2}{36} + (4-7)^2\tfrac{3}{36} + (5-7)^2\tfrac{4}{36}\\ &\quad + (6-7)^2\tfrac{5}{36} + (7-7)^2\tfrac{6}{36} + (8-7)^2\tfrac{5}{36} + (9-7)^2\tfrac{4}{36}\\ &\quad + (10-7)^2\tfrac{3}{36} + (11-7)^2\tfrac{2}{36} + (12-7)^2\tfrac{1}{36} = \tfrac{210}{36}.\end{aligned}$$ ////

EXAMPLE 15 Let X be a random variable with probability density given by $f_X(x) = \lambda e^{-\lambda x} I_{[0,\infty)}(x)$; then

$$\begin{aligned}\text{Var}[X] &= \int_{-\infty}^{\infty} (x - \mu_X)^2 f_X(x)\,dx\\ &= \int_0^{\infty} \left(x - \frac{1}{\lambda}\right)^2 \lambda e^{-\lambda x}\,dx\\ &= \frac{1}{\lambda^2}.\end{aligned}$$ ////

EXAMPLE 16 Let X be a random variable with cumulative distribution given by $F_X(x) = (1 - pe^{-\lambda x}) I_{[0,\infty)}(x)$; then

$$\begin{aligned}\text{Var}[X] &= \int_0^{\infty} 2x[1 - F(x) + F(-x)]\,dx - \mu_X^2\\ &= \int_0^{\infty} 2xpe^{-\lambda x}\,dx - \left(\frac{p}{\lambda}\right)^2\\ &= 2\frac{p}{\lambda^2} - \left(\frac{p}{\lambda}\right)^2 = \frac{p(2-p)}{\lambda^2}.\end{aligned}$$ ////

4.3 Expected Value of a Function of a Random Variable

We defined the expectation of an arbitrary random variable X, called the mean of X, in Subsec. 4.1. In this subsection, we will define the expectation of a function of a random variable for discrete or continuous random variables.

Definition 13 Expectation Let X be a random variable and $g(\cdot)$ be a function with both domain and counterdomain the real line. The *expectation* or *expected value* of the function $g(\cdot)$ of the random variable X, denoted by $\mathscr{E}[g(X)]$, is defined by:

(i)
$$\mathscr{E}[g(X)] = \sum_j g(x_j) f_X(x_j) \tag{10}$$

if X is discrete with mass points $x_1, x_2, \ldots, x_j, \ldots$ (provided this series is absolutely convergent).

(ii)
$$\mathscr{E}[g(X)] = \int_{-\infty}^{\infty} g(x) f_X(x)\, dx \tag{11}$$

if X is continuous with probability density function $f_X(x)$ (provided $\int_{-\infty}^{\infty} |g(x)| f_X(x)\, dx < \infty$).* ////

Expectation or expected value is not really a very good name since it is not necessarily what you "expect." For example, the expected value of a discrete random variable is not necessarily one of the possible values of the discrete random variable, in which case, you would not "expect" to get the expected value. A better name might be "average value" rather than "expected value."

Since $\mathscr{E}[g(X)]$ is defined in terms of the density function of X, it could be defined without reference to a random variable.

Remark If $g(x) = x$, then $\mathscr{E}[g(X)] = \mathscr{E}[X]$ is the mean of X. If $g(x) = (x - \mu_X)^2$, then $\mathscr{E}[g(X)] = \mathscr{E}[(X - \mu_X)^2] = \text{var}\,[X]$. ////

* $\mathscr{E}[g(X)]$ has been defined here for random variables that are either discrete or continuous; it can be defined for other random variables as well. For the reader who is familiar with the Stieltjes integral, $\mathscr{E}[g(X)]$ is defined as the Stieltjes integral $\int_{-\infty}^{\infty} g(x)\, dF_X(x)$ (provided this integral exists), where $F_X(\cdot)$ is the cumulative distribution function of X. If X is a random variable whose cumulative distribution function is partly discrete and partly continuous, then (according to Subsec. 3.3) $F_X(x) = (1-p)F^d(x) + pF^{ac}(x)$ for some $0 < p < 1$. Now $\mathscr{E}[g(X)]$ can be defined to be $\mathscr{E}[g(X)] = (1-p)\sum g(x_j) f^d(x_j) + p\int_{-\infty}^{\infty} g(x) f^{ac}(x)\, dx$, where $f^d(\cdot)$ is the discrete density function corresponding to $F^d(\cdot)$ and $f^{ac}(\cdot)$ is the probability density function corresponding to $F^{ac}(\cdot)$.

Theorem 3 Below are properties of expected value:

(i) $\mathscr{E}[c] = c$ for a constant c.

(ii) $\mathscr{E}[cg(X)] = c\mathscr{E}[g(X)]$ for a constant c.

(iii) $\mathscr{E}[c_1 g_1(X) + c_2 g_2(X)] = c_1\mathscr{E}[g_1(X)] + c_2\mathscr{E}[g_2(X)]$.

(iv) $\mathscr{E}[g_1(X)] \le \mathscr{E}[g_2(X)]$ if $g_1(x) \le g_2(x)$ for all x.

PROOF Assume X is continuous. To prove (i), take $g(x) = c$, then

$$\mathscr{E}[g(X)] = \mathscr{E}[c] = \int_{-\infty}^{\infty} cf_X(x)\,dx = c\int_{-\infty}^{\infty} f_X(x)\,dx = c.$$

$$\mathscr{E}[cg(X)] = \int_{-\infty}^{\infty} cg(x)f_X(x)\,dx = c\int_{-\infty}^{\infty} g(x)f_X(x)\,dx = c\mathscr{E}[g(X)],$$

which proves (ii). (iii) is given by

$$\begin{aligned}\mathscr{E}[c_1 g_1(X) + c_2 g_2(X)] &= \int_{-\infty}^{\infty} [c_1 g_1(x) + c_2 g_2(x)]f_X(x)\,dx \\ &= c_1\int_{-\infty}^{\infty} g_1(x)f_X(x)\,dx + c_2\int_{-\infty}^{\infty} g_2(x)f_X(x)\,dx \\ &= c_1\mathscr{E}[g_1(X)] + c_2\mathscr{E}[g_2(X)].\end{aligned}$$

Finally,

$$0 \le \mathscr{E}[g_2(X) - g_1(X)] = \mathscr{E}[g_2(X)] - \mathscr{E}[g_1(X)],$$

which gives (iv).

Similar proofs could be presented for the discrete random variable case. ////

Theorem 4 If X is a random variable, $\operatorname{var}[X] = \mathscr{E}[(X - \mathscr{E}[X])^2] = \mathscr{E}[X^2] - (\mathscr{E}[X])^2$ provided $\mathscr{E}[X^2]$ exists.

PROOF (We first note that if $\mathscr{E}[X^2]$ exists, then $\mathscr{E}[X]$ exists.)* By our definitions of variance and $\mathscr{E}[g(X)]$, it follows that $\operatorname{var}[X] = \mathscr{E}[(X - \mathscr{E}[X])^2]$. Now $\mathscr{E}[(X - \mathscr{E}[X])^2] = \mathscr{E}[X^2 - 2X\mathscr{E}[X] + (\mathscr{E}[X])^2] = \mathscr{E}[X^2] - 2(\mathscr{E}[X])^2 + (\mathscr{E}[X])^2 = \mathscr{E}[X^2] - (\mathscr{E}[X])^2$. ////

The above theorem provides us with two methods of calculating a variance, namely $\mathscr{E}[(X - \mu_X)^2]$ or $\mathscr{E}[X^2] - \mu_X^2$. Note that both methods require μ_X.

* Here and in the future we are not going to concern ourselves with checking existence.

$\mathscr{E}[g(X)]$ is used in each of the following three subsections. In Subsec. 4.4 and 4.5 two inequalities involving $\mathscr{E}[g(X)]$ are given. Definitions and examples of $\mathscr{E}[g(X)]$ for particular functions $g(\cdot)$ are given in Subsec. 4.6.

4.4 Chebyshev Inequality

Theorem 5 Let X be a random variable and $g(\cdot)$ a nonnegative function with domain the real line; then

$$P[g(X) \geq k] \leq \frac{\mathscr{E}[g(X)]}{k} \qquad \text{for every } k > 0. \tag{12}$$

PROOF Assume that X is a continuous random variable with probability density function $f_X(\cdot)$; then

$$\mathscr{E}[g(X)] = \int_{-\infty}^{\infty} g(x)f_X(x)\,dx = \int_{\{x:\,g(x)\geq k\}} g(x)f_X(x)\,dx$$

$$+ \int_{\{x:\,g(x)<k\}} g(x)f_X(x)\,dx \geq \int_{\{x:\,g(x)\geq k\}} g(x)f_X(x)\,dx$$

$$\geq \int_{\{x:\,g(x)\geq k\}} kf_X(x)\,dx = kP[g(X) \geq k].$$

Divide by k, and the result follows. A similar proof holds for X discrete. ////

Corollary Chebyshev inequality If X is a random variable with finite variance,

$$P[|X - \mu_X| \geq r\sigma_X] = P[(X - \mu_X)^2 \geq r^2\sigma_X^2] \leq \frac{1}{r^2} \qquad \text{for every } r > 0. \tag{13}$$

PROOF Take $g(x) = (x - \mu_X)^2$ and $k = r^2\sigma_X^2$ in Eq. (12) of Theorem 5. ////

Remark If X is a random variable with finite variance,

$$P[|X - \mu_X| < r\sigma_X] \geq 1 - \frac{1}{r^2}, \tag{14}$$

which is just a rewriting of Eq. (13). ////

The Chebyshev inequality is used in various ways. We will use it later to prove the law of large numbers. Note what Eq. (14) says:

$$P[\mu_X - r\sigma_X < X < \mu_X + r\sigma_X] \geq 1 - \frac{1}{r^2};$$

that is, the probability that X falls within $r\sigma_X$ units of μ_X is greater than or equal to $1 - 1/r^2$. For $r = 2$, one gets $P[\mu_X - 2\sigma_X < X < \mu_X + 2\sigma_X] \geq \frac{3}{4}$, or for any random variable X having finite variance at least three-fourths of the mass of X falls within two standard deviations of its mean.

Ordinarily, to calculate the probability of an event described in terms of a random variable X, the distribution or density of X is needed; the Chebyshev inequality gives a bound, which does not depend on the distribution of X, for the probability of particular events described in terms of a random variable and its mean and variance.

4.5 Jensen Inequality

Definition 14 Convex function A continuous function $g(\cdot)$ with domain and counterdomain the real line is called *convex* if for every x_0 on the real line, there exists a line which goes through the point $(x_0, g(x_0))$ and lies on or under the graph of the function $g(\cdot)$. ////

Theorem 6 Jensen inequality Let X be a random variable with mean $\mathscr{E}[X]$, and let $g(\cdot)$ be a convex function; then $\mathscr{E}[g(X)] \geq g(\mathscr{E}[X])$.

PROOF Since $g(x)$ is continuous and convex, there exists a line, say $l(x) = a + bx$, satisfying $l(x) = a + bx \leq g(x)$ and $l(\mathscr{E}[X]) = g(\mathscr{E}[X])$. $l(x)$ is a line given by the definition of continuous and convex that goes through the point $(\mathscr{E}[X], g(\mathscr{E}[X]))$. Note that $\mathscr{E}[l(X)] = \mathscr{E}[(a + bX)] = a + b\mathscr{E}[X] = l(\mathscr{E}[X])$; hence $g(\mathscr{E}[X]) = l(\mathscr{E}[X]) = \mathscr{E}[l(X)] \leq \mathscr{E}[g(X)]$ [using property (iv) of expected values (see Theorem 3) for the last inequality]. ////

The Jensen inequality can be used to prove the Rao-Blackwell theorem to appear in Chap. VII. We point out that, in general, $\mathscr{E}[g(X)] \neq g(\mathscr{E}[X])$; for example, note that $g(x) = x^2$ is convex; hence $\mathscr{E}[X^2] \geq (\mathscr{E}[X])^2$, which says that the variance of X, which is $\mathscr{E}[X^2] - (\mathscr{E}[X])^2$, is nonnegative.

4.6 Moments and Moment Generating Functions

The *moments* (or *raw moments*) of a random variable or of a distribution are the expectations of the powers of the random variable which has the given distribution.

Definition 15 Moments If X is a random variable, the *rth moment* of X, usually denoted by μ'_r, is defined as

$$\mu'_r = \mathscr{E}[X^r] \tag{15}$$

if the expectation exists. ////

Note that $\mu'_1 = \mathscr{E}[X] = \mu_X$, the mean of X.

Definition 16 Central moments If X is a random variable, the *rth central moment* of X about a is defined as $\mathscr{E}[(X - a)^r]$. If $a = \mu_X$, we have the *rth central moment* of X about μ_X, denoted by μ_r, which is

$$\mu_r = \mathscr{E}[(X - \mu_X)^r]. \tag{16}$$

////

Note that $\mu_1 = \mathscr{E}[(X - \mu_X)] = 0$ and $\mu_2 = \mathscr{E}[(X - \mu_X)^2]$, the variance of X. Also, note that all odd moments of X about μ_X are 0 if the density function of X is symmetrical about μ_X, provided such moments exist.

In the ensuing few paragraphs we will comment on how the first four moments of a random variable or density are used as measures of various characteristics of the corresponding density. For some of these characteristics, other measures can be defined in terms of *quantiles*.

Definition 17 Quantile The *qth quantile* of a random variable X or of its corresponding distribution is denoted by ξ_q and is defined as the smallest number ξ satisfying $F_X(\xi) \geq q$. ////

If X is a continuous random variable, then the *qth quantile* of X is given as the smallest number ξ satisfying $F_X(\xi) = q$. See Fig. 6.

Definition 18 Median The *median* of a random variable X, denoted by med_X, med (X), or $\xi_{.50}$, is the .5th quantile. ////

Remark In some texts the median of X is alternatively defined as any number, say med (X), satisfying $P[X \leq \text{med}(X)] \geq \frac{1}{2}$ and $P[X \geq \text{med}(X)] \geq \frac{1}{2}$. ////

If X is a continuous random variable, then the median of X satisfies

$$\int_{-\infty}^{\text{med}(X)} f_X(x)\,dx = \tfrac{1}{2} = \int_{\text{med}(X)}^{\infty} f_X(x)\,dx;$$

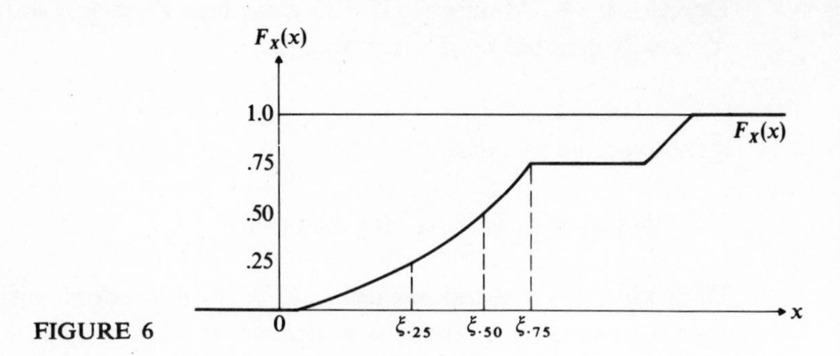

FIGURE 6

so the median of X is any number that has half the mass of X to its right and the other half to its left, which justifies use of the word "median."

We have already mentioned that $\mathscr{E}[X]$, the first moment, locates the "center" of the density of X. The median of X is also used to indicate a central location of the density of X. A third measure of location of the density of X, though not necessarily a measure of central location, is the *mode* of X, which is defined as that point (if such a point exists) at which $f_X(\cdot)$ attains its maximum. Other measures of location [for example, $\frac{1}{2}(\xi_{.25} + \xi_{.75})$] could be devised, but three, mean, median, and mode, are the ones commonly used.

We previously mentioned that the second moment about the mean, the variance of a distribution, measures the spread or dispersion of a distribution. Let us look a little further into the manner in which the variance characterizes the distribution. Suppose that $f_1(x)$ and $f_2(x)$ are two densities with the same mean μ such that

$$\int_{\mu-a}^{\mu+a} [f_1(x) - f_2(x)]\, dx \geq 0 \tag{17}$$

for every value of a. Two such densities are illustrated in Fig. 7. It can be shown that in this case the variance σ_1^2 in the first density is smaller than the

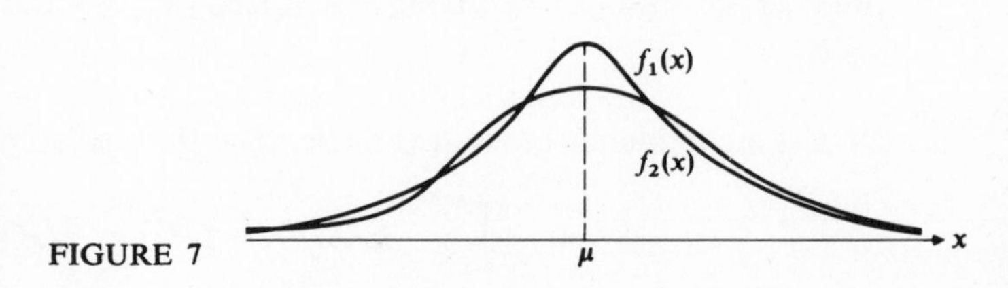

FIGURE 7

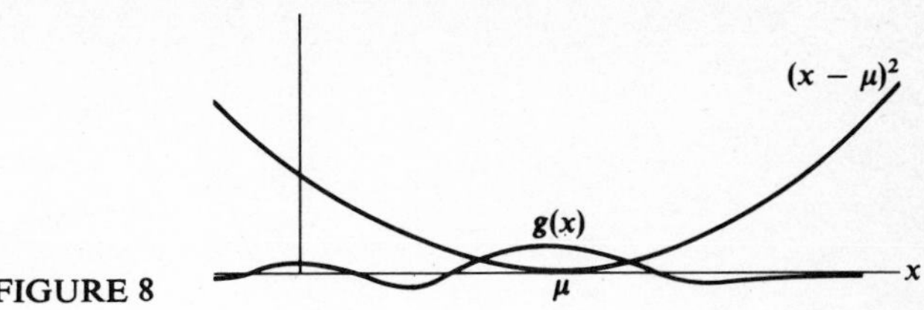

FIGURE 8

variance σ_2^2 in the second density. We shall not take the time to prove this in detail, but the argument is roughly this: Let

$$g(x) = f_1(x) - f_2(x),$$

where $f_1(x)$ and $f_2(x)$ satisfy Eq. (17). Since $\int_{-\infty}^{\infty} g(x)\,dx = 0$, the positive area between $g(x)$ and the x axis is equal to the negative area. Furthermore, in view of Eq. (17), every positive element of area $g(x')\,dx'$ may be balanced by a negative element $g(x'')\,dx''$ in such a way that x'' is further from μ than x'. When these elements of area are multiplied by $(x - \mu)^2$, the negative elements will be multiplied by larger factors than their corresponding positive elements (see Fig. 8); hence

$$\int_{-\infty}^{\infty} (x - \mu)^2 g(x)\,dx < 0$$

unless $f_1(x)$ and $f_2(x)$ are equal. Thus it follows that $\sigma_1^2 < \sigma_2^2$. The converse of these statements is not true. That is, if one is told that $\sigma_1^2 < \sigma_2^2$, he cannot conclude that the corresponding densities satisfy Eq. (17) for all values of a; although it can be shown that Eq. (17) must be true for certain values of a. Thus the condition $\sigma_1^2 < \sigma_2^2$ does not give one any precise information about the nature of the corresponding distributions, but it is evident that $f_1(x)$ has more area near the mean than $f_2(x)$, at least for certain intervals about the mean.

We indicated above how variance is used as a measure of spread or dispersion of a distribution. Alternative measures of dispersion can be defined in terms of quantiles. For example $\xi_{.75} - \xi_{.25}$, called the *interquartile range*, is a measure of spread. Also, $\xi_p - \xi_{1-p}$ for some $\frac{1}{2} < p < 1$ is a possible measure of spread.

The third moment μ_3 about the mean is sometimes called a measure of asymmetry, or *skewness*. Symmetrical distributions like those in Fig. 9 can be shown to have $\mu_3 = 0$. A curve shaped like $f_1(x)$ in Fig. 10 is said to be *skewed to the left* and can be shown to have a negative third moment about the mean; one shaped like $f_2(x)$ is called *skewed to the right* and can be shown to have a positive third moment about the mean. Actually, however, knowledge of the

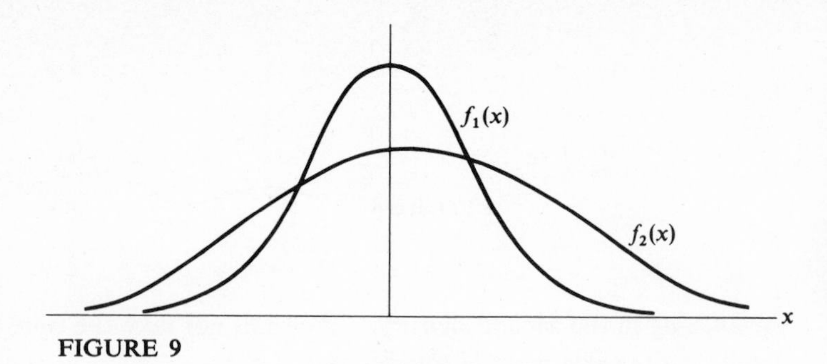

FIGURE 9

third moment gives almost no clue as to the shape of the distribution, and we mention it mainly to point out that fact. Thus, for example, the density $f_3(x)$ in Fig. 10 has $\mu_3 = 0$, but it is far from symmetrical. By changing the curve slightly we could give it either a positive or negative third moment. The ratio μ_3/σ^3, which is unitless, is called the *coefficient of skewness.*

The quantity $\mathscr{s} =$ (mean − median)/(standard deviation) provides an alternative measure of skewness. It can be proved that $-1 \leq \mathscr{s} \leq 1$.

The fourth moment about the mean is sometimes used as a measure of *excess* or *kurtosis*, which is the degree of flatness of a density near its center. Positive values of $\mu_4/\sigma^4 - 3$, called the *coefficient of excess* or *kurtosis*, are sometimes used to indicate that a density is more peaked around its center than the density of a normal curve (see Subsec. 3.2 of Chap. III), and negative values are sometimes used to indicate that a density is more flat around its center than the density of a normal curve. This measure, however, suffers from the same failing as does the measure of skewness; namely, it does not always measure what it is supposed to.

While a particular moment or a few of the moments may give little information about a distribution (see Fig. 11 for a sketch of two densities having the same first four moments. See Ref. 40. Also see Prob. 30 in Chap. III.), the entire set of moments $(\mu'_1, \mu'_2, \mu'_3, \ldots)$ will ordinarily determine the distri-

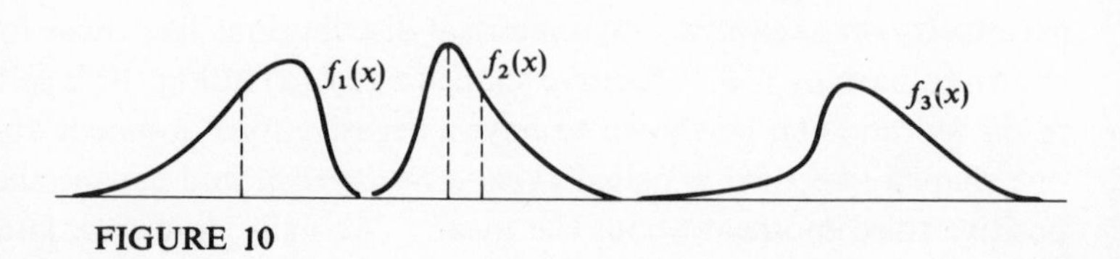

FIGURE 10

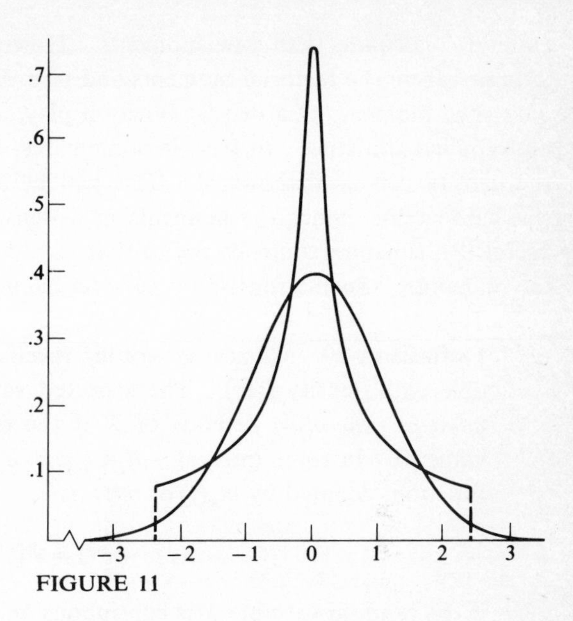

FIGURE 11

bution exactly, and for this reason we shall have occasion to use the moments in theoretical work.

In applied statistics, the first two moments are of great importance, as we shall see, but the third and higher moments are rarely useful. Ordinarily one does not know what distribution function one is working with in a practical problem, and often it makes little difference what the actual shape of the distribution is. But it is usually necessary to know at least the location of the distribution and to have some idea of its dispersion. These characteristics can be estimated by examining a sample drawn from a set of objects known to have the distribution in question. This estimation problem is probably the most important problem in applied statistics, and a large part of this book will be devoted to a study of it.

We now define another kind of moment, *factorial moment.*

Definition 19 Factorial moment If X is a random variable, the *rth factorial moment of X* is defined as (r is a positive integer):

$$\mathscr{E}[X(X-1)\cdots(X-r+1)]. \tag{18}$$

////

For some random variables (usually discrete), factorial moments are easier to calculate than raw moments. However the raw moments can be obtained from the factorial moments and vice versa.

The moments of a density function play an important role in theoretical and applied statistics. In fact, in some cases, if all the moments are known, the density can be determined. This will be discussed briefly at the end of this subsection. Since the moments of a density are important, it would be useful if a function could be found that would give us a representation of all the moments. Such a function is called a *moment generating function.*

Definition 20 Moment generating function Let X be a random variable with density $f_X(\cdot)$. The expected value of e^{tX} is defined to be the *moment generating function* of X if the expected value exists for every value of t in some interval $-h < t < h$; $h > 0$. The moment generating function, denoted by $m_X(t)$ or $m(t)$, is

$$m(t) = \mathscr{E}[e^{tX}] = \int_{-\infty}^{\infty} e^{tx} f_X(x)\, dx \tag{19}$$

if the random variable X is continuous and is

$$m(t) = \mathscr{E}[e^{tX}] = \sum_x e^{tx} f_X(x)$$

if the random variable is discrete. ////

One might note that a moment generating function is defined in terms of a density function, and since density functions were defined without reference to random variables (see Definitions 6 and 9), a moment generating function can be discussed without reference to random variables.

If a moment generating function exists, then $m(t)$ is continuously differentiable in some neighborhood of the origin. If we differentiate the moment generating function r times with respect to t, we have

$$\frac{d^r}{dt^r} m(t) = \int_{-\infty}^{\infty} x^r e^{xt} f_X(x)\, dx, \tag{20}$$

and letting $t \to 0$, we find

$$\frac{d^r}{dt^r} m(0) = \mathscr{E}[X^r] = \mu_r', \tag{21}$$

where the symbol on the left is to be interpreted to mean the rth derivative of $m(t)$ evaluated as $t \to 0$. Thus the moments of a distribution may be obtained from the moment generating function by differentiation, hence its name.

If in Eq. (19) we replace e^{xt} by its series expansion, we obtain the series expansion of $m(t)$ in terms of the moments of $f_X(\cdot)$; thus

$$m(t) = \mathscr{E}\left[1 + Xt + \frac{1}{2!}(Xt)^2 + \frac{1}{3!}(Xt)^3 + \cdots\right]$$
$$= 1 + \mu_1' t + \frac{1}{2!}\mu_2' t^2 + \cdots$$
$$= \sum_{i=0}^{\infty} \frac{1}{i!}\mu_i' t^i, \tag{22}$$

from which it is again evident that μ_r' may be obtained from $m(t)$; μ_r' is the coefficient of $t^r/r!$.

EXAMPLE 17 Let X be a random variable with probability density function given by $f_X(x) = \lambda e^{-\lambda x} I_{[0, \infty)}(x)$.

$$m_X(t) = \mathscr{E}[e^{tX}] = \int_0^\infty e^{tx} \lambda e^{-\lambda x}\, dx = \frac{\lambda}{\lambda - t} \qquad \text{for } t < \lambda.$$

$$m'(t) = \frac{dm(t)}{dt} = \frac{\lambda}{(\lambda - t)^2} \qquad \text{hence} \quad m'(0) = \mathscr{E}[X] = \frac{1}{\lambda}.$$

And $$m''(t) = \frac{2\lambda}{(\lambda - t)^3}, \qquad \text{so } m''(0) = \mathscr{E}[X^2] = \frac{2}{\lambda^2}.$$ ////

EXAMPLE 18 Consider the random variable X having probability density function $f_X(x) = x^{-2} I_{[1, \infty)}(x)$. (See Example 13.) If the moment generating function of X exists, then it is given by $\int_1^\infty x^{-2} e^{tx}\, dx$. It can be shown, however, that the integral does not exist for any $t > 0$, and hence the moment generating function does not exist for this random variable X. ////

As with moments, there is also a generating function for factorial moments.

Definition 21 Factorial moment generating function Let X be a random variable. The *factorial moment generating function* is defined as $\mathscr{E}[t^X]$ if this expectation exists. ////

The factorial moment generating function is used to generate factorial moments in the same way as the raw moments are obtained from $\mathscr{E}[e^{tX}]$ except that t approaches 1 instead of 0. It sometimes simplifies finding moments of discrete distributions.

EXAMPLE 19 Suppose X has a discrete density function given by

$$f_X(x) = \frac{e^{-\lambda}\lambda^x}{x!} \qquad \text{for } x = 0, 1, 2, \ldots.$$

Then

$$\mathscr{E}[t^X] = \sum_{x=0}^{\infty} \frac{t^x e^{-\lambda}\lambda^x}{x!} = e^{-\lambda}e^{\lambda t} = e^{\lambda(t-1)}.$$

$$\frac{d}{dt}\mathscr{E}[t^X] = \frac{d}{dt}e^{\lambda(t-1)} = \lambda e^{\lambda(t-1)}; \qquad \text{hence} \quad \frac{d}{dt}\mathscr{E}[t^X]\bigg|_{t=1} = \lambda. \qquad ////$$

In addition to raw moments, central moments, and factorial moments, there are other kinds of moments, called *cumulants*, or *semi-invariants*. Cumulants will be defined in terms of the *cumulant generating function*. We will not make use of cumulants in this book.

Definition 22 Cumulant and cumulant generating function The logarithm of the moment generating function of X is defined to be the *cumulant generating function* of X. The *rth cumulant* of X, denoted by $\kappa_r(X)$ or κ_r, is the coefficient of $t^r/r!$ in the Taylor series expansion of the cumulant generating function. ////

A moment generating function is used, as its name suggests, to generate moments. That, however, will not be its only use for us. An important use will be in determining distributions.

Theorem 7 Let X and Y be two random variables with densities $f_X(\cdot)$ and $f_Y(\cdot)$, respectively. Suppose that $m_X(t)$ and $m_Y(t)$ both exist and are equal for all t in the interval $-h < t < h$ for some $h > 0$. Then the two cumulative distribution functions $F_X(\cdot)$ and $F_Y(\cdot)$ are equal. ////

A proof of the above theorem can be obtained using certain transform theory that is beyond the scope of this book. We should note, however, what the theorem asserts. It says that if we can find the moment generating function of a random variable, then, theoretically, we can find the distribution of the random variable since there is a unique distribution function for a given moment generating function. This theorem will prove to be extremely useful in finding the distribution of certain functions of random variables. In particular, see Sec. 4 of Chap. V.

EXAMPLE 20 Suppose that a random variable X has a moment generating function $m_X(t) = 1/(1-t)$ for $-1 < t < 1$; then we know that the density of X is given by $f_X(x) = e^{-x} I_{[0,\infty)}(x)$ since we showed in Example 17 above that $\lambda e^{-\lambda x} I_{[0,\infty)}(x)$ has $\lambda/(\lambda - t)$ for its moment generating function. ////

Problem of moments We have seen that a density function determines a set of moments $\mu_1', \mu_2', \ldots$ when they exist. One of the important problems in theoretical statistics is this: Given a set of moments, what is the density function from which these moments came, and is there only one density function that has these particular moments? We shall give only partial answers. First, there exists a sequence of moments for which there is an infinite (nondenumerable) collection of different distribution functions having these same moments. In general, a sequence of moments $\mu_1', \mu_2', \ldots$ does not determine a unique distribution function. However, we did see that if the moment generating function of a random variable did exist, then this moment generating function did uniquely determine the corresponding distribution function. (See Theorem 7 above.) Hence, there are conditions (existence of the moment generating function is a sufficient condition) under which a sequence of moments does uniquely determine a distribution function. The general problem of whether or not a distribution function is determined by its sequence of moments is referred to as the *problem of moments* and will not be discussed further.

PROBLEMS

1 (*a*) Show that the following are probability density functions (p.d.f.'s):

$$f_1(x) = e^{-x} I_{(0,\infty)}(x)$$
$$f_2(x) = 2e^{-2x} I_{(0,\infty)}(x)$$
$$f(x) = (\theta + 1) f_1(x) - \theta f_2(x) \qquad 0 < \theta < 1.$$

(*b*) Prove or disprove: If $f_1(x)$ and $f_2(x)$ are p.d.f.'s and if $\theta_1 + \theta_2 = 1$, then $\theta_1 f_1(x) + \theta_2 f_2(x)$ is a p.d.f.

2 Show that the following is a density function and find its median:

$$f(x) = \frac{\alpha^2(\alpha + 2x)}{x^2(\alpha + x)^2} I_{(\alpha,\infty)}(x) + \frac{x(2\alpha + x)}{\alpha(\alpha + x)^2} I_{(0,\alpha]}(x), \text{ for } \alpha > 0$$

3 Find the constant K so that the following is a p.d.f.

$$f(x) = Kx^2 I_{(-K,K)}(x).$$

4 Suppose that the cumulative distribution function (c.d.f.) $F_X(x)$ can be written as a function of $(x - \alpha)/\beta$, where α and $\beta > 0$ are constants; that is, x, α, and β appear in $F_X(\cdot)$ only in the indicated form.

(*a*) Prove that if α is increased by $\Delta\alpha$, then so is the mean of X.

(*b*) Prove that if β is multiplied by $k(k > 0)$, then so is the standard deviation of X.

5 The experiment is to toss two balls into four boxes in such a way that each ball is equally likely to fall in any box. Let X denote the number of balls in the first box.

(*a*) What is the c.d.f. of X?

(*b*) What is the density function of X?

(*c*) Find the mean and variance of X.

6 A fair coin is tossed until a head appears. Let X denote the number of tosses required.

(*a*) Find the density function of X.

(*b*) Find the mean and variance of X.

(*c*) Find the moment generating function (m.g.f.) of X.

**7* A has two pennies; B has one. They match pennies until one of them has all three. Let X denote the number of trials required to end the game.

(*a*) What is the density function of X?

(*b*) Find the mean and variance of X.

(*c*) What is the probability that B wins the game?

8 Let $f_X(x) = (1/\beta)[1 - |(x - \alpha)/\beta|]I_{(\alpha-\beta,\, \alpha+\beta)}(x)$, where α and β are fixed constants satisfying $-\infty < \alpha < \infty$ and $\beta > 0$.

(*a*) Demonstrate that $f_X(\cdot)$ is a p.d.f., and sketch it.

(*b*) Find the c.d.f. corresponding to $f_X(\cdot)$.

(*c*) Find the mean and variance of X.

(*d*) Find the qth quantile of X.

9 Let $f_X(x) = k(1/\beta)\{1 - [(x - \alpha)/\beta]^2\}I_{(\alpha-\beta,\, \alpha+\beta)}(x)$, where $-\infty < \alpha < \infty$ and $\beta > 0$.

(*a*) Find k so that $f_X(\cdot)$ is a p.d.f., and sketch the p.d.f.

(*b*) Find the mean, median, and variance of X.

(*c*) Find $\mathscr{E}[|X - \alpha|]$.

(*d*) Find the qth quantile of X.

10 Let $f_X(x) = \frac{1}{2}\{\theta I_{(0,1)}(x) + I_{[1,2]}(x) + (1 - \theta)I_{(2,3)}(x)\}$, where θ is a fixed constant satisfying $0 \leq \theta \leq 1$.

(*a*) Find the c.d.f. of X.

(*b*) Find the mean, median, and variance of X.

11 Let $f(x;\, \theta) = \theta f(x;\, 1) + (1 - \theta)f(x;\, 0)$, where θ is a fixed constant satisfying $0 \leq \theta \leq 1$. Assume that $f(\cdot\,;\, 0)$ and $f(\cdot\,;\, 1)$ are both p.d.f.'s.

(*a*) Show that $f(\cdot\,;\, \theta)$ is also a p.d.f.

(*b*) Find the mean and variance of $f(\cdot\,;\, \theta)$ in terms of the mean and variance of $f(\cdot\,;\, 0)$ and $f(\cdot\,;\, 1)$, respectively.

(*c*) Find the m.g.f. of $f(\cdot\,;\, \theta)$ in terms of the m.g.f.'s of $f(\cdot\,;\, 0)$ and $f(\cdot\,;\, 1)$.

12 A bombing plane flies directly above a railroad track. Assume that if a large (small) bomb falls within 40 (15) feet of the track, the track will be sufficiently damaged so that traffic will be disrupted. Let X denote the perpendicular distance from the track that a bomb falls. Assume that

$$f_X(x) = \frac{100 - x}{5000} I_{[0,100)}(x).$$

(*a*) Find the probability that a large bomb will disrupt traffic.

(*b*) If the plane can carry three large (eight small) bombs and uses all three (eight), what is the probability that traffic will be disrupted?

13 (*a*) Let X be a random variable with mean μ and variance σ^2. Show that $\mathscr{E}[(X-b)^2]$, as a function of b, is minimized when $b = \mu$.

*(*b*) Let X be a continuous random variable with median m. Minimize $\mathscr{E}[|X-b|]$ as a function of b. HINT: Show that $\mathscr{E}[|X-b|] = \mathscr{E}[|X-m|] + 2\int_b^m (x-b)f_X(x)\,dx$.

14 (*a*) If X is a random variable such that $\mathscr{E}[X] = 3$ and $\mathscr{E}[X^2] = 13$, use the Chebyshev inequality to determine a lower bound for $P[-2 < X < 8]$.

(*b*) Let X be a discrete random variable with density

$$f_X(x) = \tfrac{1}{8}I_{\{-1\}}(x) + \tfrac{6}{8}I_{\{0\}}(x) + \tfrac{1}{8}I_{\{1\}}(x).$$

For $k = 2$ evaluate $P[|X - \mu_X| \geq k\sigma_X]$. (This shows that in general the Chebyshev inequality cannot be improved.)

(*c*) If X is a random variable with $\mathscr{E}[X] = \mu$ satisfying $P[X \leq 0] = 0$, show that $P[X > 2\mu] \leq \frac{1}{2}$.

15 Let X be a random variable with p.d.f. given by

$$f_X(x) = |1 - x| I_{[0,2]}(x).$$

Find the mean and variance of X.

16 Let X be a random variable having c.d.f.

$$F_X(x) = pH(x) + (1-p)G(x),$$

where p is a fixed real number satisfying $0 < p < 1$,

$$H(x) = xI_{(0,1]}(x) + I_{(1,\infty)}(x),$$

and

$$G(x) = \tfrac{1}{2}xI_{(0,2]}(x) + I_{(2,\infty)}(x).$$

(*a*) Sketch $F_X(x)$ for $p = \frac{1}{2}$.

(*b*) Give a formula for the p.d.f. of X or the discrete density function of X, whichever is appropriate.

(*c*) Evaluate $P[X \leq \frac{1}{2} | X \leq 1]$.

17 Does there exist a random variable X for which $P[\mu_X - 2\sigma_X \leq X \leq \mu_X + 2\sigma_X] = .6$?

18 An urn contains balls numbered 1, 2, 3. First a ball is drawn from the urn, and then a fair coin is tossed the number of times as the number shown on the drawn ball. Find the expected number of heads.

19 If X has distribution given by $P[X=0]=P[X=2]=p$ and $P[X=1]=1-2p$ for $0 \le p \le \frac{1}{2}$, for what p is the variance of X a maximum?

20 If X is a random variable for which $P[X \le 0]=0$ and $\mathscr{E}[X]=\mu<\infty$, prove that $P[X \le \mu t] \ge 1-1/t$ for every $t \ge 1$.

21 Given the c.d.f.

$$
\begin{aligned}
F_X(x) &= 0 && \text{for } x<0 \\
&= x^2+.2 && \text{for } 0 \le x < .5 \\
&= x && \text{for } .5 \le x < 1 \\
&= 1 && \text{for } 1 \le x.
\end{aligned}
$$

(*a*) Express $F_X(x)$ in terms of indicator functions.

(*b*) Express $F_X(x)$ in the form

$$aF^{ac}(x)+bF^{d}(x),$$

where $F^{ac}(\cdot)$ is an absolutely continuous c.d.f. and $F^{d}(\cdot)$ is a discrete c.d.f.

(*c*) Find $P[.25<X<.75]$.

(*d*) Find $P[.25<X<.5]$.

22 Let $f(x)=Ke^{-\alpha x}(1-e^{-\alpha x})I_{(0,\infty)}(x)$.

(*a*) Find K such that $f(\cdot)$ is a density function.

(*b*) Find the corresponding c.d.f.

(*c*) Find $P[X>1]$.

23 A coin is tossed four times. Let X denote the number of times a head is followed immediately by a tail. Find the distribution, mean, and variance of X.

24 Let $f_X(x;\theta)=(\theta x+\frac{1}{2})I_{(-1,1)}(x)$, where θ is a constant.

(*a*) For what range of values of θ is $f_X(\cdot;\theta)$ a density function?

(*b*) Find the mean and median of X.

(*c*) For what values of θ is var $[X]$ maximized?

25 Let X be a discrete random variable with the nonnegative integers as values. Note that $\mathscr{E}[t^X]=\sum_{j=0}^{\infty} t^j P[X=j]$. Hence, $\mathscr{E}[t^X]$ is a **probability generating function** of X, inasmuch as the coefficient of t^j gives $P[X=j]$. Find $\mathscr{E}[t^X]$ for the random variable of Probs. 6 and 7.

SPECIAL PARAMETRIC FAMILIES OF UNIVARIATE DISTRIBUTIONS

1 INTRODUCTION AND SUMMARY

The purpose of this chapter is to present certain parametric families of univariate density functions that have standard names. A *parametric family* of density functions is a collection of density functions that is *indexed* by a quantity called a *parameter*. For example, let $f(x; \lambda) = \lambda e^{-\lambda x} I_{(0, \infty)}(x)$, where $\lambda > 0$; then for each $\lambda > 0$, $f(\cdot; \lambda)$ is a probability density function. λ is the parameter, and as λ ranges over the positive numbers, the collection $\{f(\cdot; \lambda): \lambda > 0\}$ is a parametric family of density functions.

The chapter consists of three main sections: parametric families of discrete densities are given in one; parametric families of probability density functions are given in another, and comments relating the two are given in the final section. For most of the families of distributions introduced, the means, variances, and moment generating functions are presented; also, a sketch of several representative members of a presented family is often included. A table summarizing results of Secs. 2 and 3 is given in Appendix B.

2 DISCRETE DISTRIBUTIONS

In this section we list several parametric families of univariate discrete densities. Sketches of most are given; the mean and variance of each are derived, and usually examples of random experiments for which the defined parametric family might provide a realistic model are included.

The parameter (or parameters) indexes the family of densities. For each family of densities that is presented, the values that the parameter can assume will be specified. There is no uniform notation for parameters; both Greek and Latin letters are used to designate them.

2.1 Discrete Uniform Distribution

Definition 1 Discrete uniform distribution Each member of the family of discrete density functions

$$f(x) = f(x; N) = \begin{cases} \dfrac{1}{N} & \text{for } x = 1, 2, \ldots, N \\ 0 & \text{otherwise} \end{cases} = \frac{1}{N} I_{\{1, 2, \ldots, N\}}(x), \tag{1}$$

where the parameter N ranges over the positive integers, is defined to have a *discrete uniform distribution*. A random variable X having a density given in Eq. (1) is called a *discrete uniform random variable*. ////

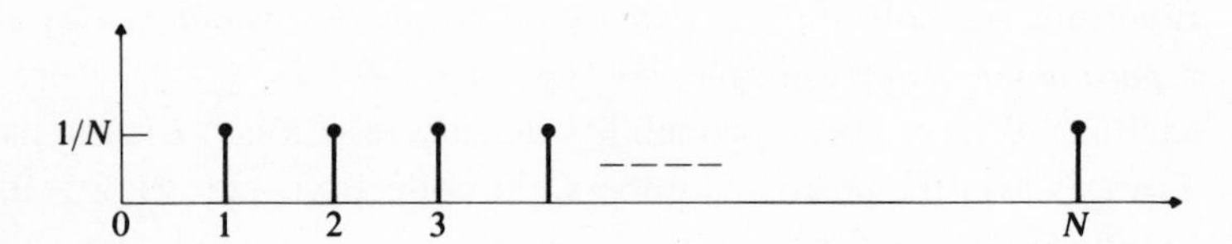

FIGURE 1
Density of discrete uniform.

Theorem 1 If X has a discrete uniform distribution, then $\mathscr{E}[X] = (N+1)/2$,

$$\text{var}[X] = \frac{(N^2 - 1)}{12}, \text{ and } m_X(t) = \mathscr{E}[e^{tX}] = \sum_{j=1}^{N} e^{jt} \frac{1}{N}.$$

PROOF

$$\mathscr{E}[X] = \sum_{j=1}^{N} j \frac{1}{N} = \frac{N+1}{2}.$$

$$\operatorname{var}[X] = \mathscr{E}[X^2] - (\mathscr{E}[X])^2 = \sum_{j=1}^{N} j^2 \frac{1}{N} - \left(\frac{N+1}{2}\right)^2$$

$$= \frac{N(N+1)(2N+1)}{6N} - \frac{(N+1)^2}{4} = \frac{(N+1)(N-1)}{12}.$$

$$\mathscr{E}[e^{tX}] = \sum_{j=1}^{N} e^{jt} \frac{1}{N}.$$ ////

Remark The discrete uniform distribution is sometimes defined in density form as $f(x; N) = [1/(N+1)]I_{\{0, 1, \ldots, N\}}(x)$, for N a nonnegative integer. If such is the case, the formulas for the mean and variance have to be modified accordingly. ////

2.2 Bernoulli and Binomial Distributions

Definition 2 Bernoulli distribution A random variable X is defined to have a *Bernoulli distribution* if the discrete density function of X is given by

$$f_X(x) = f_X(x; p)$$
$$= \left\{\begin{matrix} p^x(1-p)^{1-x} & \text{for } x = 0 \text{ or } 1 \\ & \\ 0 & \text{otherwise} \end{matrix}\right\} = p^x(1-p)^{1-x} I_{\{0, 1\}}(x), \tag{2}$$

where the parameter p satisfies $0 \le p \le 1$. $1 - p$ is often denoted by q. ////

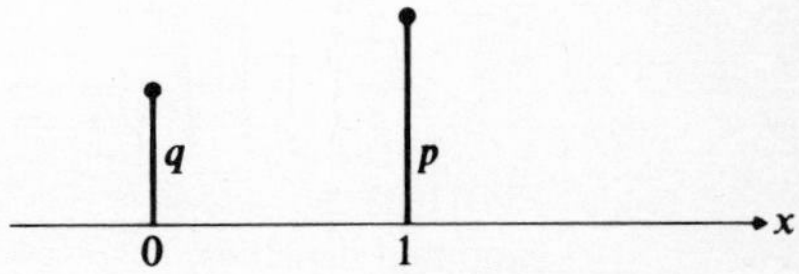

FIGURE 2
Bernoulli density.

Theorem 2 If X has a Bernoulli distribution, then

$$\mathscr{E}[X] = p, \qquad \text{var}\,[X] = pq, \qquad \text{and} \qquad m_X(t) = pe^t + q. \tag{3}$$

PROOF $\mathscr{E}[X] = 0 \cdot q + 1 \cdot p = p.$

$$\text{var}\,[X] = \mathscr{E}[X^2] - (\mathscr{E}[X])^2 = 0^2 \cdot q + 1^2 \cdot p - p^2 = pq.$$

$$m_X(t) = \mathscr{E}[e^{tX}] = q + pe^t.$$ ////

EXAMPLE 1 A random experiment whose outcomes have been classified into two categories, called "success" and "failure," represented by the letters $\mathscr{s}$ and $\mathscr{f}$, respectively, is called a *Bernoulli trial.* If a random variable X is defined as 1 if a Bernoulli trial results in success and 0 if the same Bernoulli trial results in failure, then X has a Bernoulli distribution with parameter $p = P[\text{success}]$. ////

EXAMPLE 2 For a given arbitrary probability space $(\Omega, \mathscr{A}, P[\cdot])$ and for A belonging to $\mathscr{A}$, define the random variable X to be the indicator function of A; that is, $X(\omega) = I_A(\omega)$; then X has a Bernoulli distribution with parameter $p = P[X = 1] = P[A]$. ////

Definition 3 Binomial distribution A random variable X is defined to have a *binomial distribution* if the discrete density function of X is given by

$$f_X(x) = f_X(x; n, p) = \begin{cases} \dbinom{n}{x} p^x q^{n-x} & \text{for } x = 0, 1, \ldots, n \\ 0 & \text{otherwise} \end{cases} \tag{4}$$

$$= \binom{n}{x} p^x q^{n-x} I_{\{0, 1, \ldots, n\}}(x),$$

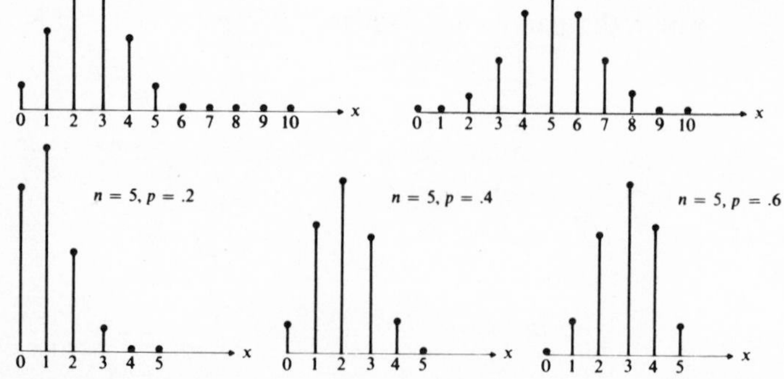

FIGURE 3
Binomial densities.

where the two parameters n and p satisfy $0 \leq p \leq 1$, n ranges over the positive integers, and $q = 1 - p$. A distribution defined by the density function given in Eq. (4) is called a *binomial distribution*. ////

Theorem 3 If X has a binomial distribution, then

$$\mathscr{E}[X] = np, \qquad \text{var}[X] = npq, \qquad \text{and} \qquad m_X(t) = (q + pe^t)^n. \tag{5}$$

PROOF

$$m_X(t) = \mathscr{E}[e^{tX}] = \sum_{x=0}^{n} e^{tx}\binom{n}{x}p^x q^{n-x} = \sum_{x=0}^{n}\binom{n}{x}(pe^t)^x q^{n-x}$$
$$= (pe^t + q)^n.$$

Now

$$m_X'(t) = npe^t(pe^t + q)^{n-1}$$

and

$$m_X''(t) = n(n-1)(pe^t)^2(pe^t + q)^{n-2} + npe^t(pe^t + q)^{n-1};$$

hence

$$\mathscr{E}[X] = m_X'(0) = np$$

and

$$\text{var}[X] = \mathscr{E}[X^2] - (\mathscr{E}[X])^2$$
$$= m_X''(0) - (np)^2 = n(n-1)p^2 + np - (np)^2 = np(1-p). \qquad ////$$

Remark The binomial distribution reduces to the Bernoulli distribution when $n = 1$. Sometimes the Bernoulli distribution is called the *point binomial*. ////

EXAMPLE 3 Consider a random experiment consisting of n repeated independent Bernoulli trials when p is the probability of success s at each individual trial. The term "repeated" is used to indicate that the probability of s remains the same from trial to trial. The sample space for such a random experiment can be represented as follows:

$$\Omega = \{(z_1, z_2, \ldots, z_n): z_i = s \text{ or } z_i = f\}.$$

z_i indicates the result of the ith trial. Since the trials are independent, the probability of any specified outcome, say $\{(f, f, s, f, s, s, \ldots, f, s)\}$, is given by $qqpqpp \cdots qp$. Let the random variable X represent the number of successes in the n repeated independent Bernoulli trials. Now $P[X = x] = P[\text{exactly } x \text{ successes and } n - x \text{ failures in } n \text{ trials}] = \binom{n}{x} p^x q^{n-x}$ for $x = 0, 1, \ldots, n$ since each outcome of the experiment that has exactly x successes has probability $p^x q^{n-x}$ and there are $\binom{n}{x}$ such outcomes. Hence X has a binomial distribution. ////

EXAMPLE 4 Consider sampling with replacement from an urn containing M balls, K of which are defective. Let X represent the number of defective balls in a sample of size n. The individual draws are Bernoulli trials where "defective" corresponds to "success," and the experiment of taking a sample of size n with replacement consists of n repeated independent Bernoulli trials where $p = P[\text{success}] = K/M$; so X has the binomial distribution

$$\binom{n}{x}\left[\frac{K}{M}\right]^x\left[1 - \frac{K}{M}\right]^{n-x} \qquad \text{for} \quad x = 0, 1, \ldots, n, \tag{6}$$

which is the same as $P[A_k]$ in Eq. (3) of Subsec. 3.5 of Chap. I, for $x = k$. ////

The sketches in Fig. 3 seem to indicate that the terms $f_X(x; n, p)$ increase monotonically and then decrease monotonically. The following theorem states that such is indeed the case.

Theorem 4 Let X have a binomial distribution with density $f_X(x; n, p)$; then $f_X(x - 1; n, p) < f_X(x; n, p)$ for $x < (n + 1)p$; $f_X(x - 1; n, p) > f_X(x; n, p)$ for $x > (n + 1)p$, and $f_X(x - 1; n, p) = f_X(x; n, p)$ if $x = (n + 1)p$ and $(n + 1)p$ is an integer, where x ranges over $1, \ldots, n$.

PROOF

$$\frac{f_X(x; n,p)}{f_X(x - 1; n,p)} = \frac{n - x + 1}{x} \cdot \frac{p}{q} = 1 + \frac{(n + 1)p - x}{xq},$$

which is greater than 1 if $x < (n + 1)p$, smaller than 1 if $x > (n + 1)p$, and equal to 1 if the integer x should equal $(n + 1)p$. ////

2.3 Hypergeometric Distribution

Definition 4 Hypergeometric distribution A random variable X is defined to have a *hypergeometric distribution* if the discrete density function of X is given by

$$f_X(x; M, K, n) = \begin{cases} \dfrac{\dbinom{K}{x}\dbinom{M-K}{n-x}}{\dbinom{M}{n}} & \text{for } x = 0, 1, \ldots, n \\ 0 & \text{otherwise} \end{cases} \tag{7}$$

$$= \frac{\dbinom{K}{x}\dbinom{M-K}{n-x}}{\dbinom{M}{n}} I_{\{0, 1, \ldots, n\}}(x)$$

where M is a positive integer, K is a nonnegative integer that is at most M, and n is a positive integer that is at most M. Any distribution function defined by the density function given in Eq. (7) above is called a *hypergeometric distribution.* ////

Theorem 5 If X is a hypergeometric distribution, then

$$\mathscr{E}[X] = n \cdot \frac{K}{M} \quad \text{and} \quad \text{var}[X] = n \cdot \frac{K}{M} \cdot \frac{M-K}{M} \cdot \frac{M-n}{M-1} \tag{8}$$

PROOF

$$\mathscr{E}[X] = \sum_{x=0}^{n} x \frac{\dbinom{K}{x}\dbinom{M-K}{n-x}}{\dbinom{M}{n}} = n \cdot \frac{K}{M} \sum_{x=1}^{n} \frac{\dbinom{K-1}{x-1}\dbinom{M-K}{n-x}}{\dbinom{M-1}{n-1}}$$

$$= n \cdot \frac{K}{M} \sum_{y=0}^{n-1} \frac{\dbinom{K-1}{y}\dbinom{M-1-K+1}{n-1-y}}{\dbinom{M-1}{n-1}}$$

$$= n \cdot \frac{K}{M},$$

using $\displaystyle\sum_{i=0}^{m} \binom{a}{i}\binom{b}{m-i} = \binom{a+b}{m}$ given in Appendix A.

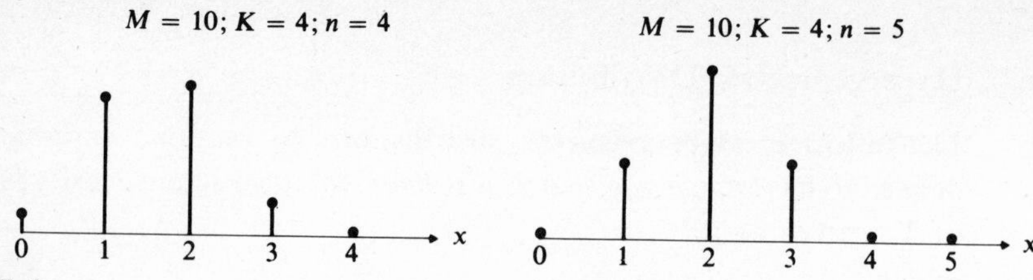

FIGURE 4
Hypergeometric densities.

$$\mathscr{E}[X(X-1)]$$

$$= \sum_{x=0}^{n} x(x-1) \frac{\binom{K}{x}\binom{M-K}{n-x}}{\binom{M}{n}}$$

$$= n(n-1)\frac{K(K-1)}{M(M-1)} \sum_{x=2}^{n} \frac{\binom{K-2}{x-2}\binom{M-K}{n-x}}{\binom{M-2}{n-2}}$$

$$= n(n-1)\frac{K(K-1)}{M(M-1)} \sum_{y=0}^{n-2} \frac{\binom{K-2}{y}\binom{M-2-K+2}{n-2-y}}{\binom{M-2}{n-2}} = n(n-1)\frac{K(K-1)}{M(M-1)}.$$

Hence

$$\begin{aligned} \text{var}[X] &= \mathscr{E}[X^2] - (\mathscr{E}[X])^2 = \mathscr{E}[X(X-1)] + \mathscr{E}[X] - (\mathscr{E}[X])^2 \\ &= n(n-1)\frac{K(K-1)}{M(M-1)} + n\frac{K}{M} - n^2\frac{K^2}{M^2} \\ &= n\frac{K}{M}\left[(n-1)\frac{K-1}{M-1} + 1 - \frac{nK}{M}\right] \\ &= \frac{nK}{M}\left[\frac{(M-K)(M-n)}{M(M-1)}\right]. \end{aligned}$$

////

Remark If we set $K/M = p$, then the mean of the hypergeometric distribution coincides with the mean of the binomial distribution, and the variance of the hypergeometric distribution is $(M-n)/(M-1)$ times the variance of the binomial distribution. ////

EXAMPLE 5 Let X denote the number of defectives in a sample of size n when sampling is done without replacement from an urn containing M balls, K of which are defective. Then X has a hypergeometric distribution. See Eq. (5) of Subsec. 3.5 in Chap. I. ////

2.4 Poisson Distribution

Definition 5 Poisson distribution A random variable X is defined to have a *Poisson distribution* if the density of X is given by

$$f_X(x) = f_X(x; \lambda) = \begin{cases} \dfrac{e^{-\lambda}\lambda^x}{x!} & \text{for } x = 0, 1, 2, \ldots \\ 0 & \text{otherwise} \end{cases} = \frac{e^{-\lambda}\lambda^x}{x!} I_{\{0, 1, \ldots\}}(x), \quad (9)$$

where the parameter λ satisfies $\lambda > 0$. The density given in Eq. (9) is called a *Poisson density*. ////

Theorem 6 Let X be a Poisson distributed random variable; then

$$\mathscr{E}[X] = \lambda, \qquad \text{var}\,[X] = \lambda, \qquad \text{and} \quad m_X(t) = e^{\lambda(e^t - 1)}. \quad (10)$$

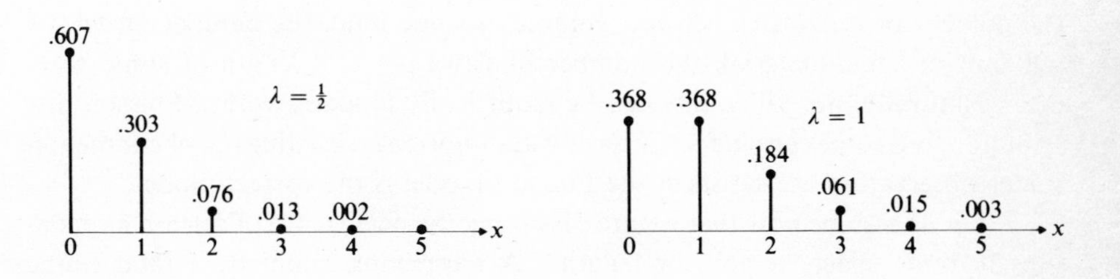

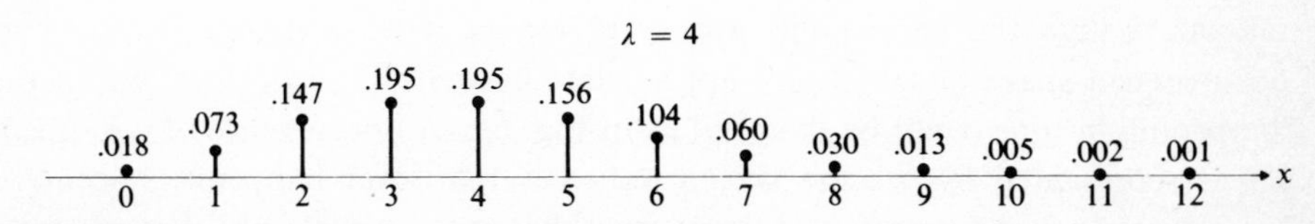

FIGURE 5
Poisson densities.

PROOF

$$m_X(t) = \mathscr{E}[e^{tX}] = \sum_{x=0}^{\infty} \frac{e^{tx}e^{-\lambda}\lambda^x}{x!}$$

$$= e^{-\lambda} \sum_{x=0}^{\infty} \frac{(\lambda e^t)^x}{x!} = e^{-\lambda}e^{\lambda e^t};$$

hence,

$$m_X'(t) = \lambda e^{-\lambda}e^t e^{\lambda e^t}$$

and

$$m_X''(t) = \lambda e^{-\lambda}e^t e^{\lambda e^t}[\lambda e^t + 1].$$

So,

$$\mathscr{E}[X] = m_X'(0) = \lambda$$

and

$$\text{var}\,[X] = \mathscr{E}[X^2] - (\mathscr{E}[X])^2 = m_X''(0) - \lambda^2 = \lambda[\lambda + 1] - \lambda^2 = \lambda. \qquad ////$$

The Poisson distribution provides a realistic model for many random phenomena. Since the values of a Poisson random variable are the nonnegative integers, any random phenomenon for which a count of some sort is of interest is a candidate for modeling by assuming a Poisson distribution. Such a count might be the number of fatal traffic accidents per week in a given state, the number of radioactive particle emissions per unit of time, the number of telephone calls per hour coming into the switchboard of a large business, the number of meteorites that collide with a test satellite during a single orbit, the number of organisms per unit volume of some fluid, the number of defects per unit of some material, the number of flaws per unit length of some wire, etc. Naturally, not all counts can be realistically modeled with a Poisson distribution, but some can; in fact, if certain assumptions regarding the phenomenon under observation are satisfied, the Poisson model is the correct model.

Let us assume now that we are observing the occurrence of certain happenings in time, space, region, or length. A happening might be a fatal traffic accident, a particle emission, the arrival of a telephone call, a meteorite collision, a defect in an area of material, a flaw in a length of wire, etc. We will talk as though the happenings are occurring in time; although happenings occurring in space or length are appropriate as well. The occurrences of the happening in time could be sketched as in Fig. 6. An occurrence of a happening is represented by $\times$; the sketch indicates that seven happenings occurred between time 0 and time t_1. Assume now that there exists a positive quantity, say v, which satisfies the following:

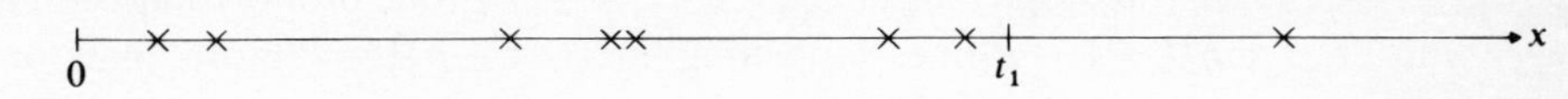

FIGURE 6

(i) The probability that exactly one happening will occur in a small time interval of length h is approximately equal to νh, or P[one happening in interval of length h] $= \nu h + o(h)$.
(ii) The probability of more than one happening in a small time interval of length h is negligible when compared to the probability of just one happening in the same time interval, or P[two or more happenings in interval of length h] $= o(h)$.
(iii) The numbers of happenings in nonoverlapping time intervals are independent.

The term $o(h)$, which is read "some function of smaller order than h," denotes an unspecified function which satisfies

$$\lim_{h \to 0} \frac{o(h)}{h} = 0.$$

The quantity ν can be interpreted as the *mean rate at which happenings occur per unit of time* and is consequently referred to as the *mean rate of occurrence.*

Theorem 7 If the above three assumptions are satisfied, the number of occurrences of a *happening* in a period of time of length t has a Poisson distribution with parameter $\lambda = \nu t$. Or if the random variable $Z(t)$ denotes the number of occurrences of the happening in a time interval of length t, then $P[Z(t) = z] = e^{-\nu t}(\nu t)^z/z!$ for $z = 0, 1, 2, \ldots$.

We will outline two different proofs, neither of which is mathematically rigorous.

PROOF For convenience, let t be a point in time after time 0; so the time interval $(0, t]$ has length t, and the time interval $(t, t + h]$ has length h. Let $P_n(s) = P[Z(s) = n] = P$[exactly n happenings in an interval of length s]; then

$$\begin{aligned}
P_0(t + h) &= P[\text{no happenings in interval } (0, t + h]] \\
&= P[\text{no happenings in } (0, t] \textit{ and } \text{no happenings in } (t, t + h]] \\
&= P[\text{no happenings in } (0, t]]P[\text{no happenings in } (t, t + h]] \\
&= P_0(t)P_0(h),
\end{aligned}$$

using (iii), the independence assumption.

Now P[no happenings in $(t, t+h]$] $= 1 - P$[one or more happenings in $(t, t+h]$] $= 1 - P$[one happening in $(t, t+h]$] $- P$[more than one happening in $(t, t+h]$] $= 1 - \nu h - o(h) - o(h)$; so $P_0(t+h) = P_0(t)[1 - \nu h - o(h) - o(h)]$, or

$$\frac{P_0(t+h) - P_0(t)}{h} = -\nu P_0(t) - P_0(t)\frac{o(h) + o(h)}{h},$$

and on passing to the limit one obtains the differential equation $P_0'(t) = -\nu P_0(t)$, whose solution is $P_0(t) = e^{-\nu t}$, using the condition $P_0(0) = 1$. Similarly, $P_1(t+h) = P_1(t)P_0(h) + P_0(t)P_1(h)$, or $P_1(t+h) = P_1(t)[1 - \nu h - o(h)] + P_0(t)[\nu h + o(h)]$, which gives the differential equation $P_1'(t) = -\nu P_1(t) + \nu P_0(t)$, the solution of which is given by $P_1(t) = \nu t e^{-\nu t}$, using the initial condition $P_1(0) = 0$. Continuing in a similar fashion one obtains $P_n'(t) = -\nu P_n(t) + \nu P_{n-1}(t)$, for $n = 2, 3, \ldots$.

It is seen that this system of differential equations is satisfied by $P_n(t) = (\nu t)^n e^{-\nu t}/n!$.

The second proof can be had by dividing the interval $(0, t)$ into, say n time subintervals, each of length $h = t/n$. The probability that k happenings occur in the interval $(0, t)$ is *approximately* equal to the probability that exactly one happening has occurred in each of k of the n subintervals that we divided the interval $(0, t)$ into. Now the probability of a happening, or "success," in a given subinterval is νh. Each subinterval provides us with a Bernoulli trial; either the subinterval has a happening, or it does not. Also, in view of the assumptions made, these Bernoulli trials are independent, repeated Bernoulli trials; hence the probability of exactly k "successes" in the n trials is given by (see Example 3)

$$\binom{n}{k}(\nu h)^k(1 - \nu h)^{n-k} = \binom{n}{k}\left[\frac{\nu t}{n}\right]^k\left[1 - \frac{\nu t}{n}\right]^{n-k},$$

which is an *approximation* to the desired probability that k happenings will occur in time interval $(0, t)$. An exact expression can be obtained by letting the number of subintervals increase to infinity, that is, by letting n tend to infinity:

$$\binom{n}{k}\left[\frac{\nu t}{n}\right]^k\left[1 - \frac{\nu t}{n}\right]^{n-k} = \frac{1}{k!}(\nu t)^k\left[1 - \frac{\nu t}{n}\right]^{n-k}\frac{(n)_k}{n^k} \to \frac{(\nu t)^k e^{-\nu t}}{k!}$$

as $n \to \infty$, noting that $\left[1 - \frac{\nu t}{n}\right]^n \to e^{-\nu t}$, $\left[1 - \frac{\nu t}{n}\right]^{-k} \to 1$, and $(n)_k/n^k \to 1$.

////

Theorem 7 gives conditions under which certain random experiments involving counts of happenings in time (or length, space, area, volume, etc.) can be realistically modeled by assuming a Poisson distribution. The parameter ν in the Poisson distribution is usually unknown. Techniques for estimating parameters such as ν will be presented in Chap. VII.

In practice great care has to be taken to avoid erroneously applying the Poisson distribution to counts. For example, in studying the distribution of insect larvae over some crop area, the Poisson model is apt to be invalid since insects lay eggs in clusters entailing that larvae are likely to be found in clusters, which is inconsistent with the assumption of independence of counts in small adjacent subareas.

EXAMPLE 6 Suppose that the average number of telephone calls arriving at the switchboard of a small corporation is 30 calls per hour. (i) What is the probability that no calls will arrive in a 3-minute period? (ii) What is the probability that more than five calls will arrive in a 5-minute interval? Assume that the number of calls arriving during any time period has a Poisson distribution. Assume that time is measured in minutes; then 30 calls per hour is equivalent to .5 calls per minute, so the *mean rate of occurrence* is .5 per minute. P[no calls in 3-minute period] $= e^{-\nu t} = e^{-(.5)(3)} = e^{-1.5} \approx .223$.

$$P[\text{more than five calls in 5-minute interval}] = \sum_{k=6}^{\infty} \frac{e^{-\nu t}(\nu t)^k}{k!}$$
$$= \sum_{k=6}^{\infty} \frac{e^{-(.5)(5)}(2.5)^k}{k!} \approx .042. \qquad ////$$

EXAMPLE 7 A merchant knows that the number of a certain kind of item that he can sell in a given period of time is Poisson distributed. How many such items should the merchant stock so that the probability will be .95 that he will have enough items to meet the customer demand for a time period of length T? Let ν denote the *mean rate of occurrence* per unit time and K the unknown number of items that the merchant should stock. Let X denote the number of demands for this kind of item during the time period of length T. The solution requires finding K so that $P[X \leq K] \geq .95$ or finding K so that $\sum_{k=0}^{K} [e^{-\nu T}(\nu T)^k/k!] \geq .95$. In particular, if the merchant sells an average of two such items per day, how many should he stock so that he will have probability at least .95 of having enough items to meet demand for a 30-day month? Find K so that

$$\sum_{k=0}^{K} \frac{e^{-(2)(30)}60^k}{k!} \geq .95,$$

or find K so that

$$\sum_{k=K+1}^{\infty} \frac{e^{-60}60^k}{k!} \leq .05.$$

The desired K can be found using an appropriate Poisson table (e.g., Molina, 1942 [45]). It is $K = 73$. ////

EXAMPLE 8 Suppose that flaws in plywood occur at random with an average of one flaw per 50 square feet. What is the probability that a 4 foot × 8 foot sheet will have no flaws? At most one flaw? To get a solution assume that the number of flaws per unit area is Poisson distributed.

$$P[\text{no flaws}] = e^{-\frac{1}{50}32} = e^{-.64} \approx .527.$$

$$P[\text{at most one flaw}] = e^{-.64} + .64e^{-.64} \approx .865.$$ ////

A Poisson density function, like the binomial density, possesses a certain monotonicity that is precisely stated in the following theorem.

Theorem 8 Consider the Poisson density

$$\frac{e^{-\lambda}\lambda^k}{k!} \qquad \text{for } k = 0, 1, 2, \ldots .$$

$$\frac{e^{-\lambda}\lambda^{k-1}}{(k-1)!} < \frac{e^{-\lambda}\lambda^k}{k!} \qquad \text{for } k < \lambda,$$

$$\frac{e^{-\lambda}\lambda^{k-1}}{(k-1)!} > \frac{e^{-\lambda}\lambda^k}{k!} \qquad \text{for } k > \lambda,$$

and

$$\frac{e^{-\lambda}\lambda^{k-1}}{(k-1)!} = \frac{e^{-\lambda}\lambda^k}{k!} \qquad \text{if } \lambda \text{ is an integer and } k = \lambda.$$

PROOF

$$\frac{e^{-\lambda}\lambda^{k-1}/(k-1)!}{e^{-\lambda}\lambda^k/k!} = \frac{k}{\lambda},$$

which is less than 1 if $k < \lambda$, greater than 1 if $k > \lambda$, and equal to 1 if λ is an integer and $k = \lambda$. ////

2.5 Geometric and Negative Binomial Distributions

Two other families of discrete distributions that play important roles in statistics are the geometric (or Pascal) and negative binomial distributions. The reason that we consider the two together is twofold; first, the geometric distribution is a special case of the negative binomial distribution, and, second, the sum of independent and identically distributed geometric random variables is negative binomially distributed, as we shall see in Chap. V. In Subsec. 3.3 of this chapter, the exponential and gamma distributions are defined. We shall see that in several respects the geometric and negative binomial distributions are discrete analogs of the exponential and gamma distributions.

Definition 6 Geometric distribution A random variable X is defined to have *geometric* (or *Pascal*) *distribution* if the density of X is given by

$$f_X(x) = f_X(x; p)$$

$$= \begin{cases} p(1-p)^x & \text{for } x = 0, 1, \ldots \\ 0 & \text{otherwise} \end{cases} = p(1-p)^x I_{\{0, 1, \ldots\}}(x), \qquad (11)$$

where the parameter p satisfies $0 < p \leq 1$. (Define $q = 1 - p$.) ////

Definition 7 Negative binomial distribution A random variable X with density

$$f_X(x) = f_X(x; r, p)$$

$$= \begin{cases} \binom{r+x-1}{x} p^r q^x = \binom{-r}{x} p^r (-q)^x & \text{for } x = 0, 1, 2, \ldots \\ 0 & \text{otherwise} \end{cases} \qquad (12)$$

$$= \binom{r+x-1}{x} p^r q^x I_{\{0, 1, \ldots\}}(x),$$

where the parameters r and p satisfy $r = 1, 2, 3, \ldots$ and $0 < p \leq 1$ $(q = 1 - p)$, is defined to have a *negative binomial distribution.* The density given by Eq. (12) is called a *negative binomial density.*

////

Remark If in the negative binomial distribution $r = 1$, then the negative binomial density specializes to the geometric density. ////

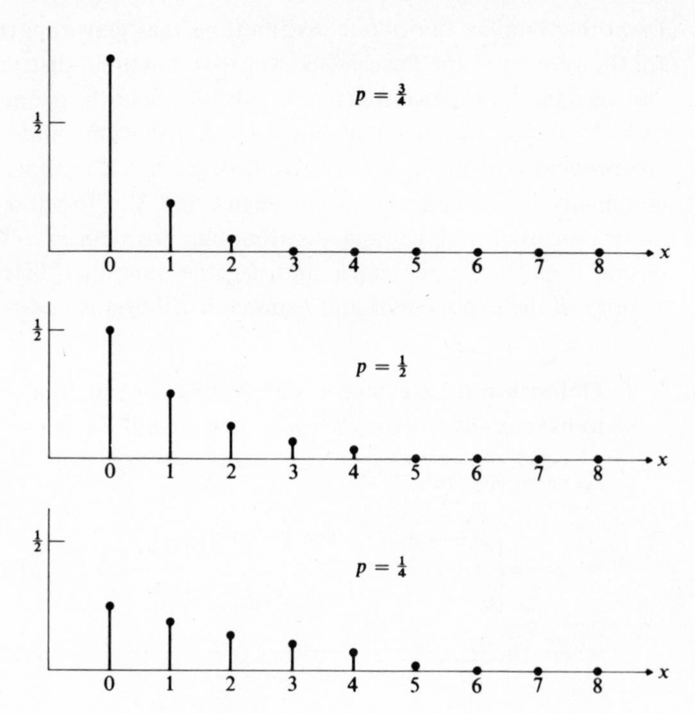

FIGURE 7
Geometric densities.

Theorem 9 If the random variable X has a geometric distribution, then

$$\mathscr{E}[X] = \frac{q}{p}, \qquad \text{var}\,[X] = \frac{q}{p^2}, \qquad \text{and} \qquad m_X(t) = \frac{p}{1 - qe^t}. \tag{13}$$

PROOF Since a geometric distribution is a special case of a negative binomial distribution, Theorem 9 is a corollary of Theorem 11. ////

The geometric distribution is well named since the values that the geometric density assumes are the terms of a geometric series. Also the mode of the geometric density is necessarily 0. A geometric density possesses one other interesting property, which is given in the following theorem.

Theorem 10 If X has the geometric density with parameter p, then

$$P[X \geq i + j \,|\, X \geq i] = P[X \geq j] \qquad \text{for } i, j = 0, 1, 2, \ldots.$$

PROOF $$P[X \geq i + j \mid X \geq i] = \frac{P[X \geq i + j]}{P[X \geq i]}$$

$$= \frac{\sum_{x=i+j}^{\infty} p(1-p)^x}{\sum_{x=i}^{\infty} p(1-p)^x} = \frac{(1-p)^{i+j}}{(1-p)^i}$$

$$= (1-p)^j$$

$$= P[X \geq j].$$ ////

Theorem 10 says that the probability that a geometric random variable is greater than or equal to $i + j$ given that it is greater than or equal to i is equal to the unconditional probability that it will be greater than or equal to j. We will comment on this again in the following example.

EXAMPLE 9 Consider a sequence of independent, repeated Bernoulli trials with p equal to the probability of success on an individual trial. Let the random variable X represent the number of trials required before the first success; then X has the geometric density given by Eq. (11). To see this, note that the first success will occur on trial $x + 1$ if this $(x + 1)$st trial results in a success and the first x trials resulted in failures; but, by independence, x successive failures followed by a success has probability $(1 - p)^x p$. In the language of this example, Theorem 10 states that the probability that at least $i + j$ trials are required before the first success, given that there have been i successive failures, is equal to the unconditional probability that at least j trials are needed before the first success. That is, the fact that one has already observed i successive failures does not change the distribution of the number of trials required to obtain the first success. ////

A random variable X that has a geometric distribution is often referred to as a discrete *waiting-time* random variable. It represents how long (in terms of the number of failures) one has to wait for a success.

Before leaving the geometric distribution, we note that some authors define the geometric distribution by assuming 1 (instead of 0) is the smallest mass point. The density then has the form

$$f(x; p) = p(1-p)^{x-1} I_{\{1, 2, \ldots\}}(x), \tag{14}$$

and the mean is $1/p$, the variance is q/p^2, and the moment generating function is $pe^t/(1 - qe^t)$.

Theorem 11 Let X have a negative binomial distribution; then

$$\mathscr{E}[X] = \frac{rq}{p}, \qquad \text{var}\,[X] = \frac{rq}{p^2}, \qquad \text{and} \qquad m_X(t) = \left[\frac{p}{1 - qe^t}\right]^r. \tag{15}$$

PROOF

$$m_X(t) = \mathscr{E}[e^{tX}] = \sum_{x=0}^{\infty} e^{tx}\binom{-r}{x}p^r(-q)^x$$

$$= \sum_{x=0}^{\infty}\binom{-r}{x}p^r(-qe^t)^x = \left[\frac{p}{1 - qe^t}\right]^r$$

[see Eq. (33) in Appendix A].

$$m_X'(t) = p^r(-r)(1 - qe^t)^{-r-1}(-qe^t)$$

and

$$m_X''(t) = rqp^r[q(r + 1)e^{2t}(1 - qe^t)^{-r-2} + e^t(1 - qe^t)^{-r-1}];$$

hence

$$\mathscr{E}[X] = m_X'(t)\bigg|_{t=0} = \frac{rq}{p}$$

and

$$\text{var}\,[X] = m_X''(t)\bigg|_{t=0} - (\mathscr{E}[X])^2 = rqp^r[qp^{-r-2}(r + 1) + p^{-r-1}] - \left(\frac{rq}{p}\right)^2$$

$$= \frac{rq^2}{p^2} + \frac{rq}{p} = \frac{rq}{p^2}. \qquad ////$$

The negative binomial distribution, like the Poisson, has the nonnegative integers for its mass points; hence, the negative binomial distribution is potentially a model for a random experiment where a count of some sort is of interest. Indeed, the negative binomial distribution has been applied in population counts, in health and accident statistics, in communications, and in other counts as well. Unlike the Poisson distribution, where the mean and variance are the same, the variance of the negative binomial distribution is greater than its mean. We will see in Subsec. 4.3 of this chapter that the negative binomial distribution can be obtained as a *contagious* distribution from the Poisson distribution.

EXAMPLE 10 Consider a sequence of independent, repeated Bernoulli trials with p equal to the probability of success on an individual trial. Let the random variable X represent the number of failures prior to the rth success; then X has the negative binomial density given by Eq. (12), as the following argument shows: The last trial must result in a success, having probability p; among the first $x + r - 1$ trials there must be $r - 1$ successes and x failures, and the probability of this is

$$\binom{x+r-1}{r-1}p^{r-1}q^x = \binom{r+x-1}{x}p^{r-1}q^x,$$

which when multiplied by p gives the desired result. ////

A random variable X having a negative binomial distribution is often referred to as a discrete *waiting-time* random variable. It represents how long (in terms of the number of failures) one waits for the rth success.

EXAMPLE 11 The negative binomial distribution is of importance in the consideration of *inverse binomial sampling*. Suppose a proportion p of individuals in a population possesses a certain characteristic. If individuals in the population are sampled until exactly r individuals with the certain characteristic are found, then the number of individuals in excess of r that are observed or sampled has a negative binomial distribution. ////

2.6 Other Discrete Distributions

In the previous five subsections we presented seven parametric families of univariate discrete density functions. Each is commonly known by the names given. There are many other families of discrete density functions. In fact, new families can be formed from the presented families by various processes. One such process is called *truncation*. We will illustrate this process by looking at the Poisson distribution truncated at 0. Suppose, as is sometimes the case, that the zero count cannot be observed yet the Poisson distribution seems a reasonable model. One might then distribute the mass ordinarily given to the mass point 0 proportionately among the other mass points obtaining the family of densities

$$f_X(x) = \begin{cases} e^{-\lambda}\lambda^x/x!(1 - e^{-\lambda}) & \text{for } x = 1, 2, \ldots \\ 0 & \text{otherwise.} \end{cases} \tag{16}$$

A random variable having density given by Eq. (16) is called a *Poisson random variable truncated at* 0.

Another process for obtaining a new family of densities from a given family can also be illustrated with the Poisson distribution. Suppose that a random variable X, representing a count of some sort, has a Poisson distribution. If the experimenter is stuck with a rather poor counter, one that cannot count beyond 2, the random variable that the experimenter actually observes has density given by

z	0	1	2
$f(z)$	$e^{-\lambda}$	$\lambda e^{-\lambda}$	$1 - e^{-\lambda} - \lambda e^{-\lambda}$

The counter counts correctly values 0 and 1 of the random variable X; but if X takes on any value 2 or more, the counter counts 2. Such a random variable is often referred to as a *censored* random variable.

The above two illustrations indicate how other families of discrete densities can be formulated from existing families. We close this section by giving two further, not so well-known, families of discrete densities.

Definition 8 Beta-binomial distribution The distribution with discrete density function

$$f(x) = f(x; n, \alpha, \beta) = \binom{n}{x} \frac{\Gamma(\alpha + \beta)}{\Gamma(\alpha)\Gamma(\beta)} \cdot \frac{\Gamma(x + \alpha)\Gamma(n + \beta - x)}{\Gamma(n + \alpha + \beta)} I_{\{0, 1, \ldots, n\}}(x) \tag{17}$$

where n is a nonnegative integer, $\alpha > 0$, and $\beta > 0$, is defined as the *beta-binomial distribution.*

$\Gamma(m)$ is the well-known gamma function $\Gamma(m) = \int_0^\infty x^{m-1} e^{-x}\, dx$ for $m > 0$. See Appendix A. The beta-binomial distribution has

$$\text{Mean} = \frac{n\alpha}{\alpha + \beta} \qquad \text{and} \qquad \text{variance} = \frac{n\alpha\beta(n + \alpha + \beta)}{(\alpha + \beta)^2(\alpha + \beta + 1)}. \tag{18}$$

It has the same mass points as the binomial distribution. If $\alpha = \beta = 1$, then the beta-binomial distribution reduces to a discrete uniform distribution over the integers 0, 1, ..., n. ////

Definition 9 Logarithmic distribution The distribution with discrete density function

$$f(x; p) = \begin{cases} \dfrac{q^x}{-x \log_e p} & \text{for } x = 1, 2, \ldots \\ 0 & \text{otherwise} \end{cases} = \frac{q^x}{-x \log_e p} I_{\{1, 2, \ldots\}}(x), \quad (19)$$

where the parameters satisfy $0 < p < 1$ and $q = 1 - p$ is defined as the *logarithmic distribution.* ////

The name is justified if one recalls the power-series expansion of $\log_e(1 - q)$. The logarithmic distribution has

$$\text{Mean} = \frac{q}{-p \log_e p} \quad \text{and} \quad \text{variance} = \frac{q(q + \log_e p)}{-(p \log_e p)^2}. \quad (20)$$

It can be derived as a limiting distribution of negative binomial distributions that have been generalized to include r, any positive number (rather than just an integer), truncated at 0. The limiting distribution is obtained by letting r approach 0.

3 CONTINUOUS DISTRIBUTIONS

In this section several parametric families of univariate probability density functions are presented. Sketches of some are included; the mean and variance (when they exist) of each are given.

3.1 Uniform or Rectangular Distribution

A very simple distribution for a continuous random variable is the uniform distribution. It is particularly useful in theoretical statistics because it is convenient to deal with mathematically.

Definition 10 Uniform distribution If the probability density function of a random variable X is given by

$$f_X(x) = f_X(x; a, b) = \frac{1}{b - a} I_{[a, b]}(x), \quad (21)$$

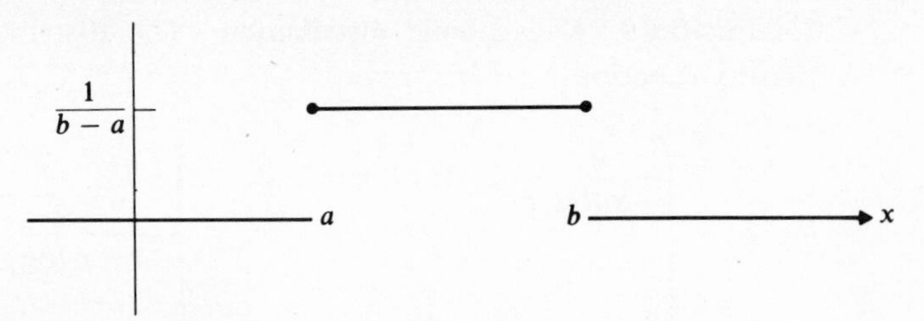

FIGURE 8
Uniform probability density.

where the parameters a and b satisfy $-\infty < a < b < \infty$, then the random variable X is defined to be *uniformly* distributed over the interval $[a, b]$, and the distribution given by Eq. (21) is called a *uniform distribution*. ////

Theorem 12 If X is uniformly distributed over $[a, b]$, then

$$\mathscr{E}[X] = \frac{a+b}{2}, \quad \text{var}[X] = \frac{(b-a)^2}{12}, \quad \text{and} \quad m_X(t) = \frac{e^{bt} - e^{at}}{(b-a)t}. \tag{22}$$

PROOF

$$\mathscr{E}[X] = \int_a^b x \frac{1}{b-a}\, dx = \frac{b^2 - a^2}{2(b-a)} = \frac{a+b}{2}.$$

$$\text{var}[X] = \mathscr{E}[X^2] - (\mathscr{E}[X])^2 = \int_a^b x^2 \frac{1}{b-a}\, dx - \left(\frac{a+b}{2}\right)^2$$

$$= \frac{b^3 - a^3}{3(b-a)} - \frac{(a+b)^2}{4} = \frac{(b-a)^2}{12}.$$

$$m_X(t) = \mathscr{E}[e^{tX}] = \int_a^b e^{tx} \frac{1}{b-a}\, dx = \frac{e^{bt} - e^{at}}{(b-a)t}.$$ ////

The uniform distribution gets its name from the fact that its density is uniform, or constant, over the interval $[a, b]$. It is also called the *rectangular* distribution—the shape of the density is rectangular.

The cumulative distribution function of a uniform random variable is given by

$$F_X(x) = \left(\frac{x-a}{b-a}\right) I_{[a,b]}(x) + I_{(b,\infty)}(x). \tag{23}$$

It provides a useful model for a few random phenomena. For instance, if it is known that the values of some random variable X can only be in a finite interval, say $[a, b]$, and if one assumes that any two subintervals of $[a, b]$ of equal length have the same probability of containing X, then X has a uniform distribution over the interval $[a, b]$. When one speaks of a *random number* from the interval $[0, 1]$, one is thinking of the value of a uniformly distributed random variable over the interval $[0, 1]$.

EXAMPLE 12 If a wheel is spun and then allowed to come to rest, the point on the circumference of the wheel that is located opposite a certain fixed marker could be considered the value of a random variable X that is uniformly distributed over the circumference of the wheel. One could then compute the probability that X will fall in any given arc. ////

Although we defined the uniform distribution as being uniformly distributed over the *closed* interval $[a, b]$, one could just as well define it over the *open* interval (a, b) [in which case $f_X(x) = (b - a)^{-1} I_{(a,b)}(x)$] or over either of the *half-open–half-closed* intervals $(a, b]$ or $[a, b)$. Note that all four of the possible densities have the same cumulative distribution function. This lack of uniqueness of probability density functions was first mentioned in Subsec. 3.2 of Chap. II.

3.2 Normal Distribution

A great many of the techniques used in applied statistics are based upon the normal distribution; it will frequently appear in the remainder of this book.

Definition 11 Normal distribution A random variable X is defined to be *normally* distributed if its density is given by

$$f_X(x) = f_X(x; \mu, \sigma) = \frac{1}{\sqrt{2\pi}\sigma} e^{-(x-\mu)^2/2\sigma^2}, \tag{24}$$

where the parameters μ and σ satisfy $-\infty < \mu < \infty$ and $\sigma > 0$. Any distribution defined by a density function given in Eq. (24) is called a *normal distribution.* ////

We have used the symbols μ and σ^2 to represent the parameters because these parameters turn out, as we shall see, to be the mean and variance, respectively, of the distribution.

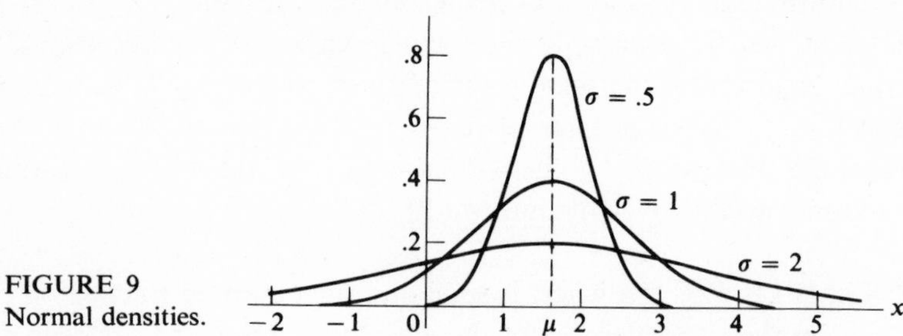

FIGURE 9
Normal densities.

One can readily check that the mode of a normal density occurs at $x = \mu$ and inflection points occur at $\mu - \sigma$ and $\mu + \sigma$. (See Fig. 9.) Since the normal distribution occurs so frequently in later chapters, special notation is introduced for it. If random variable X is normally distributed with mean μ and variance σ^2, we will write $X \sim N(\mu, \sigma^2)$. We will also use the notation $\phi_{\mu, \sigma^2}(x)$ for the density of $X \sim N(\mu, \sigma^2)$ and $\Phi_{\mu, \sigma^2}(x)$ for the cumulative distribution function.

If the normal random variable has mean 0 and variance 1, it is called a *standard* or *normalized* normal random variable. For a standard normal random variable the subscripts of the density and distribution function notations are dropped; that is,

$$\phi(x) = \frac{1}{\sqrt{2\pi}} e^{-\frac{1}{2}x^2} \qquad \text{and} \qquad \Phi(x) = \int_{-\infty}^{x} \phi(u)\, du. \tag{25}$$

Since $\phi_{\mu, \sigma^2}(x)$ is given to be a density function, it is implied that

$$\int_{-\infty}^{\infty} \phi_{\mu, \sigma^2}(x)\, dx = 1,$$

but we should satisfy ourselves that this is true. The verification is somewhat troublesome because the indefinite integral of this particular density function does not have a simple functional expression. Suppose that we represent the area under the curve by A; then

$$A = \frac{1}{\sqrt{2\pi}\sigma} \int_{-\infty}^{\infty} e^{-(x-\mu)^2/2\sigma^2}\, dx,$$

and on making the substitution $y = (x - \mu)/\sigma$, we find that

$$A = \frac{1}{\sqrt{2\pi}} \int_{-\infty}^{\infty} e^{-\frac{1}{2}y^2}\, dy.$$

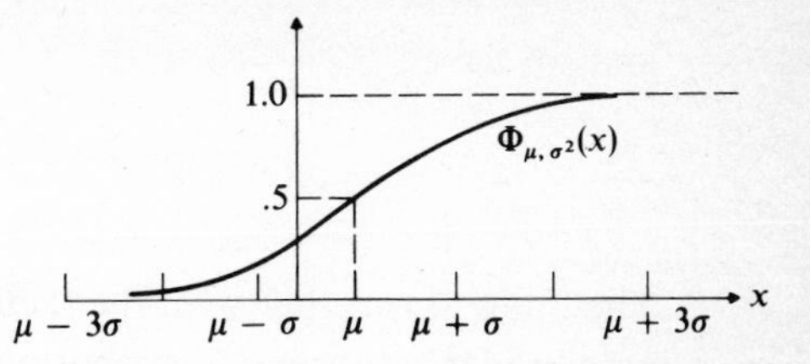

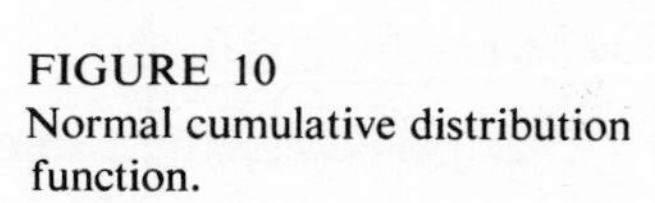
FIGURE 10
Normal cumulative distribution function.

We wish to show that $A = 1$, and this is most easily done by showing that A^2 is 1 and then reasoning that $A = 1$ since $\phi_{\mu,\sigma^2}(x)$ is positive. We may put

$$A^2 = \frac{1}{\sqrt{2\pi}} \int_{-\infty}^{\infty} e^{-\frac{1}{2}y^2}\, dy \frac{1}{\sqrt{2\pi}} \int_{-\infty}^{\infty} e^{-\frac{1}{2}z^2}\, dz$$
$$= \frac{1}{2\pi} \int_{-\infty}^{\infty} \int_{-\infty}^{\infty} e^{-\frac{1}{2}(y^2+z^2)}\, dy\, dz$$

by writing the product of two integrals as a double integral. In this integral we change the variables to polar coordinates by the substitutions

$$y = r \sin \theta$$
$$z = r \cos \theta,$$

and the integral becomes

$$A^2 = \frac{1}{2\pi} \int_0^{\infty} \int_0^{2\pi} r e^{-\frac{1}{2}r^2}\, d\theta\, dr$$
$$= \int_0^{\infty} r e^{-\frac{1}{2}r^2}\, dr$$
$$= 1.$$

Theorem 13 If X is a normal random variable,

$$\mathscr{E}[X] = \mu, \qquad \operatorname{var}[X] = \sigma^2, \qquad \text{and} \qquad m_X(t) = e^{\mu t + \sigma^2 t^2/2}. \tag{26}$$

PROOF

$$m_X(t) = \mathscr{E}[e^{tX}] = e^{t\mu}\mathscr{E}[e^{t(X-\mu)}]$$
$$= e^{t\mu} \int_{-\infty}^{\infty} \frac{1}{\sqrt{2\pi}} e^{t(x-\mu)} e^{-(1/2\sigma^2)(x-\mu)^2}\, dx$$
$$= e^{t\mu} \frac{1}{\sqrt{2\pi}} \int_{-\infty}^{\infty} e^{-(1/2\sigma^2)[(x-\mu)^2 - 2\sigma^2 t(x-\mu)]}\, dx.$$

If we complete the square inside the bracket, it becomes

$$(x-\mu)^2 - 2\sigma^2 t(x-\mu) = (x-\mu)^2 - 2\sigma^2 t(x-\mu) + \sigma^4 t^2 - \sigma^4 t^2$$
$$= (x - \mu - \sigma^2 t)^2 - \sigma^4 t^2,$$

and we have

$$m_X(t) = e^{t\mu}e^{\sigma^2 t^2/2} \frac{1}{\sqrt{2\pi}\sigma} \int_{-\infty}^{\infty} e^{-(x-\mu-\sigma^2 t)^2/2\sigma^2}\, dx.$$

The integral together with the factor $1/\sqrt{2\pi}\sigma$ is necessarily 1 since it is the area under a normal distribution with mean $\mu + \sigma^2 t$ and variance σ^2. Hence,

$$m_X(t) = e^{\mu t + \sigma^2 t^2/2}.$$

On differentiating $m_X(t)$ twice and substituting $t = 0$, we find

$$\mathscr{E}[X] = m'_X(0) = \mu$$

and

$$\text{var}\,[X] = \mathscr{E}[X^2] - (\mathscr{E}[X])^2 = m''_X(0) - \mu^2 = \sigma^2,$$

thus justifying our use of the symbols μ and σ^2 for the parameters. ////

Since the indefinite integral of $\phi_{\mu,\sigma^2}(x)$ does not have a simple functional form, one can only exhibit the cumulative distribution function as

$$\Phi_{\mu,\sigma^2}(x) = \int_{-\infty}^{x} \phi_{\mu,\sigma^2}(u)\, du. \tag{27}$$

The following theorem shows that we can find the probability that a normally distributed random variable, with mean μ and variance σ^2, falls in any interval in terms of the standard normal cumulative distribution function, and this standard normal cumulative distribution function is tabled in Table 2 of Appendix D.

Theorem 14 If $X \sim N(\mu, \sigma^2)$, then

$$P[a < X < b] = \Phi\left(\frac{b-\mu}{\sigma}\right) - \Phi\left(\frac{a-\mu}{\sigma}\right). \tag{28}$$

PROOF

$$P[a < X < b] = \int_a^b \frac{1}{\sqrt{2\pi}\sigma} e^{-\frac{1}{2}[(x-\mu)/\sigma]^2}\, dx$$

$$= \int_{(a-\mu)/\sigma}^{(b-\mu)/\sigma} \frac{1}{\sqrt{2\pi}} e^{-\frac{1}{2}z^2}\, dz$$

$$= \Phi\left(\frac{b-\mu}{\sigma}\right) - \Phi\left(\frac{a-\mu}{\sigma}\right). \qquad ////$$

Remark $\Phi(x) = 1 - \Phi(-x)$. ////

The normal distribution appears to be a reasonable model of the behavior of certain random phenomena. It also is the limiting form of many other probability distributions. Some such limits are given in Subsec. 4.1 of this chapter. The normal distribution is also the limiting distribution in the famous *central-limit theorem*, which is discussed in Sec. 4 of Chap. V and again in Sec. 3 of Chap. VI.

Most students are already somewhat familiar with the normal distribution because of their experience with "grading on the curve." This notion is covered in the following example.

EXAMPLE 13 Suppose that an instructor assumes that a student's final score is the value of a normally distributed random variable. If the instructor decides to award a grade of A to those students whose score exceeds $\mu + \sigma$, a B to those students whose score falls between μ and $\mu + \sigma$, a C if a score falls between $\mu - \sigma$ and μ, a D if a score falls between $\mu - 2\sigma$ and $\mu - \sigma$, and an F if the score falls below $\mu - 2\sigma$, then the proportions of each grade given can be calculated. For example, since

$$P[X > \mu + \sigma] = 1 - P[X < \mu + \sigma] = 1 - \Phi\left(\frac{\mu + \sigma - \mu}{\sigma}\right)$$
$$= 1 - \Phi(1) \approx .1587,$$

one would expect 15.87 percent of the students to receive A's. ////

EXAMPLE 14 Suppose that the diameters of shafts manufactured by a certain machine are normal random variables with mean 10 centimeters and standard deviation .1 centimeter. If for a given application the shaft must meet the requirement that its diameter fall between 9.9 and 10.2 centimeters, what proportion of the shafts made by this machine will meet the requirement?

$$P[9.9 < X < 10.2] = \Phi\left(\frac{10.2 - 10}{.1}\right) - \Phi\left(\frac{9.9 - 10}{.1}\right)$$
$$= \Phi(2) - \Phi(-1) \approx .9772 - .1587 = .8185. \qquad ////$$

3.3 Exponential and Gamma Distributions

Two other families of distributions that play important roles in statistics are the (negative) exponential and gamma distributions, which are defined in this subsection. The reason that the two are considered together is twofold; first, the exponential is a special case of the gamma, and, second, the sum of independent identically distributed exponential random variables is gamma-distributed, as we shall see in Chap. V.

Definition 12 Exponential distribution If a random variable X has a density given by

$$f_X(x; \lambda) = \lambda e^{-\lambda x} I_{[0, \infty)}(x), \tag{29}$$

where $\lambda > 0$, then X is defined to have an (negative) *exponential distribution.* ////

Definition 13 Gamma distribution If a random variable X has density given by

$$f_X(x; r, \lambda) = \frac{\lambda}{\Gamma(r)} (\lambda x)^{r-1} e^{-\lambda x} I_{[0, \infty)}(x), \tag{30}$$

where $r > 0$ and $\lambda > 0$, then X is defined to have a *gamma distribution.* $\Gamma(\cdot)$ is the gamma function and it is discussed in Appendix A. ////

Remark If in the gamma density $r = 1$, the gamma density specializes to the exponential density. ////

Theorem 15 If X has an exponential distribution, then

$$\mathscr{E}[X] = \frac{1}{\lambda}, \qquad \text{var}\,[X] = \frac{1}{\lambda^2}, \qquad \text{and} \qquad m_X(t) = \frac{\lambda}{\lambda - t} \qquad \text{for} \qquad t < \lambda. \tag{31}$$

PROOF The exponential distribution was the distribution used as an example for some definitions given in Chap. II, and derivations of the above appear there. Also, Theorem 15 is a corollary to the following theorem. ////

Theorem 16 If X has a gamma distribution with parameters r and λ, then

$$\mathscr{E}[X] = \frac{r}{\lambda}, \qquad \text{var}\,[X] = \frac{r}{\lambda^2}, \qquad \text{and} \qquad m_X(t) = \left(\frac{\lambda}{\lambda - t}\right)^r \qquad \text{for } t < \lambda. \tag{32}$$

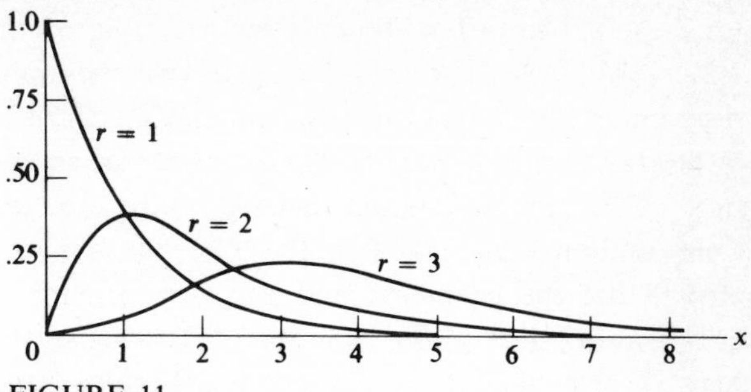

FIGURE 11
Gamma densities ($\lambda = 1$).

PROOF

$$m_X(t) = \mathscr{E}[e^{tX}]$$
$$= \int_0^\infty \frac{\lambda^r}{\Gamma(r)} e^{tx} x^{r-1} e^{-\lambda x}\, dx$$
$$= \left(\frac{\lambda}{\lambda - t}\right)^r \int_0^\infty \frac{(\lambda - t)^r}{\Gamma(r)} x^{r-1} e^{-(\lambda - t)x}\, dx = \left(\frac{\lambda}{\lambda - t}\right)^r.$$
$$m_X'(t) = r\lambda^r(\lambda - t)^{-r-1}$$

and

$$m_X''(t) = r(r+1)\lambda^r(\lambda - t)^{-r-2};$$

hence

$$\mathscr{E}[X] = m_X'(0) = \frac{r}{\lambda}$$

and

$$\text{var}\,[X] = \mathscr{E}[X^2] - (\mathscr{E}[X])^2$$
$$= m_X''(0) - \left(\frac{r}{\lambda}\right)^2 = \frac{r(r+1)}{\lambda^2} - \left(\frac{r}{\lambda}\right)^2 = \frac{r}{\lambda^2}. \qquad ////$$

The exponential distribution has been used as a model for lifetimes of various things. When we introduced the Poisson distribution, we spoke of certain happenings, for example, particle emissions, occurring in time. The length of the time interval between successive happenings can be shown to have an exponential distribution provided that the number of happenings in a fixed time interval has a Poisson distribution. We comment on this again in Subsec. 4.2 below. Also, if we assume again that the number of happenings in a fixed time interval is Poisson distributed, the length of time between time 0 and the instant when the rth happening occurs can be shown to have a gamma distribution. So a gamma random variable can be thought of as a continuous waiting-time random variable. It is the time one has to wait for the rth happening. Recall that the geometric and negative binomial random variables were discrete waiting-time random variables. In a sense, they are discrete analogs of the negative exponential and gamma distributions, respectively.

Theorem 17 If the random variable X has a gamma distribution with parameters r and λ, where r is a positive integer, then

$$F_X(x) = 1 - \sum_{j=0}^{r-1} \frac{e^{-\lambda x}(\lambda x)^j}{j!}. \tag{33}$$

PROOF The proof can be obtained by successive integrations by parts. ////

For $\lambda = 1$, $F_X(x)$ given in Eq. (33) is called the *incomplete gamma function* and has been extensively tabulated.

Theorem 18 If the random variable X has an exponential distribution with parameter λ, then

$$P[X > a + b \mid X > a] = P[X > b], \qquad \text{for } a > 0 \text{ and } b > 0.$$

PROOF

$$P[X > a + b \mid X > a] = \frac{P[X > a + b]}{P[X > a]} = \frac{e^{-\lambda(a+b)}}{e^{-\lambda a}}$$

$$= e^{-\lambda b} = P[X > b].$$

////

Let X represent the lifetime of a given component; then, in words, Theorem 18 states that the conditional probability that the component will last $a + b$ time units given that it has lasted a time units is the same as its initial probability of lasting b time units. Another way of saying this is to say that an "old" functioning component has the same lifetime distribution as a "new" functioning component or that the component is not subject to fatigue or to wear.

3.4 Beta Distribution

A family of probability densities of continuous random variables taking on values in the interval (0, 1) is the family of beta distributions.

Definition 14 Beta distribution If a random variable X has a density given by

$$f_X(x) = f_X(x; a, b) = \frac{1}{\text{B}(a, b)} x^{a-1}(1-x)^{b-1} I_{(0,1)}(x), \tag{34}$$

where $a > 0$ and $b > 0$, then X is defined to have a *beta distribution.* ////

The function $\text{B}(a, b) = \int_0^1 x^{a-1}(1-x)^{b-1}\,dx$, called the *beta function*, is mentioned briefly in Appendix A.

Remark The beta distribution reduces to the uniform distribution over (0, 1) if $a = b = 1$. ////

Remark The cumulative distribution function of a beta-distributed random variable is

$$F_X(x; a, b) = I_{(0,1)}(x) \int_0^x \frac{1}{\text{B}(a, b)} u^{a-1}(1-u)^{b-1}\,du + I_{[1,\infty)}(x); \tag{35}$$

it is often called the *incomplete beta* and has been extensively tabulated. ////

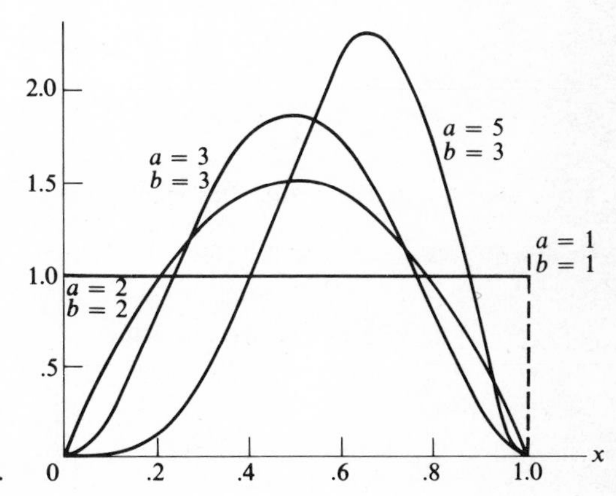

FIGURE 12
Beta densities.

The moment generating function for the beta distribution does not have a simple form; however the moments are readily found by using their definition.

Theorem 19 If X is a beta-distributed random variable, then

$$\mathscr{E}[X] = \frac{a}{a+b} \qquad \text{and} \qquad \text{var}\,[X] = \frac{ab}{(a+b+1)(a+b)^2}.$$

PROOF

$$\begin{aligned}\mathscr{E}[X^k] &= \frac{1}{\mathrm{B}(a, b)} \int_0^1 x^{k+a-1}(1-x)^{b-1}\,dx \\ &= \frac{\mathrm{B}(k+a, b)}{\mathrm{B}(a, b)} = \frac{\Gamma(k+a)\Gamma(b)}{\Gamma(k+a+b)} \cdot \frac{\Gamma(a+b)}{\Gamma(a)\Gamma(b)} \\ &= \frac{\Gamma(k+a)\Gamma(a+b)}{\Gamma(a)\Gamma(k+a+b)};\end{aligned}$$

hence,

$$\mathscr{E}[X] = \frac{\Gamma(a+1)\Gamma(a+b)}{\Gamma(a)\Gamma(a+b+1)} = \frac{a}{a+b},$$

and

$$\begin{aligned}\text{var}\,[X] = \mathscr{E}[X^2] - (\mathscr{E}[X])^2 &= \frac{\Gamma(a+2)\Gamma(a+b)}{\Gamma(a)\Gamma(a+b+2)} - \left(\frac{a}{a+b}\right)^2 \\ &= \frac{(a+1)a}{(a+b+1)(a+b)} - \left(\frac{a}{a+b}\right)^2 = \frac{ab}{(a+b+1)(a+b)^2}.\end{aligned}$$

////

The family of beta densities is a two-parameter family of densities that is positive on the interval (0, 1) and can assume quite a variety of different shapes, and, consequently, the beta distribution can be used to model an experiment for which one of the shapes is appropriate.

3.5 Other Continuous Distributions

In this subsection other parametric families of probability density functions that will appear later in this book are briefly introduced; many other families exist. The introductions of the three families of distributions, that go by the names of Student's t distribution, chi-square distribution, and F distribution, are deferred until Chap. VI. These three families, as we shall see, are very important when sampling from normal distributions.

Cauchy distribution A distribution which we shall find useful for illustrative purposes is the *Cauchy*, which has the density

$$f_X(x; \alpha, \beta) = \frac{1}{\pi\beta\{1 + [(x - \alpha)/\beta]^2\}}, \tag{36}$$

where $-\infty < \alpha < \infty$ and $\beta > 0$.

Although the Cauchy density is symmetrical about the parameter α, its mean and higher moments do not exist. The cumulative distribution function is

$$\begin{aligned} F_X(x) &= \frac{1}{\pi}\int_{-\infty}^{x} \frac{du}{\pi\beta\{1 + [(u - a)/\beta]^2\}} \\ &= \frac{1}{2} + \frac{1}{\pi} \arctan \frac{x - \alpha}{\beta}. \end{aligned} \tag{37}$$

Lognormal distribution Let X be a positive random variable, and let a new random variable Y be defined as $Y = \log_e X$. If Y has a normal distribution, then X is said to have a *lognormal* distribution. The density of a lognormal distribution is given by

$$f\ (x; \mu, \sigma^2) = \frac{1}{x\sqrt{2\pi}\sigma} \exp\left[-\frac{1}{2\sigma^2}(\log_e x - \mu)^2\right] I_{(0,\infty)}(x), \tag{38}$$

where $-\infty < \mu < \infty$ and $\sigma > 0$.

$$\mathscr{E}[X] = e^{\mu + \frac{1}{2}\sigma^2} \qquad \text{and} \qquad \text{var}\,[X] = e^{2\mu + 2\sigma^2} - e^{2\mu + \sigma^2} \tag{39}$$

for a lognormal random variable X. Also, if X has a lognormal distribution, then $\mathscr{E}[\log_e (X)] = \mu$, and $\text{var}\,[\log_e (X)] = \sigma^2$.

Double exponential or Laplace distribution A random variable X is said to have a *double exponential*, or *Laplace*, distribution if the density function of X is given by

$$f_X(x) = f_X(x; \alpha, \beta) = \frac{1}{2\beta} \exp\left(-\frac{|x - \alpha|}{\beta}\right), \tag{40}$$

where $-\infty < \alpha < \infty$ and $\beta > 0$. If X has a Laplace distribution, then

$$\mathscr{E}[X] = \alpha \qquad \text{and} \qquad \text{var}\,[X] = 2\beta^2. \tag{41}$$

Weibull distribution The density

$$f(x; a, b) = abx^{b-1}e^{-ax^b}I_{(0,\infty)}(x) \tag{42}$$

where $a > 0$ and $b > 0$, is called the *Weibull* density, a distribution that has been successfully used in reliability theory. For $b = 1$, the Weibull density reduces to the exponential density. It has mean $(1/a)^{1/b}\Gamma(1 + b^{-1})$ and variance $(1/a)^{2/b}[\Gamma(1 + 2b^{-1}) - \Gamma^2(1 + b^{-1})]$.

Logistic distribution The *logistic* distribution is given in cumulative distribution form by

$$F(x; \alpha, \beta) = \frac{1}{1 + e^{-(x-\alpha)/\beta}}, \tag{43}$$

where $-\infty < \alpha < \infty$ and $\beta > 0$. The mean of the logistic distribution is given by α. The variance is given by $\beta^2\pi^2/3$. Note that $F(\alpha - d;\ \alpha,\ \beta) = 1 - F(\alpha + d;\ \alpha,\ \beta)$, and so the density of the logistic is symmetrical about α. This distribution has been used to model tolerance levels in bioassay problems.

Pareto distribution The *Pareto* distribution is given in density-function form by

$$f_X(x; x_0, \theta) = \frac{\theta}{x_0}\left(\frac{x_0}{x}\right)^{\theta+1} I_{(x_0, \infty)}(x), \tag{44}$$

where $\theta > 0$ and $x_0 > 0$. The mean and variance respectively of the Pareto distribution are given by

$$\frac{\theta x_0}{\theta - 1} \quad \text{for} \quad \theta > 1 \qquad \text{and} \qquad \frac{\theta x_0^2}{\theta - 2} - \left(\frac{\theta x_0}{\theta - 1}\right)^2 \quad \text{for} \quad \theta > 2.$$

This distribution has found application in modeling problems involving distributions of incomes when incomes exceed a certain limit x_0.

Gumbel distribution The cumulative distribution function

$$F(x;\ \alpha,\ \beta) = \exp(-e^{-(x-\alpha)/\beta}), \tag{45}$$

where $-\infty < \alpha < \infty$ and $\beta > 0$ is called the *Gumbel* distribution. It appears as a limiting distribution in the theory of *extreme-value* statistics.

Pearsonian system of distributions Consider a density function $f_X(x)$ which satisfies the differential equation

$$\frac{1}{f_X(x)} \frac{df_X(x)}{dx} = \frac{x + a}{b_0 + b_1 x + b_2 x^2} \tag{46}$$

for constants $a, b_0, b_1,$ and b_2. Such a density is said to belong to the *Pearsonian system of density functions.* Many of the probability density functions that we have considered are special cases of the Pearsonian system. For example, if

$$f_X(x) = \frac{\lambda^r x^{r-1} e^{-\lambda x}}{\Gamma(r)} I_{[0,\infty)}(x),$$

then

$$\frac{1}{f_X(x)} \frac{df_X(x)}{dx} = -\lambda + \frac{r-1}{x} = \frac{x - (r-1)/\lambda}{-x/\lambda}$$

for $x > 0$; so the gamma distribution is a member of the Pearsonian system with $a = -(r-1)/\lambda$, $b_1 = -1/\lambda$, and $b_0 = b_2 = 0$.

4 COMMENTS

We conclude this chapter by making several comments that tie together some of the density functions defined in Secs. 2 and 3 of this chapter.

4.1 Approximations

Although many approximations of one distribution by another exist, we will give only three here. Others will be given along with the central-limit theorem in Chaps. V and VI.

Binomial by Poisson We defined the binomial discrete density function, with parameters n and p, as

$$\binom{n}{x} p^x (1-p)^{n-x} \qquad \text{for } x = 0, 1, \ldots, n.$$

If the parameter n approaches infinity and p approaches 0 in such a way that np remains constant, say equal to λ, then

$$\binom{n}{x} p^x (1-p)^{n-x} \to \frac{e^{-\lambda}\lambda^x}{x!} \tag{47}$$

for fixed integer x. The above follows immediately from the following consideration:

$$\begin{aligned}\binom{n}{x} p^x (1-p)^{n-x} &= \frac{(n)_x}{x!}\left(\frac{\lambda}{n}\right)^x \left(1 - \frac{\lambda}{n}\right)^{n-x} \\ &= \frac{\lambda^x}{x!}\frac{(n)_x}{n^x}\left(1 - \frac{\lambda}{n}\right)^n \left(1 - \frac{\lambda}{n}\right)^{-x} \to \frac{e^{-\lambda}\lambda^x}{x!},\end{aligned}$$

since

$$\frac{(n)_x}{n^x} \to 1, \qquad \left(1 - \frac{\lambda}{n}\right)^{-x} \to 1, \qquad \text{and} \qquad \left(1 - \frac{\lambda}{n}\right)^n \to e^{-\lambda} \qquad \text{as} \quad n \to \infty.$$

Thus, for large n and small p the binomial probability $\binom{n}{x} p^x(1-p)^{n-x}$ can be approximated by the Poisson probability $e^{-np}(np)^x/x!$. The utility of this approximation is evident if one notes that the binomial probability involves two parameters and the Poisson only one.

Binomial and Poisson by normal

Theorem 20 Let random variable X have a Poisson distribution with parameter λ; then for fixed $a < b$

$$P\left[a < \frac{X - \lambda}{\sqrt{\lambda}} < b\right]$$
$$= P[\lambda + a\sqrt{\lambda} < X < \lambda + b\sqrt{\lambda}] \to \Phi(b) - \Phi(a) \quad \text{as} \quad \lambda \to \infty. \tag{48}$$

PROOF Omitted. [Eq. (48) can be proved using Stirling's formula, which is given in Appendix A. It also follows from the central-limit theorem.] ////

Theorem 21 De Moivre–Laplace limit theorem Let a random variable X have a binomial distribution with parameters n and p; then for fixed $a < b$

$$P\left[a \le \frac{X - np}{\sqrt{npq}} \le b\right] = P[np + a\sqrt{npq} \le X \le np + b\sqrt{npq}] \to$$
$$\Phi(b) - \Phi(a) \qquad \text{as } n \to \infty. \tag{49}$$

PROOF Omitted. (This is a special case of the central-limit theorem, given in Chaps. V and VI.) ////

Remark We approximated the binomial distribution with a Poisson distribution in Eq. (47) for large n and small p. Theorem 21 gives a normal approximation of the binomial distribution for large n. ////

The usefulness of Theorems 20 and 21 rests in the approximations that they give. For instance, Eq. (49) states that $P[np + a\sqrt{npq} \le X \le np + b\sqrt{npq}]$

is approximately equal to $\Phi(b) - \Phi(a)$ for large n. Or if $c = np + a\sqrt{npq}$ and $d = np + b\sqrt{npq}$, then Eq. (49) gives that $P[c \le X \le d]$ is approximately equal to

$$\Phi\left(\frac{d - np}{\sqrt{npq}}\right) - \Phi\left(\frac{c - np}{\sqrt{npq}}\right)$$

for large n, and, so, an approximate value for the probability that a binomial random variable falls in an interval can be obtained from the standard normal distribution. Note that the binomial distribution is discrete and the approximating normal distribution is continuous.

EXAMPLE 15 Suppose that two fair dice are tossed 600 times. Let X denote the number of times a total of 7 occurs. Then X has a binomial distribution with parameters $n = 600$ and $p = \frac{1}{6}$. $\mathscr{E}[X] = 100$. Find $P[90 \le X \le 110]$.

$$P[90 \le X \le 110] = \sum_{j=90}^{110} \binom{600}{j} \left(\frac{1}{6}\right)^j \left(\frac{5}{6}\right)^{600-j},$$

a sum that is tedious to evaluate. Using the approximation given by Eq. (49), we have

$$P[90 \le X \le 110] \approx \Phi\left(\frac{110 - 100}{\sqrt{\frac{500}{6}}}\right) - \Phi\left(\frac{90 - 100}{\sqrt{\frac{500}{6}}}\right)$$

$$= \Phi(\sqrt{\tfrac{6}{5}}) - \Phi(-\sqrt{\tfrac{6}{5}}) \approx \Phi(1.095) - \Phi(-1.095) \approx .726.$$

////

4.2 Poisson and Exponential Relationship

When the Poisson distribution was introduced in Subsec. 2.4, an experiment consisting of the counting of the number of happenings of a certain phenomenon in time was given special consideration. We argued that under certain conditions the count of the number of happenings in a fixed time interval was Poisson distributed with parameter, the mean, proportional to the length of the interval. Suppose now that one of these happenings has just occurred; what then is the distribution of the length of time, say X, that one will have to wait until the next happening? $P[X > t] = P$[no happenings in time interval of length t] $= e^{-\nu t}$, where ν is the *mean occurrence rate*; so

$$F_X(t) = P[X \le t] = 1 - P[X > t] = 1 - e^{-\nu t} \qquad \text{for } t > 0;$$

that is, X has an exponential distribution. On the other hand, it can be proved, under an independence assumption, that if the happenings are occurring in time in such a way that the distribution of the lengths of time between successive happenings is exponential, then the distribution of the number of happenings in a fixed time interval is Poisson distributed. Thus the exponential and Poisson distributions are related.

4.3 Contagious Distributions and Truncated Distributions

A brief introduction to the concept of *contagious distributions* is given here. If $f_0(\cdot), f_1(\cdot), \ldots, f_n(\cdot), \ldots$ is a sequence of density functions which are either all discrete density functions or all probability density functions which may or may not depend on parameters, and $p_0, p_1, \ldots, p_n, \ldots$ is a sequence of parameters satisfying $p_i \geq 0$ and $\sum_{i=0}^{\infty} p_i = 1$, then $\sum_{i=0}^{\infty} p_i f_i(x)$ is a density function, which is sometimes called a *contagious* distribution or a *mixture*. For example, if $f_0(x) = \phi_{\mu_0, \sigma_0^2}(x)$ (a normal with mean μ_0 and variance σ_0^2) and $f_1(x) = \phi_{\mu_1, \sigma_1^2}(x)$, then

$$p_0 \phi_{\mu_0,\sigma_0^2}(x) + p_1\, \phi_{\mu_1,\sigma_1^2}(x)$$

$$= (1-p)\frac{1}{\sqrt{2\pi}\sigma_0} e^{-\frac{1}{2}[(x-\mu_0)/\sigma_0]^2} + p\frac{1}{\sqrt{2\pi}\sigma_1} e^{-\frac{1}{2}[(x-\mu_1)/\sigma_1]^2} \tag{50}$$

where $p_1 = p$ and $p_0 = 1 - p$, is a *mixture* of two normal densities. Equation (50) is also sometimes referred to as a *contaminated* normal. A random variable X has distribution given by Eq. (50) if it is normally distributed with mean μ_1 and variance σ_1^2 with probability p and normally distributed with mean μ_0 and variance σ_0^2 with probability $1 - p$. Contagious distributions or mixtures can be useful models for certain experiments. For instance, the mixture of two normal distributions given in Eq. (50) has five parameters, namely, p, μ_0, μ_1, σ_0, and σ_1. If we vary these five parameters, the density can be forced to assume a variety of different shapes, some of which are bimodal; that is, the density has two distinct local maximums.

Physical considerations of the random experiment at hand can sometimes persuade one to consider modeling the experiment with a mixture. The experimenter may know that the phenomena that he is observing are a mixture; for example, the radioactive particle emissions under observation might be a mixture of the particle emissions of two, or several, different types of radioactive materials.

The concept of mixing can be extended. Let $\{f(x;\theta)\}$ be a family of density functions parameterized or indexed by θ. Let the totality of values that the parameter θ can assume be denoted by $\overline{\underline{\Theta}}$. If $\overline{\underline{\Theta}}$ is an interval (possibly infinite) and $g(\theta)$ is a probability density function which is 0 for all arguments not in $\overline{\underline{\Theta}}$, then

$$\int_{\overline{\underline{\Theta}}} f(x;\theta)g(\theta)\,d\theta \tag{51}$$

is again a density function, called a *contagious distribution* or a *mixture*. For example, suppose $f(x;\theta) = e^{-\theta}\theta^x/x!$ for $x = 0, 1, 2, \ldots$ and $f(x;\theta) = 0$ otherwise and

$$g(\theta) = \frac{\lambda^r}{\Gamma(r)}\,\theta^{r-1}e^{-\lambda\theta}I_{(0,\infty)}(\theta),$$

a gamma density. Then

$$\begin{aligned}
\int_0^\infty f(x;\theta)\cdot g(\theta)\,d\theta &= \int_0^\infty \frac{e^{-\theta}\theta^x}{x!}\cdot\frac{\lambda^r}{\Gamma(r)}\,\theta^{r-1}e^{-\theta\lambda}\,d\theta \\
&= \frac{\lambda^r}{x!\Gamma(r)}\int_0^\infty \theta^{r+x-1}e^{-(\lambda+1)\theta}\,d\theta \\
&= \frac{\lambda^r}{x!\Gamma(r)}\cdot\frac{\Gamma(r+x)}{(\lambda+1)^{r+x}}\int_0^\infty \frac{[(\lambda+1)\theta]^{r+x-1}e^{-(\lambda+1)\theta}\,d[(\lambda+1)\theta]}{\Gamma(r+x)} \\
&= \left(\frac{\lambda}{\lambda+1}\right)^r \frac{\Gamma(r+x)}{(x!)\Gamma(r)}\,\frac{1}{(\lambda+1)^x} \\
&= \binom{r+x-1}{x}\left(\frac{\lambda}{\lambda+1}\right)^r\left(\frac{1}{\lambda+1}\right)^x \qquad \text{for} \quad x = 0, 1, \ldots,
\end{aligned}$$

which is the density function of a negative binomial distribution with parameters r and $p = \lambda/(\lambda+1)$. We say that the derived negative binomial distribution is the *gamma mixture of Poissons*.

$$\int_0^\infty \frac{e^{-\theta}\theta^x}{x!}\,g(\theta)\,d\theta$$

is sometimes called a *compound* Poisson, where $g(\theta)I_{(0,\infty)}(\theta)$ is a probability density function.

We have sketchily illustrated above how new parametric families of densities can be obtained from existing families by the technique of mixing. In Subsec. 2.6 we indicated how truncation could be employed to generate new families of discrete densities. Truncation can also be utilized to form other families of continuous distributions. For instance, the family of beta distributions provides densities that are useful in modeling an experiment for which it is known that the values that the random variable can assume are between 0 and 1. A truncated normal or gamma distribution would also provide a useful model for such an experiment. A normal distribution that is truncated at 0 on the left and at 1 on the right is defined in density form as

$$f(x) = f(x; \mu, \sigma) = \frac{\phi_{\mu,\sigma^2}(x)I_{(0,1)}(x)}{\Phi_{\mu,\sigma^2}(1) - \Phi_{\mu,\sigma^2}(0)}. \tag{52}$$

This truncated normal distribution, like the beta distribution, assumes values between 0 and 1.

Truncation can be defined in general. If X is a random variable with density $f_X(\cdot)$ and cumulative distribution $F_X(\cdot)$, then the density of X truncated on the left at a and on the right at b is given by

$$\frac{f_X(x)I_{(a,b)}(x)}{F_X(b) - F_X(a)}. \tag{53}$$

PROBLEMS

1 (*a*) Let X be a random variable having a binomial distribution with parameters $n = 25$ and $p = .2$. Evaluate $P[X < \mu_X - 2\sigma_X]$.

(*b*) If X is a random variable with Poisson distribution satisfying $P[X = 0] = P[X = 1]$, what is $\mathscr{E}[X]$?

(*c*) If X is uniformly distributed over (1, 2), find z such that $P[X > z + \mu_X] = \frac{1}{4}$.

(*d*) If X is normally distributed with mean 2 and variance 1, find $P[|X - 2| < 1]$.

(*e*) Suppose X is binomially distributed with parameters n and p; further suppose that $\mathscr{E}[X] = 5$ and var $[X] = 4$. Find n and p.

(*f*) If $\mathscr{E}[X] = 10$ and $\sigma_X = 3$, can X have a negative binomial distribution?

(*g*) If X has a negative exponential distribution with mean 2, find $P[X < 1 \mid X < 2]$.

(*h*) Name three distributions for which $P[X \le \mu_X] = \frac{1}{2}$.

(*i*) Let X be a random variable having binomial distribution with parameters $n = 100$ and $p = .1$. Evaluate $P[X \le \mu_X - 3\sigma_X]$.

(*j*) If X has a Poisson distribution and $P[X = 0] = \frac{1}{2}$, what is $\mathscr{E}[X]$?

(*k*) Suppose X has a binomial distribution with parameters n and p. For what p is var $[X]$ maximized if we assumed n is fixed?

(*l*) Suppose X has a negative exponential distribution with parameter λ. If $P[X \le 1] = P[X > 1]$, what is var $[X]$?

(*m*) Suppose X is a continuous random variable with uniform distribution having mean 1 and variance $\frac{4}{3}$. What is $P[X < 0]$?

(*n*) If X has a beta distribution, can $\mathscr{E}[1/X]$ be unity?

(*o*) Can X ever have the same distribution as $-X$? If so, when?

(*p*) If X is a random variable having moment generating function exp $(e^t - 1)$, what is $\mathscr{E}[X]$?

2 (*a*) Find the mode of the beta distribution.

(*b*) Find the mode of the gamma distribution.

3 Name a parametric family of distributions which satisfies:
(*a*) The mean must be greater than or equal to the variance.
(*b*) The mean must be equal to the variance.
(*c*) The mean must be less than or equal to the variance.
(*d*) The mean can be less than, equal to, or greater than the variance (for different parameter values).

4 (*a*) If X is normally distributed with mean 2 and variance 2, express $P[|X-1|\leq 2]$ in terms of the standard normal cumulative distribution function.
(*b*) If X is normally distributed with mean $\mu > 0$ and variance $\sigma^2 = \mu^2$, express $P[X < -\mu \mid X < \mu]$ in terms of the standard normal cumulative distribution function.
(*c*) Let X be normally distributed with mean μ and variance σ^2. Suppose σ^2 is some function of μ, say $\sigma^2 = h(\mu)$. Pick $h(\cdot)$ so that $P[X \leq 0]$ does not depend on μ for $\mu > 0$.

5 Use the alternate definition of the median as given in the remark following Definition 18 of Chap. II. Find the median in each of the following cases:
(*a*) $f_X(x) = \lambda e^{-\lambda x} I_{(0,\infty)}(x)$.
(*b*) X is uniformly distributed on the interval (θ_1, θ_2).
(*c*) X has a binomial distribution with $n = 4$, $p = .5$.
(*d*) X has a binomial distribution with $n = 5$, $p = .5$.
(*e*) X has a binomial distribution with $n = 2$, $p = .9$.

**6* A contractor has found through experience that the low bid for a job (excluding his own bid) is a random variable that is uniformly distributed over the interval $(\frac{3}{4}C, 2C)$, where C is the contractor's cost estimate (no profit or loss) of the job. If profit is defined as 0 if the contractor does not get the job (his bid is greater than the low bid) and as the difference between his bid and his cost estimate C if he gets the job, what should he bid (in terms of C) in order to maximize his expected profit?

7 A merchant has found that the number of items of brand XYZ that he can sell in a day is a Poisson random variable with mean 4.
(*a*) How many items of brand XYZ should the merchant stock to be 95 percent certain that he will have enough to last for 25 days? (Give a numerical answer.)
(*b*) What is the expected number of days out of 25 that the merchant will sell no items of brand XYZ?

8 (*a*) If X is binomially distributed with parameters n and p, what is the distribution of $Y = n - X$?
(*b*) Two dice are thrown n times. Let X denote the number of throws in which the number on the first die exceeds the number on the second die. What is the distribution of X?
*(*c*) A drunk performs a "random walk" over positions $0, \pm 1, \pm 2, \ldots$ as follows: He starts at 0. He takes successive one-unit steps, going to the right with probability p and to the left with probability $1 - p$. His steps are inde-

pendent. Let X denote his position after n steps. Find the distribution of $(X+n)/2$, and then find $\mathscr{E}[X]$.

*(*d*) Let X_1 (X_2) have a binomial distribution with parameters n and p_1 (n and p_2). If $p_1 < p_2$, show that $P[X_1 \le k] \ge P[X_2 \le k]$ for $k = 0, 1, \ldots, n$. (This result says that the smaller the p, the more the binomial distribution is shifted to the left.)

9 In a town with 5000 adults, a sample of 100 is asked their opinion of a proposed municipal project; 60 are found to favor it, and 40 oppose it. If, in fact, the adults of the town were equally divided on the proposal, what would be the probability of obtaining a majority of 60 or more favoring it in a sample of 100?

10 A distributor of bean seeds determines from extensive tests that 5 percent of a large batch of seeds will not germinate. He sells the seeds in packages of 200 and guarantees 90 percent germination. What is the probability that a given package will violate the guarantee?

**11* (*a*) A manufacturing process is intended to produce electrical fuses with no more than 1 percent defective. It is checked every hour by trying 10 fuses selected at random from the hour's production. If 1 or more of the 10 fail, the process is halted and carefully examined. If, in fact, its probability of producing a defective fuse is .01, what is the probability that the process will needlessly be examined in a given instance?

(*b*) Referring to part (*a*), how many fuses (instead of 10) should be tested if the manufacturer desires that the probability be about .95 that the process will be examined when it is producing 10 percent defectives?

12 An insurance company finds that .005 percent of the population die from a certain kind of accident each year. What is the probability that the company must pay off on more than 3 of 10,000 insured risks against such accidents in a given year?

13 (*a*) If X has a Poisson distribution with $P[X=1] = P[X=2]$, what is $P[X = 1 \text{ or } 2]$?

(*b*) If X has a Poisson distribution with mean 1, show that $\mathscr{E}[|X-1|] = 2\sigma_X/e$.

**14* Recall Theorems 4 and 8. Formulate, and then prove or disprove a similar theorem for the negative binomial distribution.

**15* Let X be normally distributed with mean μ and variance σ^2. Truncate the density of X on the left at a and on the right at b, and then calculate the mean of the truncated distribution. (Note that the mean of the truncated distribution should fall between a and b. Furthermore, if $a = \mu - c$ and $b = \mu + c$, then the mean of the truncated distribution should equal μ.)

**16* Show that the hypergeometric distribution can be approximated by the binomial distribution for large M and K; i.e., show that

$$\lim_{\substack{M\to\infty \\ K\to\infty \\ K/M\to p}} \frac{\binom{K}{x}\binom{M-K}{n-x}}{\binom{M}{n}} = \binom{n}{x} p^x(1-p)^{n-x}$$

17 Let X be the life in hours of a radio tube. Assume that X is normally distributed with mean 20 and variance σ^2. If a purchaser of such radio tubes requires that at least 90 percent of the tubes have lives exceeding 150 hours, what is the largest value σ can be and still have the purchaser satisfied?

18 Assume that the number of fatal car accidents in a certain state obeys a Poisson distribution with an average of one per day.

(*a*) What is the probability of more than ten such accidents in a week?

(*b*) What is the probability that more than 2 days will lapse between two such accidents?

19 The distribution given by

$$f(x; \beta) = \frac{1}{\beta^2} x e^{-\frac{1}{2}(x/\beta)^2} I_{(0, \infty)}(x) \qquad \text{for } \beta > 0$$

is called the *Rayleigh* distribution.

(*a*) Show that the mean and variance exist, and find them.

(*b*) Does the Rayleigh distribution belong to the Pearsonian system?

20 The distribution given by

$$f(x; \beta) = \frac{4}{\beta^3 \sqrt{\pi}} x^2 e^{-x^2/\beta^2} I_{(0, \infty)}(x) \qquad \text{for } \beta > 0$$

is called the *Maxwell* distribution.

(*a*) Show that the mean and variance exist, and find them.

(*b*) Does this distribution belong to the Pearsonian system?

21 The distribution given by

$$f(x; n) = \frac{1}{B(\frac{1}{2}, [n-2]/2)} (1 - x^2)^{(n-4)/2} I_{[-1, 1]}(x)$$

is called the *r distribution*.

(*a*) Show that the mean and variance exist, and find them.

(*b*) Does this distribution belong to the Pearsonian system?

22 A die is cast until a 6 appears. What is the probability that it must be cast more than five times?

23 Red-blood-cell deficiency may be determined by examining a specimen of the blood under a microscope. Suppose that a certain small fixed volume contains, on an average, 20 red cells for normal persons. What is the probability that a specimen from a normal person will contain less than 15 red cells?

24 A telephone switchboard handles 600 calls, on an average, during a rush hour. The board can make a maximum of 20 connections per minute. Use the Poisson distribution to evaluate the probability that the board will be overtaxed during any given minute.

25 Suppose that a particle is equally likely to release one, two, or three other particles, and suppose that these second-generation particles are in turn each equally likely to release one, two, or three third-generation particles. What is the density of the number of third-generation particles?

26 Find the mean of the Gumbel distribution.

27 Derive the mean and variance of the Weibull distribution.

**28* Show that

$$P[X \geq k] = \sum_{j=k}^{n} \binom{n}{j} p^j q^{n-j} = \frac{1}{B(k, n-k+1)} \int_0^p u^{k-1}(1-u)^{n-k}\, du$$

for X a binomially distributed random variable. That is, if X is binomially distributed with parameters n and p and Y is beta-distributed with parameters k and $n - k + 1$, then $F_Y(p) = 1 - F_X(k-1)$.

**29* Suppose that X has a binomial distribution with parameters n and p and Y has a negative binomial distribution with parameters r and p. Show that $F_X(r-1) = 1 - F_Y(n-r)$.

**30* If U is a random variable that is uniformly distributed over the interval [0, 1], then the random variable $Z_\lambda = [U^\lambda - (1-U)^\lambda]/\lambda$ is said to have *Tukey's symmetrical lambda distribution*. Find the first four moments of Z_λ. Find two different λ's, say λ_1 and λ_2, such that Z_{λ_1} and Z_{λ_2} have the same first four moments and unit standard deviations.

IV

JOINT AND CONDITIONAL DISTRIBUTIONS, STOCHASTIC INDEPENDENCE, MORE EXPECTATION

1 INTRODUCTION AND SUMMARY

The purpose of this chapter is to introduce the concepts of k-dimensional distribution functions, conditional distributions, joint and conditional expectation, and independence of random variables. It, like Chap. II, is primarily a "definitions-and-their-understanding" chapter.

The chapter is divided into four main sections in addition to the present one. In Sec. 2, joint distributions, both in cumulative and density-function form, are introduced. The important k-dimensional discrete distribution, called the *multinomial*, is included as an example. Conditional distributions and independence of random variables are the subject of Sec. 3. Section 4 deals with expectation with respect to k-variate distributions. Definitions of covariance, the correlation coefficient, and joint moment generating functions, all of which are special expectations, are given. The important concept of conditional expectation is discussed in Subsec. 4.3. Results relating independence and expectation are presented in Subsec. 4.5, and the famous Cauchy-Schwarz inequality is proved in Subsec. 4.6. The last main section, Sec. 5, is devoted to the important bivariate normal distribution, which gives one unified example of many of the terms defined in the preceding sections.

This chapter is the multidimensional analog of Chap. II. It provides definitions needed to understand distributional-theory results of Chap. V.

2 JOINT DISTRIBUTION FUNCTIONS

In the study of many random experiments, there are, or can be, more than one random variable of interest; hence we are compelled to extend our definitions of the distribution and density function of one random variable to those of several random variables. Such definitions are the essence of this section, which is the multivariate counterpart of Secs. 2 and 3 of Chap. II. As in the univariate case we will first define, in Subsec. 2.1, the cumulative distribution function. Although it is not as convenient to work with as density functions, it does exist for any set of k random variables. Density functions for jointly discrete and jointly continuous random variables will be given in Subsecs. 2.2 and 2.3, respectively.

2.1 Cumulative Distribution Function

Definition 1 Joint cumulative distribution function Let $X_1, X_2, \ldots, X_k$ be k random variables all defined on the same probability space $(\Omega, \mathscr{A}, P[\cdot])$. The *joint cumulative distribution function* of $X_1, \ldots, X_k$, denoted by $F_{X_1, \ldots, X_k}(\cdot, \ldots, \cdot)$, is defined as $P[X_1 \leq x_1; \ldots; X_k \leq x_k]$ for all $(x_1, x_2, \ldots, x_k)$. ////

Thus a joint cumulative distribution function is a function with domain euclidean k space and counterdomain the interval $[0, 1]$. If $k = 2$, the joint cumulative distribution function is a function of two variables, and so its domain is just the xy plane.

EXAMPLE 1 Consider the experiment of tossing two tetrahedra (regular four-sided polyhedron) each with sides labeled 1 to 4. Let X denote the number on the downturned face of the first tetrahedron and Y the larger of the downturned numbers. The goal is to find $F_{X, Y}(\cdot, \cdot)$, the joint cumulative distribution function of X and Y. Observe first that the random variables X and Y jointly take on only the values

$$\begin{aligned}&(1, 1), (1, 2), (1, 3), (1, 4),\\&(2, 2), (2, 3), (2, 4),\\&(3, 3), (3, 4),\\&(4, 4).\end{aligned}$$

(The first component is the value of X, and the second the value of Y.)

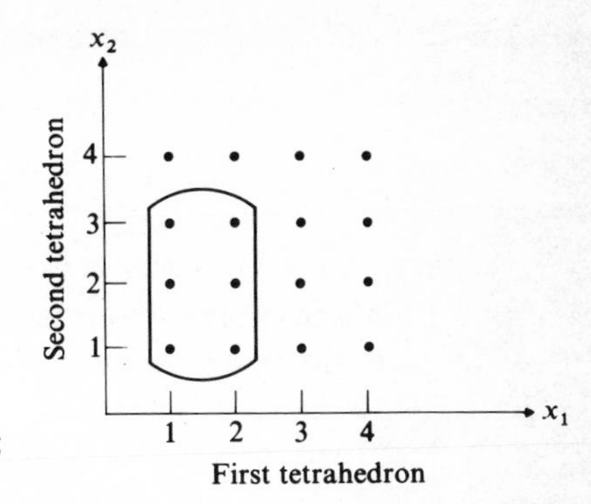

FIGURE 1
Sample space for experiment of tossing two tetrahedra.

The sample space for this experiment is displayed in Fig. 1. The 16 sample points are assumed to be equally likely. Our objective is to find $F_{X,Y}(x, y)$ for each point (x, y). As an example let $(x, y) = (2, 3)$, and find $F_{X,Y}(2, 3) = P[X \le 2;\ Y \le 3]$. Now the event $\{X \le 2 \text{ and } Y \le 3\}$ corresponds to the encircled sample points in Fig. 1; hence $F_{X,Y}(2, 3) = \frac{6}{16}$. Similarly, $F_{X,Y}(x, y)$ can be found for other values of x and y. $F_{X,Y}(x, y)$ is tabled in Fig. 2. ////

We saw that the cumulative distribution function of a unidimensional random variable had certain properties; the same is true of a joint cumulative. We shall list these properties for the joint cumulative distribution function of two random variables; the generalization to k dimensions is straightforward.

TABLE OF VALUES OF $F_{X,Y}(x,y)$

$4 \le y$	0	$\frac{4}{16}$	$\frac{8}{16}$	$\frac{12}{16}$	1
$3 \le y < 4$	0	$\frac{3}{16}$	$\frac{6}{16}$	$\frac{9}{16}$	$\frac{9}{16}$
$2 \le y < 3$	0	$\frac{2}{16}$	$\frac{4}{16}$	$\frac{4}{16}$	$\frac{4}{16}$
$1 \le y < 2$	0	$\frac{1}{16}$	$\frac{1}{16}$	$\frac{1}{16}$	$\frac{1}{16}$
$y < 1$	0	0	0	0	0
	$x < 1$	$1 \le x < 2$	$2 \le x < 3$	$3 \le x < 4$	$4 \le x$

FIGURE 2

Properties of bivariate cumulative distribution function $F(\cdot, \cdot)$

(i) $F(-\infty, y) = \lim_{x \to -\infty} F(x, y) = 0$ for all y, $F(x, -\infty) = \lim_{y \to -\infty} F(x, y) = 0$ for all x, and $\lim_{\substack{x \to \infty \\ y \to \infty}} F(x, y) = F(\infty, \infty) = 1$.

(ii) If $x_1 < x_2$ and $y_1 < y_2$, then $P[x_1 < X \le x_2; y_1 < Y \le y_2] = F(x_2, y_2) - F(x_2, y_1) - F(x_1, y_2) + F(x_1, y_1) \ge 0$.

(iii) $F(x, y)$ is right continuous in each argument; that is, $\lim_{0<h\to 0} F(x + h, y) = \lim_{0<h\to 0} F(x, y + h) = F(x, y)$.

We will not prove these properties. Property (ii) is a *monotonicity* property of sorts; it is not equivalent to $F(x_1, y_1) \le F(x_2, y_2)$ for $x_1 \le x_2$ and $y_1 \le y_2$. Consider, for example, the bivariate function $G(x, y)$ defined as in Fig. 3. Note that $G(x_1, y_1) \le G(x_2, y_2)$ for $x_1 \le x_2$ and $y_1 \le y_2$, yet $G(1 + \varepsilon, 1 + \varepsilon) - G(1 + \varepsilon, 1 - \varepsilon) - G(1 - \varepsilon, 1 + \varepsilon) + G(1 - \varepsilon, 1 - \varepsilon) = 1 - (1 - \varepsilon) - (1 - \varepsilon) = 2\varepsilon - 1 < 0$ for $\varepsilon < \frac{1}{2}$; so $G(x, y)$ does not satisfy property (ii) and consequently is not a bivariate cumulative distribution function.

Definition 2 Bivariate cumulative distribution function Any function satisfying properties (i) to (iii) is defined to be a *bivariate cumulative distribution function* without reference to any random variables. ////

Definition 3 Marginal cumulative distribution function If $F_{X,Y}(\cdot, \cdot)$ is the joint cumulative distribution function of X and Y, then the cumulative distribution functions $F_X(\cdot)$ and $F_Y(\cdot)$ are called *marginal cumulative distribution functions*. ////

TABLE OF $G(x, y)$

$1 \le y$	0	x	1
$0 \le y < 1$	0	0	y
$y < 0$	0	0	0
	$x < 0$	$0 \le x < 1$	$1 \le x$

FIGURE 3

Remark $F_X(x) = F_{X,Y}(x, \infty)$, and $F_Y(y) = F_{X,Y}(\infty, y)$; that is, knowledge of the joint cumulative distribution function of X and Y implies knowledge of the two marginal cumulative distribution functions. ////

The converse of the above remark is not generally true; in fact, an example (Example 8) will be given in Subsec. 2.3 below that gives an entire family of joint cumulative distribution functions, and each member of the family has the same marginal distributions.

We will conclude this section with a remark that gives an inequality involving the joint cumulative distribution and marginal distributions. The proof is left as an exercise.

Remark $F_X(x) + F_Y(y) - 1 \le F_{X,Y}(x, y) \le \sqrt{F_X(x)F_Y(y)}$ for all x, y. ////

2.2 Joint Density Functions for Discrete Random Variables

If $X_1, X_2, \ldots, X_k$ are random variables defined on the same probability space, then $(X_1, X_2, \ldots, X_k)$ is called a *k-dimensional random variable*.

Definition 4 Joint discrete random variables The k-dimensional random variable $(X_1, X_2, \ldots, X_k)$ is defined to be a *k-dimensional discrete random variable* if it can assume values only at a countable number of points $(x_1, x_2, \ldots, x_k)$ in k-dimensional real space. We also say that the random variables $X_1, X_2, \ldots, X_k$ are *joint discrete random variables*. ////

Definition 5 Joint discrete density function If $(X_1, X_2, \ldots, X_k)$ is a k-dimensional discrete random variable, then the *joint discrete density function* of $(X_1, X_2, \ldots, X_k)$, denoted by $f_{X_1, X_2, \ldots, X_k}(\cdot, \cdot, \ldots, \cdot)$, is defined to be

$$f_{X_1, X_2, \ldots, X_k}(x_1, x_2, \ldots, x_k) = P[X_1 = x_1; X_2 = x_2; \ldots; X_k = x_k]$$

for $(x_1, x_2, \ldots, x_k)$, a value of $(X_1, X_2, \ldots, X_k)$ and is defined to be 0 otherwise. ////

Remark $\sum f_{X_1, \ldots, X_k}(x_1, \ldots, x_k) = 1$, where the summation is over all possible values of $(X_1, \ldots, X_k)$. ////

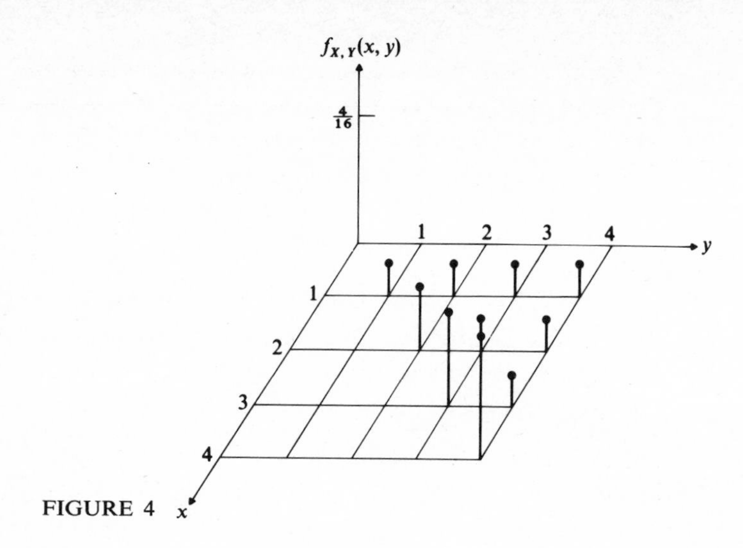

FIGURE 4

EXAMPLE 2 Let X denote the number on the downturned face of the first tetrahedron and Y the larger of the downturned numbers in the experiment of tossing two tetrahedra. The values that (X, Y) can take on are (1, 1), (1, 2), (1, 3), (1, 4), (2, 2), (2, 3), (2, 4), (3, 3), (3, 4), and (4, 4); hence X and Y are jointly discrete. The joint discrete density function of X and Y is given in Fig. 4.

In tabular form it is given as

(x, y)	(1, 1)	(1, 2)	(1, 3)	(1, 4)	(2, 2)	(2, 3)	(2, 4)	(3, 3)	(3, 4)	(4, 4)
$f_{X,Y}(x,y)$	$\frac{1}{16}$	$\frac{1}{16}$	$\frac{1}{16}$	$\frac{1}{16}$	$\frac{2}{16}$	$\frac{1}{16}$	$\frac{1}{16}$	$\frac{3}{16}$	$\frac{1}{16}$	$\frac{4}{16}$

or in another tabular form as

4	$\frac{1}{16}$	$\frac{1}{16}$	$\frac{1}{16}$	$\frac{4}{16}$
3	$\frac{1}{16}$	$\frac{1}{16}$	$\frac{3}{16}$	
2	$\frac{1}{16}$	$\frac{2}{16}$		
1	$\frac{1}{16}$			
y / x	1	2	3	4

////

Theorem 1 If X and Y are jointly discrete random variables, then knowledge of $F_{X,Y}(\cdot, \cdot)$ is equivalent to knowledge of $f_{X,Y}(\cdot, \cdot)$. Also, the statement extends to k-dimensional discrete random variables.

PROOF Let $(x_1, y_1), (x_2, y_2), \ldots$ be the possible values of (X, Y). If $f_{X,Y}(\cdot, \cdot)$ is given, then $F_{X,Y}(x, y) = \sum f_{X,Y}(x_i, y_i)$, where the summation is over all i for which $x_i \leq x$ and $y_i \leq y$. Conversely, if $F_{X,Y}(\cdot, \cdot)$ is given, then for (x_i, y_i), a possible value of (X, Y),

$$\begin{aligned} f_{X,Y}(x_i, y_i) = F_{X,Y}(x_i, y_i) &- \lim_{0<h\to 0} F_{X,Y}(x_i - h, y_i) \\ &- \lim_{0<h\to 0} F_{X,Y}(x_i, y_i - h) \\ &+ \lim_{0<h\to 0} F_{X,Y}(x_i - h, y_i - h). \end{aligned}$$ ////

Definition 6 Marginal discrete density If X and Y are jointly discrete random variables, then $f_X(\cdot)$ and $f_Y(\cdot)$ are called *marginal* discrete density functions. More generally, let $X_{i_1}, \ldots, X_{i_m}$ be any subset of the jointly discrete random variables $X_1, \ldots, X_k$; then $f_{X_{i_1}, \ldots, X_{i_m}}(x_{i_1}, \ldots, x_{i_m})$ is also called a *marginal density*. ////

Remark If $X_1, \ldots, X_k$ are jointly discrete random variables, then any marginal discrete density can be found from the joint density, but not conversely. For example, if X and Y are jointly discrete with values $(x_1, y_1), (x_2, y_2), \ldots$, then

$$f_X(x_k) = \sum_{\{i: x_i = x_k\}} f_{X,Y}(x_i, y_i) \quad \text{and} \quad f_Y(y_k) = \sum_{\{i: y_i = y_k\}} f_{X,Y}(x_i, y_i).$$ ////

Heretofore we have indexed the values of (X, Y) with a single index, namely i. That is, we listed values as $(x_1, y_1), (x_2, y_2), \ldots, (x_i, y_i), \ldots$. The values of (X, Y) could also be indexed by using separate indices for the X and Y values. For instance, we could let i index the possible X values, say $x_1, \ldots, x_i, \ldots$, and j index the possible Y values, say $y_1, \ldots, y_j, \ldots$. Then the values of (X, Y) would be a subset of the points (x_i, y_j) for $i = 1, 2, \ldots$ and $j = 1, 2, \ldots$. If this latter method of indexing is used, then the marginal density of X is obtained as follows:

$$f_X(x_k) = \sum_j f_{X,Y}(x_k, y_j),$$

where the summation is over all y_j for the fixed x_k. The marginal density of Y is analogously obtained. The following example may help to clarify these two different methods of indexing the values of (X, Y).

EXAMPLE 3 Return to the experiment of tossing two tetrahedra, and define X as the number on the downturned face of the first tetrahedron and Y as the larger of the numbers on the two downturned faces. The joint density of X and Y is given in Fig. 4. The values of (X, Y) can be listed as (1, 1), (1, 2), (1, 3), (1, 4), (2, 2), (2, 3), (2, 4), (3, 3), (3, 4), and (4, 4), 10 points in all. Or, if we note that X has values 1, 2, 3, and 4; Y has values 1, 2, 3, and 4; and Y is greater than or equal X, the values of (X, Y) are $\{(i, j): i = 1, \ldots, 4; j = 1, \ldots, 4; \text{ and } i \le j\}$. Let us use each of these methods of indexing to evaluate $F_{X,Y}(2, 3)$ from the joint density. Under the first method of indexing,

$$\begin{aligned} F_{X,Y}(2, 3) &= \sum_{\{i: x_i \le 2, y_i \le 3\}} f_{X,Y}(x_i, y_i) \\ &= f_{X,Y}(1, 1) + f_{X,Y}(1, 2) \\ &\quad + f_{X,Y}(1, 3) + f_{X,Y}(2, 2) + f_{X,Y}(2, 3) = \tfrac{6}{16} \end{aligned}$$

Under the second method of indexing,

$$F_{X,Y}(2, 3) = \sum_{i=1}^{2} \sum_{j=i}^{3} f_{X,Y}(i, j) = \tfrac{6}{16}.$$

Similarly, all other values of $F_{X,Y}(\cdot, \cdot)$ could be obtained. Also

$$\begin{aligned} f_Y(3) &= \sum_{\{i: y_i = 3\}} f_{X,Y}(x_i, y_i) = f_{X,Y}(1, 3) + f_{X,Y}(2, 3) + f_{X,Y}(3, 3) \\ &= \tfrac{1}{16} + \tfrac{1}{16} + \tfrac{3}{16} = \tfrac{5}{16}. \end{aligned}$$

Similarly $f_Y(1) = \frac{1}{16}$, $f_Y(2) = \frac{3}{16}$, and $f_Y(4) = \frac{7}{16}$, which together with $f_Y(3) = \frac{5}{16}$ give the marginal discrete density function of Y. ////

EXAMPLE 4 We mentioned that marginal densities can be obtained from the joint density, but not conversely. The following is an example of a family of joint densities that all have the same marginals, and hence we see that in general the joint density is not uniquely determined from knowledge of the marginals. Consider altering the joint density given in the previous examples as follows:

y \ x	1	2	3	4
4	$\frac{1}{16} + \varepsilon$	$\frac{1}{16} - \varepsilon$	$\frac{1}{16}$	$\frac{4}{16}$
3	$\frac{1}{16} - \varepsilon$	$\frac{1}{16} + \varepsilon$	$\frac{3}{16}$	
2	$\frac{1}{16}$	$\frac{2}{16}$		
1	$\frac{1}{16}$			

For each $0 \le \varepsilon \le \frac{1}{16}$, the above table defines a joint density. Note that the marginal densities are independent of ε, and hence each of the joint densities (there is a different joint density for each $0 \le \varepsilon \le \frac{1}{16}$) has the same marginals. ////

We saw that the binomial distribution was associated with independent, repeated Bernoulli trials; we shall see in the example below that the *multinomial* distribution is associated with independent, repeated trials that generalize from Bernoulli trials with two outcomes to more than two outcomes.

EXAMPLE 5 Suppose that there are $k + 1$ (distinct) possible outcomes of a trial. Denote these outcomes by $\mathscr{s}_1, \mathscr{s}_2, \ldots, \mathscr{s}_{k+1}$, and let $p_i = P[\mathscr{s}_i]$, $i = 1, \ldots, k + 1$. Obviously we must have $\sum_{i=1}^{k+1} p_i = 1$, just as $p + q = 1$ in the binomial case. Suppose that we repeat the trial n times. Let X_i denote the number of times outcome $\mathscr{s}_i$ occurs in the n trials, $i = 1, \ldots, k + 1$. If the trials are repeated and independent, then the discrete density function of the random variables $X_1, \ldots, X_k$ is

$$f_{X_1, \ldots, X_k}(x_1, \ldots, x_k) = \frac{n!}{\prod_{i=1}^{k+1} x_i!} \prod_{i=1}^{k+1} p_i^{x_i}, \tag{1}$$

where $x_i = 0, \ldots, n$ and $\sum_{i=1}^{k+1} x_i = n$. Note that $X_{k+1} = n - \sum_{i=1}^{k} X_i$.

To justify Eq. (1), note that the left-hand side is $P[X_1 = x_1; X_2 = x_2; \ldots; X_{k+1} = x_{k+1}]$; so, we want the probability that the n trials result in exactly x_1 outcomes $\mathscr{s}_1$, exactly x_2 outcomes $\mathscr{s}_2, \ldots,$ exactly x_{k+1} outcomes $\mathscr{s}_{k+1}$, where $\sum_1^{k+1} x_i = n$. Any specific ordering of these n outcomes has probability $p_1^{x_1} \cdot p_2^{x_2} \cdots p_{k+1}^{x_{k+1}}$ by the assumption of independent trials, and there are $n!/x_1!\,x_2! \cdots x_{k+1}!$ such orderings. ////

Definition 7 Multinomial distribution The joint discrete density function given in Eq. (1) is called the *multinomial distribution*. ////

The multinomial distribution is a $(k + 1)$ parameter family of distributions, the parameters being n and $p_1, p_2, \ldots, p_k$. p_{k+1} is, like q in the binomial distribution, exactly determined by $p_{k+1} = 1 - p_1 - p_2 - \cdots - p_k$. A

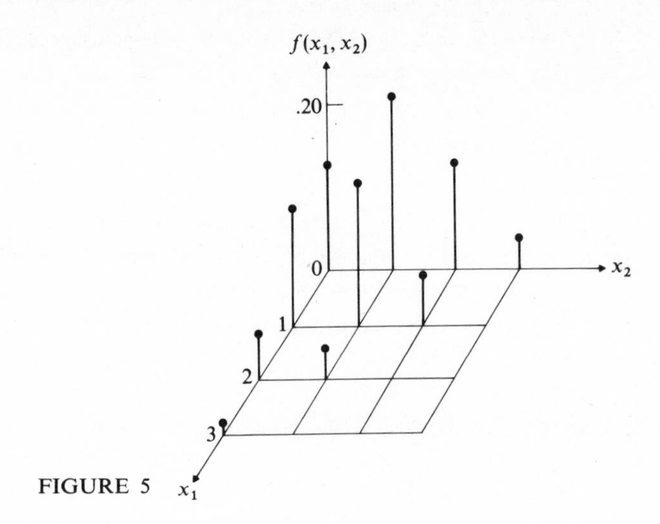

FIGURE 5

particular case of a multinomial distribution is obtained by putting, for example, $n = 3$, $k = 2$, $p_1 = .2$, and $p_2 = .3$, to get

$$f_{X_1, X_2}(x_1, x_2) = f(x_1, x_2) = \frac{3!}{x_1! x_2!(3 - x_1 - x_2)!}(.2)^{x_1}(.3)^{x_2}(.5)^{3 - x_1 - x_2}.$$

This density is plotted in Fig. 5.

We might observe that if $X_1, X_2, \ldots, X_k$ have the multinomial distribution given in Eq. (1), then the marginal distribution of X_i is a binomial distribution with parameters n and p_i. This observation can be verified by recalling the experiment of repeated, independent trials. Each trial can be thought of as resulting either in outcome $\mathscr{O}_i$ or not in outcome $\mathscr{O}_i$, in which case the trial is Bernoulli, implying that X_i has a binomial distribution with parameters n and p_i.

2.3 Joint Density Functions for Continuous Random Variables

Definition 8 Joint continuous random variables and density function The k-dimensional random variable $(X_1, X_2, \ldots, X_k)$ is defined to be a k-dimensional *continuous random variable* if and only if there exists a function $f_{X_1, \ldots, X_k}(\cdot, \ldots, \cdot) \geq 0$ such that

$$F_{X_1, \ldots, X_k}(x_1, \ldots, x_k) = \int_{-\infty}^{x_k} \cdots \int_{-\infty}^{x_1} f_{X_1, \ldots, X_k}(u_1, \ldots, u_k)\, du_1 \ldots du_k \tag{2}$$

for all $(x_1, \ldots, x_k)$. $f_{X_1, \ldots, X_k}(\cdot, \ldots, \cdot)$ is defined to be a *joint probability density function.* ////

As in the unidimensional case, a joint probability density function has two properties:

(i) $f_{X_1, \ldots, X_k}(x_1, \ldots, x_k) \geq 0$.

(ii) $\int_{-\infty}^{\infty} \cdots \int_{-\infty}^{\infty} f_{X_1, \ldots, X_k}(x_1, \ldots, x_k)\, dx_1 \ldots dx_k = 1$.

A unidimensional probability density function was used to find probabilities. For example, for X a continuous random variable with probability density $f_X(\cdot)$, $P[a < X < b] = \int_a^b f_X(x)\, dx$; that is, the *area* under $f_X(\cdot)$ over the interval (a, b) gave $P[a < X < b]$; and, more generally, $P[X \in B] = \int_B f_X(x)\, dx$; that is, the *area* under $f_X(\cdot)$ over the set B gave $P[X \in B]$. In the two-dimensional case, *volume* gives probabilities. For instance, let (X_1, X_2) be jointly continuous random variables with joint probability density function $f_{X_1, X_2}(x_1, x_2)$, and let R be some region in the $x_1 x_2$ plane; then $P[(X_1, X_2) \in R] = \iint_R f_{X_1, X_2}(x_1, x_2)\, dx_1\, dx_2$; that is, the probability that (X_1, X_2) falls in the region R is given by the *volume* under $f_{X_1, X_2}(\cdot, \cdot)$ over the region R. In particular if $R = \{(x_1, x_2): a_1 < x_1 \leq b_1; a_2 < x_2 \leq b_2\}$, then

$$P[a_1 < X_1 \leq b_1; a_2 < X_2 \leq b_2] = \int_{a_2}^{b_2} \left[\int_{a_1}^{b_1} f_{X_1, X_2}(x_1, x_2)\, dx_1 \right] dx_2 .$$

A joint probability density function is defined as any nonnegative integrand satisfying Eq. (2) and hence is not uniquely defined.

EXAMPLE 6 Consider the bivariate function

$$f(x, y) = K(x + y) I_{(0,1)}(x) I_{(0,1)}(y) = K(x + y) I_U(x, y),$$

where $U = \{(x, y): 0 < x < 1 \text{ and } 0 < y < 1\}$, a unit square. Can the constant K be selected so that $f(x, y)$ will be a joint probability density function? If K is positive, $f(x, y) \geq 0$.

$$\begin{aligned}
\int_{-\infty}^{\infty} \int_{-\infty}^{\infty} Kf(x, y)\, dx\, dy &= \int_0^1 \int_0^1 K(x + y)\, dx\, dy \\
&= K \int_0^1 \int_0^1 (x + y)\, dx\, dy \\
&= K \int_0^1 (\tfrac{1}{2} + y)\, dy \\
&= K(\tfrac{1}{2} + \tfrac{1}{2}) \\
&= 1
\end{aligned}$$

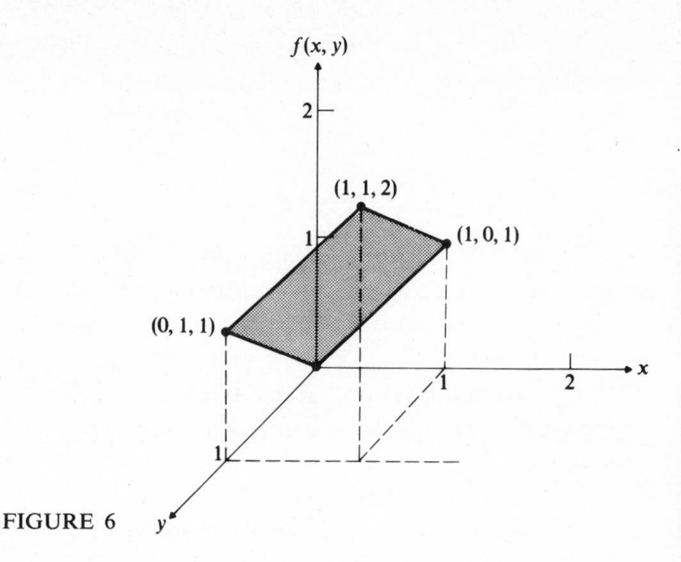

FIGURE 6

for $K = 1$. So $f(x, y) = (x + y)I_{(0,1)}(x)I_{(0,1)}(y)$ is a joint probability density function. It is sketched in Fig. 6.

Probabilities of events defined in terms of the random variables can be obtained by integrating the joint probability density function over the indicated region; for example

$$\begin{aligned} P[0 < X < \tfrac{1}{2}; 0 < Y < \tfrac{1}{4}] &= \int_0^{\frac{1}{4}} \int_0^{\frac{1}{2}} (x + y)\, dx\, dy \\ &= \int_0^{\frac{1}{4}} \left(\tfrac{1}{8} + \frac{y}{2}\right) dy \\ &= \tfrac{1}{32} + \tfrac{1}{64} \\ &= \tfrac{3}{64}, \end{aligned}$$

which is the volume under the surface $z = x + y$ over the region $\{(x, y): 0 < x < \frac{1}{2}; 0 < y < \frac{1}{4}\}$ in the xy plane. ////

Theorem 2 If X and Y are jointly continuous random variables, then knowledge of $F_{X,Y}(\cdot, \cdot)$ is equivalent to knowledge of an $f_{X,Y}(\cdot, \cdot)$. The remark extends to k-dimensional continuous random variables.

PROOF For a given $f_{X,Y}(\cdot, \cdot)$, $F_{X,Y}(x, y)$ is obtained for any (x, y) by

$$F_{X,Y}(x, y) = \int_{-\infty}^{y} \int_{-\infty}^{x} f_{X,Y}(u, v)\, du\, dv.$$

For given $F_{X,Y}(\cdot, \cdot)$, an $f_{X,Y}(x, y)$ can be obtained by

$$f_{X,Y}(x, y) = \frac{\partial^2 F_{X,Y}(x, y)}{\partial x \, \partial y}$$

for x, y points, where $F_{X,Y}(x, y)$ is differentiable. ////

Definition 9 Marginal probability density functions If X and Y are jointly continuous random variables, then $f_X(\cdot)$ and $f_Y(\cdot)$ are called *marginal probability density functions.* More generally, let $X_{i_1}, \ldots, X_{i_m}$ be any subset of the jointly continuous random variables $X_1, \ldots, X_k$. $f_{X_{i_1}, \ldots, X_{i_m}}(x_{i_1}, \ldots, x_{i_m})$ is called a *marginal density of the* m-dimensional random variable $(X_{i_1}, \ldots, X_{i_m})$. ////

Remark If $X_1, \ldots, X_k$ are jointly continuous random variables, then any marginal probability density function can be found. (However, knowledge of all marginal densities does not, in general, imply knowledge of the joint density, as Example 8 below shows.) If X and Y are jointly continuous, then

$$f_X(x) = \int_{-\infty}^{\infty} f_{X,Y}(x, y) \, dy \qquad \text{and} \qquad f_Y(y) = \int_{-\infty}^{\infty} f_{X,Y}(x, y) \, dx \tag{3}$$

since

$$f_X(x) = \frac{dF_X(x)}{dx} = \frac{d}{dx}\left[\int_{-\infty}^{x} \left(\int_{-\infty}^{\infty} f_{X,Y}(u, y) \, dy\right) du\right] = \int_{-\infty}^{\infty} f_{X,Y}(x, y) \, dy.$$

////

EXAMPLE 7 Consider the joint probability density

$$f_{X,Y}(x, y) = (x + y)I_{(0,1)}(x)I_{(0,1)}(y).$$

$$\begin{aligned}
F_{X,Y}(x, y) &= I_{(0,1)}(x)I_{(0,1)}(y) \int_0^y \int_0^x (u + v) \, du \, dv \\
&\quad + I_{(0,1)}(x)I_{[1,\infty)}(y) \int_0^1 \int_0^x (u + v) \, du \, dv \\
&\quad + I_{[1,\infty)}(x)I_{(0,1)}(y) \int_0^y \int_0^1 (u + v) \, du \, dv \\
&\quad + I_{[1,\infty)}(x)I_{[1,\infty)}(y) \\
&= \tfrac{1}{2}\{(x^2y + xy^2)I_{(0,1)}(x)I_{(0,1)}(y) + (x^2 + x)I_{(0,1)}(x)I_{[1,\infty)}(y) \\
&\quad + (y + y^2)I_{[1,\infty)}(x)I_{(0,1)}(y)\} + I_{[1,\infty)}(x)I_{[1,\infty)}(y).
\end{aligned}$$

$$\begin{aligned} f_X(x) &= \int_{-\infty}^{\infty} f_{X,Y}(x, y)\, dy \\ &= I_{(0,1)}(x) \int_0^1 (x + y)\, dy \\ &= (x + \tfrac{1}{2}) I_{(0,1)}(x); \end{aligned}$$

or,

$$\begin{aligned} f_X(x) &= \frac{\partial F_{X,Y}(x, \infty)}{\partial x} \\ &= \frac{\partial F_X(x)}{\partial x} \\ &= I_{(0,1)}(x) \frac{\partial}{\partial x}\left(\frac{x^2 + x}{2}\right) \\ &= (x + \tfrac{1}{2}) I_{(0,1)}(x). \end{aligned}$$

////

EXAMPLE 8 Let $f_X(x)$ and $f_Y(y)$ be two probability density functions with corresponding cumulative distribution functions $F_X(x)$ and $F_Y(y)$, respectively. For $-1 \le \alpha \le 1$, define

$$f_{X,Y}(x, y; \alpha) = f_X(x) f_Y(y)\{1 + \alpha[2F_X(x) - 1][2F_Y(y) - 1]\}. \tag{4}$$

We will show (i) that for each α satisfying $-1 \le \alpha \le 1$, $f_{X,Y}(x, y; \alpha)$ is a joint probability density function and (ii) that the marginals of $f_{X,Y}(x, y; \alpha)$ are $f_X(x)$ and $f_Y(y)$, respectively. Thus, $\{f_{X,Y}(x, y; \alpha): -1 \le \alpha \le 1\}$ will be an infinite family of joint probability density functions, each having the same two given marginals. To verify (i) we must show that $f_{X,Y}(x, y; \alpha)$ is nonnegative and, if integrated over the xy plane, integrates to 1.

$$f_X(x) f_Y(y)\{1 + \alpha[2F_X(x) - 1][2F_Y(y) - 1]\} \ge 0 \qquad \text{if } 1 \ge -\alpha[2F_X(x) - 1][2F_Y(y) - 1];$$

but α, $2F_X(x) - 1$, and $2F_Y(y) - 1$ are all between -1 and 1, and hence also their product, which implies $f_{X,Y}(x, y; \alpha)$ is nonnegative. Since

$$1 = \int_{-\infty}^{\infty} f_X(x)\, dx = \int_{-\infty}^{\infty} \left(\int_{-\infty}^{\infty} f_{X,Y}(x, y; \alpha)\, dy \right) dx,$$

it suffices to show that $f_X(x)$ and $f_Y(y)$ are the marginals of $f_{X,Y}(x, y; \alpha)$.

$$\int_{-\infty}^{\infty} f_{X,Y}(x, y; \alpha)\, dy$$

$$= \int_{-\infty}^{\infty} f_X(x) f_Y(y)\{1 + \alpha[2F_X(x) - 1][2F_Y(y) - 1]\}\, dy$$

$$= f_X(x) \int_{-\infty}^{\infty} f_Y(y)\, dy + \alpha f_X(x)[2F_X(x) - 1] \int_{-\infty}^{\infty} [2F_Y(y) - 1] f_Y(y)\, dy$$

$$= f_X(x), \qquad \text{noting that} \quad \int_{-\infty}^{\infty} [2F_Y(y) - 1] f_Y(y)\, dy$$

$$= \int_0^1 (2u - 1)\, du = 0$$

by making the transformation $u = F_Y(y)$. ////

3 CONDITIONAL DISTRIBUTIONS AND STOCHASTIC INDEPENDENCE

In the preceding section we defined the joint distribution and joint density functions of several random variables; in this section we define conditional distributions and the related concept of stochastic independence. Most definitions will be given first for only two random variables and later extended to k random variables.

3.1 Conditional Distribution Functions for Discrete Random Variables

Definition 10 Conditional discrete density function Let X and Y be jointly discrete random variables with joint discrete density function $f_{X,Y}(\cdot, \cdot)$. The *conditional discrete density function* of Y given $X = x$, denoted by $f_{Y|X}(\cdot \mid x)$, is defined to be

$$f_{Y|X}(y \mid x) = \frac{f_{X,Y}(x, y)}{f_X(x)}, \tag{5}$$

if $f_X(x) > 0$, where $f_X(x)$ is the marginal density of X evaluated at x. $f_{Y|X}(\cdot \mid x)$ is undefined for $f_X(x) = 0$. Similarly,

$$f_{X|Y}(x \mid y) = \frac{f_{X,Y}(x, y)}{f_Y(y)}, \tag{6}$$

if $f_Y(y) > 0$. ////

Since X and Y are discrete, they have mass points, say $x_1, x_2, \ldots$ for X and $y_1, y_2, \ldots$ for Y. If $f_X(x) > 0$, then $x = x_i$ for some i, and $f_X(x_i) = P[X = x_i]$. The numerator of the right-hand side of Eq. (5) is $f_{X,Y}(x_i, y_j) = P[X = x_i; Y = y_j]$; so

$$f_{Y|X}(y_j|x_i) = \frac{f_{X,Y}(x_i, y_j)}{f_X(x_i)} = \frac{P[X = x_i; Y = y_j]}{P[X = x_i]} = P[Y = y_j | X = x_i],$$

for y_j a mass point of Y and x_i a mass point of X; hence $f_{Y|X}(\cdot|x)$ is a conditional probability as defined in Subsec. 3.6 of Chap. I. $f_{Y|X}(\cdot|x)$ is called a conditional discrete density function and hence should possess the properties of a discrete density function. To see that it does, consider x as some fixed mass point of X. Then $f_{Y|X}(y|x)$ is a function with argument y, and to be a discrete density function must be nonnegative and, if summed over the possible values (mass points) of Y, must sum to 1. $f_{Y|X}(y|x)$ is nonnegative since $f_{X,Y}(x, y)$ is nonnegative and $f_X(x)$ is positive.

$$\sum_j f_{Y|X}(y_j|x) = \sum_j \frac{f_{X,Y}(x, y_j)}{f_X(x)} = \frac{1}{f_X(x)} \sum_j f_{X,Y}(x, y_j) = \frac{f_X(x)}{f_X(x)} = 1,$$

where the summation is over all the mass points of Y. (We used the fact that the marginal discrete density of X is obtained by summing the joint density of X and Y over the possible values of Y.) So $f_{Y|X}(\cdot|x)$ is indeed a density; it tells us how the values of Y are distributed for a given value x of X.

The conditional cumulative distribution of Y given $X = x$ can be defined for two jointly discrete random variables by recalling the close relationship between discrete density functions and cumulative distribution functions.

Definition 11 Conditional discrete cumulative distribution If X and Y are jointly discrete random variables, the *conditional cumulative distribution* of Y given $X = x$, denoted by $F_{Y|X}(\cdot|x)$, is defined to be $F_{Y|X}(y|x) = P[Y \le y | X = x]$ for $f_X(x) > 0$. ////

Remark $F_{Y|X}(y|x) = \sum_{\{j: y_j \le y\}} f_{Y|X}(y_j|x)$. ////

EXAMPLE 9 Return to the experiment of tossing two tetrahedra. Let X denote the number on the downturned face of the first and Y the larger of the downturned numbers. What is the density of Y given that $X = 2$?

$$f_{Y|X}(2|2) = \frac{f_{X,Y}(2, 2)}{f_X(2)} = \frac{\frac{2}{16}}{\frac{4}{16}} = \frac{1}{2}$$

$$f_{Y|X}(3|2) = \frac{f_{X,Y}(2, 3)}{f_X(2)} = \frac{\frac{1}{16}}{\frac{4}{16}} = \frac{1}{4}$$

$$f_{Y|X}(4|2) = \frac{f_{X,Y}(2, 4)}{f_X(2)} = \frac{\frac{1}{16}}{\frac{4}{16}} = \frac{1}{4}.$$

Also,

$$f_{Y|X}(y|3) = \begin{cases} \frac{3}{4} & \text{for } y = 3 \\ \frac{1}{4} & \text{for } y = 4. \end{cases} \qquad ////$$

Definition 12 Conditional discrete density function Let $(X_1, \ldots, X_k)$ be a k-dimensional discrete random variable, and let $X_{i_1}, \ldots, X_{i_r}$ and $X_{j_1}, \ldots, X_{j_s}$ be two disjoint subsets of the random variables $X_1, \ldots, X_k$. The *conditional density* of the r-dimensional random variable $(X_{i_1}, \ldots, X_{i_r})$ given the value $(x_{j_1}, \ldots, x_{j_s})$ of $(X_{j_1}, \ldots, X_{j_s})$ is defined to be

$$f_{X_{i_1}, \ldots, X_{i_r}|X_{j_1}, \ldots, X_{j_s}}(x_{i_1}, \ldots, x_{i_r}|x_{j_1}, \ldots, x_{j_s})$$
$$= \frac{f_{X_{i_1}, \ldots, X_{i_r}, X_{j_1}, \ldots, X_{j_s}}(x_{i_1}, \ldots, x_{i_r}, x_{j_1}, \ldots, x_{j_s})}{f_{X_{j_1}, \ldots, X_{j_s}}(x_{j_1}, \ldots, x_{j_s})}. \qquad ////$$

EXAMPLE 10 Let $X_1, \ldots, X_5$ be jointly discrete random variables. Take $r = s = 2$, $(X_{i_1}, X_{i_2}) = (X_1, X_2)$, and $(X_{j_1}, X_{j_2}) = (X_3, X_5)$; then

$$f_{X_1, X_2|X_3, X_5}(x_1, x_2|x_3, x_5) = \frac{f_{X_1, X_2, X_3, X_5}(x_1, x_2, x_3, x_5)}{f_{X_3, X_5}(x_3, x_5)}. \qquad ////$$

EXAMPLE 11 Suppose 12 cards are drawn without replacement from an ordinary deck of playing cards. Let X_1 be the number of aces drawn, X_2 be the number of 2s, X_3 be the number of 3s, and X_4 be the number of 4s. The joint density of these four random variables is given by

$$f_{X_1, X_2, X_3, X_4}(x_1, x_2, x_3, x_4)$$
$$= \frac{\binom{4}{x_1}\binom{4}{x_2}\binom{4}{x_3}\binom{4}{x_4}\binom{36}{12 - x_1 - x_2 - x_3 - x_4}}{\binom{52}{12}},$$

where $x_i = 0, 1, 2, 3$, or 4 and $i = 1, \ldots, 4$, subject to the restriction that $\sum x_i \le 12$. There are a large number of conditional densities associated with this density; an example is

$$f_{X_2, X_4 | X_1, X_3}(x_2, x_4 | x_1, x_3)$$

$$= \frac{\binom{4}{x_1}\binom{4}{x_2}\binom{4}{x_3}\binom{4}{x_4}\binom{36}{12 - x_1 - x_2 - x_3 - x_4}\Big/\binom{52}{12}}{\binom{4}{x_1}\binom{4}{x_3}\binom{44}{12 - x_1 - x_3}\Big/\binom{52}{12}}$$

$$= \frac{\binom{4}{x_2}\binom{4}{x_4}\binom{36}{12 - x_1 - x_2 - x_3 - x_4}}{\binom{44}{12 - x_1 - x_3}},$$

where $x_i = 0, 1, \ldots, 4$ and $x_2 + x_4 \le 12 - x_1 - x_3$. ////

3.2 Conditional Distribution Functions for Continuous Random Variables

Definition 13 Conditional probability density function Let X and Y be jointly continuous random variables with joint probability density function $f_{X,Y}(x, y)$. The *conditional probability density function* of Y given $X = x$, denoted by $f_{Y|X}(\cdot | x)$, is defined to be

$$f_{Y|X}(y | x) = \frac{f_{X,Y}(x, y)}{f_X(x)} \tag{7}$$

if $f_X(x) > 0$, where $f_X(x)$ is the marginal probability density of X, and is undefined at points when $f_X(x) = 0$.
Similarly,

$$f_{X|Y}(x | y) = \frac{f_{X,Y}(x, y)}{f_Y(y)} \qquad \text{if } f_Y(y) > 0, \tag{8}$$

and is undefined if $f_Y(y) = 0$ ////

$f_{Y|X}(\cdot | x)$ is called a (conditional) probability density function and hence should possess the properties of a probability density function. $f_{Y|X}(\cdot | x)$ is clearly nonnegative, and

$$\int_{-\infty}^{\infty} f_{Y|X}(y | x)\, dy = \int_{-\infty}^{\infty} \frac{f_{X,Y}(x, y)}{f_X(x)}\, dy$$

$$= \frac{1}{f_X(x)} \int_{-\infty}^{\infty} f_{X,Y}(x, y)\, dy = \frac{f_X(x)}{f_X(x)} = 1.$$

The density $f_{Y|X}(\cdot|x)$ is a density of the random variable Y given that x is the value of the random variable X. In the conditional density $f_{Y|X}(\cdot|x)$, x is fixed and could be thought of as a parameter. Consider $f_{Y|X}(\cdot|x_0)$, that is, the density of Y given that X was observed to be x_0. Now $f_{X,Y}(x, y)$ plots as a surface over the xy plane. A plane perpendicular to the xy plane which intersects the xy plane on the line $x = x_0$ will intersect the surface in the curve $f_{X,Y}(x_0, y)$. The area under this curve is

$$\int_{-\infty}^{\infty} f_{X,Y}(x_0, y)\, dy = f_X(x_0).$$

Hence, if we divide $f_{X,Y}(x_0, y)$ by $f_X(x_0)$, we obtain a density which is precisely $f_{Y|X}(y|x_0)$.

Again, the conditional cumulative distribution can be defined in the natural way.

Definition 14 Conditional continuous cumulative distribution If X and Y are jointly continuous, then the *conditional cumulative distribution* of Y given $X = x$ is defined as

$$F_{Y|X}(y|x) = \int_{-\infty}^{y} f_{Y|X}(z|x)\, dz$$

for all x such that $f_X(x) > 0$. ////

EXAMPLE 12 Suppose $f_{X,Y}(x, y) = (x + y)I_{(0,1)}(x)I_{(0,1)}(y)$.

$$f_{Y|X}(y|x) = \frac{(x + y)I_{(0,1)}(x)I_{(0,1)}(y)}{(x + \frac{1}{2})I_{(0,1)}(x)} = \frac{x + y}{x + \frac{1}{2}} I_{(0,1)}(y)$$

for $0 < x < 1$. Note that

$$\begin{aligned} F_{Y|X}(y|x) &= \int_{-\infty}^{y} f_{Y|X}(z|x)\, dz \\ &= \int_{0}^{y} \frac{x + z}{x + \frac{1}{2}}\, dz = \frac{1}{x + \frac{1}{2}} \int_{0}^{y} (x + z)\, dz \\ &= \frac{1}{x + \frac{1}{2}} (xy + y^2/2) \qquad \text{for } 0 < y < 1. \end{aligned}$$

////

Conditional probability density functions can be analogously defined for k-dimensional continuous random variables. For instance,

$$f_{X_1,X_2,X_4|X_3,X_5}(x_1, x_2, x_4|x_3, x_5) = \frac{f_{X_1,X_2,X_3,X_4,X_5}(x_1, x_2, x_3, x_4, x_5)}{f_{X_3,X_5}(x_3, x_5)}$$

for $f_{X_3,X_5}(x_3, x_5) > 0$.

3.3 More on Conditional Distribution Functions

We have defined the conditional cumulative distribution $F_{Y|X}(y|x)$ for either jointly continuous or jointly discrete random variables. If X is discrete and Y is any random variable, then $F_{Y|X}(y|x)$ can be defined as $P[Y \leq y | X = x]$ if x is a mass point of X. We would like to define $P[Y \leq y | X = x]$ and more generally $P[A | X = x]$, where A is any event, for X either a discrete or continuous random variable. Thus we seek to define *the conditional probability of an event A given a random variable $X = x$.*

We start by assuming that the event A and the random variable X are both defined on the same probability space. We want to define $P[A | X = x]$. If X is discrete, either x is a mass point of X, or it is not; and if x is a mass point of X,

$$P[A | X = x] = \frac{P[A;\ X = x]}{P[X = x]},$$

which is well defined; on the other hand, if x is not a mass point of X, we are not interested in $P[A | X = x]$. Now if X is continuous, $P[A | X = x]$ cannot be analogously defined since $P[X = x] = 0$; however, if x is such that the events $\{x - h < X < x + h\}$ have positive probability for every $h > 0$, then $P[A | X = x]$ could be defined as

$$P[A | X = x] = \lim_{0 < h \to 0} P[A | x - h < X < x + h] \tag{9}$$

provided that the limit exists. We will take Eq. (9) as our definition of $P[A | X = x]$ if the indicated limit exists, and leave $P[A | X = x]$ undefined otherwise. (It is, in fact, possible to give $P[A | X = x]$ meaning even if $P[X = x] = 0$, and such is done in advanced probability theory.)

We will seldom be interested in $P[A | X = x]$ per se, but will be interested in using it to calculate certain probabilities. We note the following formulas:

(i)
$$P[A] = \sum_{i=1}^{\infty} P[A | X = x_i] f_X(x_i) \tag{10}$$

if X is discrete with mass points $x_1, x_2, \ldots$.

(ii)
$$P[A] = \int_{-\infty}^{\infty} P[A | X = x] f_X(x)\, dx \tag{11}$$

if X is continuous.

(iii)
$$P[A;\ X \in B] = \sum_{\{i:\ x_i \in B\}} P[A | X = x_i] f_X(x_i) \tag{12}$$

if X is discrete with mass points $x_1, x_2, \ldots$.

(iv) $$P[A; X \in B] = \int_B P[A \mid X = x] f_X(x)\, dx \tag{13}$$

if X is continuous.

Although we will not prove the above formulas, we note that Eq. (10) is just the theorem of total probabilities given in Subsec. 3.6 of Chap. I and the others are generalizations of the same. Some problems are of such a nature that it is easy to find $P[A \mid X = x]$ and difficult to find $P[A]$. If, however, $f_X(\cdot)$ is known, then $P[A]$ can be easily obtained using the appropriate one of the above formulas.

Remark $F_{X,Y}(x, y) = \int_{-\infty}^{x} F_{Y|X}(y \mid x') f_X(x')\, dx'$ results from Eq. (13) by taking $A = \{Y \le y\}$ and $B = (-\infty, x]$; and $F_Y(y) = \int_{-\infty}^{\infty} F_{Y|X}(y \mid x) f_X(x)\, dx$ is obtained from Eq. (11) by taking $A = \{Y \le y\}$. ////

We add one other formula, whose proof is also omitted. Suppose $A = \{h(X, Y) \le z\}$, where $h(\cdot, \cdot)$ is some function of two variables; then

(v) $$P[A \mid X = x] = P[h(X, Y) \le z \mid X = x] = P[h(x, Y) \le z \mid X = x]. \tag{14}$$

The following is a classical example that uses Eq. (11); another example utilizing Eqs. (14) and (11) appears at the end of the next subsection.

EXAMPLE 13 Three points are selected randomly on the circumference of a circle. What is the probability that there will be a semicircle on which all three points will lie? By selecting a point "randomly," we mean that the point is equally likely to be any point on the circumference of the circle; that is, the point is uniformly distributed over the circumference of the circle. Let us use the first point to orient the circle; for example, orient the circle (assumed centered at the origin) so that the first point falls on the positive x axis. Let X denote the position of the second point, and let A denote the event that all three points lie on the same half circle. X is uniformly distributed over the interval $(0, 2\pi)$. According to Eq. (11), $P[A] = \int P[A \mid X = x] f_X(x)\, dx$. Note that for $0 < x < \pi$, $P[A \mid X = x] = (\pi - x + \pi)/2\pi$ since, given $X = x$, event A occurs if and only if the third point falls between $x - \pi$ and π. Similarly, $P[A \mid X = x] = (x + \pi - \pi)/2\pi$ for $\pi \le x < 2\pi$. Hence $P[A] = \int_0^{2\pi} P[A \mid X = x](1/2\pi)\, dx = (1/2\pi)\{\int_0^{\pi}[(2\pi - x)/2\pi]\, dx + \int_{\pi}^{2\pi}(x/2\pi)\, dx\} = \frac{3}{4}$. ////

3.4 Independence

When we defined the conditional probability of two events in Chap. I, we also defined independence of events. We have now defined the conditional distribution of random variables; so we should define independence of random variables as well.

Definition 15 Stochastic independence Let $(X_1, X_2, \ldots, X_k)$ be a k-dimensional random variable. $X_1, X_2, \ldots, X_k$ are defined to be *stochastically independent* if and only if

$$F_{X_1, \ldots, X_k}(x_1, \ldots, x_k) = \prod_{i=1}^{k} F_{X_i}(x_i) \tag{15}$$

for all $x_1, x_2, \ldots, x_k$. ////

Definition 16 Stochastic independence Let $(X_1, X_2, \ldots, X_k)$ be a k-dimensional discrete random variable with joint discrete density function $f_{X_1, \ldots, X_k}(\cdot, \ldots, \cdot)$. $X_1, \ldots, X_k$ are *stochastically independent* if and only if

$$f_{X_1, \ldots, X_k}(x_1, \ldots, x_k) = \prod_{i=1}^{k} f_{X_i}(x_i) \tag{16}$$

for all values $(x_1, \ldots, x_k)$ of $(X_1, \ldots, X_k)$. ////

Definition 17 Stochastic independence Let $(X_1, \ldots, X_k)$ be a k-dimensional continuous random variable with joint probability density function $f_{X_1, \ldots, X_k}(\cdot, \ldots, \cdot)$. $X_1, \ldots, X_k$ are *stochastically independent* if and only if

$$f_{X_1, \ldots, X_k}(x_1, \ldots, x_k) = \prod_{i=1}^{k} f_{X_i}(x_i) \tag{17}$$

for all $x_1, \ldots, x_k$. ////

Remark Often the word "stochastically" will be omitted. ////

We saw that independence of events was closely related to conditional probability; likewise independence of random variables is closely related to conditional distributions of random variables. For example, suppose X and Y are two independent random variables; then $f_{X,Y}(x, y) = f_X(x)f_Y(y)$ by definition of independence; however, $f_{X,Y}(x, y) = f_{Y|X}(y|x)f_X(x)$ by definition of conditional density, which implies that $f_{Y|X}(y|x) = f_Y(y)$; that is, the conditional density of Y given x is the unconditional density of Y. So to show that two random variables are *not* independent, it suffices to show that $f_{Y|X}(y|x)$ depends on x.

EXAMPLE 14 Let X be the number on the downturned face of the first tetrahedron and Y the larger of the two downturned numbers in the experiment of tossing two tetrahedra. Are X and Y independent? Obviously not, since $f_{Y|X}(2|3) = P[Y = 2 | X = 3] = 0 \neq f_Y(2) = P[Y = 2] = \frac{3}{16}$. ////

EXAMPLE 15 Let $f_{X,Y}(x, y) = (x + y)I_{(0,1)}(x)I_{(0,1)}(y)$. Are therefore X and Y independent? No, since $f_{Y|X}(y|x) = [(x + y)/(x + \frac{1}{2})]I_{(0,1)}(y)$ for $0 < x < 1$, $f_{Y|X}(y|x)$ depends on x and hence cannot equal $f_Y(y)$. ////

EXAMPLE 16 Let $f_{X,Y}(x, y) = e^{-(x+y)}I_{(0,\infty)}(x)I_{(0,\infty)}(y)$. X and Y are independent since

$$f_{X,Y}(x, y) = [e^{-x}I_{(0,\infty)}(x)][e^{-y}I_{(0,\infty)}(y)] = f_X(x)f_Y(y)$$

for all (x, y). ////

It can be proved that if $X_1, \ldots, X_k$ are jointly continuous random variables, then Definitions 15 and 17 are equivalent. Similarly, for jointly discrete random variables, Definitions 15 and 16 are equivalent. It can also be proved that Eq. (15) is equivalent to $P[X_1 \in B_1; \ldots; X_k \in B_k] = \prod_{i=1}^{k} P[X_i \in B_i]$ for sets $B_1, \ldots, B_k$. The following important result is easily derived using the above equivalent notions of independence.

Theorem 3 If $X_1, \ldots, X_k$ are independent random variables and $g_1(\cdot), \ldots, g_k(\cdot)$ are k functions such that $Y_j = g_j(X_j)$, $j = 1, \ldots, k$ are random variables, then $Y_1, \ldots, Y_k$ are independent.

PROOF Note that if $g_j^{-1}(B_j) = \{z: g_j(z) \in B_j\}$, then the events $\{Y_j \in B_j\}$ and $\{X_j \in g_j^{-1}(B_j)\}$ are equivalent; consequently, $P[Y_1 \in B_1; \ldots; Y_k \in B_k] = P[X_1 \in g_1^{-1}(B_1); \ldots; X_k \in g_k^{-1}(B_k)] = \prod_{j=1}^{k} P[X_j \in g_j^{-1}(B_j)]$ $= \prod_{j=1}^{k} P[Y_j \in B_j]$. ////

For $k = 2$, the above theorem states that if two random variables, say X and Y, are independent, then a function of X is independent of a function of Y. Such a result is certainly intuitively plausible.

We will return to independence of random variables in Subsec. 4.5.

Equation (14) of the previous subsection states that $P[h(X, Y) \leq z | X = x] = P[h(x, Y) \leq z | X = x]$. Now if X and Y are assumed to be independent, then $P[h(x, Y) \leq z | X = x] = P[h(x, Y) \leq z]$, which is a probability that may be easy to calculate for certain problems.

EXAMPLE 17 Let a random variable Y represent the diameter of a shaft and a random variable X represent the inside diameter of the housing that is intended to support the shaft. By design the shaft is to have diameter 99.5 units and the housing inside diameter 100 units. If the manufacturing process of each of the items is imperfect, so that in fact Y is uniformly distributed over the interval (98.5, 100.5) and X is uniformly distributed over (99, 101), what is the probability that a particular shaft can be successfully paired with a particular housing, when "successfully paired" is taken to mean that $X - h < Y < X$ for some small positive quantity h? Assume that X and Y are independent; then

$$P[X - h < Y < X] = \int_{-\infty}^{\infty} P[X - h < Y < X | X = x] f_X(x)\, dx$$
$$= \int_{99}^{101} P[x - h < Y < x]\tfrac{1}{2}\, dx.$$

Suppose now that $h = 1$; then

$$P[x - 1 < Y < x] = \begin{cases} \dfrac{x - 98.5}{2} & \text{for } 99 < x \leq 99.5 \\[2ex] \dfrac{1}{2} & \text{for } 99.5 < x < 100.5 \\[2ex] \dfrac{100.5 - (x - 1)}{2} & \text{for } 100.5 < x \leq 101. \end{cases}$$

Hence,

$$P[X - 1 < Y < X] = \int_{99}^{101} P[x - 1 < Y < x]\tfrac{1}{2}\, dx$$
$$= \int_{99}^{99.5} \tfrac{1}{2}(x - 98.5)\tfrac{1}{2}\, dx$$
$$+ \int_{99.5}^{100.5} \tfrac{1}{2}(\tfrac{1}{2})\, dx + \int_{100.5}^{101} (\tfrac{1}{2})(100.5 - x + 1)\tfrac{1}{2}\, dx = \tfrac{7}{16}.$$

////

4 EXPECTATION

When we introduced the concept of expectation for univariate random variables in Sec. 4 of Chap. II, we first defined the mean and variance as particular expectations and then defined the expectation of a general function of a random variable. Here, we will commence, in Subsec. 4.1, with the definition of the expectation of a general function of a k-dimensional random variable. The definition will be given for only those k-dimensional random variables which have densities.

4.1 Definition

Definition 18 Expectation Let $(X_1, \ldots, X_k)$ be a k-dimensional random variable with density $f_{X_1, \ldots, X_k}(\cdot, \ldots, \cdot)$. The *expected value* of a function $g(\cdot, \ldots, \cdot)$ of the k-dimensional random variable, denoted by $\mathscr{E}[g(X_1, \ldots, X_k)]$, is defined to be

$$\mathscr{E}[g(X_1, \ldots, X_k)] = \sum g(x_1, \ldots, x_k) f_{X_1, \ldots, X_k}(x_1, \ldots, x_k) \tag{18}$$

if the random variable $(X_1, \ldots, X_k)$ is discrete where the summation is over all possible values of $(X_1, \ldots, X_k)$, and

$$\mathscr{E}[g(X_1, \ldots, X_k)]$$
$$= \int_{-\infty}^{\infty} \int_{-\infty}^{\infty} \cdots \int_{-\infty}^{\infty} g(x_1, \ldots, x_k) f_{X_1, \ldots, X_k}(x_1, \ldots, x_k)\, dx_1 \ldots dx_k \tag{19}$$

if the random variable $(X_1, \ldots, X_k)$ is continuous. ////

In order for the above to be defined, it is understood that the sum and multiple integral, respectively, exist.

Theorem 4 In particular, if $g(x_1, \ldots, x_k) = x_i$, then

$$\mathscr{E}[g(X_1, \ldots, X_k)] = \mathscr{E}[X_i] = \mu_{X_i}. \tag{20}$$

PROOF Assume that $(X_1, \ldots, X_k)$ is continuous. [The proof for $(X_1, \ldots, X_k)$ discrete is similar.]

$$\mathscr{E}[g(X_1, \ldots, X_k)] = \int_{-\infty}^{\infty} \int_{-\infty}^{\infty} \cdots \int_{-\infty}^{\infty} x_i f_{X_1, \ldots, X_k}(x_1, \ldots, x_k)\, dx_1 \ldots dx_k$$
$$= \int_{-\infty}^{\infty} x_i f_{X_i}(x_i)\, dx_i = \mathscr{E}[X_i]$$

using the fact that the marginal density $f_{X_i}(x_i)$ is obtained from the joint density by

$$\int_{-\infty}^{\infty} \cdots \int_{-\infty}^{\infty} f_{X_1, \ldots, X_k}(x_1, \ldots, x_k)\, dx_1 \ldots dx_{i-1} \cdot dx_{i+1} \ldots dx_k. \qquad ////$$

Similarly, the following theorem can be proved.

Theorem 5 If $g(x_1, \ldots, x_k) = (x_i - \mathscr{E}[X_i])^2$, then

$$\mathscr{E}[g(X_1, \ldots, X_k)] = \mathscr{E}[(X_i - \mathscr{E}[X_i])^2] = \text{var}\,[X_i]. \qquad ////$$

We might note that the "expectation" in the notation $\mathscr{E}[X_i]$ of Eq. (20) has two different interpretations; one is that the expectation is taken over the joint distribution of $X_1, \ldots, X_k$, and the other is that the expectation is taken over the marginal distribution of X_i. What Theorem 4 really says is that these two expectations are equivalent, and hence we are justified in using the same notation for both.

EXAMPLE 18 Consider the experiment of tossing two tetrahedra. Let X be the number on the first and Y the larger of the two numbers. We gave the joint discrete density function of X and Y in Example 2.

$$\begin{aligned}
\mathscr{E}[XY] &= \sum xy f_{X,Y}(x, y) \\
&= 1 \cdot 1(\tfrac{1}{16}) + 1 \cdot 2(\tfrac{1}{16}) + 1 \cdot 3(\tfrac{1}{16}) + 1 \cdot 4(\tfrac{1}{16}) \\
&\quad + 2 \cdot 2(\tfrac{2}{16}) + 2 \cdot 3(\tfrac{1}{16}) + 2 \cdot 4(\tfrac{1}{16}) + 3 \cdot 3(\tfrac{3}{16}) \\
&\quad + 3 \cdot 4(\tfrac{1}{16}) + 4 \cdot 4(\tfrac{4}{16}) = \tfrac{135}{16}. \\
\mathscr{E}[X + Y] &= (1 + 1)\tfrac{1}{16} + (1 + 2)\tfrac{1}{16} + (1 + 3)\tfrac{1}{16} + (1 + 4)\tfrac{1}{16} \\
&\quad + (2 + 2)\tfrac{2}{16} + (2 + 3)\tfrac{1}{16} + (2 + 4)\tfrac{1}{16} + (3 + 3)\tfrac{3}{16} \\
&\quad + (3 + 4)\tfrac{1}{16} + (4 + 4)\tfrac{4}{16} = \tfrac{90}{16}.
\end{aligned}$$

$\mathscr{E}[X] = \frac{5}{2}$, and $\mathscr{E}[Y] = \frac{50}{16}$; hence $\mathscr{E}[X + Y] = \mathscr{E}[X] + \mathscr{E}[Y]$. ////

EXAMPLE 19 Suppose $f_{X,Y}(x, y) = (x + y)I_{(0,1)}(x)I_{(0,1)}(y)$.

$$\mathscr{E}[XY] = \int_0^1 \int_0^1 xy(x + y)\, dx\, dy = \tfrac{1}{3}.$$

$$\mathscr{E}[X + Y] = \int_0^1 \int_0^1 (x + y)(x + y)\, dx\, dy = \tfrac{7}{6}.$$

$$\mathscr{E}[X] = \mathscr{E}[Y] = \tfrac{7}{12}. \qquad ////$$

EXAMPLE 20 Let the three-dimensional random variable (X_1, X_2, X_3) have the density

$$f_{X_1,X_2,X_3}(x_1, x_2, x_3) = 8x_1x_2x_3 I_{(0,1)}(x_1)I_{(0,1)}(x_2)I_{(0,1)}(x_3).$$

Suppose we want to find (i) $\mathscr{E}[3X_1 + 2X_2 + 6X_3]$, (ii) $\mathscr{E}[X_1X_2X_3]$, and (iii) $\mathscr{E}[X_1X_2]$. For (i) we have $g(x_1, x_2, x_3) = 3x_1 + 2x_2 + 6x_3$ and obtain

$$\begin{aligned}\mathscr{E}[g(X_1, X_2, X_3)] &= \mathscr{E}[3X_1 + 2X_2 + 6X_3] \\ &= \int_0^1 \int_0^1 \int_0^1 (3x_1 + 2x_2 + 6x_3)8x_1x_2x_3\, dx_1\, dx_2\, dx_3 = \tfrac{22}{3}.\end{aligned}$$

For (ii), we get

$$\mathscr{E}[X_1X_2X_3] = \int_0^1 \int_0^1 \int_0^1 8x_1^2x_2^2x_3^2\, dx_1\, dx_2\, dx_3 = \tfrac{8}{27},$$

and for (iii) we get $\mathscr{E}[X_1X_2] = \frac{4}{9}$. ////

The following remark, the proof of which is left to the reader, displays a property of joint expectation. It is a generalization of (ii) in Theorem 3 of Chap. II.

Remark $\mathscr{E}\left[\sum_1^m c_i g_i(X_1, \ldots, X_k)\right] = \sum_1^m c_i \mathscr{E}[g_i(X_1, \ldots, X_k)]$ for constants $c_1, c_2, \ldots, c_m$. ////

4.2 Covariance and Correlation Coefficient

Definition 19 Covariance Let X and Y be any two random variables defined on the same probability space. The *covariance* of X and Y, denoted by cov $[X, Y]$ or $\sigma_{X,Y}$, is defined as

$$\text{cov}\,[X, Y] = \mathscr{E}[(X - \mu_X)(Y - \mu_Y)] \tag{21}$$

provided that the indicated expectation exists. ////

Definition 20 Correlation coefficient The *correlation coefficient*, denoted by $\rho[X, Y]$ or $\rho_{X,Y}$, of random variables X and Y is defined to be

$$\rho_{X,Y} = \frac{\text{cov}\,[X, Y]}{\sigma_X \sigma_Y} \tag{22}$$

provided that cov $[X, Y]$, σ_X, and σ_Y exist, and $\sigma_X > 0$ and $\sigma_Y > 0$. ////

Both the *covariance* and the *correlation coefficient* of random variables X and Y are measures of a *linear relationship* of X and Y in the following sense: $\text{cov}[X, Y]$ will be positive when $X - \mu_X$ and $Y - \mu_Y$ tend to have the same sign with high probability, and $\text{cov}[X, Y]$ will be negative when $X - \mu_X$ and $Y - \mu_Y$ tend to have opposite signs with high probability. $\text{cov}[X, Y]$ tends to measure the linear relationship of X and Y; however, its actual magnitude does not have much meaning since it depends on the variability of X and Y. The correlation coefficient removes, in a sense, the individual variability of each X and Y by dividing the covariance by the product of the standard deviations, and thus the correlation coefficient is a better measure of the linear relationship of X and Y than is the covariance. Also, the correlation coefficient is unitless and, as we shall see in Subsec. 4.6 below, satisfies $-1 \le \rho_{X,Y} \le 1$.

Remark $\text{cov}[X, Y] = \mathscr{E}[(X - \mu_X)(Y - \mu_Y)] = \mathscr{E}[XY] - \mu_X \mu_Y$.

PROOF
$$\begin{aligned}\mathscr{E}[(X - \mu_X)(Y - \mu_Y)] &= \mathscr{E}[XY - \mu_X Y - \mu_Y X + \mu_X \mu_Y] \\ &= \mathscr{E}[XY] - \mu_X \mathscr{E}[Y] - \mu_Y \mathscr{E}[X] + \mu_X \mu_Y \\ &= \mathscr{E}[XY] - \mu_X \mu_Y. \end{aligned}$$ ////

EXAMPLE 21 Find $\rho_{X,Y}$ for X, the number on the first, and Y, the larger of the two numbers, in the experiment of tossing two tetrahedra. We would expect that $\rho_{X,Y}$ is positive since when X is large, Y tends to be large too. We calculated $\mathscr{E}[XY]$, $\mathscr{E}[X]$, and $\mathscr{E}[Y]$ in Example 18 and obtained $\mathscr{E}[XY] = \frac{135}{16}$, $\mathscr{E}[X] = \frac{5}{2}$, and $\mathscr{E}[Y] = \frac{50}{16}$. Thus $\text{cov}[X, Y] = \frac{135}{16} - \frac{5}{2} \cdot \frac{50}{16} = \frac{10}{16}$. Now $\mathscr{E}[X^2] = \frac{30}{4}$ and $\mathscr{E}[Y^2] = \frac{170}{16}$; hence $\text{var}[X] = \frac{5}{4}$ and $\text{var}[Y] = \frac{55}{64}$. So,

$$\rho_{X,Y} = \frac{\frac{10}{16}}{\sqrt{\frac{5}{4}}\sqrt{\frac{55}{64}}} = \frac{2}{\sqrt{11}}.$$ ////

EXAMPLE 22 Find $\rho_{X,Y}$ for X and Y if $f_{X,Y}(x, y) = (x + y)I_{(0,1)}(x)I_{(0,1)}(y)$. We saw that $\mathscr{E}[XY] = \frac{1}{3}$ and $\mathscr{E}[X] = \mathscr{E}[Y] = \frac{7}{12}$ in Example 19. Now $\mathscr{E}[X^2] = \mathscr{E}[Y^2] = \frac{5}{12}$; hence $\text{var}[X] = \text{var}[Y] = \frac{11}{144}$. Finally

$$\rho_{X,Y} = \frac{\frac{1}{3} - \frac{49}{144}}{\frac{11}{144}} = -\frac{1}{11}.$$

Does a negative correlation coefficient seem right? ////

4.3 Conditional Expectations

In the following chapters we shall have occasion to find the expected value of random variables in conditional distributions, or the expected value of one random variable given the value of another.

Definition 21 Conditional expectation Let (X, Y) be a two-dimensional random variable and $g(\cdot, \cdot)$, a function of two variables. The *conditional expectation* of $g(X, Y)$ given $X = x$, denoted by $\mathscr{E}[g(X, Y) | X = x]$, is defined to be

$$\mathscr{E}[g(X, Y) | X = x] = \int_{-\infty}^{\infty} g(x, y) f_{Y|X}(y | x) \, dy \qquad (23)$$

if (X, Y) are jointly continuous, and

$$\mathscr{E}[g(X, Y) | X = x] = \sum g(x, y_j) f_{Y|X}(y_j | x) \qquad (24)$$

if (X, Y) are jointly discrete, where the summation is over all possible values of Y. ////

In particular, if $g(x, y) = y$, we have defined $\mathscr{E}[Y | X = x] = \mathscr{E}[Y | x]$. $\mathscr{E}[Y | x]$ and $\mathscr{E}[g(X, Y) | x]$ are functions of x. Note that this definition can be generalized to more than two dimensions. For example, let $(X_1, \ldots, X_k, Y_1, \ldots, Y_m)$ be a $(k + m)$-dimensional continuous random variable with density $f_{X_1, \ldots, X_k, Y_1, \ldots, Y_m}(x_1, \ldots, x_k, y_1, \ldots, y_m)$; then

$$\begin{aligned} &\mathscr{E}[g(X_1, \ldots, X_k, Y_1, \ldots, Y_m) | x_1, \ldots, x_k] \\ &\quad = \int_{-\infty}^{\infty} \cdots \int_{-\infty}^{\infty} g(x_1, \ldots, x_k, y_1, \ldots, y_m) \\ &\quad \times f_{Y_1, \ldots, Y_m | X_1, \ldots, X_k}(y_1, \ldots, y_m | x_1, \ldots, x_k) \, dy_1 \ldots dy_m. \end{aligned}$$ ////

EXAMPLE 23 In the experiment of tossing two tetrahedra with X, the number on the first, and Y, the larger of the two numbers, we found that

$$f_{Y|X}(y | 2) = \begin{cases} \frac{1}{2} & \text{for } y = 2 \\ \frac{1}{4} & \text{for } y = 3 \\ \frac{1}{4} & \text{for } y = 4 \end{cases}$$

in Example 9. Hence $\mathscr{E}[Y | X = 2] = \sum y f_{Y|X}(y | X = 2) = 2 \cdot \frac{1}{2} + 3 \cdot \frac{1}{4} + 4 \cdot \frac{1}{4} = \frac{11}{4}$. ////

EXAMPLE 24 For $f_{X,Y}(x, y) = (x + y)I_{(0,1)}(x)I_{(0,1)}(y)$, we found that

$$f_{Y|X}(y|x) = \frac{x+y}{x+\frac{1}{2}} I_{(0,1)}(y)$$

for $0 < x < 1$ in Example 12. Hence

$$\mathscr{E}[Y|X = x] = \int_0^1 y \frac{x+y}{x+\frac{1}{2}} \, dy = \frac{1}{x+\frac{1}{2}} \left(\frac{x}{2} + \frac{1}{3}\right)$$

for $0 < x < 1$. ////

As we stated above, $\mathscr{E}[g(Y)|x]$ is, in general, a function of x, Let us denote it by $h(x)$; that is, $h(x) = \mathscr{E}[g(Y)|x]$. Now we can evaluate the expectation of $h(X)$, a function of X, and will have $\mathscr{E}[h(X)] = \mathscr{E}[\mathscr{E}[g(Y)|X]]$. This gives us

$$\begin{aligned}
\mathscr{E}[\mathscr{E}[g(Y)|X]] = \mathscr{E}[h(X)] &= \int_{-\infty}^{\infty} h(x) f_X(x) \, dx \\
&= \int_{-\infty}^{\infty} \mathscr{E}[g(Y)|x] f_X(x) \, dx \\
&= \int_{-\infty}^{\infty} \left[\int_{-\infty}^{\infty} g(y) f_{Y|X}(y|x) \, dy\right] f_X(x) \, dx \\
&= \int_{-\infty}^{-\infty} \int_{-\infty}^{\infty} g(y) f_{Y|X}(y|x) f_X(x) \, dy \, dx \\
&= \int_{-\infty}^{\infty} \int_{-\infty}^{\infty} g(y) f_{X,Y}(x, y) \, dy \, dx \\
&= \mathscr{E}[g(Y)].
\end{aligned}$$

Thus we have proved for jointly continuous random variables X and Y (the proof for X and Y jointly discrete is similar) the following simple yet very useful theorem.

Theorem 6 Let (X, Y) be a two-dimensional random variable; then

$$\mathscr{E}[g(Y)] = \mathscr{E}[\mathscr{E}[g(Y)|X]], \tag{25}$$

and in particular

$$\mathscr{E}[Y] = \mathscr{E}[\mathscr{E}[Y|X]]. \tag{26}$$

////

Definition 22 Regression curve $\mathscr{E}[Y|X = x]$ is called the *regression curve* of Y on x. It is also denoted by $\mu_{Y|X=x} = \mu_{Y|x}$. ////

Definition 23 Conditional variance The *variance* of Y given $X = x$ is defined by $\text{var}[Y|X = x] = \mathscr{E}[Y^2|X = x] - (\mathscr{E}[Y|X = x])^2$. ////

Theorem 7 $\text{var}[Y] = \mathscr{E}[\text{var}[Y|X]] + \text{var}[\mathscr{E}[Y|X]]$.

PROOF

$$\begin{aligned}\mathscr{E}[\text{var}[Y|X]] &= \mathscr{E}[\mathscr{E}[Y^2|X]] - \mathscr{E}[(\mathscr{E}[Y|X])^2] \\ &= \mathscr{E}[Y^2] - (\mathscr{E}[Y])^2 - \mathscr{E}[(\mathscr{E}[Y|X])^2] + (\mathscr{E}[Y])^2 \\ &= \text{var}[Y] - \mathscr{E}[(\mathscr{E}[Y|X])^2] + (\mathscr{E}[\mathscr{E}[Y|X]])^2 \\ &= \text{var}[Y] - \text{var}[\mathscr{E}[Y|X]].\end{aligned}$$ ////

Let us note in words what the two theorems say. Equation (26) states that the mean of Y is the mean or expectation of the conditional mean of Y, and Theorem 7 states that the variance of Y is the mean or expectation of the conditional variance of Y, plus the variance of the conditional mean of Y.

We will conclude this subsection with one further theorem. The proof can be routinely obtained from Definition 21 and is left as an exercise. Also, the theorem can be generalized to more than two dimensions.

Theorem 8 Let (X, Y) be a two-dimensional random variable and $g_1(\cdot)$ and $g_2(\cdot)$ functions of one variable. Then

(i) $\mathscr{E}[g_1(Y) + g_2(Y)|X = x] = \mathscr{E}[g_1(Y)|X = x] + \mathscr{E}[g_2(Y)|X = x]$.
(ii) $\mathscr{E}[g_1(Y)g_2(X)|X = x] = g_2(x)\mathscr{E}[g_1(Y)|X = x]$. ////

4.4 Joint Moment Generating Function and Moments

We will use our definition of the expectation of a function of several variables to define joint moments and the joint moment generating function.

Definition 24 Joint moments The *joint raw moments* of $X_1, \ldots, X_k$ are defined by $\mathscr{E}[X_1^{r_1}X_2^{r_2} \cdots X_k^{r_k}]$, where the r_i's are 0 or any positive integer; the *joint moments* about the means are defined by

$$\mathscr{E}[(X_1 - \mu_{X_1})^{r_1} \cdots (X_k - \mu_{X_k})^{r_k}].$$ ////

Remark If $r_i = r_j = 1$ and all other r_m's are 0, then that particular joint moment about the means becomes $\mathscr{E}[(X_i - \mu_{X_i})(X_j - \mu_{X_j})]$, which is just the covariance between X_i and X_j. ////

Definition 25 Joint moment generating function The *joint moment generating function* of $(X_1, \ldots, X_k)$ is defined by

$$m_{X_1, \ldots, X_k}(t_1, \ldots, t_k) = \mathscr{E}\left[\exp \sum_{j=1}^{k} t_j X_j\right], \tag{27}$$

if the expectation exists for all values of $t_1, \ldots, t_k$ such that $-h < t_j < h$ for some $h > 0$, $j = 1, \ldots, k$. ////

The rth moment of X_j may be obtained from $m_{X_1, \ldots, X_k}(t_1, \ldots, t_k)$ by differentiating it r times with respect to t_j and then taking the limit as all the t's approach 0. Also $\mathscr{E}[X_i^r X_j^s]$ can be obtained by differentiating the joint moment generating function r times with respect to t_i and s times with respect to t_j and then taking the limit as all the t's approach 0. Similarly other joint raw moments can be generated.

Remark $m_X(t_1) = m_{X,Y}(t_1, 0) = \lim_{t_2 \to 0} m_{X,Y}(t_1, t_2)$, and $m_Y(t_2) = m_{X,Y}(0, t_2) = \lim_{t_1 \to 0} m_{X,Y}(t_1, t_2)$; that is, the marginal moment generating functions can be obtained from the joint moment generating function. ////

An example of a joint moment generating function will appear in Sec. 5 of this chapter.

4.5 Independence and Expectation

We have already defined independence and expectation; in this section we will relate the two concepts.

Theorem 9 If X and Y are independent and $g_1(\cdot)$ and $g_2(\cdot)$ are two functions, each of a single argument, then

$$\mathscr{E}[g_1(X)g_2(Y)] = \mathscr{E}[g_1(X)] \cdot \mathscr{E}[g_2(Y)].$$

PROOF We will give the proof for jointly continuous random variables.

$$\begin{aligned}
\mathscr{E}[g_1(X)g_2(Y)] &= \int_{-\infty}^{\infty} \int_{-\infty}^{\infty} g_1(x)g_2(y)f_{X,Y}(x, y)\, dx\, dy \\
&= \int_{-\infty}^{\infty} \int_{-\infty}^{\infty} g_1(x)g_2(y)f_X(x)f_Y(y)\, dx\, dy \\
&= \int_{-\infty}^{\infty} g_1(x)f_X(x)\, dx \cdot \int_{-\infty}^{\infty} g_2(y)f_Y(y)\, dy \\
&= \mathscr{E}[g_1(X)] \cdot \mathscr{E}[g_2(Y)].
\end{aligned}$$

////

Corollary If X and Y are independent, then $\text{cov}[X, Y] = 0$.

PROOF Take $g_1(x) = x - \mu_X$ and $g_2(y) = y - \mu_Y$; by Theorem 9,

$$\begin{aligned}\text{cov}[X, Y] &= \mathscr{E}[(X - \mu_X)(Y - \mu_Y)] = \mathscr{E}[g_1(X)g_2(Y)] \\ &= \mathscr{E}[g_1(X)]\mathscr{E}[g_2(Y)] \\ &= \mathscr{E}[X - \mu_X] \cdot \mathscr{E}[Y - \mu_Y] = 0 \qquad \text{since } \mathscr{E}[X - \mu_X] = 0.\end{aligned}$$ ////

Definition 26 **Uncorrelated random variables** Random variables X and Y are defined to be *uncorrelated* if and only if $\text{cov}[X, Y] = 0$. ////

Remark The converse of the above corollary is not always true; that is, $\text{cov}[X, Y] = 0$ does not always imply that X and Y are independent, as the following example shows. ////

EXAMPLE 25 Let U be a random variable which is uniformly distributed over the interval (0, 1). Define $X = \sin 2\pi U$ and $Y = \cos 2\pi U$. X and Y are clearly not independent since if a value of X is known, then U is one of two values, and so Y is also one of two values; hence the conditional distribution of Y is not the same as the marginal distribution. $\mathscr{E}[Y] = \int_0^1 \cos 2\pi u \, du = 0$, and $\mathscr{E}[X] = \int_0^1 \sin 2\pi u \, du = 0$; so $\text{cov}[X, Y] = \mathscr{E}[XY] = \int_0^1 \sin 2\pi u \cos 2\pi u \, du = \frac{1}{2} \int_0^1 \sin 4\pi u \, du = 0$. ////

Theorem 10 Two jointly distributed random variables X and Y are independent if and only if $m_{X,Y}(t_1, t_2) = m_X(t_1)m_Y(t_2)$ for all t_1, t_2 for which $-h < t_i < h$, $i = 1, 2$, for some $h > 0$.

PROOF [Recall that $m_X(t_1)$ is the moment generating function of X. Also note that $m_X(t_1) = m_{X,Y}(t_1, 0)$.] X and Y independent imply that the joint moment generating function factors into the product of the marginal moment generating functions by Theorem 9 by taking $g_1(x) = e^{t_1 x}$ and $g_2(y) = e^{t_2 y}$. The proof in the other direction will be omitted. ////

Remark Both Theorems 9 and 10 can be generalized from two random variables to k random variables. ////

4.6 Cauchy-Schwarz Inequality

Theorem 11 Cauchy-Schwarz inequality Let X and Y have finite second moments; then $(\mathscr{E}[XY])^2 = |\mathscr{E}[XY]|^2 \le \mathscr{E}[X^2]\mathscr{E}[Y^2]$, with equality if and only if $P[Y = cX] = 1$ for some constant c.

PROOF The existence of expectations $\mathscr{E}[X]$, $\mathscr{E}[Y]$, and $\mathscr{E}[XY]$ follows from the existence of expectations $\mathscr{E}[X^2]$ and $\mathscr{E}[Y^2]$. Define $0 \le h(t) = \mathscr{E}[(tX - Y)^2] = \mathscr{E}[X^2]t^2 - 2\mathscr{E}[XY]t + \mathscr{E}[Y^2]$. Now $h(t)$ is a quadratic function in t which is greater than or equal to 0. If $h(t) > 0$, then the roots of $h(t)$ are not real; so $4(\mathscr{E}[XY])^2 - 4\mathscr{E}[X^2]\mathscr{E}[Y^2] < 0$, or $(\mathscr{E}[XY])^2 < \mathscr{E}[X^2]\mathscr{E}[Y^2]$. If $h(t) = 0$ for some t, say t_0, then $\mathscr{E}[(t_0 X - Y)^2] = 0$, which implies $P[t_0 X = Y] = 1$. ////

Corollary $|\rho_{X,Y}| \le 1$, with equality if and only if one random variable is a linear function of the other with probability 1.

PROOF Rewrite the Cauchy-Schwarz inequality as $|\mathscr{E}[UV]| \le \sqrt{\mathscr{E}[U^2]\mathscr{E}[V^2]}$, and set $U = X - \mu_X$ and $V = Y - \mu_Y$. ////

5 BIVARIATE NORMAL DISTRIBUTION

One of the important multivariate densities is the *multivariate normal density*, which is a generalization of the normal distribution for a unidimensional random variable. In this section we shall discuss a special case, the case of the bivariate normal. In our discussion we will include the joint density, marginal densities, conditional densities, conditional means and variances, covariance, and the moment generating function. This section, then, will give an example of many of the concepts defined in the preceding sections of this chapter.

5.1 Density Function

Definition 27 Bivariate normal distribution Let the two-dimensional random variable (X, Y) have the joint probability density function

$$f_{X,Y}(x, y) = f(x, y) = \frac{1}{2\pi\sigma_X\sigma_Y\sqrt{1-\rho^2}}$$

$$\times \exp\left\{-\frac{1}{2(1-\rho^2)}\left[\left(\frac{x-\mu_X}{\sigma_X}\right)^2 - 2\rho\frac{x-\mu_X}{\sigma_X}\frac{y-\mu_Y}{\sigma_Y} + \left(\frac{y-\mu_Y}{\sigma_Y}\right)^2\right]\right\} \qquad (28)$$

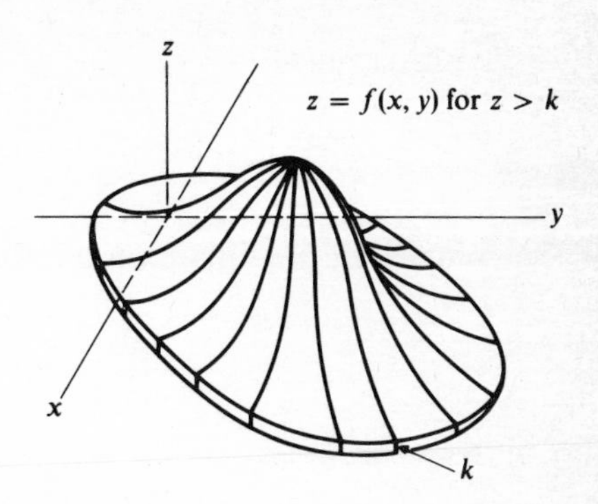

FIGURE 7

for $-\infty < x < \infty$, $-\infty < y < \infty$, where σ_Y, σ_X, μ_X, μ_Y, and ρ are constants such that $-1 < \rho < 1$, $0 < \sigma_Y$, $0 < \sigma_X$, $-\infty < \mu_X < \infty$, and $-\infty < \mu_Y < \infty$. Then the random variable (X, Y) is defined to have a *bivariate normal distribution.* ////

The density in Eq. (28) may be represented by a bell-shaped surface $z = f(x, y)$ as in Fig. 7. Any plane parallel to the xy plane which cuts the surface will intersect it in an elliptic curve, while any plane perpendicular to the xy plane will cut the surface in a curve of the normal form. The probability that a point (X, Y) will lie in any region R of the xy plane is obtained by integrating the density over that region:

$$P[(X, Y) \text{ is in } R] = \iint_R f(x, y)\, dy\, dx. \tag{29}$$

The density might, for example, represent the distribution of hits on a vertical target, where x and y represent the horizontal and vertical deviations from the central lines. And in fact the distribution closely approximates the distribution of this as well as many other bivariate populations encountered in practice.

We must first show that the function actually represents a density by showing that its integral over the whole plane is 1; that is,

$$\int_{-\infty}^{\infty} \int_{-\infty}^{\infty} f(x, y)\, dy\, dx = 1. \tag{30}$$

The density is, of course, positive. To simplify the integral, we shall substitute

$$u = \frac{x - \mu_X}{\sigma_X} \qquad \text{and} \qquad v = \frac{y - \mu_Y}{\sigma_Y}, \tag{31}$$

so that it becomes

$$\int_{-\infty}^{\infty}\int_{-\infty}^{\infty}\frac{1}{2\pi\sqrt{1-\rho^2}}e^{-[\frac{1}{2}/(1-\rho^2)](u^2-2\rho uv+v^2)}\,dv\,du.$$

On completing the square on u in the exponent, we have

$$\int_{-\infty}^{\infty}\int_{-\infty}^{\infty}\frac{1}{2\pi\sqrt{1-\rho^2}}e^{-[\frac{1}{2}/(1-\rho^2)][(u-\rho v)^2+(1-\rho^2)v^2]}\,dv\,du,$$

and if we substitute

$$w=\frac{u-\rho v}{\sqrt{1-\rho^2}} \qquad \text{and} \qquad dw=\frac{du}{\sqrt{1-\rho^2}},$$

the integral may be written as the product of two simple integrals

$$\int_{-\infty}^{\infty}\frac{1}{\sqrt{2\pi}}e^{-w^2/2}\,dw\int_{-\infty}^{\infty}\frac{1}{\sqrt{2\pi}}e^{-v^2/2}\,dv, \tag{32}$$

both of which are 1, as we have seen in studying the univariate normal distribution. Equation (30) is thus verified.

Remark The cumulative bivariate normal distribution

$$F(x, y)=\int_{-\infty}^{y}\left(\int_{-\infty}^{x}f(x', y')\,dx'\right)dy'$$

may be reduced to a form involving only the parameter ρ by making the substitution in Eq. (31). ////

5.2 Moment Generating Function and Moments

To obtain the moments of X and Y, we shall find their joint moment generating function, which is given by

$$m_{X,Y}(t_1, t_2)=m(t_1, t_2)=\mathscr{E}[e^{t_1X+t_2Y}]=\int_{-\infty}^{\infty}\int_{-\infty}^{\infty}e^{t_1x+t_2y}f(x, y)\,dy\,dx.$$

Theorem 12 The moment generating function of the bivariate normal distribution is

$$m(t_1, t_2)=\exp[t_1\mu_X+t_2\mu_Y+\tfrac{1}{2}(t_1^2\sigma_X^2+2\rho t_1t_2\sigma_X\sigma_Y+t_2^2\sigma_Y^2)]. \tag{33}$$

PROOF Let us again substitute for x and y in terms of u and v to obtain

$$m(t_1, t_2) = e^{t_1\mu_X + t_2\mu_Y} \int_{-\infty}^{\infty} \int_{-\infty}^{\infty} e^{t_1\sigma_X u + t_2\sigma_Y v} \frac{1}{2\pi\sqrt{1-\rho^2}} e^{-[\frac{1}{2}/(1-\rho^2)](u^2 - 2\rho uv + v^2)}\, dv\, du. \tag{34}$$

The combined exponents in the integrand may be written

$$-\frac{1}{2(1-\rho^2)}[u^2 - 2\rho uv + v^2 - 2(1-\rho^2)t_1\sigma_X u - 2(1-\rho^2)t_2\sigma_Y v],$$

and on completing the square first on u and then on v, we find this expression becomes

$$-\frac{1}{2(1-\rho^2)}\{[u - \rho v - (1-\rho^2)t_1\sigma_X]^2 + (1-\rho^2)(v - \rho t_1\sigma_X - t_2\sigma_Y)^2 - (1-\rho^2)(t_1^2\sigma_X^2 + 2\rho t_1 t_2\sigma_X\sigma_Y + t_2^2\sigma_Y^2)\},$$

which, if we substitute

$$w = \frac{u - \rho v - (1-\rho^2)t_1\sigma_X}{\sqrt{1-\rho^2}} \qquad \text{and} \qquad z = v - \rho t_1\sigma_X - t_2\sigma_Y,$$

becomes

$$-\tfrac{1}{2}w^2 - \tfrac{1}{2}z^2 + \tfrac{1}{2}(t_1^2\sigma_X^2 + 2\rho t_1 t_2\sigma_X\sigma_Y + t_2^2\sigma_Y^2),$$

and the integral in Eq. (34) may be written

$$\begin{aligned} m(t_1, t_2) &= e^{t_1\mu_X + t_2\mu_Y} \exp[\tfrac{1}{2}(t_1^2\sigma_X^2 + 2\rho t_1 t_2\sigma_X\sigma_Y + t_2^2\sigma_Y^2)] \\ &\quad \times \int_{-\infty}^{\infty} \int_{-\infty}^{\infty} \frac{1}{2\pi} e^{-w^2/2 - z^2/2}\, dw\, dz \\ &= \exp[t_1\mu_X + t_2\mu_Y + \tfrac{1}{2}(t_1^2\sigma_X^2 + 2\rho t_1 t_2\sigma_X\sigma_Y + t_2^2\sigma_Y^2)] \end{aligned}$$

since the double integral is equal to unity. ////

Theorem 13 If (X, Y) has bivariate normal distribution, then

$$\begin{aligned} \mathscr{E}[X] &= \mu_X, \\ \mathscr{E}[Y] &= \mu_Y, \\ \text{var}[X] &= \sigma_X^2, \\ \text{var}[Y] &= \sigma_Y^2, \\ \text{cov}[X, Y] &= \rho\sigma_X\sigma_Y, \end{aligned}$$

and

$$\rho_{X,Y} = \rho.$$

PROOF The moments may be obtained by evaluating the appropriate derivative of $m(t_1, t_2)$ at $t_1 = 0$, $t_2 = 0$. Thus,

$$\mathscr{E}[X] = \left.\frac{\partial m}{\partial t_1}\right|_{t_1, t_2 = 0} = \mu_X$$

$$\mathscr{E}[X^2] = \left.\frac{\partial^2 m}{\partial t_1^2}\right|_{t_1, t_2 = 0} = \mu_X^2 + \sigma_X^2.$$

Hence the variance of X is

$$\mathscr{E}[(X - \mu_X)^2] = \mathscr{E}[X^2] - \mu_X^2 = \sigma_X^2.$$

Similarly, on differentiating with respect to t_2, one finds the mean and variance of Y to be μ_Y and σ_Y^2. We can also obtain joint moments

$$\mathscr{E}[X^r Y^s]$$

by differentiating $m(t_1, t_2)$ r times with respect to t_1 and s times with respect to t_2 and then putting t_1 and t_2 equal to 0. The covariance of X and Y is

$$\begin{aligned}\mathscr{E}[(X - \mu_X)(Y - \mu_Y)] &= \mathscr{E}[XY - X\mu_Y - Y\mu_X + \mu_X \mu_Y] \\ &= \mathscr{E}[XY] - \mu_X \mu_Y \\ &= \left.\frac{\partial^2}{\partial t_1\, \partial t_2} m(t_1, t_2)\right|_{t_1 = t_2 = 0} - \mu_X \mu_Y \\ &= \rho \sigma_X \sigma_Y.\end{aligned}$$

Hence, the parameter ρ is the correlation coefficient of X and Y. ////

Theorem 14 If (X, Y) has a bivariate normal distribution, then X and Y are independent if and only if X and Y are uncorrelated.

PROOF X and Y are uncorrelated if and only if $\text{cov}[X, Y] = 0$ or, equivalently, if and only if $\rho_{X,Y} = \rho = 0$. It can be observed that if $\rho = 0$, the joint density $f(x, y)$ becomes the product of two univariate normal distributions; so that $\rho = 0$ implies X and Y are independent. We know that, in general, independence of X and Y implies that X and Y are uncorrelated. ////

5.3 Marginal and Conditional Densities

Theorem 15 If (X, Y) has a bivariate normal distribution, then the marginal distributions of X and Y are univariate normal distributions; that is, X is normally distributed with mean μ_X and variance σ_X^2, and Y is normally distributed with mean μ_Y and variance σ_Y^2.

PROOF The marginal density of one of the variables X, for example, is by definition

$$f_X(x) = \int_{-\infty}^{\infty} f(x, y)\, dy;$$

and again substituting

$$v = \frac{y - \mu_Y}{\sigma_Y}$$

and completing the square on v, one finds that

$$f_X(x) = \int_{-\infty}^{\infty} \frac{1}{2\pi\sigma_X\sqrt{1-\rho^2}} \times \exp\left[-\frac{1}{2}\left(\frac{x-\mu_X}{\sigma_X}\right)^2 - \frac{1}{2(1-\rho^2)}\left(v - \rho\frac{x-\mu_X}{\sigma_X}\right)^2\right] dv.$$

Then the substitutions

$$w = \frac{v - \rho(x-\mu_X)/\sigma_X}{\sqrt{1-\rho^2}} \quad \text{and} \quad dw = \frac{dv}{\sqrt{1-\rho^2}}$$

show at once that

$$f_X(x) = \frac{1}{\sqrt{2\pi\sigma_X^2}} \exp\left[-\frac{1}{2}\left(\frac{x-\mu_X}{\sigma_X}\right)^2\right],$$

the univariate normal density. Similarly the marginal density of Y may be found to be

$$f_Y(y) = \frac{1}{\sqrt{2\pi\sigma_Y^2}} \exp\left[-\frac{1}{2}\left(\frac{y-\mu_Y}{\sigma_Y}\right)^2\right]. \qquad ////$$

Theorem 16 If (X, Y) has a bivariate normal distribution, then the conditional distribution of X given $Y = y$ is normal with mean $\mu_X + (\rho\sigma_X/\sigma_Y)(y - \mu_Y)$ and variance $\sigma_X^2(1-\rho^2)$. Also, the conditional distribution of Y given $X = x$ is normal with mean $\mu_Y + (\rho\sigma_Y/\sigma_X)(x - \mu_X)$ and variance $\sigma_Y^2(1-\rho^2)$.

PROOF The conditional distributions are obtained from the joint and marginal distributions. Thus, the conditional density of X for fixed values of Y is

$$f_{X|Y}(x|y) = \frac{f(x, y)}{f_Y(y)},$$

and, after substituting, the expression may be put in the form

$$f_{X|Y}(x|y) = \frac{1}{\sqrt{2\pi}\sigma_X\sqrt{1-\rho^2}} \exp\left\{-\frac{1}{2\sigma_X^2(1-\rho^2)}\left[x - \mu_X - \frac{\rho\sigma_X}{\sigma_Y}(y - \mu_Y)\right]^2\right\}, \tag{35}$$

which is a univariate normal density with mean $\mu_X + (\rho\sigma_X/\sigma_Y)(y - \mu_Y)$ and with variance $\sigma_X^2(1 - \rho^2)$. The conditional distribution of Y may be obtained by interchanging x and y throughout Eq. (35) to get

$$f_{Y|X}(y|x) = \frac{1}{\sqrt{2\pi}\sigma_Y\sqrt{1-\rho^2}} \exp\left\{-\frac{1}{2\sigma_Y^2(1-\rho^2)}\left[y - \mu_Y - \frac{\rho\sigma_Y}{\sigma_X}(x - \mu_X)\right]^2\right\}. \tag{36}$$

////

As we already noted, the mean value of a random variable in a conditional distribution is called a *regression curve* when regarded as a function of the fixed variable in the conditional distribution. Thus the regression for X on $Y = y$ in Eq. (35) is $\mu_X + (\rho\sigma_X/\sigma_Y)(y - \mu_Y)$, which is a linear function of y in the present case. For bivariate distributions in general, the mean of X in the conditional density of X given $Y = y$ will be some function of y, say $g(\cdot)$, and the equation

$$x = g(y)$$

when plotted in the xy plane gives the regression curve for X. It is simply a curve which gives the location of the mean of X for various values of Y in the conditional density of X given $Y = y$.

For the bivariate normal distribution, the regression curve is the straight line obtained by plotting

$$x = \mu_X + \frac{\rho\sigma_X}{\sigma_Y}(y - \mu_Y),$$

as shown in Fig. 8. The conditional density of X given $Y = y$, $f_{X|Y}(x|y)$, is also plotted in Fig. 8 for two particular values y_0 and y_1 of Y.

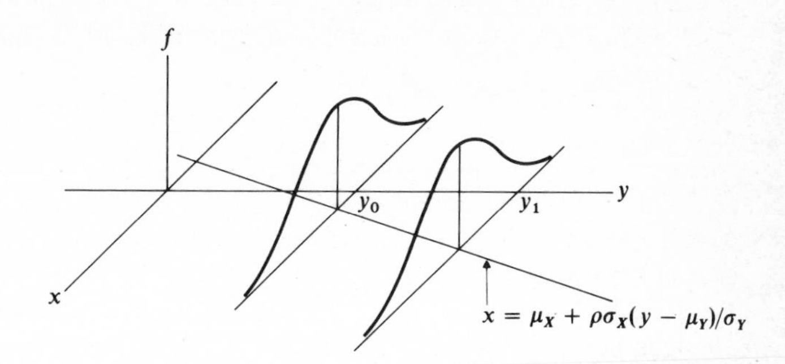

FIGURE 8

PROBLEMS

1 Prove or disprove:

(*a*) If $P[X > Y] = 1$, then $\mathscr{E}[X] > \mathscr{E}[Y]$.

(*b*) If $\mathscr{E}[X] > \mathscr{E}[Y]$, then $P[X > Y] = 1$.

(*c*) If $\mathscr{E}[X] > \mathscr{E}[Y]$, then $P[X > Y] > 0$.

2 Prove or disprove:

(*a*) If $F_X(z) > F_Y(z)$ for all z, then $\mathscr{E}[Y] > \mathscr{E}[X]$.

(*b*) If $\mathscr{E}[Y] > \mathscr{E}[X]$, then $F_X(z) > F_Y(z)$ for all z.

(*c*) If $\mathscr{E}[Y] > \mathscr{E}[X]$, then $F_X(z) > F_Y(z)$ for some z.

(*d*) If $F_X(z) = F_Y(z)$ for all z, then $P[X = Y] = 1$.

(*e*) If $F_X(z) > F_Y(z)$ for all z, then $P[X < Y] > 0$.

(*f*) If $Y = X + 1$, then $F_X(z) = F_Y(z + 1)$ for all z.

3 If X_1 and X_2 are independent random variables with distribution given by $P[X_i = -1] = P[X_i = 1] = \frac{1}{2}$ for $i = 1, 2$, then are X_1 and $X_1 X_2$ independent?

4 A penny and dime are tossed. Let X denote the number of heads up. Then the penny is tossed again. Let Y denote the number of heads up on the dime (from the first toss) and the penny from the second toss.

(*a*) Find the conditional distribution of Y given $X = 1$.

(*b*) Find the covariance of X and Y.

5 If X and Y have joint distribution given by

$$f_{X,Y}(x, y) = 2I_{(0,y)}(x)I_{(0,1)}(y).$$

(*a*) Find cov $[X, Y]$.

(*b*) Find the conditional distribution of Y given $X = x$.

6 Consider a sample of size 2 drawn without replacement from an urn containing three balls, numbered 1, 2, and 3. Let X be the number on the first ball drawn and Y the larger of the two numbers drawn.

(*a*) Find the joint discrete density function of X and Y.

(*b*) Find $P[X = 1 \mid Y = 3]$.

(*c*) Find cov $[X, Y]$.

7 Consider two random variables X and Y having a joint probability density function

$$f_{X,Y}(x, y) = \tfrac{1}{2}xyI_{(0,x)}(y)I_{(0,2)}(x).$$

(*a*) Find the marginal distributions of X and Y.
(*b*) Are X and Y independent?

8 If X has a Bernoulli distribution with parameter p (that is, $P[X=1]=p=1-P[X=0]$), $\mathscr{E}[Y|X=0]=1$, and $\mathscr{E}[Y|X=1]=2$, what is $\mathscr{E}[Y]$?

9 Consider a sample of size 2 drawn without replacement from an urn containing three balls, numbered 1, 2, and 3. Let X be the smaller of the two numbers drawn and Y the larger.
(*a*) Find the joint discrete density function of X and Y.
(*b*) Find the conditional distribution of Y given $X=1$.
(*c*) Find cov $[X, Y]$.

10 Let X and Y be independent random variables, each having the same geometric distribution. Find $P[X=Y]$.

11 If $F(\cdot)$ is a cumulative distribution function:
(*a*) Is $F(x, y)=F(x)+F(y)$ a joint cumulative distribution function?
(*b*) Is $F(x, y)=F(x)F(y)$ a joint cumulative distribution function?
(*c*) Is $F(x, y)=\max[F(x), F(y)]$ a joint cumulative distribution function?
(*d*) Is $F(x, y)=\min[F(x), F(y)]$ a joint cumulative distribution function?

12 Prove

$$F_X(x)+F_Y(y)-1 \le F_{X,Y}(x, y) \le \sqrt{F_X(x)F_Y(y)} \qquad \text{for all } x, y.$$

13 Three fair coins are tossed. Let X denote the number of heads on the first two coins, and let Y denote the number of tails on the last two coins.
(*a*) Find the joint distribution of X and Y.
(*b*) Find the conditional distribution of Y given that $X=1$.
(*c*) Find cov $[X, Y]$.

14 Let random variable X have a density function $f(\cdot)$, cumulative distribution function $F(\cdot)$, mean μ, and variance σ^2. Define $Y=\alpha+\beta X$, where α and β are constants satisfying $-\infty<\alpha<\infty$ and $\beta>0$.
(*a*) Select α and β so that Y has mean 0 and variance 1.
(*b*) What is the correlation coefficient between X and Y?
(*c*) Find the cumulative distribution function of Y in terms of α, β, and $F(\cdot)$.
(*d*) If X is symmetrically distributed about μ, is Y necessarily symmetrically distributed about its mean? (HINT: Z is symmetrically distributed about constant C if $Z-C$ and $-(Z-C)$ have the same distribution.)

15 Suppose that random variable X is uniformly distributed over the interval (0, 1); that is, $f_X(x)=I_{(0,1)}(x)$. Assume that the conditional distribution of Y given $X=x$ has a binomial distribution with parameters n and $p=x$; i.e.,

$$P[Y=y|X=x]=\binom{n}{y}x^y(1-x)^{n-y} \qquad \text{for } y=0, 1, \ldots, n.$$

(*a*) Find $\mathscr{E}[Y]$.
(*b*) Find the distribution of Y.

16 Suppose that the joint probability density function of (X, Y) is given by

$$f_{X,Y}(x, y) = [1 - \alpha(1 - 2x)(1 - 2y)]I_{(0,1)}(x)I_{(0,1)}(y),$$

where the parameter α satisfies $-1 \le \alpha \le 1$.
(*a*) Prove or disprove: X and Y are independent if and only if X and Y are uncorrelated.

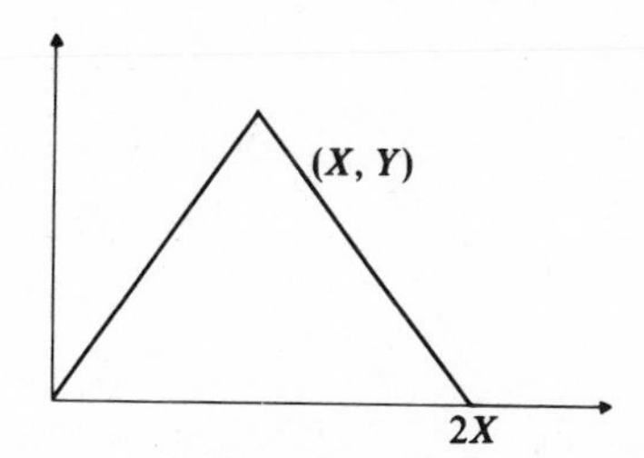

An isosceles triangle is formed as indicated in the sketch.
(*b*) If (X, Y) has the joint density given above, pick α to maximize the expected area of the triangle.
(*c*) What is the probability that the triangle falls within the unit square with corners at (0, 0), (1, 0), (1, 1), and (0, 1)?
*(*d*) Find the expected length of the perimeter of the triangle.

17 Consider tossing two tetrahedra with sides numbered 1 to 4. Let Y_1 denote the smaller of the two downturned numbers and Y_2 the larger.
(*a*) Find the joint density function of Y_1 and Y_2.
(*b*) Find $P[Y_1 \ge 2, Y_2 \ge 2]$.
(*c*) Find the mean and variance of Y_1 and Y_2.
(*d*) Find the conditional distribution of Y_2 given Y_1 for each of the possible values of Y_1.
(*e*) Find the correlation coefficient of Y_1 and Y_2.

18 Let $f_{X,Y}(x, y) = e^{-(x+y)}I_{(0,\infty)}(x)I_{(0,\infty)}(y)$
(*a*) Find $P[X > 1]$.
(*b*) Find $P[1 < X + Y < 2]$.
(*c*) Find $P[X < Y \mid X < 2Y]$.
(*d*) Find m such that $P[X + Y < m] = \frac{1}{2}$.
(*e*) Find $P[0 < X < 1 \mid Y = 2]$.
(*f*) Find the correlation coefficient of X and Y.

**19* Let $f_{X,Y}(x, y) = e^{-y}(1 - e^{-x})I_{(0,y]}(x)I_{[0,\infty)}(y) + e^{-x}(1 - e^{-y})I_{(0,x]}(y)I_{[0,\infty)}(x)$.
(*a*) Show that $f_{X,Y}(\cdot, \cdot)$ is a probability density function.
(*b*) Find the marginal distributions of X and Y.
(*c*) Find $\mathscr{E}[Y \mid X = x]$ for $0 < x$.
(*d*) Find $P[X \le 2, Y \le 2]$.
(*e*) Find the correlation coefficient of X and Y.
(*f*) Find another joint probability density function having the same marginals.

*20 Suppose X and Y are independent and identically distributed random variables with probability density function $f(\cdot)$ that is symmetrical about 0.
(a) Prove that $P[|X+Y| \leq 2|X|] > \frac{1}{2}$.
(b) Select some symmetrical probability density function $f(\cdot)$, and evaluate $P[|X+Y| \leq 2|X|]$.

*21 Prove or disprove: If $\mathscr{E}[Y|X] = X$, $\mathscr{E}[X|Y] = Y$, and both $\mathscr{E}[X^2]$ and $\mathscr{E}[Y^2]$ are finite, then $P[X = Y] = 1$. (Possible HINT: $P[X = Y] = 1$ if var $[X - Y] = 0$.)

22 A multivariate Chebyshev inequality: Let $(X_1, \ldots, X_m)$ be jointly distributed with $\mathscr{E}[X_j] = \mu_j$ and var $[X_j] = \sigma_j^2$ for $j = 1, \ldots, m$. Define $A_j = \{|X_j - \mu_j| \leq \sqrt{m}t\sigma_j\}$. Show that $P[\bigcap_{j=1}^{m} A_j] \geq 1 - t^{-2}$, for $t > 0$.

23 Let $f_X(\cdot)$ be a probability density function with corresponding cumulative distribution function $F_X(\cdot)$. In terms of $f_X(\cdot)$ and/or $F_X(\cdot)$:
(a) Find $P[X > x_0 + \Delta x | X > x_0]$.
(b) Find $P[x_0 < X < x_0 + \Delta x | X > x_0]$.
(c) Find the limit of the above divided by Δx as Δx goes to 0.
(d) Evaluate the quantities in parts (a) to (c) for $f_X(x) = \lambda e^{-\lambda x} I_{(0, \infty)}(x)$.

24 Let N equal the number of times a certain device may be used before it breaks. The probability is p that it will break on any one try given that it did not break on any of the previous tries.
(a) Express this in terms of conditional probabilities.
(b) Express it in terms of a density function, and find the density function.

25 Player A tosses a coin with sides numbered 1 and 2. B spins a spinner evenly graduated from 0 to 3. B's spinner is fair, but A's coin is not; it comes up 1 with a probability p, not necessarily equal to $\frac{1}{2}$. The payoff X of this game is the difference in their numbers (A's number minus B's). Find the cumulative distribution function of X.

26 An urn contains four balls; two of the balls are numbered with a 1, and the other two are numbered with a 2. Two balls are drawn from the urn without replacement. Let X denote the smaller of the numbers on the drawn balls and Y the larger.
(a) Find the joint density of X and Y.
(b) Find the marginal distribution of Y.
(c) Find the cov $[X, Y]$.

27 The joint probability density function of X and Y is given by

$$f_{X,Y}(x, y) = 3(x + y)I_{(0,1)}(x + y)I_{(0,1)}(x)I_{(0,1)}(y).$$

(Note the symmetry in x and y.)
(a) Find the marginal density of X.
(b) Find $P[X + Y < .5]$.
(c) Find $\mathscr{E}[Y|X = x]$.
(d) Find cov $[X, Y]$.

28 The discrete density of X is given by $f_X(x) = x/3$ for $x = 1, 2$, and $f_{Y|X}(y|x)$ is binomial with parameters x and $\frac{1}{2}$; that is,

$$f_{Y|X}(y|x) = P[Y = y | X = x] = \binom{x}{y} (\tfrac{1}{2})^x$$

for $y = 0, \ldots, x$ and $x = 1, 2$.

(*a*) Find $\mathscr{E}[X]$ and var $[X]$.

(*b*) Find $\mathscr{E}[Y]$.

(*c*) Find the joint distribution of X and Y.

29 Let the joint density function of X and Y be given by $f_{X,Y}(x, y) = 8xy$ for $0 < x < y < 1$ and be 0 elsewhere.

(*a*) Find $\mathscr{E}[Y | X = x]$.

(*b*) Find $\mathscr{E}[XY | X = x]$.

(*c*) Find var $[Y | X = x]$.

30 Let Y be a random variable having a Poisson distribution with parameter λ. Assume that the conditional distribution of X given $Y = y$ is binomially distributed with parameters y and p. Find the distribution of X, if $X = 0$ when $Y = 0$.

31 Assume that X and Y are independent random variables and X (Y) has binomial distribution with parameters 3 and $\frac{1}{3}$ (2 and $\frac{1}{2}$). Find $P[X = Y]$.

32 Let X and Y have bivariate normal distribution with parameters $\mu_X = 5$, $\mu_Y = 10$, $\sigma_X^2 = 1$, and $\sigma_Y^2 = 25$.

(*a*) If $\rho > 0$, find ρ when $P[4 < Y < 16 | X = 5] = .954$.

*(*b*) If $\rho = 0$, find $P[X + Y \leq 16]$.

33 Two dice are cast 10 times. Let X be the number of times no 1s appear, and let Y be the number of times two 1s appear.

(*a*) What is the probability that X and Y will each be less than 3?

(*b*) What is the probability that $X + Y$ will be 4?

34 Three coins are tossed n times.

(*a*) Find the joint density of X, the number of times no heads appear; Y, the number of times one head appears; and Z, the number of times two heads appear.

(*b*) Find the conditional density of X and Z given Y.

35 Six cards are drawn without replacement from an ordinary deck.

(*a*) Find the joint density of the number of aces X and the number of kings Y.

(*b*) Find the conditional density of X given Y.

36 Let the two-dimensional random variable (X, Y) have the joint density

$$f_{X,Y}(x, y) = \tfrac{1}{8}(6 - x - y)I_{(0,2)}(x)I_{(2,4)}(y).$$

(*a*) Find $\mathscr{E}[Y | X = x]$. (*b*) Find $\mathscr{E}[Y^2 | X = x]$.

(*c*) Find var $[Y | X = x]$. (*d*) Show that $\mathscr{E}[Y] = \mathscr{E}[\mathscr{E}[Y | X]]$.

(*e*) Find $\mathscr{E}[XY | X = x]$.

37 The trinomial distribution (multinomial with $k + 1 = 3$) of two random variables X and Y is given by

$$f_{X,Y}(x, y) = \frac{n!}{x!y!(n - x - y)!} p^x q^y (1 - p - q)^{n-x-y}$$

for $x, y = 0, 1, \ldots, n$ and $x + y \leq n$, where $0 \leq p$, $0 \leq q$, and $p + q \leq 1$.

(*a*) Find the marginal distribution of Y.
(*b*) Find the conditional distribution of X given Y, and obtain its expected value.
(*c*) Find $\rho[X, Y]$.

38 Let (X, Y) have probability density function $f_{X,Y}(x, y)$, and let $u(X)$ and $v(Y)$ be functions of X and Y, respectively. Show that

$$\mathscr{E}[u(X)v(Y)|X=x] = u(x)\mathscr{E}[v(Y)|X=x].$$

39 If X and Y are two random variables and $\mathscr{E}[Y|X=x]=\mu$, where μ does not depend on x, show that $\text{var}[Y] = \mathscr{E}[\text{var}[Y|X]]$.

40 If X and Y are two independent random variables, does $\mathscr{E}[Y|X=x]$ depend on x?

41 If the joint moment generating function of (X, Y) is given by $m_{X,Y}(t_1, t_2) = \exp[\frac{1}{2}(t_1^2 + t_2^2)]$ what is the distribution of Y?

42 Define the moment generating function of $Y|X=x$. Does $m_Y(t) = \mathscr{E}[m_{Y|X}(t)]$?

43 Toss three coins. Let X denote the number of heads on the first two and Y denote the number of heads on the last two.
(*a*) Find the joint distribution of X and Y.
(*b*) Find $\mathscr{E}[Y|X=1]$.
(*c*) Find $\rho_{X,Y}$.
(*d*) Give a joint distribution that is not the joint distribution given in part (*a*) yet has the same marginal distributions as the joint distribution given in part (*a*).

44 Suppose that X and Y are jointly continuous random variables, $f_{Y|X}(y|x) = I_{(x,x+1)}(y)$, and $f_X(x) = I_{(0,1)}(x)$.
(*a*) Find $\mathscr{E}[Y]$. (*b*) Find $\text{cov}[X, Y]$.
(*c*) Find $P[X+Y<1]$. (*d*) Find $f_{X|Y}(x|y)$.

45 Let (X, Y) have a joint discrete density function

$$f_{X,Y}(x, y)$$
$$= p_1^x(1-p_1)^{1-x}p_2^y(1-p_2)^{1-y}[1+\alpha(x-p_1)(y-p_2)]I_{\{0,1\}}(x)I_{\{0,1\}}(y),$$

where $0<p_1<1$, $0<p_2<1$, and $-1 \le \alpha \le 1$. Prove or disprove: X and Y are independent if and only if they are uncorrelated.

**46* Let (X, Y) be jointly discrete random variables such that each X and Y have at most two mass points. Prove or disprove: X and Y are independent if and only if they are uncorrelated.

V

DISTRIBUTIONS OF FUNCTIONS OF RANDOM VARIABLES

1 INTRODUCTION AND SUMMARY

As the title of this chapter indicates, we are interested in finding the distributions of functions of random variables. More precisely, for given random variables, say $X_1, X_2, \ldots, X_n$, and given functions of the n given random variables, say $g_1(\cdot, \ldots, \cdot), g_2(\cdot, \ldots, \cdot), \ldots, g_k(\cdot, \ldots, \cdot)$, we want, in general, to find the joint distribution of $Y_1, Y_2, \ldots, Y_k$, where $Y_j = g_j(X_1, \ldots, X_n), j = 1, 2, \ldots, k$. If the joint density of the random variables $X_1, X_2, \ldots, X_n$ is given, then theoretically at least, we can find the joint distribution of $Y_1, Y_2, \ldots, Y_k$. This follows since the joint cumulative distribution function of $Y_1, \ldots, Y_k$ satisfies the following:

$$\begin{aligned} F_{Y_1, \ldots, Y_k}(y_1, \ldots, y_k) &= P[Y_1 \le y_1; \ldots; Y_k \le y_k] \\ &= P[g_1(X_1, \ldots, X_n) \le y_1; \ldots; g_k(X_1, \ldots, X_n) \le y_k] \end{aligned}$$

for fixed $y_1, \ldots, y_k$, which is the probability of an event described in terms of $X_1, \ldots, X_n$, and theoretically such a probability can be determined by integrating or summing the joint density over the region corresponding to the event. The problem is that in general one cannot easily evaluate the desired probability for each $y_1, \ldots, y_k$. One of the important problems of statistical inference, the estimation of parameters, provides us with an example of a problem in which it is useful to be able to find the distribution of a function of joint random variables.

In this chapter three techniques for finding the distribution of functions of random variables will be presented. These three techniques are called (i) the *cumulative-distribution-function technique*, alluded to above and discussed in Sec. 3, (ii) the *moment-generating-function technique*, considered in Sec. 4, and (iii) the *transformation technique*, considered in Secs. 5 and 6. A number of important examples are given, including the distribution of sums of independent random variables (in Subsec. 4.2) and the distribution of the minimum and maximum (in Subsec. 3.2). Presentation of other important derived distributions is deferred until later chapters. For instance, the distributions of *chi-square*, *Student's t*, and *F*, all derived from sampling from a normal distribution, are given in Sec. 4 of the next chapter.

Preceding the presentation of the techniques for finding the distribution of functions of random variables is a discussion, given in Sec. 2, of expectations of functions of random variables. As one might suspect, an expectation, for example, the mean or the variance, of a function of given random variables can sometimes be expressed in terms of expectations of the given random variables. If such is the case and one is only interested in certain expectations, then it is not necessary to solve the problem of finding the distribution of the function of the given random variables. One important function of given random variables is their sum, and in Subsec. 2.2 the mean and variance of a sum of given random variables are derived.

We have remarked several times in past chapters that our intermediate objective was the understanding of *distribution theory*. This chapter provides us with a presentation of *distribution theory* at a level that is deemed adequate for the understanding of the statistical concepts that are given in the remainder of this book.

2 EXPECTATIONS OF FUNCTIONS OF RANDOM VARIABLES

2.1 Expectation Two Ways

An expectation of a function of a set of random variables can be obtained two different ways. To illustrate, consider a function of just one random variable, say X. Let $g(\cdot)$ be the function, and set $Y = g(X)$. Since Y is a random variable, $\mathscr{E}[Y]$ is defined (if it exists), and $\mathscr{E}[g(X)]$ is defined (if it exists). For instance, if X and $Y = g(X)$ are continuous random variables, then by definition

$$\mathscr{E}[Y] = \int_{-\infty}^{\infty} y f_Y(y)\, dy, \tag{1}$$

and

$$\mathscr{E}[g(X)] = \int_{-\infty}^{\infty} g(x) f_X(x)\, dx; \tag{2}$$

but $Y = g(X)$, so it seems reasonable that $\mathscr{E}[Y] = \mathscr{E}[g(X)]$. This can, in fact, be proved; although we will not bother to do it. Thus we have two ways of calculating the expectation of $Y = g(X)$; one is to average Y with respect to the density of Y, and the other is to average $g(X)$ with respect to the density of X.

In general, for given random variables $X_1, \ldots, X_n$, let $Y = g(X_1, \ldots, X_n)$; then $\mathscr{E}[Y] = \mathscr{E}[g(X_1, \ldots, X_n)]$, where (for jointly continuous random variables)

$$\mathscr{E}[Y] = \int_{-\infty}^{\infty} y f_Y(y)\, dy \tag{3}$$

and

$$\mathscr{E}[g(X_1, \ldots, X_n)] = \int_{-\infty}^{\infty} \cdots \int_{-\infty}^{\infty} g(x_1, \ldots, x_n) f_{X_1, \ldots, X_n}(x_1, \ldots, x_n)\, dx_1 \ldots dx_n. \tag{4}$$

In practice, one would naturally select that method which makes the calculations easier. One might suspect that Eq. (3) gives the better method of the two since it involves only a single integral whereas Eq. (4) involves a multiple integral. On the other hand, Eq. (3) involves the density of Y, a density that may have to be obtained before integration can proceed.

EXAMPLE 1 Let X be a standard normal random variable, and let $g(x) = x^2$. For $Y = g(X) = X^2$,

$$\mathscr{E}[Y] = \int_{-\infty}^{\infty} y f_Y(y)\, dy,$$

and

$$\mathscr{E}[g(X)] = \mathscr{E}[X^2] = \int_{-\infty}^{\infty} x^2 f_X(x)\, dx.$$

Now

$$\mathscr{E}[X^2] = \int_{-\infty}^{\infty} x^2 \frac{1}{\sqrt{2\pi}} e^{-\frac{1}{2}x^2}\, dx = 1$$

and

$$\mathscr{E}[Y] = \int_0^\infty y \frac{1}{\Gamma(1/2)} (1/2)^{\frac{1}{2}} y^{-\frac{1}{2}} e^{-\frac{1}{2}y} \, dy = 1,$$

using the fact that Y has a gamma distribution with parameters $r = \frac{1}{2}$ and $\lambda = \frac{1}{2}$. (See Example 2 in Subsec. 3.1 below.) ////

2.2 Sums of Random Variables

A simple, yet important, function of several random variables is their sum.

Theorem 1 For random variables $X_1, \ldots, X_n$

$$\mathscr{E}\left[\sum_1^n X_i\right] = \sum_1^n \mathscr{E}[X_i], \tag{5}$$

and

$$\operatorname{var}\left[\sum_1^n X_i\right] = \sum_1^n \operatorname{var}[X_i] + 2 \sum_{i<j} \sum \operatorname{cov}[X_i, X_j]. \tag{6}$$

PROOF That $\mathscr{E}\left[\sum_1^n X_i\right] = \sum_1^n \mathscr{E}[X_i]$ follows from a property of expectation (see the last Remark in Subsec. 4.1 of Chap. IV).

$$\begin{aligned}
\operatorname{var}\left[\sum_1^n X_i\right] &= \mathscr{E}\left[\left(\sum_1^n X_i - \mathscr{E}\left[\sum_1^n X_i\right]\right)^2\right] = \mathscr{E}\left[\left(\sum_1^n (X_i - \mathscr{E}[X_i])\right)^2\right] \\
&= \mathscr{E}\left[\sum_{i=1}^n \sum_{j=1}^n (X_i - \mathscr{E}[X_i])(X_j - \mathscr{E}[X_j])\right] \\
&= \sum_{i=1}^n \sum_{j=1}^n \mathscr{E}[(X_i - \mathscr{E}[X_i])(X_j - \mathscr{E}[X_j])] \\
&= \sum_{i=1}^n \operatorname{var}[X_i] + 2 \sum_{i<j} \sum \operatorname{cov}[X_i, X_j].
\end{aligned}$$ ////

Corollary If $X_1, \ldots, X_n$ are uncorrelated random variables, then

$$\operatorname{var}\left[\sum_1^n X_i\right] = \sum_1^n \operatorname{var}[X_i].$$ ////

The following theorem gives a result that is somewhat related to the above theorem inasmuch as its proof, which is left as an exercise, is similar.

Theorem 2 Let $X_1, \ldots, X_n$ and $Y_1, \ldots, Y_m$ be two sets of random variables, and let $a_1, \ldots, a_n$ and $b_1, \ldots, b_m$ be two sets of constants; then

$$\operatorname{cov}\left[\sum_1^n a_i X_i, \sum_1^m b_j Y_j\right] = \sum_{i=1}^n \sum_{j=1}^m a_i b_j \operatorname{cov}[X_i, Y_j]. \tag{7}$$

////

Corollary If $X_1, \ldots, X_n$ are random variables and $a_1, \ldots, a_n$ are constants, then

$$\begin{aligned} \operatorname{var}\left[\sum_1^n a_i X_i\right] &= \sum_{i=1}^n \sum_{j=1}^n a_i a_j \operatorname{cov}[X_i, X_j] \\ &= \sum_{i=1}^n a_i^2 \operatorname{var}[X_i] + \sum_{i \neq j}\sum a_i a_j \operatorname{cov}[X_i, X_j]. \end{aligned} \tag{8}$$

In particular, if $X_1, \ldots, X_n$ are independent and identically distributed random variables with mean μ_X and variance σ_X^2 and if $\bar{X}_n = (1/n) \sum_1^n X_i$, then

$$\mathscr{E}[\bar{X}_n] = \mu_X, \qquad \text{and} \qquad \operatorname{var}[\bar{X}_n] = \frac{\sigma_X^2}{n}. \tag{9}$$

PROOF Let $m = n$, $Y_i = X_i$, and $b_i = a_i$, $i = 1, \ldots, n$ in the above theorem; then

$$\operatorname{var}\left[\sum_1^n a_i X_i\right] = \operatorname{cov}\left[\sum_1^n a_i X_i, \sum_1^m b_j Y_j\right],$$

and Eq. (8) follows from Eq. (7). To obtain the variance part of Eq. (9) from Eq. (8), set $a_i = 1/n$ and $\sigma_X^2 = \operatorname{var}[X_i]$. The mean part of Eq. (9) is routinely derived as

$$\mathscr{E}[\bar{X}_n] = \frac{1}{n}\mathscr{E}\left[\sum_1^n X_i\right] = \frac{1}{n}\sum_1^n \mathscr{E}[X_i] = \frac{1}{n}\sum_1^n \mu_X = \mu_X. \qquad ////$$

Corollary If X_1 and X_2 are two random variables, then

$$\operatorname{var}[X_1 \pm X_2] = \operatorname{var}[X_1] + \operatorname{var}[X_2] \pm 2\operatorname{cov}[X_1, X_2]. \tag{10}$$

////

Equation (10) gives the variance of the sum or the difference of two random variables. Clearly

$$\mathscr{E}[X_1 \pm X_2] = \mathscr{E}[X_1] \pm \mathscr{E}[X_2]. \tag{11}$$

2.3 Product and Quotient

In the above subsection the mean and variance of the sum and difference of two random variables were obtained. It was found that the mean and variance of the sum or difference of random variables X and Y could be expressed in terms of the means, variances, and covariance of X and Y. We consider now the problem of finding the first two moments of the product and quotient of X and Y.

Theorem 3 Let X and Y be two random variables for which var $[XY]$ exists; then

$$\mathscr{E}[XY] = \mu_X \mu_Y + \text{cov}\,[X, Y], \tag{12}$$

and

$$\begin{aligned}\text{var}\,[XY] &\\ = \mu_Y^2 \,\text{var}\,[X] &+ \mu_X^2 \,\text{var}\,[Y] + 2\mu_X \mu_Y \,\text{cov}\,[X, Y]\\ &- (\text{cov}\,[X, Y])^2 + \mathscr{E}[(X - \mu_X)^2(Y - \mu_Y)^2] +\\ &2\mu_Y \mathscr{E}[(X - \mu_X)^2(Y - \mu_Y)] + 2\mu_X \mathscr{E}[(X - \mu_X)(Y - \mu_Y)^2].\end{aligned} \tag{13}$$

PROOF

$$XY = \mu_X \mu_Y + (X - \mu_X)\mu_Y + (Y - \mu_Y)\mu_X + (X - \mu_X)(Y - \mu_Y).$$

Calculate $\mathscr{E}[XY]$ and $\mathscr{E}[(XY)^2]$ to get the desired results. ////

Corollary If X and Y are independent, $\mathscr{E}[XY] = \mu_X \mu_Y$, and var $[XY] = \mu_Y^2$ var $[X] + \mu_X^2$ var $[Y]$ + var $[X]$ var $[Y]$.

PROOF If X and Y are independent,

$$\begin{aligned}\mathscr{E}[(X - \mu_X)^2(Y - \mu_Y)^2] &= \mathscr{E}[(X - \mu_X)^2]\mathscr{E}[(Y - \mu_Y)^2]\\ &= \text{var}\,[X]\,\text{var}\,[Y],\\ \mathscr{E}[(X - \mu_X)^2(Y - \mu_Y)] &= \mathscr{E}[(X - \mu_X)^2]\mathscr{E}[Y - \mu_Y] = 0,\end{aligned}$$

and

$$\mathscr{E}[(X - \mu_X)(Y - \mu_Y)^2] = 0. \qquad ////$$

Note that the mean of the product can be expressed in terms of the means and covariance of X and Y but the variance of the product requires higher-order moments.

In general, there are no simple exact formulas for the mean and variance of the quotient of two random variables in terms of moments of the two random variables; however, there are approximate formulas which are sometimes useful.

Theorem 4

$$\mathscr{E}\left[\frac{X}{Y}\right] \approx \frac{\mu_X}{\mu_Y} - \frac{1}{\mu_Y^2}\operatorname{cov}[X, Y] + \frac{\mu_X}{\mu_Y^3}\operatorname{var}[Y], \tag{14}$$

and

$$\operatorname{var}\left[\frac{X}{Y}\right] \approx \left(\frac{\mu_X}{\mu_Y}\right)^2 \left(\frac{\operatorname{var}[X]}{\mu_X^2} + \frac{\operatorname{var}[Y]}{\mu_Y^2} - \frac{2\operatorname{cov}[X, Y]}{\mu_X \mu_Y}\right). \tag{15}$$

PROOF To find the approximate formula for $\mathscr{E}[X/Y]$, consider the Taylor series expansion of x/y expanded about (μ_X, μ_Y); drop all terms of order higher than 2, and then take the expectation of both sides. The approximate formula for var $[X/Y]$ is similarly obtained by expanding in a Taylor series and retaining only second-order terms. ////

Two comments are in order: First, it is not unusual that the mean and variance of the quotient X/Y do not exist even though the moments of X and Y do exist. (See Examples 5, 23, and 24.) Second, the method of proof of Theorem 4 can be used to find approximate formulas for the mean and variance of functions of X and Y other than the quotient. For example,

$$\mathscr{E}[g(X, Y)] \approx g(\mu_X, \mu_Y) + \frac{1}{2}\operatorname{var}[X] \frac{\partial^2}{\partial x^2} g(x, y)\bigg|_{\mu_X, \mu_Y}$$
$$+ \frac{1}{2}\operatorname{var}[Y] \frac{\partial^2}{\partial y^2} g(x, y)\bigg|_{\mu_X, \mu_Y} + \operatorname{cov}[X, Y] \frac{\partial^2}{\partial y\, \partial x} g(x, y)\bigg|_{\mu_X, \mu_Y}, \tag{16}$$

and

$$\operatorname{var}[g(X, Y)] \approx \operatorname{var}[X]\left\{\frac{\partial}{\partial x} g(x, y)\bigg|_{\mu_X, \mu_Y}\right\}^2 + \operatorname{var}[Y]\left\{\frac{\partial}{\partial y} g(x, y)\bigg|_{\mu_X, \mu_Y}\right\}^2$$
$$+ 2\operatorname{cov}[X, Y]\left\{\frac{\partial}{\partial x} g(x, y)\bigg|_{\mu_X, \mu_Y} \cdot \frac{\partial}{\partial y} g(x, y)\bigg|_{\mu_X, \mu_Y}\right\}. \tag{17}$$

3 CUMULATIVE-DISTRIBUTION-FUNCTION TECHNIQUE

3.1 Description of Technique

If the joint distribution of random variables $X_1, \ldots, X_n$ is given, then, theoretically, the joint distribution of random variables of $Y_1, \ldots, Y_k$ can be determined, where $Y_j = g_j(X_1, \ldots, X_n)$, $j = 1, \ldots, k$ for given functions $g_1(\cdot, \ldots, \cdot), \ldots,$

$g_k(\cdot, \ldots, \cdot)$. By definition, the joint cumulative distribution function of $Y_1, \ldots, Y_k$ is $F_{Y_1, \ldots, Y_k}(y_1, \ldots, y_k) = P[Y_1 \leq y_1; \ldots; Y_k \leq y_k]$. But for each $y_1, \ldots, y_k$ the event $\{Y_1 \leq y_1; \ldots; Y_k \leq y_k\} \equiv \{g_1(X_1, \ldots, X_n) \leq y_1; \ldots; g_k(X_1, \ldots, X_n) \leq y_k\}$. This latter event is an event described in terms of the given functions $g_1(\cdot, \ldots, \cdot), \ldots, g_k(\cdot, \ldots, \cdot)$ and the given random variables $X_1, \ldots, X_n$. Since the joint distribution of $X_1, \ldots, X_n$ is assumed given, presumably the probability of event $\{g_1(X_1, \ldots, X_n) \leq y_1; \ldots; g_k(X_1, \ldots, X_n) \leq y_k\}$ can be calculated and consequently $F_{Y_1, \ldots, Y_k}(\cdot, \ldots, \cdot)$ determined. The above described technique for deriving the joint distribution of $Y_1, \ldots, Y_k$ will be called the *cumulative-distribution-function technique.*

An important special case arises if $k = 1$; then there is only one function, say $g(X_1, \ldots, X_n)$, of the given random variables for which one needs to derive the distribution.

EXAMPLE 2 Let there be only one given random variable, say X, which has a standard normal distribution. Suppose the distribution of $Y = g(X) = X^2$ is desired.

$$
\begin{aligned}
&F_Y(y) \\
&= P[Y \leq y] = P[X^2 \leq y] = P[-\sqrt{y} \leq X \leq \sqrt{y}] = \Phi(\sqrt{y}) - \Phi(-\sqrt{y}) \\
&= 2\int_0^{\sqrt{y}} \phi(u)\,du = 2\int_0^{\sqrt{y}} \frac{1}{\sqrt{2\pi}} e^{-\frac{1}{2}u^2}\,du \\
&= \frac{2}{\sqrt{2\pi}} \int_0^y \frac{1}{2\sqrt{z}} e^{-\frac{1}{2}z}\,dz = \int_0^y \frac{1}{\Gamma(\frac{1}{2})} \frac{1}{\sqrt{2z}} e^{-\frac{1}{2}z}\,dz, \quad \text{for } y > 0,
\end{aligned}
$$

which can be recognized as the cumulative distribution function of a gamma distribution with parameters $r = \frac{1}{2}$ and $\lambda = \frac{1}{2}$. ////

Other applications of the cumulative-distribution-function technique expounded above are given in the following three subsections.

3.2 Distribution of Minimum and Maximum

Let $X_1, \ldots, X_n$ be n given random variables. Define $Y_1 = \min[X_1, \ldots, X_n]$ and $Y_n = \max[X_1, \ldots, X_n]$. To be certain to understand the meaning of $Y_n = \max[X_1, \ldots, X_n]$, recall that each X_i is a function with domain Ω, the sample space of a random experiment. For each $\omega \in \Omega$, $X_i(\omega)$ is some real number. Now Y_n is to be a random variable; that is, for each ω, $Y_n(\omega)$ is to be some real number. As defined, $Y_n(\omega) = \max[X_1(\omega), \ldots, X_n(\omega)]$; that is, for a given ω, $Y_n(\omega)$ is the largest of the real numbers $X_1(\omega), \ldots, X_n(\omega)$.

The distributions of Y_1 and Y_n are desired. $F_{Y_n}(y) = P[Y_n \le y] = P[X_1 \le y; \ldots; X_n \le y]$ since the largest of the X_i's is less than or equal to y if and only if all the X_i's are less than or equal to y. Now, if the X_i's are assumed independent, then

$$P[X_1 \le y; \ldots; X_n \le y] = \prod_{i=1}^{n} P[X_i \le y] = \prod_{i=1}^{n} F_{X_i}(y);$$

so the distribution of $Y_n = \max[X_1, \ldots, X_n]$ can be expressed in terms of the marginal distributions of $X_1, \ldots, X_n$. If in addition it is assumed that all the $X_1, \ldots, X_n$ have the same cumulative distribution, say $F_X(\cdot)$, then

$$\prod_{i=1}^{n} F_{X_i}(y) = [F_X(y)]^n.$$

We have proved Theorem 5.

Theorem 5 If $X_1, \ldots, X_n$ are independent random variables and $Y_n = \max[X_1, \ldots, X_n]$, then

$$F_{Y_n}(y) = \prod_{i=1}^{n} F_{X_i}(y). \tag{18}$$

If $X_1, \ldots, X_n$ are independent and identically distributed with common cumulative distribution function $F_X(\cdot)$, then

$$F_{Y_n}(y) = [F_X(y)]^n. \tag{19}$$

////

Corollary If $X_1, \ldots, X_n$ are independent identically distributed continuous random variables with common probability density function $f_X(\cdot)$ and cumulative distribution function $F_X(\cdot)$, then

$$f_{Y_n}(y) = n[F_X(y)]^{n-1} f_X(y). \tag{20}$$

PROOF

$$f_{Y_n}(y) = \frac{d}{dy} F_{Y_n}(y) = n[F_X(y)]^{n-1} f_X(y). \qquad ////$$

Similarly,

$$F_{Y_1}(y) = P[Y_1 \le y] = 1 - P[Y_1 > y] = 1 - P[X_1 > y; \ \ldots; \ X_n > y]$$

since Y_1 is greater than y if and only if every $X_i > y$. And if $X_1, \ldots, X_n$ are independent, then

$$1 - P[X_1 > y; \ldots; X_n > y] = 1 - \prod_{i=1}^{n} P[X_i > y] = 1 - \prod_{i=1}^{n} [1 - F_{X_i}(y)].$$

If further it is assumed that $X_1, \ldots, X_n$ are identically distributed with common cumulative distribution function $F_X(\cdot)$, then

$$1 - \prod_{i=1}^{n} [1 - F_{X_i}(y)] = 1 - [1 - F_X(y)]^n,$$

and we have proved Theorem 6.

Theorem 6 If $X_1, \ldots, X_n$ are independent random variables and $Y_1 = \min [X_1, \ldots, X_n]$, then

$$F_{Y_1}(y) = 1 - \prod_{i=1}^{n} [1 - F_{X_i}(y)]. \tag{21}$$

And if $X_1, \ldots, X_n$ are independent and identically distributed with common cumulative distribution function $F_X(\cdot)$, then

$$F_{Y_1}(y) = 1 - [1 - F_X(y)]^n. \tag{22}$$

////

Corollary If $X_1, \ldots, X_n$ are independent identically distributed continuous random variables with common probability density $f_X(\cdot)$ and cumulative distribution $F_X(\cdot)$, then

$$f_{Y_1}(y) = n[1 - F_X(y)]^{n-1} f_X(y). \tag{23}$$

PROOF

$$f_{Y_1}(y) = \frac{d}{dy} F_{Y_1}(y) = n[1 - F_X(y)]^{n-1} f_X(y).$$ ////

EXAMPLE 3 Suppose that the life of a certain light bulb is exponentially distributed with mean 100 hours. If 10 such light bulbs are installed simultaneously, what is the distribution of the life of the light bulb that fails first, and what is its expected life? Let X_i denote the life of the ith light bulb; then $Y_1 = \min [X_1, \ldots, X_{10}]$ is the life of the light bulb that fails first. Assume that the X_i's are independent.

Now $f_{X_i}(x) = \frac{1}{100} e^{-\frac{1}{100}x} I_{(0, \infty)}(x)$, and

$$F_{X_i}(x) = (1 - e^{-\frac{1}{100}x}) I_{(0, \infty)}(x);$$

so

$$\begin{aligned} f_{Y_1}(y) &= 10(e^{-\frac{1}{100}y})^{10-1}(\tfrac{1}{100} e^{-\frac{1}{100}y}) I_{(0, \infty)}(y) \\ &= \tfrac{10}{100} e^{-\frac{10}{100}y} I_{(0, \infty)}(y), \end{aligned}$$

which is an exponential distribution with parameter $\lambda = \frac{1}{10}$; hence $\mathscr{E}[Y_1] = 1/\lambda = 10$. ////

3.3 Distribution of Sum and Difference of Two Random Variables

Theorem 7 Let X and Y be jointly distributed continuous random variables with density $f_{X,Y}(x, y)$, and let $Z = X + Y$ and $V = X - Y$. Then,

$$f_Z(z) = \int_{-\infty}^{\infty} f_{X,Y}(x, z - x)\, dx = \int_{-\infty}^{\infty} f_{X,Y}(z - y, y)\, dy, \tag{24}$$

and

$$f_V(v) = \int_{-\infty}^{\infty} f_{X,Y}(x, x - v)\, dx = \int_{-\infty}^{\infty} f_{X,Y}(v + y, y)\, dy. \tag{25}$$

PROOF We will prove only the first part of Eq. (24); the others are proved in an analogous manner.

$$\begin{aligned} F_Z(z) = P[Z \leq z] = P[X + Y \leq z] &= \iint\limits_{x+y \leq z} f_{X,Y}(x, y)\, dx\, dy \\ &= \int_{-\infty}^{\infty} \left[\int_{-\infty}^{z-x} f_{X,Y}(x, y)\, dy\right] dx \\ &= \int_{-\infty}^{\infty} \left[\int_{-\infty}^{z} f_{X,Y}(x, u - x)\, du\right] dx \end{aligned}$$

by making the substitution $y = u - x$.

Now

$$\begin{aligned} f_Z(z) = \frac{dF_Z(z)}{dz} &= \frac{d}{dz}\left\{\int_{-\infty}^{z} \left[\int_{-\infty}^{\infty} f_{X,Y}(x, u - x)\, dx\right] du\right\} \\ &= \int_{-\infty}^{\infty} f_{X,Y}(x, z - x)\, dx. \end{aligned}$$

////

Corollary If X and Y are independent continuous random variables and $Z = X + Y$, then

$$f_Z(z) = f_{X+Y}(z) = \int_{-\infty}^{\infty} f_Y(z-x)f_X(x)\,dx = \int_{-\infty}^{\infty} f_X(z-y)f_Y(y)\,dy. \tag{26}$$

PROOF Equation (26) follows immediately from independence and Eq. (24); however, we will give a direct proof using a conditional distribution formula. [See Eq. (11) of Chap. IV.]

$$\begin{aligned} P[Z \le z] = P[X + Y \le z] &= \int_{-\infty}^{\infty} P[X + Y \le z \mid X = x] f_X(x)\,dx \\ &= \int_{-\infty}^{\infty} P[x + Y \le z] f_X(x)\,dx \\ &= \int_{-\infty}^{\infty} F_Y(z-x) f_X(x)\,dx. \end{aligned}$$

Hence,

$$\begin{aligned} f_Z(z) = \frac{dF_Z(z)}{dz} &= \frac{d}{dz}\left[\int_{-\infty}^{\infty} F_Y(z-x)f_X(x)\,dx\right] \\ &= \int_{-\infty}^{\infty} \frac{dF_Y(z-x)}{dz} f_X(x)\,dx \\ &= \int_{-\infty}^{\infty} f_Y(z-x)f_X(x)\,dx. \end{aligned}$$ ////

Remark The formula given in Eq. (26) is often called the *convolution* formula. In mathematical analysis, the function $f_Z(\cdot)$ is called the *convolution* of the functions $f_Y(\cdot)$ and $f_X(\cdot)$. ////

EXAMPLE 4 Suppose that X and Y are independent and identically distributed with density $f_X(x) = f_Y(x) = I_{(0,1)}(x)$. Note that since both X and Y assume values between 0 and 1, $Z = X + Y$ assumes values between 0 and 2.

$$\begin{aligned} f_Z(z) &= \int_{-\infty}^{\infty} f_Y(z-x)f_X(x)\,dx = \int_{-\infty}^{\infty} I_{(0,1)}(z-x)I_{(0,1)}(x)\,dx \\ &= \int_{-\infty}^{\infty} \{I_{(0,z)}(x)I_{(0,1)}(z) + I_{(z-1,1)}(x)I_{[1,2)}(z)\}\,dx \\ &= I_{(0,1)}(z)\int_0^z dx + I_{[1,2)}(z)\int_{z-1}^{1} dx \\ &= zI_{(0,1)}(z) + (2-z)I_{[1,2)}(z). \end{aligned}$$ ////

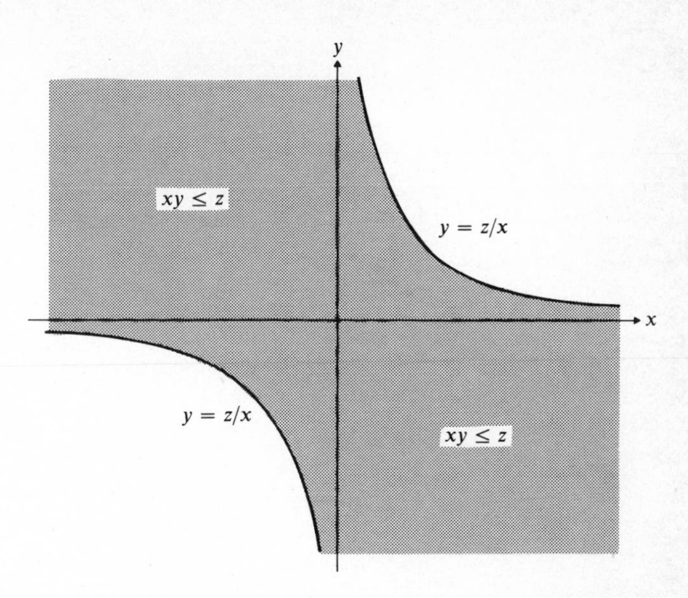

FIGURE 1

3.4 Distribution of Product and Quotient

Theorem 8 Let X and Y be jointly distributed continuous random variables with density $f_{X,Y}(x, y)$, and let $Z = XY$ and $U = X/Y$; then

$$f_Z(z) = \int_{-\infty}^{\infty} \frac{1}{|x|} f_{X,Y}\left(x, \frac{z}{x}\right) dx = \int_{-\infty}^{\infty} \frac{1}{|y|} f_{X,Y}\left(\frac{z}{y}, y\right) dy, \tag{27}$$

and

$$f_U(u) = \int_{-\infty}^{\infty} |y| f_{X,Y}(uy, y)\, dy. \tag{28}$$

PROOF Again, only the first part of Eq. (27) will be proved. (See Fig. 1 for $z > 0$.)

$$F_Z(z) = P[Z \le z] = \iint_{xy \le z} f_{X,Y}(x, y)\, dx\, dy$$

$$= \int_{-\infty}^{0} \left[\int_{z/x}^{\infty} f_{X,Y}(x, y)\, dy\right] dx + \int_{0}^{\infty} \left[\int_{-\infty}^{z/x} f_{X,Y}(x, y)\, dy\right] dx,$$

which on making the substitution $u = xy$

$$= \int_{-\infty}^{0} \left[\int_{z}^{-\infty} f_{X,Y}\left(x, \frac{u}{x}\right) \frac{du}{x} \right] dx + \int_{0}^{\infty} \left[\int_{-\infty}^{z} f_{X,Y}\left(x, \frac{u}{x}\right) \frac{du}{x} \right] dx$$

$$= \int_{-\infty}^{z} \left[\int_{-\infty}^{0} \frac{1}{-x} f_{X,Y}\left(x, \frac{u}{x}\right) dx \right] du + \int_{-\infty}^{z} \left[\int_{0}^{\infty} \frac{1}{x} f_{X,Y}\left(x, \frac{u}{x}\right) dx \right] du$$

$$= \int_{-\infty}^{z} \left[\int_{-\infty}^{\infty} \frac{1}{|x|} f_{X,Y}\left(x, \frac{u}{x}\right) dx \right] du;$$

hence

$$f_Z(z) = \frac{dF_Z(z)}{dz}$$

$$= \int_{-\infty}^{\infty} \frac{1}{|x|} f_{X,Y}\left(x, \frac{z}{x}\right) dx. \qquad ////$$

EXAMPLE 5 Suppose X and Y are independent random variables, each uniformly distributed over the interval (0, 1). Let $Z = XY$ and $U = X/Y$.

$$f_Z(z) = \int_{-\infty}^{\infty} \frac{1}{|x|} f_{X,Y}\left(x, \frac{z}{x}\right) dx$$

$$= \int_{-\infty}^{\infty} \frac{1}{|x|} I_{(0,1)}(x) I_{(0,1)}\left(\frac{z}{x}\right) dx$$

$$= I_{(0,1)}(z) \int_{0}^{1} \frac{1}{x} I_{(z,1)}(x)\, dx$$

$$= I_{(0,1)}(z) \int_{z}^{1} \frac{1}{x}\, dx = -\log z\, I_{(0,1)}(z).$$

$$f_U(u) = \int_{-\infty}^{\infty} |y| f_{X,Y}(uy, y)\, dy$$

$$= \int_{-\infty}^{\infty} |y| I_{(0,1)}(uy) I_{(0,1)}(y)\, dy \qquad \text{(see Fig. 2)}$$

$$= \int_{-\infty}^{\infty} |y| \{I_{(0,1)}(u) I_{(0,1)}(y) + I_{[1,\infty)}(u) I_{(0,1/u)}(y)\}\, dy$$

$$= I_{(0,1)}(u) \int_{0}^{1} y\, dy + I_{[1,\infty)}(u) \int_{0}^{1/u} y\, dy$$

$$= \frac{1}{2} I_{(0,1)}(u) + \frac{1}{2}\left(\frac{1}{u}\right)^2 I_{[1,\infty)}(u).$$

Note that $\mathscr{E}[X/Y] = \mathscr{E}[U] = \frac{1}{2}\int_0^1 u\, du + \frac{1}{2}\int_1^\infty (1/u)\, du = \infty$, quite different from $\mathscr{E}[X]/\mathscr{E}[Y] = 1$. ////

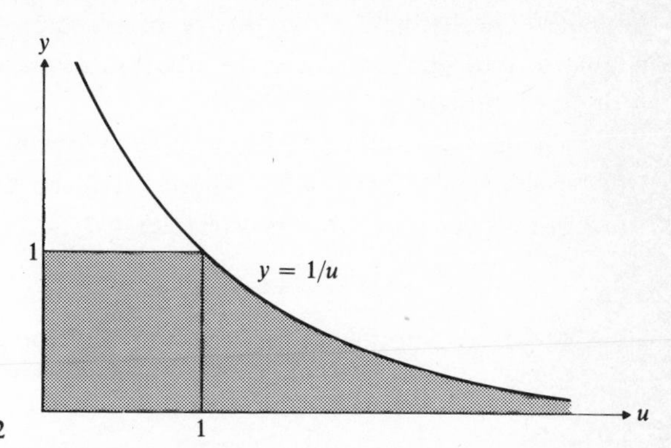

FIGURE 2

4 MOMENT-GENERATING-FUNCTION TECHNIQUE

4.1 Description of Technique

There is another method of determining the distribution of functions of random variables which we shall find to be particularly useful in certain instances. This method is built around the concept of the moment generating function and will be called the *moment-generating-function technique.*

The statement of the problem remains the same. For given random variables $X_1, \ldots, X_n$ with given density $f_{X_1, \ldots, X_n}(x_1, \ldots, x_n)$ and given functions $g_1(\cdot, \ldots, \cdot), \ldots, g_k(\cdot, \ldots, \cdot)$, find the joint distribution of $Y_1 = g_1(X_1, \ldots, X_n)$, $\ldots$, $Y_k = g_k(X_1, \ldots, X_n)$. Now the joint moment generating function of $Y_1, \ldots, Y_k$, if it exists, is

$$m_{Y_1, \ldots, Y_k}(t_1, \ldots, t_k) = \mathscr{E}[e^{t_1 Y_1 + \cdots + t_k Y_k}]$$

$$= \int \cdots \int e^{t_1 g_1(x_1, \ldots, x_n) + \cdots + t_k g_k(x_1, \ldots, x_n)}$$

$$\times f_{X_1, \ldots, X_n}(x_1, \ldots, x_n) \prod_{i=1}^{n} dx_i. \tag{29}$$

If after the integration of Eq. (29) is performed, the resulting function of $t_1, \ldots, t_k$ can be recognized as the joint moment generating function of some known joint distribution, it will follow that $Y_1, \ldots, Y_k$ has that joint distribution by virtue of the fact that a moment generating function, when it exists, is unique and uniquely determines its distribution function.

For $k > 1$, this method will be of limited use to us because we can recognize only a few joint moment generating functions. For $k = 1$, the moment generating function is a function of a single argument, and we should have a better chance of recognizing the resulting moment generating function.

This method is quite powerful in connection with certain techniques of advanced mathematics (the theory of transforms) which, in many instances, enable one to determine the distribution associated with the derived moment generating function.

The most useful application of the moment-generating-function technique will be given in Subsec. 4.2. There it will be used to find the distribution of sums of independent random variables.

EXAMPLE 6 Suppose X has a normal distribution with mean 0 and variance 1. Let $Y = X^2$, and find the distribution of Y.

$$\begin{aligned} m_Y(t) = \mathscr{E}[e^{tY}] &= \int_{-\infty}^{\infty} e^{tx^2} \frac{1}{\sqrt{2\pi}} e^{-\frac{1}{2}x^2} \, dx \\ &= \frac{1}{\sqrt{2\pi}} \int_{-\infty}^{\infty} e^{-\frac{1}{2}x^2(1-2t)} \, dx \\ &= \frac{1}{\sqrt{2\pi}} \cdot \frac{(1-2t)^{-\frac{1}{2}}}{(1-2t)^{-\frac{1}{2}}} \int_{-\infty}^{\infty} e^{-\frac{1}{2}x^2(1-2t)} \, dx \\ &= (1-2t)^{-\frac{1}{2}} = \left(\frac{\frac{1}{2}}{\frac{1}{2}-t}\right)^{\frac{1}{2}} \qquad \text{for} \quad t < \frac{1}{2}, \end{aligned}$$

which we recognize as the moment generating function of a gamma with parameters $r = \frac{1}{2}$ and $\lambda = \frac{1}{2}$. (It is also called a chi-square distribution with one degree of freedom. See Subsec. 4.3 of Chap. VI.) ////

EXAMPLE 7 Let X_1 and X_2 be two independent standard normal random variables. Let $Y_1 = g_1(X_1, X_2) = X_1 + X_2$ and $Y_2 = g_2(X_1, X_2) = X_2 - X_1$. Find the joint distribution of Y_1 and Y_2.

$$\begin{aligned} m_{Y_1, Y_2}(t_1, t_2) &= \mathscr{E}[e^{Y_1 t_1 + Y_2 t_2}] \\ &= \mathscr{E}[e^{(X_1+X_2)t_1 + (X_2 - X_1)t_2}] \\ &= \mathscr{E}[e^{X_1(t_1-t_2) + X_2(t_1+t_2)}] \\ &= \mathscr{E}[e^{X_1(t_1-t_2)}]\mathscr{E}[e^{X_2(t_1+t_2)}] \\ &= m_{X_1}(t_1 - t_2) m_{X_2}(t_1 + t_2) \\ &= \exp\frac{(t_1-t_2)^2}{2} \exp\frac{(t_1+t_2)^2}{2} \\ &= \exp(t_1^2 + t_2^2) = \exp\frac{2t_1^2}{2} \exp\frac{2t_2^2}{2} \\ &= m_{Y_1}(t_1) m_{Y_2}(t_2). \end{aligned}$$

We note that Y_1 and Y_2 are independent random variables (by Theorem 10 of Chap. IV) and each has a normal distribution with mean 0 and variance 2. ////

In the above example we were able to manipulate expectations and avoid performing an integration to find the desired joint moment generating function. In the following example the integration will have to be performed.

EXAMPLE 8 Let X_1 and X_2 be two independent standard normal random variables. Let $Y = (X_2 - X_1)^2/2$, and find the distribution of Y.

$$
\begin{aligned}
m_Y(t) &= \mathscr{E}[\exp Yt] = \mathscr{E}\left[\exp \frac{(X_2 - X_1)^2}{2} t\right] \\
&= \int_{-\infty}^{\infty} \int_{-\infty}^{\infty} \frac{1}{2\pi} \exp\left[\frac{(x_2 - x_1)^2}{2} t - \frac{x_1^2 + x_2^2}{2}\right] dx_1\, dx_2 \\
&= \int_{-\infty}^{\infty} \int_{-\infty}^{\infty} \frac{1}{2\pi} \exp\left\{-\frac{1}{2}[x_1^2(1 - t) + 2x_1 x_2 t + x_2^2(1 - t)]\right\} dx_1\, dx_2 \\
&= \int_{-\infty}^{\infty} \frac{1}{2\pi} \exp\left[-\frac{1}{2} x_2^2(1 - t)\right] \\
&\quad \times \left\{\int_{-\infty}^{\infty} \exp\left[-\frac{1 - t}{2}\left(x_1^2 + \frac{2x_1 x_2 t}{1 - t}\right)\right] dx_1\right\} dx_2 \\
&= \int_{-\infty}^{\infty} \frac{1}{\sqrt{2\pi}} \exp\left[-\frac{x_2^2(1 - t)}{2}\right] \exp \frac{x_2^2 t^2}{2(1 - t)} \\
&\quad \times \frac{1}{\sqrt{1 - t}} \left\{\int_{-\infty}^{\infty} \frac{\sqrt{1 - t}}{\sqrt{2\pi}} \exp\left[-\frac{1 - t}{2}\left(x_1 + \frac{x_2 t}{1 - t}\right)^2\right] dx_1\right\} dx_2 \\
&= \frac{1}{\sqrt{1 - t}} \int_{-\infty}^{\infty} \frac{1}{\sqrt{2\pi}} \exp\left[-\frac{1}{2}\left(1 - t - \frac{t^2}{1 - t}\right) x_2^2\right] dx_2 \\
&= \frac{1}{\sqrt{1 - t}} \cdot \frac{\sqrt{1 - t}}{\sqrt{1 - 2t}} \cdot \frac{\sqrt{1 - 2t}}{\sqrt{1 - t}} \frac{1}{\sqrt{2\pi}} \int_{-\infty}^{\infty} \exp\left(-\frac{1}{2} \frac{1 - 2t}{1 - t} x_2^2\right) dx_2 \\
&= (1 - 2t)^{-\frac{1}{2}} = \left(\frac{\frac{1}{2}}{\frac{1}{2} - t}\right)^{\frac{1}{2}}, \text{ for } t < 1/2,
\end{aligned}
$$

which is the moment generating function of a gamma distribution with parameters $r = \frac{1}{2}$ and $\lambda = \frac{1}{2}$; hence,

$$f_Y(y) = [\sqrt{\tfrac{1}{2}}/\Gamma(\tfrac{1}{2})] y^{-\frac{1}{2}} e^{-y/2} I_{(0, \infty)}(y). \qquad ////$$

4.2 Distribution of Sums of Independent Random Variables

In this subsection we employ the moment-generating-function technique to find the distribution of the sum of independent random variables.

Theorem 9 If $X_1, \ldots, X_n$ are independent random variables and the moment generating function of each exists for all $-h < t < h$ for some $h > 0$, let $Y = \sum_1^n X_i$; then

$$m_Y(t) = \mathscr{E}[\exp \sum X_i t] = \prod_{i=1}^n m_{X_i}(t) \qquad \text{for} \quad -h < t < h.$$

PROOF

$$m_Y(t) = \mathscr{E}[\exp \sum X_i t] = \mathscr{E}\left[\prod_{i=1}^n e^{X_i t}\right]$$
$$= \prod_{i=1}^n \mathscr{E}[e^{X_i t}] = \prod_{i=1}^n m_{X_i}(t)$$

using Theorem 10 of Chap. IV. ////

The power and utility of Theorem 9 becomes apparent if we recall Theorem 7 of Chap. II, which says that a moment generating function, when it exists, determines the distribution function. Thus, if we can recognize $\prod_{i=1}^n m_{X_i}(t)$ as the moment generating function corresponding to a particular distribution, then we have found the distribution of $\sum_1^n X_i$. In the following examples, we will be able to do just that.

EXAMPLE 9 Suppose that $X_1, \ldots, X_n$ are independent Bernoulli random variables; that is, $P[X_i = 1] = p$, and $P[X_i = 0] = 1 - p$. Now

$$m_{X_i}(t) = pe^t + q.$$

So

$$m_{\sum X_i}(t) = \prod_{i=1}^n m_{X_i}(t) = (pe^t + q)^n,$$

the moment generating function of a binomial random variable; hence $\sum_1^n X_i$ has a binomial distribution with parameters n and p. ////

EXAMPLE 10 Suppose that $X_1, \ldots, X_n$ are independent Poisson distributed random variables, X_i having parameter λ_i. Then

$$m_{X_i}(t) = \mathscr{E}[e^{tX_i}] = \exp \lambda_i(e^t - 1),$$

and hence

$$m_{\Sigma X_i}(t) = \prod_{i=1}^{n} m_{X_i}(t) = \prod_{i=1}^{n} \exp \lambda_i(e^t - 1) = \exp \sum \lambda_i(e^t - 1),$$

which is again the moment generating function of a Poisson distributed random variable having parameter $\sum \lambda_i$. So the distribution of a sum of independent Poisson distributed random variables is again a Poisson distributed random variable with a parameter equal to the sum of the individual parameters. ////

EXAMPLE 11 Assume that $X_1, \ldots, X_n$ are independent and identically distributed exponential random variables; then

$$m_{X_i}(t) = \frac{\lambda}{\lambda - t}.$$

So

$$m_{\Sigma X_i}(t) = \prod_{i=1}^{n} m_{X_i}(t) = \left(\frac{\lambda}{\lambda - t}\right)^n,$$

which is the moment generating function of a gamma distribution with parameters n and λ; hence,

$$f_{\Sigma X_i}(x) = \frac{\lambda^n}{\Gamma(n)} x^{n-1} e^{-\lambda x} I_{(0, \infty)}(x),$$

the density of a gamma distribution with parameters n and λ. ////

EXAMPLE 12 Assume that $X_1, \ldots, X_n$ are independent random variables and

$$X_i \sim N(\mu_i, \sigma_i^2);$$

then

$$a_i X_i \sim N(a_i \mu_i, a_i^2 \sigma_i^2),$$

and

$$m_{a_i X_i}(t) = \exp (a_i \mu_i t + \tfrac{1}{2} a_i^2 \sigma_i^2 t^2).$$

Hence

$$m_{\sum a_i X_i}(t) = \prod_{i=1}^{n} m_{a_i X_i}(t) = \exp[(\sum a_i \mu_i)t + \tfrac{1}{2}(\sum a_i^2 \sigma_i^2)t^2],$$

which is the moment generating function of a normal random variable; so

$$\sum_1^n a_i X_i \sim N\left(\sum_1^n a_i \mu_i, \sum_1^n a_i^2 \sigma_i^2\right).$$

The above says that any linear combination (that is, $\sum a_i X_i$) of independent normal random variables is itself a normally distributed random variable. (Actually, any linear combination of jointly normally distributed random variables is normally distributed. Independence is not required.) In particular, if

$$X \sim N(\mu_X, \sigma_X^2), \qquad Y \sim N(\mu_Y, \sigma_Y^2),$$

and X and Y are independent, then

$$X + Y \sim N(\mu_X + \mu_Y, \sigma_X^2 + \sigma_Y^2),$$

and

$$X - Y \sim N(\mu_X - \mu_Y, \sigma_X^2 + \sigma_Y^2).$$

If $X_1, \ldots, X_n$ are independent and identically distributed random variables distributed $N(\mu, \sigma^2)$, then

$$\bar{X}_n = \frac{1}{n} \sum X_i \sim N\left(\mu, \frac{\sigma^2}{n}\right);$$

that is, the sample mean has a (not approximate) normal distribution. ////

In the above examples we found the exact distribution of the sums of certain independent random variables. Other examples, including the important result that the sum of independent identically distributed geometric random variables has a negative binomial distribution, are given in the Problems. One is often more interested in the *average*, that is, $(1/n) \sum_1^n X_i$, than in the sum. Note, however, that if the distribution of the sum is known, then the distribution of the average is readily derivable since

$$F_{(1/n)\sum X_i}(z) = P\left[\frac{1}{n} \sum X_i \le z\right] = P[\sum X_i \le nz] = F_{\sum X_i}(nz). \tag{30}$$

In Examples 9 to 12 above, where we derived the distribution of a sum, we have in essence also derived the distribution of the corresponding average. One of the most important theorems of all probability theory, the central-limit theorem, gives an *approximate* distribution of an average. We will state this theorem next and then again in our discussion of sampling in Chap. VI, where we will outline its proof.

Theorem 10 Central-limit theorem If for each positive integer n, $X_1, \ldots, X_n$ are independent and identically distributed random variables with mean μ_X and variance σ_X^2, then for each z

$$F_{Z_n}(z) \text{ converges to } \Phi(z) \text{ as } n \text{ approaches } \infty, \tag{31}$$

where

$$Z_n = \frac{(\bar{X}_n - \mathscr{E}[\bar{X}_n])}{\sqrt{\operatorname{var}[\bar{X}_n]}} = \frac{\bar{X}_n - \mu_X}{\sigma_X/\sqrt{n}}. \qquad ////$$

We have made use of Eq. (9), which stated that $\mathscr{E}[\bar{X}_n] = \mu_X$ and $\operatorname{var}[\bar{X}_n] = \sigma_X^2/n$. Equation (31) states that for each fixed argument z the value of the cumulative distribution function of Z_n, for $n = 1, 2, \ldots$, converges to the value $\Phi(z)$. [Recall that $\Phi(\cdot)$ is the cumulative distribution function of the standard normal distribution.]

Note what the central-limit theorem says: If you have independent random variables $X_1, \ldots, X_n, \ldots$, each with the same distribution which has a mean and variance, then $\bar{X}_n = (1/n) \sum X_i$ "standardized" by subtracting its mean and then dividing by its standard deviation has a distribution that approaches a standard normal distribution. The key thing to note is that it does not make any difference what common distribution the $X_1, \ldots, X_n, \ldots$ have, as long as they have a mean and variance. A number of useful approximations can be garnered from the central-limit theorem, and they are listed as a corollary.

Corollary If $X_1, \ldots, X_n$ are independent and identically distributed random variables with common mean μ_X and variance σ_X^2, then

$$P\left[a < \frac{\bar{X}_n - \mu_X}{\sigma_X/\sqrt{n}} < b\right] \approx \Phi(b) - \Phi(a), \tag{32}$$

$$P[c < \bar{X}_n < d] \approx \Phi\left(\frac{d - \mu_X}{\sigma_X/\sqrt{n}}\right) - \Phi\left(\frac{c - \mu_X}{\sigma_X/\sqrt{n}}\right), \tag{33}$$

or

$$P\left[r < \sum_1^n X_i < s\right] \approx \Phi\left(\frac{s - n\mu_X}{\sqrt{n}\sigma_X}\right) - \Phi\left(\frac{r - n\mu_X}{\sqrt{n}\sigma_X}\right). \tag{34}$$

////

Equations (32) to (34) give approximate values for the probabilities of certain events described in terms of averages or sums. The practical utility of the central-limit theorem is inherent in these approximations.

At this stage we can conveniently discuss and contrast two terms that are a vital part of a statistician's vocabulary. These two terms are *limiting distribution* and *asymptotic distribution.* A distribution is called a limiting distribution function if it is the limit distribution function of a sequence of distribution functions. Equation (31) provides us with an example; $\Phi(z)$ is the limiting distribution function of the sequence of distribution functions $F_{Z_1}(\cdot)$, $F_{Z_2}(\cdot)$, ..., $F_{Z_n}(\cdot)$, ... Also $\Phi(z)$ is called the limiting distribution of the sequence of random variables $Z_1, Z_2, \ldots, Z_n, \ldots$. On the other hand, an asymptotic distribution of a random variable, say Y_n, in a sequence of random variables $Y_1, Y_2, \ldots Y_n, \ldots$ is any distribution that is approximately equal to the actual distribution of Y_n for large n. As an example [see Eq. (33)], we say that $\bar{X}_n$ has an asymptotic distribution that is a normal distribution with mean μ_X and variance σ_X^2/n. Note that an asymptotic distribution may depend on n whereas a limiting distribution does not (for a limiting distribution the dependence on n was removed in taking the limit). Yet the two terms are closely related since it was precisely the fact that the sequence $Z_1, Z_2, \ldots, Z_n, \ldots$ had limiting standard normal distribution that allowed us to say that $\bar{X}_n$ had an asymptotic normal distribution with mean μ_X and variance σ_X^2/n. The idea is that if the distribution of Z_n is converging to $\Phi(z)$, then for large n the distribution of Z_n must be approximately distributed $N(0, 1)$. But if $Z_n = (\bar{X}_n - \mu_X)/(\sigma_X/\sqrt{n})$ is approximately distributed $N(0, 1)$, then $\bar{X}_n$ is approximately distributed $N(\mu_X, \sigma_X^2/n)$.

In concluding this section we give two further examples concerning sums. The first shows how expressing one random variable as a sum of other simpler random variables is often a useful ploy. The second shows how the distribution of a sum can be obtained even though the number of terms in the sum is also a random variable, something that occasionally occurs in practice.

EXAMPLE 13 Consider n repeated independent trials, each of which has possible outcomes $\mathscr{s}_1, \ldots, \mathscr{s}_{k+1}$. Let p_j denote the probability of outcome $\mathscr{s}_j$ on a particular trial, and let X_j denote the number of the n trials resulting in outcome $\mathscr{s}_j$, $j = 1, \ldots, k+1$. We saw that $(X_1, \ldots, X_k)$ had a multinomial distribution. Now let

$$Z_{j\alpha} = \begin{cases} 1 & \text{if } \alpha\text{th trial results in outcome } \mathscr{s}_j \\ 0 & \text{otherwise;} \end{cases}$$

then $X_j = \sum_{\alpha=1}^{n} Z_{j\alpha}$. Now suppose we want to find $\text{cov}[X_i, X_j]$. Intuitively, we might suspect that such covariance is negative since when one of the random variables is large another tends to be small.

$$\text{cov}[X_i, X_j] = \text{cov}\left[\sum_{\beta=1}^{n} Z_{i\beta}, \sum_{\alpha=1}^{n} Z_{j\alpha}\right] = \sum_{\beta=1}^{n}\sum_{\alpha=1}^{n} \text{cov}[Z_{i\beta}, Z_{j\alpha}]$$

by Theorem 2. Now if $\alpha \neq \beta$, then $Z_{i\beta}$ and $Z_{j\alpha}$ are independent since they correspond to different trials, which are independent. Hence

$$\sum_{\beta=1}^{n}\sum_{\alpha=1}^{n} \text{cov}[Z_{i\beta}, Z_{j\alpha}] = \sum_{\alpha=1}^{n} \text{cov}[Z_{i\alpha}, Z_{j\alpha}].$$

But $\text{cov}[Z_{i\alpha}, Z_{j\alpha}] = \mathscr{E}[Z_{i\alpha}Z_{j\alpha}] - \mathscr{E}[Z_{i\alpha}]\mathscr{E}[Z_{j\alpha}]$, and $\mathscr{E}[Z_{i\alpha}Z_{j\alpha}] = 0$ since at least one of $Z_{i\alpha}$ and $Z_{j\alpha}$ must be 0. Now $\mathscr{E}[Z_{i\alpha}] = p_i$, and $\mathscr{E}[Z_{j\alpha}] = p_j$; so $\text{cov}[X_i, X_j] = -np_i p_j$. ////

EXAMPLE 14 Let $X_1, \ldots, X_n, \ldots$ be a sequence of independent and identically distributed random variables with mean μ_X and variance σ_X^2. Let N be an integer-valued random variable, and define $S_N = \sum_{i=1}^{N} X_i$; that is, S_N is the sum of the first N X_i's, where N is a random variable as are the X_i's. Thus S_N is a sum of a *random number* of *random variables*. Let us assume that N is independent of the X_i's. Then $\mathscr{E}[S_N] = \mathscr{E}[\mathscr{E}[S_N | N]]$ by Eq. (26) of Chap. IV. But $\mathscr{E}[S_N | N = n] = \mathscr{E}[X_1 + \cdots + X_n] = n\mu_X$; so $\mathscr{E}[S_N | N] = N\mu_X$, and $\mathscr{E}[\mathscr{E}[S_N | N]] = \mathscr{E}[N\mu_X] = \mu_X \mathscr{E}[N] = \mu_N \mu_X$. Similarly, using Theorem 7 of Chap. IV,

$$\begin{aligned}
\text{var}[S_N] &= \mathscr{E}[\text{var}[S_N | N]] + \text{var}[\mathscr{E}[S_N | N]] \\
&= \mathscr{E}[N\sigma_X^2] + \text{var}[N\mu_X] \\
&= \sigma_X^2 \mathscr{E}[N] + \mu_X^2 \text{var}[N] \\
&= \mu_N \cdot \sigma_X^2 + \sigma_N^2 \cdot \mu_X^2.
\end{aligned}$$

Suppose now that N has a geometric distribution [see Eq. (14) of Chap. III] with parameter p, X_i has an exponential distribution with parameter λ, and we are interested in the distribution of S_N. Further assume independence of N and the X_i's. Now, for $z > 0$,

$$P[S_N \leq z] = \sum_{n=1}^{\infty} P[S_N \leq z | N = n] P[N = n]$$

(by using the fact that a sum of independent and identically distributed exponential random variables has a gamma distribution)

$$= \sum_{n=1}^{\infty} \left[\int_0^z \frac{1}{\Gamma(n)} \lambda^n u^{n-1} e^{-\lambda u} \, du \right] p(1-p)^{n-1}$$

$$= \int_0^z \lambda p e^{-\lambda u} \sum_{n=1}^{\infty} \frac{[\lambda u(1-p)]^{n-1}}{(n-1)!} \, du$$

$$= \int_0^z \lambda p e^{-\lambda u} e^{(1-p)\lambda u} \, du = \lambda p \int_0^z e^{-\lambda p u} \, du = 1 - e^{-\lambda p z}.$$

That is, S_N has an exponential distribution with parameter $p\lambda$. Recall that [see Eq. (14) of Chap. III] $\mathscr{E}[N] = 1/p$ and var $[N] = (1-p)/p^2$; also, $\mathscr{E}[X] = 1/\lambda$, and var $[X] = 1/\lambda^2$. So, as a check of the formulas for the mean and variance derived above, note that

$$\mathscr{E}[S_N] = \mu_N \mu_X = \frac{1}{p} \cdot \frac{1}{\lambda} = \frac{1}{p\lambda}$$

and

$$\text{var } [S_N] = \mu_N \sigma_X^2 + \sigma_N^2 \mu_X^2 = \frac{1}{p} \frac{1}{\lambda^2} + \frac{1-p}{p^2} \frac{1}{\lambda^2} = \frac{1}{(p\lambda)^2},$$

which are the mean and variance, respectively, of an exponential distribution with parameter $p\lambda$. ////

5 THE TRANSFORMATION $Y = g(X)$

The last of our three techniques for finding the distribution of functions of given random variables is the *transformation technique*. It is discussed in this section for the special case of finding the distribution of a function of a unidimensional random variable. That is, for a given random variable X we seek the distribution of $Y = g(X)$ for some function $g(\cdot)$. Discussion of the general case is deferred until Sec. 6 below. Both the notation $Y = g(X)$ and the notation $y = g(x)$ will appear in the ensuing paragraphs; $y = g(x)$ is the usual notation for the function or transformation specified by $g(\cdot)$, and $Y = g(X)$ defines the random variable Y as the function $g(\cdot)$ of the random variable X.

5.1 Distribution of $Y = g(X)$

A random variable X may be transformed by some function $g(\cdot)$ to define a new random variable Y. The density of Y, $f_Y(y)$, will be determined by the transformation $g(\cdot)$ together with the density $f_X(x)$ of X.

First, if X is a discrete random variable with mass points $x_1, x_2, \ldots$, then the distribution of $Y = g(X)$ is determined directly by the laws of probability. If X takes on the values $x_1, x_2, \ldots$ with probabilities $f_X(x_1), f_X(x_2), \ldots$, then the possible values of Y are determined by substituting the successive values of X in $g(\cdot)$. It may be that several values of X give rise to the same value of Y. The probability that Y takes on a given value, say y_j, is

$$f_Y(y_j) = \sum_{\{i:g(x_i)=y_j\}} f_X(x_i). \tag{35}$$

EXAMPLE 15 Suppose X takes on the values 0, 1, 2, 3, 4, 5 with probabilities $f_X(0), f_X(1), f_X(2), f_X(3), f_X(4)$, and $f_X(5)$. If $Y = g(X) = (X - 2)^2$, note that Y can take on values 0, 1, 4, and 9; then $f_Y(0) = f_X(2)$, $f_Y(1) = f_X(1) + f_X(3)$, $f_Y(4) = f_X(0) + f_X(4)$, and $f_Y(9) = f_X(5)$. ////

Second, if X is a continuous random variable, then the cumulative distribution function of $Y = g(X)$ can be found by integrating $f_X(x)$ over the appropriate region; that is,

$$F_Y(y) = P[Y \leq y] = P[g(X) \leq y] = \int_{\{x:g(x)\leq y\}} f_X(x)\,dx. \tag{36}$$

This is just the cumulative-distribution-function technique.

EXAMPLE 16 Let X be a random variable with uniform distribution over the interval (0, 1) and let $Y = g(X) = X^2$. The density of Y is desired. Now

$$F_Y(y) = P[Y \leq y] = P[X^2 \leq y] = \int_{\{x:\,x^2\leq y\}} f_X(x)\,dx = \int_0^{\sqrt{y}} dx = \sqrt{y}$$

for $0 < y < 1$; so

$$F_Y(y) = \sqrt{y} I_{(0,1)}(y) + I_{[1,\infty)}(y),$$

and therefore

$$f_Y(y) = \frac{1}{2}\frac{1}{\sqrt{y}} I_{(0,1)}(y). \qquad ////$$

Application of the cumulative-distribution-function technique to find the density of $Y = g(X)$, as in the above example, produces the transformation technique, the result of which is given in the following theorem.

Theorem 11 Suppose X is a continuous random variable with probability density function $f_X(\cdot)$. Set $\mathfrak{X} = \{x : f_X(x) > 0\}$. Assume that:

(i) $y = g(x)$ defines a one-to-one transformation of $\mathfrak{X}$ onto $\mathfrak{Y}$.
(ii) The derivative of $x = g^{-1}(y)$ with respect to y is continuous and nonzero for $y \in \mathfrak{Y}$, where $g^{-1}(y)$ is the inverse function of $g(x)$; that is, $g^{-1}(y)$ is that x for which $g(x) = y$.

Then $Y = g(X)$ is a continuous random variable with density

$$f_Y(y) = \left|\frac{d}{dy} g^{-1}(y)\right| f_X(g^{-1}(y)) I_{\mathfrak{Y}}(y).$$

PROOF The above is a standard theorem from calculus on the change of variable in a definite integral; so we will only sketch the proof. Consider the case when $\mathfrak{X}$ is an interval. Let us suppose that $g(x)$ is a monotone increasing function over $\mathfrak{X}$; that is, $g'(x) > 0$, which is true if and only if $(d/dy)g^{-1}(y) > 0$ over $\mathfrak{Y}$. For $y \in \mathfrak{Y}$, $F_Y(y) = P[g(X) \leq y] = P[X \leq g^{-1}(y)] = F_X(g^{-1}(y))$, and hence $f_Y(y) = (d/dy)F_Y(y) = [(d/dy)g^{-1}(y)] f_X(g^{-1}(y))$ by chain rule of differentiation. On the other hand, if $g(x)$ is a monotone decreasing function over $\mathfrak{X}$, so that $g'(x) < 0$ and $(d/dy)g^{-1}(y) < 0$, then $F_Y(y) = P[g(X) \leq y] = P[X \geq g^{-1}(y)] = 1 - F_X(g^{-1}(y))$, and therefore $f_Y(y) = -[(d/dy)g^{-1}(y)] f_X(g^{-1}(y)) = |(d/dy)g^{-1}(y)| f_X(g^{-1}(y))$ for $y \in \mathfrak{Y}$. ////

EXAMPLE 17 Suppose X has a beta distribution. What is the distribution of $Y = -\log_e X$? $\mathfrak{X} = \{x : f_X(x) > 0\} = \{x : 0 < x < 1\}$. $y = g(x) = -\log_e x$ defines a one-to-one transformation of $\mathfrak{X}$ onto $\mathfrak{Y} = \{y : y > 0\}$. $x = g^{-1}(y) = e^{-y}$, so $(d/dy)g^{-1}(y) = -e^{-y}$, which is continuous and nonzero for $y \in \mathfrak{Y}$. By Theorem 11,

$$\begin{aligned} f_Y(y) &= \left|\frac{d}{dy} g^{-1}(y)\right| f_X(g^{-1}(y)) I_{\mathfrak{Y}}(y) \\ &= e^{-y} \frac{1}{\mathrm{B}(a, b)} (e^{-y})^{a-1}(1 - e^{-y})^{b-1} I_{(0, \infty)}(y) \\ &= \frac{1}{\mathrm{B}(a, b)} e^{-ay}(1 - e^{-y})^{b-1} I_{(0, \infty)}(y). \end{aligned}$$

In particular, if $b = 1$, then $\mathrm{B}(a, b) = 1/a$; so $f_Y(y) = ae^{-ay} I_{(0, \infty)}(y)$, an exponential distribution with parameter a. ////

EXAMPLE 18 Suppose X has the Pareto density $f_X(x) = \theta x^{-\theta-1} I_{[1,\infty)}(x)$ and the distribution of $Y = \log_e X$ is desired.

$$f_Y(y) = \left| \frac{d}{dy} g^{-1}(y) \right| f_X(g^{-1}(y)) I_{\mathfrak{Y}}(y)$$
$$= e^y \theta (e^y)^{-\theta-1} I_{[1,\infty)}(e^y) = \theta e^{-\theta y} I_{[0,\infty)}(y). \qquad ////$$

The condition that $g(x)$ be a one-to-one transformation of $\mathfrak{X}$ onto $\mathfrak{Y}$ is unnecessarily restrictive. For the transformation $y = g(x)$, each point in $\mathfrak{X}$ will correspond to just one point in $\mathfrak{Y}$; but to a point in $\mathfrak{Y}$ there may correspond more than one point in $\mathfrak{X}$, which says that the transformation is not one-to-one, and consequently Theorem 11 is not directly applicable. If, however, $\mathfrak{X}$ can be decomposed into a finite (or even countable) number of disjoint sets, say $\mathfrak{X}_1, \ldots, \mathfrak{X}_m$, so that $y = g(x)$ defines a one-to-one transformation of each $\mathfrak{X}_i$ into $\mathfrak{Y}$, then the joint density of $Y = g(X)$ can be found. Let $x = g_i^{-1}(y)$ denote the inverse of $y = g(x)$ for $x \in \mathfrak{X}_i$. Then the density of $Y = g(X)$ is given by

$$f_Y(y) = \sum \left| \frac{d}{dy} g_i^{-1}(y) \right| f_X(g_i^{-1}(y)) I_{\mathfrak{Y}}(y), \qquad (37)$$

where the summation is over those values of i for which $g(x) = y$ for some value of x in $\mathfrak{X}_i$.

EXAMPLE 19 Let X be a continuous random variable with density $f_X(\cdot)$, and let $Y = g(X) = X^2$. Note that if $\mathfrak{X}$ is an interval containing both negative and positive points, then $y = g(x) = x^2$ is not one-to-one. However, if $\mathfrak{X}$ is decomposed into $\mathfrak{X}_1 = \{x\colon x \in \mathfrak{X}, x < 0\}$ and $\mathfrak{X}_2 = \{x\colon x \in \mathfrak{X}, x \geq 0\}$, then $y = g(x)$ defines a one-to-one transformation on each $\mathfrak{X}_i$. Note that $g_1^{-1}(y) = -\sqrt{y}$ and $g_2^{-1}(y) = \sqrt{y}$. By Eq. (37),

$$f_Y(y) = \left[\frac{1}{2} \frac{1}{\sqrt{y}} f_X(-\sqrt{y}) + \frac{1}{2} \frac{1}{\sqrt{y}} f_X(\sqrt{y}) \right] I_{(0,\infty)}(y).$$

In particular, if

$$f_X(x) = (\tfrac{1}{2}) e^{-|x|},$$

then

$$f_Y(y) = \frac{1}{2} \frac{1}{\sqrt{y}} e^{-\sqrt{y}} I_{(0,\infty)}(y);$$

or, if

$$f_X(x) = \frac{2}{9}(x+1)I_{(-1,2)}(x),$$

then

$$f_Y(y) = \left[\frac{1}{2}\frac{1}{\sqrt{y}}\frac{2}{9}(-\sqrt{y}+1) + \frac{1}{2}\frac{1}{\sqrt{y}}\frac{2}{9}(1+\sqrt{y})\right]I_{(0,1)}(y) + \left[\frac{1}{2}\frac{1}{\sqrt{y}}\frac{2}{9}(1+\sqrt{y})\right]I_{[1,4)}(y). \qquad ////$$

5.2 Probability Integral Transform

If X is a random variable with cumulative distribution $F_X(\cdot)$, then $F_X(\cdot)$ is a candidate for $g(\cdot)$ in the transformation $Y = g(X)$. The following theorem gives the distribution of $Y = F_X(X)$ if $F_X(\cdot)$ is continuous. Since $F_X(\cdot)$ is a nondecreasing function, the inverse function $F_X^{-1}(\cdot)$ may be defined for any value of y between 0 and 1 as: $F_X^{-1}(y)$ is the smallest x satisfying $F_X(x) \geq y$.

Theorem 12 If X is a random variable with continuous cumulative distribution function $F_X(x)$, then $U = F_X(X)$ is uniformly distributed over the interval (0, 1). Conversely, if U is uniformly distributed over the interval (0, 1), then $X = F_X^{-1}(U)$ has cumulative distribution function $F_X(\cdot)$.

PROOF $P[U \leq u] = P[F_X(X) \leq u] = P[X \leq F_X^{-1}(u)] = F_X(F_X^{-1}(u)) = u$ for $0 < u < 1$. Conversely, $P[X \leq x] = P[F_X^{-1}(U) \leq x] = P[U \leq F_X(x)] = F_X(x)$. ////

In various statistical applications, particularly in simulation studies, it is often desired to generate values of some random variable X. To generate a value of a random variable X having continuous cumulative distribution function $F_X(\cdot)$, it suffices to generate a value of a random variable U that is uniformly distributed over the interval (0, 1). This follows from Theorem 12 since if U is a random variable with a uniform distribution over the interval (0, 1), then $X = F_X^{-1}(U)$ is a random variable having distribution $F_X(\cdot)$. So to get a value, say x, of a random variable X, obtain a value, say u, of a random variable U, compute $F_X^{-1}(u)$, and set it equal to x. A value u of a random variable U is called a *random number*. Many computer-oriented random-number generators are available.

EXAMPLE 20 $F_X(x) = (1 - e^{-\lambda x})I_{(0,\infty)}(x)$. $F_X^{-1}(y) = -(1/\lambda)\log_e(1-y)$; so $-(1/\lambda)\log_e(1-U)$ is a random variable having distribution $(1 - e^{-\lambda x}) I_{(0,\infty)}(x)$ if U is a random variable uniformly distributed over the interval $(0, 1)$. ////

The transformation $Y = F_X(X)$ is called the *probability integral transformation.* It plays an important role in the theory of distribution-free statistics and goodness-of-fit tests.

6 TRANSFORMATIONS

In Sec. 5 we considered the problem of obtaining the distribution of a function of a given random variable. It is natural to consider next the problem of obtaining the joint distribution of several random variables which are functions of a given set of random variables.

6.1 Discrete Random Variables

Suppose that the discrete density function $f_{X_1,\ldots,X_n}(x_1, \ldots, x_n)$ of the n-dimensional discrete random variable $(X_1, \ldots, X_n)$ is given. Let $\mathfrak{X}$ denote the mass points of $(X_1, \ldots, X_n)$; that is,

$$\mathfrak{X} = \{(x_1, \ldots, x_n) : f_{X_1,\ldots,X_n}(x_1, \ldots, x_n) > 0\}.$$

Suppose that the joint density of $Y_1 = g_1(X_1, \ldots, X_n), \ldots, Y_k = g_k(X_1, \ldots, X_n)$ is desired. It can be observed that $Y_1, \ldots, Y_k$ are jointly discrete and $P[Y_1 = y_1; \ldots; Y_k = y_k] = f_{Y_1,\ldots,Y_k}(y_1, \ldots, y_k) = \sum f_{X_1,\ldots,X_n}(x_1, \ldots, x_n)$, where the summation is over those $(x_1, \ldots, x_n)$ belonging to $\mathfrak{X}$ for which $(y_1, \ldots, y_k) = (g_1(x_1, \ldots, x_n), \ldots, g_k(x_1, \ldots, x_n))$.

EXAMPLE 21 Let (X_1, X_2, X_3) have a joint discrete density function given by

(x_1, x_2, x_3)	(0, 0, 0)	(0, 0, 1)	(0, 1, 1)	(1, 0, 1)	(1, 1, 0)	(1, 1, 1)
$f_{X_1,X_2,X_3}(x_1, x_2, x_3)$	$\frac{1}{8}$	$\frac{3}{8}$	$\frac{1}{8}$	$\frac{1}{8}$	$\frac{1}{8}$	$\frac{1}{8}$

Find the joint density of $Y_1 = g_1(X_1, X_2, X_3) = X_1 + X_2 + X_3$ and $Y_2 = g_2(X_1, X_2, X_3) = |X_3 - X_2|$.

$$\mathfrak{X} = \{(0, 0, 0), (0, 0, 1), (0, 1, 1), (1, 0, 1), (1, 1, 0), (1, 1, 1)\}.$$

$$f_{Y_1, Y_2}(0, 0) = f_{X_1, X_2, X_3}(0, 0, 0) = \tfrac{1}{8},$$
$$f_{Y_1, Y_2}(1, 1) = f_{X_1, X_2, X_3}(0, 0, 1) = \tfrac{3}{8},$$
$$f_{Y_1, Y_2}(2, 0) = f_{X_1, X_2, X_3}(0, 1, 1) = \tfrac{1}{8},$$
$$f_{Y_1, Y_2}(2, 1) = f_{X_1, X_2, X_3}(1, 0, 1) + f_{X_1, X_2, X_3}(1, 1, 0) = \tfrac{2}{8},$$

and

$$f_{Y_1, Y_2}(3, 0) = f_{X_1, X_2, X_3}(1, 1, 1) = \tfrac{1}{8}. \qquad ////$$

6.2 Continuous Random Variables

Suppose now that we are given the joint probability density function $f_{X_1, \ldots, X_n}(x_1, \ldots, x_n)$ of the n-dimensional continuous random variable $(X_1, \ldots, X_n)$. Let

$$\mathfrak{X} = \{(x_1, \ldots, x_n): f_{X_1, \ldots, X_n}(x_1, \ldots, x_n) > 0\}. \tag{38}$$

Again assume that the joint density of the random variables $Y_1 = g_1(X_1, \ldots, X_n)$, $\ldots$, $Y_k = g_k(X_1, \ldots, X_n)$ is desired, where k is some integer satisfying $1 \leq k \leq n$. If $k < n$, we will introduce additional, new random variables $Y_{k+1} = g_{k+1}(X_1, \ldots, X_n), \ldots, Y_n = g_n(X_1, \ldots, X_n)$ for judiciously selected functions $g_{k+1}, \ldots, g_n$; then we will find the joint distribution of $Y_1, \ldots, Y_n$, and finally we will find the desired marginal distribution of $Y_1, \ldots, Y_k$ from the joint distribution of $Y_1, \ldots, Y_n$. This use of possibly introducing additional random variables makes the transformation $y_1 = g_1(x_1, \ldots, x_n), \ldots, y_n = g_n(x_1, \ldots, x_n)$ a transformation from an n-dimensional space to an n-dimensional space. Henceforth we will assume that we are seeking the joint distribution of $Y_1 = g_1(X_1, \ldots, X_n), \ldots, Y_n = g_n(X_1, \ldots, X_n)$ (rather than the joint distribution of $Y_1, \ldots, Y_k$) when we have given the joint probability density of $X_1, \ldots, X_n$.

We will state our results first for $n = 2$ and later generalize to $n > 2$. Let $f_{X_1, X_2}(x_1, x_2)$ be given. Set $\mathfrak{X} = \{(x_1, x_2): f_{X_1, X_2}(x_1, x_2) > 0\}$. We want to find the joint distribution of $Y_1 = g_1(X_1, X_2)$ and $Y_2 = g_2(X_1, X_2)$ for known functions $g_1(\cdot, \cdot)$ and $g_2(\cdot, \cdot)$. Now suppose that $y_1 = g_1(x_1, x_2)$ and $y_2 = g_2(x_1, x_2)$ defines a one-to-one transformation which maps $\mathfrak{X}$ onto, say, $\mathfrak{Y}$. x_1 and x_2 can be expressed in terms of y_1 and y_2; so we can write, say, $x_1 = g_1^{-1}(y_1, y_2)$ and $x_2 = g_2^{-1}(y_1, y_2)$. Note that $\mathfrak{X}$ is a subset of the x_1x_2 plane and $\mathfrak{Y}$ is a subset of the y_1y_2 plane. The determinant

$$\begin{vmatrix} \dfrac{\partial x_1}{\partial y_1} & \dfrac{\partial x_1}{\partial y_2} \\ \dfrac{\partial x_2}{\partial y_1} & \dfrac{\partial x_2}{\partial y_2} \end{vmatrix} \tag{39}$$

will be called the *Jacobian* of the transformation and will be denoted by J. The above discussion permits us to state Theorem 13.

Theorem 13 Let X_1 and X_2 be jointly continuous random variables with density function $f_{X_1, X_2}(x_1, x_2)$. Set $\mathfrak{X} = \{(x_1, x_2): f_{X_1, X_2}(x_1, x_2) > 0\}$. Assume that:

(i) $y_1 = g_1(x_1, x_2)$ and $y_2 = g_2(x_1, x_2)$ defines a one-to-one transformation of $\mathfrak{X}$ onto $\mathfrak{Y}$.

(ii) The first partial derivatives of $x_1 = g_1^{-1}(y_1, y_2)$ and $x_2 = g_2^{-1}(y_1, y_2)$ are continuous over $\mathfrak{Y}$.

(iii) The Jacobian of the transformation is nonzero for $(y_1, y_2) \in \mathfrak{Y}$.

Then the joint density of $Y_1 = g_1(X_1, X_2)$ and $Y_2 = g_2(X_1, X_2)$ is given by

$$f_{Y_1, Y_2}(y_1, y_2) = |J| f_{X_1, X_2}(g_1^{-1}(y_1, y_2), g_2^{-1}(y_1, y_2)) I_{\mathfrak{Y}}(y_1, y_2). \tag{40}$$

PROOF We omit the proof; it is essentially the same as the derivation of the formulas for transforming variables in double integrals, which may be found in many advanced calculus textbooks. $\mathfrak{Y}$ is that subset of the $y_1 y_2$ plane consisting of points (y_1, y_2) for which there exists a $(x_1, x_2) \in \mathfrak{X}$ such that $(y_1, y_2) = (g_1(x_1, x_2), g_2(x_1, x_2))$. $I_{\mathfrak{Y}}(y_1, y_2) = I_{\mathfrak{X}}(g_1^{-1}(y_1, y_2), g_2^{-1}(y_1, y_2))$. ////

EXAMPLE 22 Suppose that X_1 and X_2 are independent random variables, each uniformly distributed over the interval (0, 1). Then $f_{X_1, X_2}(x_1, x_2) = I_{(0,1)}(x_1) I_{(0,1)}(x_2)$. $\mathfrak{X} = \{(x_1, x_2): 0 < x_1 < 1 \text{ and } 0 < x_2 < 1\}$. Let $y_1 = g_1(x_1, x_2) = x_1 + x_2$ and $y_2 = g_2(x_1, x_2) = x_2 - x_1$; then $x_1 = \frac{1}{2}(y_1 - y_2) = g_1^{-1}(y_1, y_2)$, and $x_2 = \frac{1}{2}(y_1 + y_2) = g_2^{-1}(y_1, y_2)$.

$$J = \begin{vmatrix} \dfrac{\partial x_1}{\partial y_1} & \dfrac{\partial x_1}{\partial y_2} \\ \dfrac{\partial x_2}{\partial y_1} & \dfrac{\partial x_2}{\partial y_2} \end{vmatrix} = \begin{vmatrix} \frac{1}{2} & -\frac{1}{2} \\ \frac{1}{2} & \frac{1}{2} \end{vmatrix} = \frac{1}{2}.$$

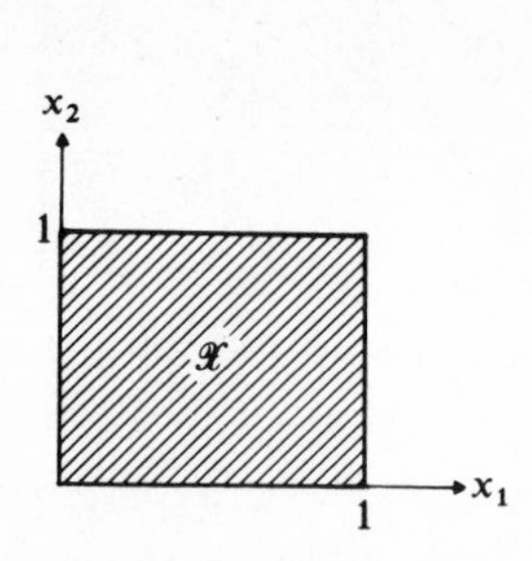

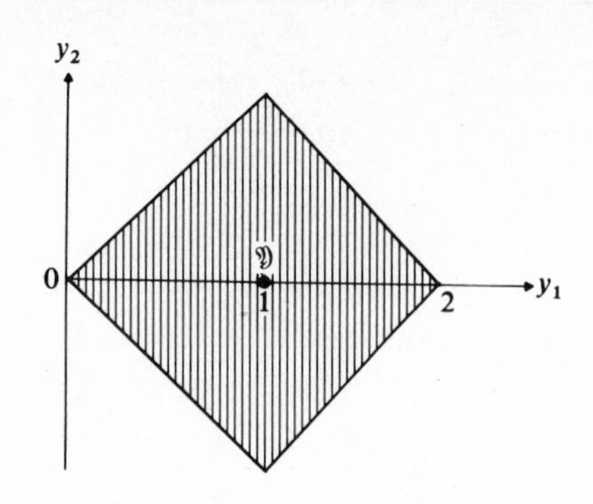

FIGURE 3

$\mathfrak{X}$ and $\mathfrak{Y}$ are sketched in Fig. 3. Note that the boundary $x_1 = 0$ of $\mathfrak{X}$ goes into the boundary $\frac{1}{2}(y_1 - y_2) = 0$ of $\mathfrak{Y}$, the boundary $x_2 = 0$ of $\mathfrak{X}$ goes into the boundary $\frac{1}{2}(y_1 + y_2) = 0$ of $\mathfrak{Y}$, the boundary $x_1 = 1$ of $\mathfrak{X}$ goes into the boundary $\frac{1}{2}(y_1 - y_2) = 1$ of $\mathfrak{Y}$, and the boundary $x_2 = 1$ of $\mathfrak{X}$ goes into the boundary $\frac{1}{2}(y_1 + y_2) = 1$ of $\mathfrak{Y}$. Now the transformation is one-to-one, the first partial derivatives of g_1^{-1} and g_2^{-1} are continuous, and the Jacobian is nonzero; so

$$\begin{aligned} f_{Y_1, Y_2}(y_1, y_2) &= |J| f_{X_1, X_2}(g_1^{-1}(y_1, y_2), g_2^{-1}(y_1, y_2)) \\ &= \tfrac{1}{2} I_{(0,1)}\left(\frac{y_1 - y_2}{2}\right) I_{(0,1)}\left(\frac{y_1 + y_2}{2}\right) \\ &= \begin{cases} \frac{1}{2} & \text{for } (y_1, y_2) \in \mathfrak{Y} \\ 0 & \text{otherwise.} \end{cases} \end{aligned}$$

////

EXAMPLE 23 Let X_1 and X_2 be two independent standard normal random variables. Let $Y_1 = X_1 + X_2$ and $Y_2 = X_1/X_2$. Then

$$x_1 = g_1^{-1}(y_1, y_2) = \frac{y_1 y_2}{1 + y_2} \quad \text{and} \quad x_2 = g_2^{-1}(y_1, y_2) = \frac{y_1}{1 + y_2}.$$

$$J = \begin{vmatrix} \dfrac{y_2}{1 + y_2} & \dfrac{y_1}{(1 + y_2)^2} \\ \dfrac{1}{1 + y_2} & -\dfrac{y_1}{(1 + y_2)^2} \end{vmatrix} = -\frac{y_1(y_2 + 1)}{(1 + y_2)^3} = -\frac{y_1}{(1 + y_2)^2}.$$

$$f_{Y_1, Y_2}(y_1, y_2)$$

$$= \frac{|y_1|}{(1 + y_2)^2} \frac{1}{2\pi} \exp\left\{-\frac{1}{2}\left[\frac{(y_1 y_2)^2}{(1 + y_2)^2} + \frac{y_1^2}{(1 + y_2)^2}\right]\right\}$$

$$= \frac{1}{2\pi} \frac{|y_1|}{(1 + y_2)^2} \exp\left[-\frac{1}{2}\frac{(1 + y_2^2)y_1^2}{(1 + y_2)^2}\right].$$

To find the marginal distribution of, say, Y_2, we must integrate out y_1; that is

$$f_{Y_2}(y_2) = \int_{-\infty}^{\infty} f_{Y_1, Y_2}(y_1, y_2)\, dy_1$$

$$= \frac{1}{2\pi} \frac{1}{(1 + y_2)^2} \int_{-\infty}^{\infty} |y_1| \exp\left[-\frac{1}{2}\frac{(1 + y_2^2)y_1^2}{(1 + y_2)^2}\right] dy_1.$$

Let

$$u = \frac{1}{2}\frac{(1 + y_2^2)}{(1 + y_2)^2} y_1^2;$$

then

$$du = \frac{(1 + y_2^2)}{(1 + y_2)^2} y_1\, dy_1$$

and so

$$f_{Y_2}(y_2) = \frac{1}{2\pi} \cdot \frac{1}{(1 + y_2)^2} \cdot \frac{(1 + y_2)^2}{1 + y_2^2} (2) \int_0^{\infty} e^{-u}\, du = \frac{1}{\pi} \cdot \frac{1}{1 + y_2^2},$$

a Cauchy density. That is, the *ratio of two independent standard normal random variables has a Cauchy distribution.* ////

EXAMPLE 24 Let X_i have a gamma density with parameters n_i and λ for $i = 1, 2$. Assume that X_1 and X_2 are independent. Again, we seek the joint distribution of $Y_1 = X_1 + X_2$ and $Y_2 = X_1/X_2$.

$$x_1 = g_1^{-1}(y_1, y_2) = \frac{y_1 y_2}{1 + y_2} \quad \text{and} \quad x_2 = g_2^{-1}(y_1, y_2) = \frac{y_1}{1 + y_2};$$

hence

$$|J| = \frac{y_1}{(1+y_2)^2}.$$

$$f_{Y_1, Y_2}(y_1, y_2)$$

$$= \frac{y_1}{(1+y_2)^2} \cdot \frac{1}{\Gamma(n_1)} \cdot \frac{1}{\Gamma(n_2)} \lambda^{n_1+n_2} \left(\frac{y_1 y_2}{1+y_2}\right)^{n_1-1} \left(\frac{y_1}{1+y_2}\right)^{n_2-1} e^{-\lambda y_1} I_{\mathfrak{Y}}(y_1, y_2)$$

$$= \frac{\lambda^{n_1+n_2}}{\Gamma(n_1)\Gamma(n_2)} y_1^{n_1+n_2-1} e^{-\lambda y_1} \frac{y_2^{n_1-1}}{(1+y_2)^{n_1+n_2}} I_{(0,\infty)}(y_1) I_{(0,\infty)}(y_2)$$

$$= \left[\frac{\lambda^{n_1+n_2}}{\Gamma(n_1+n_2)} y_1^{n_1+n_2-1} e^{-\lambda y_1} I_{(0,\infty)}(y_1)\right]$$

$$\times \left[\frac{1}{B(n_1, n_2)} \frac{y_2^{n_1-1}}{(1+y_2)^{n_1+n_2}} I_{(0,\infty)}(y_2)\right].$$

We see that $f_{Y_1, Y_2}(y_1, y_2) = f_{Y_1}(y_1) f_{Y_2}(y_2)$; so Y_1 and Y_2 are independent. Also, we see that the distribution of $Y_1 = X_1 + X_2$ is a gamma distribution with parameters $n_1 + n_2$ and λ. If $n_1 = n_2 = 1$, then Y_2 is the ratio of two independent exponentially distributed random variables and has density

$$f_{Y_2}(y_2) = \frac{1}{(1+y_2)^2} I_{(0,\infty)}(y_2),$$

a density which has an infinite mean. ////

EXAMPLE 25 Let X_i have a gamma distribution with parameters n_i and λ for $i = 1, 2$, and assume X_1 and X_2 are independent. Suppose now that the distribution of $Y_1 = X_1/(X_1 + X_2)$ is desired. We have only the one function $y_1 = g_1(x_1, x_2) = x_1/(x_1 + x_2)$; so we have to select the other to use the transformation technique. Since x_1 and x_2 occur in the exponent of their joint density as their sum, $x_1 + x_2$ is a good choice. Let $y_2 = x_1 + x_2$; then $x_1 = y_1 y_2$, $x_2 = y_2 - y_1 y_2$, and

$$J = \begin{vmatrix} y_2 & y_1 \\ -y_2 & 1 - y_1 \end{vmatrix} = y_2.$$

Hence

$$f_{Y_1, Y_2}(y_1, y_2)$$

$$= y_2 \frac{1}{\Gamma(n_1)} \frac{1}{\Gamma(n_2)} \lambda^{n_1+n_2}(y_1 y_2)^{n_1-1}(y_2 - y_1 y_2)^{n_2-1} e^{-\lambda y_2} I_{\mathfrak{Y}}(y_1, y_2)$$

$$= \frac{\lambda^{n_1+n_2}}{\Gamma(n_1)\Gamma(n_2)} y_1^{n_1-1}(1 - y_1)^{n_2-1} y_2^{n_1+n_2-1} e^{-\lambda y_2} I_{(0, 1)}(y_1) I_{(0, \infty)}(y_2)$$

$$= \left[\frac{1}{\mathrm{B}(n_1, n_2)} y_1^{n_1-1}(1 - y_1)^{n_2-1} I_{(0, 1)}(y_1)\right]$$

$$\times \left[\frac{\lambda^{n_1+n_2}}{\Gamma(n_1 + n_2)} y_2^{n_1+n_2-1} e^{-\lambda y_2} I_{(0, \infty)}(y_2)\right].$$

It turns out that Y_1 and Y_2 are independent and Y_1 has a beta distribution with parameters n_1 and n_2. ////

Of the three conditions that are imposed on the transformation $y_1 = g_1(x_1, x_2)$ and $y_2 = g_2(x_1, x_2)$, the sometimes restrictive condition that the transformation be one-to-one can be relaxed. For the transformation $y_1 = g_1(x_1, x_2)$ and $y_2 = g_2(x_1, x_2)$, each point in $\mathfrak{X}$ will correspond to just one point in $\mathfrak{Y}$; but to a point in $\mathfrak{Y}$ there may correspond more than one point in $\mathfrak{X}$, which says that the transformation is not one-to-one and consequently Theorem 13 as stated is not applicable. If, however, $\mathfrak{X}$ can be decomposed into a finite number of disjoint sets, say $\mathfrak{X}_1, \ldots, \mathfrak{X}_m$, so that $y_1 = g_1(x_1, x_2)$ and $y_2 = g_2(x_1, x_2)$ define a one-to-one transformation of each $\mathfrak{X}_i$ onto $\mathfrak{Y}$ then the joint density of $Y_1 = g_1(X_1, X_2)$ and $Y_2 = g_2(X_1, X_2)$ can be found. Let $x_1 = g_{1i}^{-1}(y_1, y_2)$ and $x_2 = g_{2i}^{-1}(y_1, y_2)$ denote the inverse transformation of $\mathfrak{Y}$ onto $\mathfrak{X}_i$ for $i = 1, \ldots, m$, and set

$$J_i = \begin{vmatrix} \dfrac{\partial g_{1i}^{-1}}{\partial y_1} & \dfrac{\partial g_{1i}^{-1}}{\partial y_2} \\ \dfrac{\partial g_{2i}^{-1}}{\partial y_1} & \dfrac{\partial g_{2i}^{-1}}{\partial y_2} \end{vmatrix}.$$

Theorem 14 Let X_1 and X_2 be two jointly continuous random variables with density function $f_{X_1, X_2}(x_1, x_2)$. Assume that $\mathfrak{X}$ can be decomposed into sets $\mathfrak{X}_1, \ldots, \mathfrak{X}_m$ such that the transformation $y_1 = g_1(x_1, x_2)$ and $y_2 = g_2(x_1, x_2)$ is one-to-one from $\mathfrak{X}_i$ onto $\mathfrak{Y}$. Let $x_1 = g_{1i}^{-1}(y_1, y_2)$ and $x_2 = g_{2i}^{-1}(y_1, y_2)$ denote the inverse transformation of $\mathfrak{Y}$ onto $\mathfrak{X}_i$, $i = 1, \ldots, m$. Assume that all first partial derivatives of g_{1i}^{-1} and g_{2i}^{-1} are continuous on $\mathfrak{Y}$ and that J_i does not vanish on $\mathfrak{Y}$, $i = 1, \ldots, m$. Then

$$f_{Y_1, Y_2}(y_1, y_2) = \sum_{i=1}^{m} |J_i| f_{X_1, X_2}(g_{1i}^{-1}(y_1, y_2), g_{2i}^{-1}(y_1, y_2)) I_{\mathfrak{Y}}(y_1, y_2). \tag{41}$$

////

We illustrate this theorem with Example 26.

EXAMPLE 26 Assume that X_1 and X_2 are independent standard normal random variables. Consider the transformation $y_1 = x_1^2 + x_2^2$ and $y_2 = x_2$, which implies $x_1 = \pm\sqrt{y_1 - y_2^2}$ and $x_2 = y_2$ so that the transformation is not one-to-one. Here $\mathfrak{X} = \{(x_1, x_2): -\infty < x_1 < \infty, -\infty < x_2 < \infty\}$, and $\mathfrak{Y} = \{(y_1, y_2): 0 \leq y_1 < \infty, -\sqrt{y_1} < y_2 < \sqrt{y_1}\}$. If $\mathfrak{X}$ is decomposed into $\mathfrak{X}_1$ and $\mathfrak{X}_2$, where $\mathfrak{X}_1 = \{(x_1, x_2): 0 \leq x_1 < \infty, -\infty < x_2 < \infty\}$ and $\mathfrak{X}_2 = \{(x_1, x_2): -\infty < x_1 < 0, -\infty < x_2 < \infty\}$ (in the terminology of Theorem 14, $m = 2$), then our transformation is one-to-one for $\mathfrak{X}_i$ onto $\mathfrak{Y}$, $i = 1, 2$. $g_{11}^{-1}(y_1, y_2) = \sqrt{y_1 - y_2^2}$, and $g_{21}^{-1}(y_1, y_2) = y_2$; $g_{12}^{-1}(y_1, y_2) = -\sqrt{y_1 - y_2^2}$, and $g_{22}^{-1}(y_1, y_2) = y_2$; so

$$J_1 = \begin{vmatrix} \frac{1}{2}(y_1 - y_2^2)^{-\frac{1}{2}} & \dfrac{\partial g_{11}^{-1}}{\partial y_2} \\ 0 & 1 \end{vmatrix} = \tfrac{1}{2}(y_1 - y_2^2)^{-\frac{1}{2}},$$

and

$$J_2 = \begin{vmatrix} -\frac{1}{2}(y_1 - y_2^2)^{-\frac{1}{2}} & \dfrac{\partial g_{12}^{-1}}{\partial y_2} \\ 0 & 1 \end{vmatrix} = -\frac{1}{2}(y_1 - y_2^2)^{-\frac{1}{2}}.$$

Hence,

$$\begin{aligned} f_{Y_1, Y_2}(y_1, y_2) &= [|J_1| f_{X_1, X_2}(g_{11}^{-1}(y_1, y_2), g_{21}^{-1}(y_1, y_2)) \\ &\quad + |J_2| f_{X_1, X_2}(g_{12}^{-1}(y_1, y_2), g_{22}^{-1}(y_1, y_2))] I_{\mathfrak{Y}}(y_1, y_2) \\ &= \frac{1}{\sqrt{y_1 - y_2^2}} \cdot \frac{1}{2\pi} \cdot e^{-\frac{1}{2}y_1} \end{aligned}$$

for $y_1 \geq 0$ and $-\sqrt{y_1} < y_2 < \sqrt{y_1}$. Now

$$\begin{aligned} f_{Y_1}(y_1) &= \int_{-\infty}^{\infty} f_{Y_1, Y_2}(y_1, y_2)\, dy_2 \\ &= \frac{1}{2\pi} e^{-\frac{1}{2}y_1} \int_{-\sqrt{y_1}}^{\sqrt{y_1}} \frac{1}{\sqrt{y_1 - y_2^2}}\, dy_2 \\ &= \frac{1}{2\pi} e^{-\frac{1}{2}y_1} \left\{ \arc\sin \frac{y_2}{\sqrt{y_1}} \Big|_{-\sqrt{y_1}}^{\sqrt{y_1}} \right\} \\ &= \frac{1}{2\pi} e^{-\frac{1}{2}y_1} \left(\frac{\pi}{2} + \frac{\pi}{2} \right) = \frac{1}{2} e^{-\frac{1}{2}y_1} \qquad \text{for} \quad y_1 > 0, \end{aligned}$$

an exponential distribution. ////

Theorems 13 and 14 can be generalized from $n = 2$ to $n > 2$. We will state the generalization of Theorem 14. (Theorem 13 is a special case of Theorem 14.)

Theorem 15 Let $X_1, X_2, \ldots, X_n$ be jointly continuous random variables with density function $f_{X_1, \ldots, X_n}(x_1, \ldots, x_n)$. Let $\mathfrak{X} = \{(x_1, \ldots, x_n)\colon f_{X_1, \ldots, X}(x_1 \ldots, x_n) > 0\}$. Assume that $\mathfrak{X}$ can be decomposed into sets $\mathfrak{X}_1, \ldots, \mathfrak{X}_m$ such that $y_1 = g_1(x_1, \ldots, x_n)$, $y_2 = g_2(x_1, \ldots, x_n), \ldots, y_n = g_n(x_1, \ldots, x_n)$ is a one-to-one transformation of $\mathfrak{X}_i$ onto $\mathfrak{Y}$, $i = 1, \ldots, m$. Let $x_1 = g_{1i}^{-1}(y_1, \ldots, y_n), \ldots, x_n = g_{ni}^{-1}(y_1, \ldots, y_n)$ denote the inverse transformation of $\mathfrak{Y}$ onto $\mathfrak{X}_i$, $i = 1, \ldots, m$. Define

$$J_i = \begin{vmatrix} \dfrac{\partial g_{1i}^{-1}}{\partial y_1} & \dfrac{\partial g_{1i}^{-1}}{\partial y_2} & \cdots & \cdots & \dfrac{\partial g_{1i}^{-1}}{\partial y_n} \\ \dfrac{\partial g_{2i}^{-1}}{\partial y_1} & \dfrac{\partial g_{2i}^{-1}}{\partial y_2} & \cdots & \cdots & \dfrac{\partial g_{2i}^{-1}}{\partial y_n} \\ \cdots & \cdots & \cdots & \cdots & \cdots \\ \dfrac{\partial g_{ni}^{-1}}{\partial y_1} & \dfrac{\partial g_{ni}^{-1}}{\partial y_2} & \cdots & \cdots & \dfrac{\partial g_{ni}^{-1}}{\partial y_n} \end{vmatrix}$$

for $i = 1, \ldots, m$.

Assume that all the partial derivatives in J_i are continuous over $\mathfrak{Y}$ and the determinant J_i is nonzero, $i = 1, \ldots, m$. Then

$$f_{Y_1, \ldots, Y_n}(y_1, \ldots, y_n) = \sum_{i=1}^{m} |J_i| f_{X_1, \ldots, X_n}(g_{1i}^{-1}(y_1, \ldots, y_n), \ldots, g_{ni}^{-1}(y_1, \ldots, y_n)) \tag{42}$$

for $(y_1, \ldots, y_n)$ in $\mathfrak{Y}$. ////

EXAMPLE 27 Let X_1, X_2, and X_3 be independent standard normal random variables, $y_1 = x_1$, $y_2 = (x_1 + x_2)/2$, and $y_3 = (x_1 + x_2 + x_3)/3$. Then $x_1 = y_1$, $x_2 = 2y_2 - y_1$, and $x_3 = 3y_3 - 2y_2$; so the transformation is one-to-one. ($m = 1$ in Theorem 15.)

$$J = \begin{vmatrix} 1 & 0 & 0 \\ -1 & 2 & 0 \\ 0 & -2 & 3 \end{vmatrix} = 6.$$

$$f_{Y_1,Y_2,Y_3}(y_1, y_2, y_3)$$
$$= |J| f_{X_1,X_2,X_3}(x_1, x_2, x_3)$$
$$= 6\left(\frac{1}{\sqrt{2\pi}}\right)^3 \exp\{-\tfrac{1}{2}[y_1^2 + (2y_2 - y_1)^2 + (3y_3 - 2y_2)^2]\}$$
$$= 6\left(\frac{1}{\sqrt{2\pi}}\right)^3 \exp[-\tfrac{1}{2}(2y_1^2 - 4y_1y_2 + 8y_2^2 - 12y_2y_3 + 9y_3^2)].$$

The marginal distributions can be obtained from the joint distribution; for instance,

$$f_{Y_3}(y_3)$$
$$= \int_{-\infty}^{\infty}\int_{-\infty}^{\infty} f_{Y_1,Y_2,Y_3}(y_1, y_2, y_3)\, dy_1\, dy_2$$
$$= 6\left(\frac{1}{\sqrt{2\pi}}\right)^3 \int_{-\infty}^{\infty} \exp[-\tfrac{1}{2}(6y_2^2 - 12y_2y_3 + 9y_3^2)]$$
$$\times\left(\int_{-\infty}^{\infty} \exp[-\tfrac{1}{2}(2y_1^2 - 4y_1y_2 + 2y_2^2)]\, dy_1\right) dy_2$$
$$= \frac{6}{\sqrt{2}}\left(\frac{1}{\sqrt{2\pi}}\right)^2 \int_{-\infty}^{\infty} \exp[-\tfrac{1}{2}(6y_2^2 - 12y_2y_3 + 6y_3^2)]\exp[-\tfrac{1}{2}(3y_3^2)]\, dy_2$$
$$= (\sqrt{3}/\sqrt{2\pi}) \exp[-\tfrac{3}{2}y_3^2];$$

that is, Y_3 is normally distributed with mean 0 and variance $\frac{1}{3}$. ////

PROBLEMS

1 (*a*) Let X_1, X_2, and X_3 be uncorrelated random variables with common variance σ^2. Find the correlation coefficient between $X_1 + X_2$ and $X_2 + X_3$.

(*b*) Let X_1 and X_2 be uncorrelated random variables. Find the correlation coefficient between $X_1 + X_2$ and $X_2 - X_1$ in terms of var $[X_1]$ and var $[X_2]$.

(*c*) Let X_1, X_2, and X_3 be independently distributed random variables with common mean μ and common variance σ^2. Find the correlation coefficient between $X_2 - X_1$ and $X_3 - X_1$.

2 Prove Theorem 2.

3 Let X have c.d.f. $F_X(\cdot) = F(\cdot)$. What in terms of $F(\cdot)$ is the distribution of $XI_{[0,\infty)}(X) = \max[0, X]$?

4 Consider drawing balls, one at a time, without replacement, from an urn containing M balls, K of which are defective. Let the random variable $X(Y)$ denote the number of the draw on which the first defective (nondefective) ball is obtained. Let Z denote the number of the draw on which the rth defective ball is obtained.

(*a*) Find the distribution of X.

(*b*) Find the distribution of Z. (Such distribution is often called the *negative hypergeometric* distribution.)

(*c*) Set $M = 5$ and $K = 2$. Find the joint distribution of X and Y.

5 Let $X_1, \ldots, X_n$ be independent and identically distributed with common density

$$f_X(x) = x^{-2} I_{(1, \infty)}(x).$$

Set $Y = \min [X_1, \ldots, X_n]$. Does $\mathscr{E}[X_1]$ exist? If so, find it. Does $\mathscr{E}[Y]$ exist? If so, find it.

6 Let X and Y be two random variables having finite means.

(*a*) Prove or disprove: $\mathscr{E}[\max [X, Y]] \geq \max [\mathscr{E}[X], \mathscr{E}[Y]]$.

(*b*) Prove or disprove: $\mathscr{E}[\max [X, Y] + \min [X, Y]] = \mathscr{E}[X] + \mathscr{E}[Y]$.

7 The area of a rectangle is obtained by first measuring the length and width and then multiplying the two measurements together. Let X denote the measured length, Y the measured width. Assume that the measurements X and Y are random variables with joint probability density function given by $f_{X,Y}(x, y) = kI_{[.9L,\,1.1L]}(x)I_{[.8W,\,1.2W]}(y)$, where L and W are parameters satisfying $L \geq W > 0$ and k is a constant which may depend on L and W.

(*a*) Find $\mathscr{E}[XY]$ and var $[XY]$.

(*b*) Find the distribution of XY.

8 If X and Y are independent random variables with (negative) exponential distributions having respective parameters λ_1 and λ_2, find $\mathscr{E}[\max [X, Y]]$.

9 Projectiles are fired at the origin of an xy coordinate system. Assume that the point which is hit, say (X, Y), consists of a pair of independent standard normal random variables. For two projectiles fired independently of one another, let (X_1, Y_1) and (X_2, Y_2) represent the points which are hit, and let Z be the distance between them. Find the distribution of Z^2. HINT: What is the distribution of $(X_2 - X_1)^2$? Of $(Y_2 - Y_1)^2$? Is $(X_2 - X_1)^2$ independent of $(Y_2 - Y_1)^2$?

10 A certain explosive device will detonate if any one of n short-lived fuses lasts longer than .8 seconds. Let X_i represent the life of the ith fuse. It can be assumed that each X_i is uniformly distributed over the interval 0 to 1 second. Furthermore, it can be assumed that the X_i's are independent.

(*a*) How many fuses are needed (i.e., how large should n be) if one wants to be 95 percent certain that the device will detonate?

(*b*) If the device has nine fuses, what is the average life of the fuse that lasts the longest?

11 Suppose that random variable X_n has a c.d.f. given by $[(n-1)/n]\Phi(x) + (1/n)F_n(x)$, where $\Phi(\cdot)$ is the c.d.f. of a standard normal and for each n $F_n(\cdot)$ is a c.d.f. What is the limiting distribution of X_n?

12 Let X and Y be independent random variables each having a geometric distribution.

*(*a*) Find the distribution of $X/(X + Y)$. [Define $X/(X + Y)$ to be zero if $X + Y = 0$.]

(*b*) Find the joint moment generating function of X and $X + Y$.

13 Let X_1 and X_2 be independent standard normal random variables. Let $Y_1 = X_1 + X_2$ and $Y_2 = X_1^2 + X_2^2$.

(*a*) Show that the joint moment generating function of Y_1 and Y_2 is

$$\frac{\exp\,[t_1^2/(1-2t_2)]}{1-2t_2} \qquad \text{for } -\infty < t_1 < \infty \text{ and } -\infty < t_2 < \tfrac{1}{2}.$$

(*b*) Find the correlation coefficient of Y_1 and Y_2.

14 Let X and Y be independent standard normal random variables. Find the m.g.f. of XY.

15 Suppose that X_1 and X_2 are independent random variables, each having a standard normal distribution.

(*a*) Find the joint distribution of $(X_1 + X_2)/\sqrt{2}$ and $(X_2 - X_1)/\sqrt{2}$.

(*b*) Argue that $2X_1X_2$ and $X_2^2 - X_1^2$ have the same distribution. HINT:

$$X_2^2 - X_1^2 = 2\,\frac{X_1 + X_2}{\sqrt{2}}\,\frac{X_2 - X_1}{\sqrt{2}}$$

16 A dry-bean supplier fills bean bags with a machine that does not work very well, and he advertises that each bag contains 1 pound of beans. In fact, the weight of the beans that the machine puts into a bag is a random variable with mean 16 ounces and standard deviation 1 ounce. If a box contains 16 bags of beans:

(*a*) Find the mean and variance of the weight of the beans in a box.

(*b*) Find approximately the probability that the weight of the beans in a box exceeds 250 ounces.

(*c*) Find the probability that two or fewer underweight (less than 16 ounce) bags are in the box if the weight of beans in a bag is assumed to be normally distributed.

17 Numbers are selected at random from the interval (0, 1).

(*a*) If 10 numbers are selected, what is the probability that exactly 5 are less than $\frac{1}{2}$?

(*b*) If 10 numbers are selected, on the average how many are less than $\frac{1}{2}$?

(*c*) If 100 numbers are selected, what is the probability that the average of the numbers is less than $\frac{1}{2}$?

18 Let X_i denote the number of meteors that collide with a test satellite during the ith orbit. Let $S_n = \sum_{i=1}^{n} X_i$; that is, S_n is the total number of meteors that collides with the satellite during n orbits. Assume that the X_i's are independent and identically distributed Poisson random variables having mean λ.

(*a*) Find $\mathscr{E}[S_n]$ and var $[S_n]$.

(*b*) If $n = 100$ and $\lambda = 4$, find approximately $P[S_{100} > 440]$.

19 How many light bulbs should you buy if you want to be 95 percent certain that you will have 1000 hours of light if each of the bulbs is known to have a lifetime that is (negative) exponentially distributed with an average life of 100 hours?

(*a*) Assume that all the bulbs are burning simultaneously.

(*b*) Assume that one bulb is used until it burns out and then it is replaced, etc.

20 (*a*) If $X_1, \ldots, X_n$ are independent and identically distributed gamma random variables, what is the distribution of $X_1 + \cdots + X_n$?

(*b*) If $X_1, \ldots, X_n$ are independent gamma random variables and if X_i has parameters r_i and λ, $i = 1, \ldots, n$, what is the distribution of $X_1 + \cdots + X_n$?

21 (*a*) If $X_1, \ldots, X_n$ are independent identically distributed geometric random variables, what is the distribution of $X_1 + \cdots + X_n$?

(*b*) If $X_1, \ldots, X_n$ are independent identically distributed geometric random variables with density $\theta(1-\theta)^{x-1} I_{\{1, 2, \ldots\}}(x)$, what is the distribution of $X_1 + \cdots + X_n$?

(*c*) If $X_1, \ldots, X_n$ are independent identically distributed negative binomial random variables, what is the distribution of $X_1 + \cdots + X_n$?

(*d*) If $X_1, \ldots, X_n$ are independent negative binomial random variables and if X_i has parameters r_i and p, what is the distribution of $X_1 + \cdots + X_n$?

******22* Kitty Oil Co. has decided to drill for oil in 10 different locations; the cost of drilling at each location is \$10,000. (Total cost is then \$100,000.) The probability of finding oil in a given location is only $\frac{1}{5}$, but if oil is found at a given location, then the amount of money the company will get selling oil (excluding the initial \$10,000 drilling cost) from that location is an exponential random variable with mean \$50,000. Let Y be the random variable that denotes the number of locations where oil is found, and let Z denote the total amount of money received from selling oil from all the locations.

(*a*) Find $\mathscr{E}[Z]$.

(*b*) Find $P[Z > 100{,}000 \mid Y = 1]$ and $P[Z > 100{,}000 \mid Y = 2]$.

(*c*) How would you find $P[Z > 100{,}000]$? Is $P[Z > 100{,}000] > \frac{1}{2}$?

23 If $X_1, \ldots, X_k$ are independent Poisson distributed random variables, show that the conditional distribution of X_1, given $X_1 + \cdots + X_k$, is binomial.

******24* Assume that $X_1, \ldots, X_{k+1}$ are independent Poisson distributed random variables with respective parameters $\lambda_1, \ldots, \lambda_{k+1}$. Show that the conditional distribution of $X_1, \ldots, X_k$ given that $X_1 + \cdots + X_{k+1} = n$ has a multinomial distribution with parameters $n, \lambda_1/\lambda, \ldots, \lambda_k/\lambda$, where $\lambda = \lambda_1 + \cdots + \lambda_{k+1}$.

25 If X has a uniform distribution over the interval $(-\pi/2, \pi/2)$, find the distribution of $Y = \tan X$.

26 If X has a normal distribution with mean μ and variance σ^2, find the distribution, mean, and variance of $Y = e^X$.

27 Suppose X has c.d.f. $F_X(x) = \exp[-e^{-(x-\alpha)/\beta}]$. What is the distribution of $Y = \exp[-(X - \alpha)/\beta]$?

28 Let X have density

$$f_X(x; a, b) = \frac{1}{B(a, b)} \frac{x^{a-1}}{(1+x)^{a+b}} I_{(0, \infty)}(x),$$

where $a > 0$ and $b > 0$. (This density is often called a *beta distribution of the second kind*.) Find the distribution of $Y = 1/(1 + X)$.

29 If X has a uniform distribution on the interval $(0, 1)$, find the distribution of $1/X$. Does $\mathscr{E}[1/X]$ exist? If so, find it.

30 (*a*) Give an example of a distribution of a random variable X for which $\mathscr{E}[1/X]$ is not finite.

(*b*) Give an example of a distribution of a random variable X for which $\mathscr{E}[1/X]$ is finite, and evaluate $\mathscr{E}[1/X]$.

31 If $f_X(x) = 2xe^{-x^2}I_{(0, \infty)}(x)$, find the density of $Y = X^2$.

32 If X has a beta distribution, what is the distribution of $1 - X$?

33 If $f_X(x) = e^{-x}I_{(0, \infty)}(x)$, find the distribution of $X/(1 + X)$.

34 If $f_X(x) = 1/\pi(1 + x^2)$, find the distribution of $1/X$.

35 If $f_X(x) = 0$ for $x \leq 0$, find the density of $Y = aX^2 + b$ in terms of $f_X(\cdot)$ for $a > 0$.

36 If X has the Weibull distribution as given in Eq. (42) of Chap. III, what is the distribution of $Y = aX^b$?

37 (*a*) Let $Y = X^2$ and $f_X(x) = (1/\theta)I_{(0, \theta)}(x)$, $\theta > 0$. Find the c.d.f. of X and Y. Find the density of Y.

(*b*) Let $Y = X^2$ and $f_X(x) = (\frac{1}{2}\theta)I_{(-\theta, \theta)}(x)$, $\theta > 0$. Find the c.d.f. and density of Y.

38 If X and Y are independent random variables, each having the same geometric distribution, find the distribution of $Y - X$.

39 If X and Y are independent random variables, each having the same negative exponential distribution, find the distribution of $Y - X$.

40 If X, Y, and Z are independent random variables, each uniformly distributed over (0, 1), what is the distribution of XY/Z?

41 Assume that X and Y are independent random variables, where X has a p.d.f. given by $f_X(x) = 2xI_{(0, 1)}(x)$ and Y has a p.d.f. given by $f_Y(y) = 2(1 - y)I_{(0, 1)}(y)$. Find the distribution of $X + Y$.

***42* Let X and Y be independent Poisson distributed random variables. Find the distribution of $Y - X$.

43 If $f_X(x) = I_{(0, 1)}(x)$, find the density of $Y = 3X + 1$.

***44* Let X and Y be two independent beta-distributed random variables. Is XY always beta-distributed? If not, find conditions on the parameters of X and Y that will imply that XY is beta-distributed.

45 If $f_{X, Y}(x, y) = e^{-(x+y)}I_{(0, \infty)}(x)I_{(0, \infty)}(y)$, find the density of $Z = (X + Y)/2$.

46 If $f_{X, Y}(x, y) = 4xye^{-(x^2+y^2)}I_{(0, \infty)}(x)I_{(0, \infty)}(y)$, find the density of $\sqrt{X^2 + Y^2}$.

47 If $f_{X, Y}(x, y) = 4xyI_{(0, 1)}(x)I_{(0, 1)}(y)$, find the joint density of X^2 and Y^2.

48 If $f_{X, Y}(x, y) = 3xI_{(0,x)}(y)I_{(0, 1)}(x)$, find the density of $Z = X - Y$.

49 If $f_X(x) = [(1 + x)/2]I_{(-1, 1)}(x)$, find the density of $Y = X^2$.

50 If $f_{X, Y}(x, y) = I_{(0, 1)}(x)I_{(0, 1)}(y)$, find the density of Z, where

$$Z = (X + Y)I_{(-\infty, 1]}(X + Y) + (X + Y - 1)I_{(1, \infty)}(X + Y).$$

51 If $f_{X, Y}(x, y) = e^{-(x+y)}I_{(0, \infty)}(x)I_{(0, \infty)}(y)$, find the joint density of X and $X + Y$.

52 If $f_{X, Y, Z}(x, y, z) = e^{-(x+y+z)}I_{(0, \infty)}(x)I_{(0, \infty)}(y)I_{(0, \infty)}(z)$, find the density of their average $(X + Y + Z)/3$.

53 If X_1 and X_2 are independent and each has probability density given by $\lambda e^{-\lambda x}I_{(0, \infty)}(x)$, find the joint distribution of $Y_1 = X_1/X_2$ and $Y_2 = X_1 + X_2$ and the marginal distributions of Y_1 and Y_2.

*54 Let X_1 and X_2 be independent random variables, each normally distributed with parameters $\mu = 0$ and $\sigma^2 = 1$. Find the joint distribution of $Y_1 = X_1^2 + X_2^2$ and $Y_2 = X_1/X_2$. Find the marginal distribution of Y_1 and of Y_2. Are Y_1 and Y_2 independent?

55 If the joint distribution of X and Y is given by

$$f_{X,Y}(x, y) = 2e^{-(x+y)}I_{[0,y]}(x)I_{[0,\infty)}(y),$$

find the joint distribution of X and $X + Y$. Find the marginal distributions of X and $X + Y$.

56 Let $f_{X,Y}(x, y) = K(x + y)I_{(0,1)}(x)I_{(0,1)}(y)I_{(0,1)}(x + y)$.
(a) Find $f_X(\cdot)$.
(b) Find the joint and marginal distributions of $X + Y$ and $Y - X$.

57 Suppose $f_{(X,Y)|Z}(x, y|z) = [z + (1 - z)(x + y)]I_{(0,1)}(x)I_{(0,1)}(y)$ for $0 \le z \le 2$, and $f_Z(z) = \frac{1}{2}I_{[0,2]}(z)$.
(a) Find $\mathscr{E}[X + Y]$.
(b) Are X and Y independent? Verify.
(c) Are X and Z independent? Verify.
(d) Find the joint distribution of X and $X + Y$.
(e) Find the distribution of $\max[X, Y]|Z = z$.
(f) Find the distribution of $(X + Y)|Z = z$.

58 A system will function as long as at least one of three components functions. When all three components are functioning, the distribution of the life of each is exponential with parameter $\frac{1}{3}\lambda$. When only two are functioning, the distribution of the life of each of the two is exponential with parameter $\frac{1}{2}\lambda$; and when only one is functioning, the distribution of its life is exponential with parameter λ.
(a) What is the distribution of the lifetime of the system?
(b) Suppose now that only one component (of the three components) is used at a time and it is replaced when it fails. What is the distribution of the lifetime of such a system?

59 The system in the sketch will function as long as component C_1 and at least one of the components C_2 and C_3 functions. Let X_i be the random variable denoting the lifetime of component C_i, $i = 1, 2$, and 3. Let $Y = \max[X_2, X_3]$ and $Z = \min[X_1, Y]$. Assume that the X_i's are independent (negative) exponential random variables with mean 1.
(a) Find $\mathscr{E}[Z]$ and var $[Z]$.
(b) Find the distribution of the lifetime of the system.

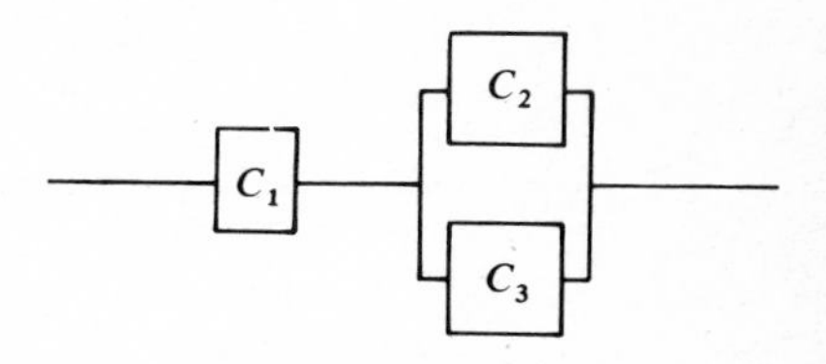

60 A system, which is composed of two components, will function as long as at least one of the two components functions. When both components are operating, the lifetime distribution of each is exponential with mean 1. However, the distribution of the remaining lifetime of the good component, after one fails, is exponential with mean $\frac{1}{2}$. (The idea is that after one component fails the other component carries twice the load and hence has only half the expected lifetime.) Find the lifetime distribution of the system.

**61* Suppose that (X, Y) has a bivariate normal distribution. Find the joint distribution of $aX + bY$ and $cX + dY$ for constants a, b, c, and d satisfying $ad - bc \neq 0$. Find the distribution of $aX + bY$. HINT: Use the moment-generating-function technique and see Example 7.

62 Let X_1 and X_2 be independent standard normal random variables. Let U be independent of X_1 and X_2, and assume that U is uniformly distributed over $(0, 1)$. Define $Z = UX_1 + (1 - U)X_2$.

(*a*) Find the conditional distribution of Z given $U = u$.

(*b*) Find $\mathscr{E}[Z]$ and var $[Z]$.

* (*c*) Find the distribution of Z.

VI

SAMPLING AND SAMPLING DISTRIBUTIONS

1 INTRODUCTION AND SUMMARY

The purpose of this chapter is to introduce the concept of *sampling* and to present some distribution theoretical results that are engendered by sampling. It is a connecting chapter—it merges the distribution theory of the first five chapters into the statistical theory of the last five chapters. The intent is to present here in one location some of the laborious derivations of distributions that are associated with sampling and that will be necessary in our future study of the theory of statistics, especially estimation and testing hypotheses. Our thinking is that by deriving these results now, our later presentation of the statistical theory will not have to be interrupted by their derivations. The nature of the material to be given here is such that it is not easily motivated.

Section 2 begins with a discussion of populations and samples. It ends with the definitions of *statistic* and of *sample moments*. Sample moments are important and useful statistics. Section 3 is devoted to the consideration of various results associated with the *sample mean*. The law of large numbers and the central-limit theorem are given, and then the exact distribution of the sample means from several of the different parametric families of distributions introduced in Chap. III is given. Sampling from the normal distribution is considered in Sec. 4, where the chi-square, F, and t distributions are defined. *Order statistics* are discussed in the final section; they, like sample moments, are important and useful statistics.

2 SAMPLING

2.1 Inductive Inference

Up to now we have been concerned with certain aspects of the theory of probability, including distribution theory. Now the subject of sampling brings us to the theory of statistics proper, and here we shall consider briefly one important area of the theory of statistics and its relation to sampling.

Progress in science is often ascribed to experimentation. The research worker performs an experiment and obtains some data. On the basis of the data, certain conclusions are drawn. The conclusions usually go beyond the materials and operations of the particular experiment. In other words, the scientist may generalize from a particular experiment to the class of all similar experiments. This sort of extension from the particular to the general is called *inductive inference*. It is one way in which new knowledge is found.

Inductive inference is well known to be a hazardous process. In fact, it is a theorem of logic that in inductive inference uncertainty is present. One simply cannot make absolutely certain generalizations. However, uncertain inferences can be made, and the degree of uncertainty can be measured if the experiment has been performed in accordance with certain principles. One function of statistics is the provision of techniques for making inductive inferences and for measuring the degree of uncertainty of such inferences. Uncertainty is measured in terms of probability, and that is the reason we have devoted so much time to the theory of probability.

Before proceeding further we shall say a few words about another kind of inference—*deductive* inference. While conclusions which are reached by inductive inference are only probable, those reached by deductive inference are conclusive. To illustrate deductive inference, consider the following two statements:

(i) One of the interior angles of each right triangle equals 90°.
(ii) Triangle A is a right triangle.

If we accept these two statements, then we are forced to the conclusion:

(iii) One of the angles of triangle A equals 90°.

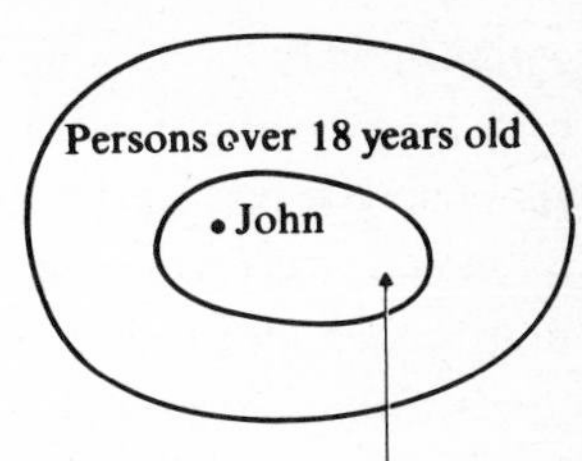

FIGURE 1

This is an example of deductive inference, which can be described as a method of deriving information [statement (iii)] from accepted facts [statements (i) and (ii)]. Statement (i) is called the *major premise*, statement (ii) the *minor premise*, and statement (iii) the *conclusion*. For another example, consider the following:

(i) Major premise: All West Point graduates are over 18 years of age.
(ii) Minor premise: John is a West Point graduate.
(iii) Conclusion: John is over 18 years of age.

West Point graduates is a subset of all persons over 18 years old, and John is an element in the subset of West Point graduates; hence John is also an element in the set of persons who are over 18 years old.

While deductive inference is extremely important, much of the new knowledge in the real world comes about by the process of inductive inference. In the science of mathematics, for example, deductive inference is used to prove theorems, while in the empirical sciences inductive inference is used to find new knowledge.

Let us illustrate inductive inference by a simple example. Suppose that we have a storage bin which contains (let us say) 10 million flower seeds which we know will each produce either white or red flowers. The information which we want is: How many (or what percent) of these 10 million seeds will produce white flowers? Now the only way in which we can be sure that this question is answered correctly is to plant every seed and observe the number producing white flowers. However, this is not feasible since we want to sell the seeds. Even if we did not want to sell the seeds, we would prefer to obtain an answer without expending so much effort. Of course, without planting each seed and observing the color of flower that each produces we cannot be certain of the number of seeds producing white flowers. Another thought which occurs is: Can we plant a few of the seeds and, on the basis of the colors of these few flowers, make a statement as to how many of the 10 million seeds will produce white flowers? The answer is that we cannot make an exact prediction as to how many white flowers the seeds will produce but we can make a probabilistic statement if we select the few seeds in a certain fashion. This is inductive inference. We select a few of the 10 million seeds, plant them, observe the number which produce white flowers, and on the basis of these few we make a prediction as to how many of the 10 million will produce white flowers; from a knowledge of the color of a few we generalize to the whole 10 million. We cannot be certain of our answer, but we can have confidence in it in a frequency-ratio-probability sense.

2.2 Populations and Samples

We have seen in the previous subsection that a central problem in discovering new knowledge in the real world consists of observing a few of the elements under discussion, and on the basis of these few we make a statement about the totality of elements. We shall now investigate this procedure in more detail.

> **Definition 1 Target population** The totality of elements which are under discussion and about which information is desired will be called the *target population.* ////

In the example in the previous subsection the 10 million seeds in the storage bin form the target population. The target population may be all the dairy cattle in Wisconsin on a certain date, the prices of bread in New York City on a certain date, the hypothetical sequence of heads and tails obtained by tossing a certain coin an infinite number of times, the hypothetical set of an infinite number of measurements of the velocity of light, and so forth. The important thing is that the target population must be capable of being quite well defined; it may be real or hypothetical.

The problem of inductive inference is regarded as follows from the point of view of statistics: The object of an investigation is to find out something about a certain target population. It is generally impossible or impractical to examine the entire population, but one may examine a part of it (a sample from it) and, on the basis of this limited investigation, make inferences regarding the entire target population.

The problem immediately arises as to how the sample of the population should be selected. We stated in the previous section that we could make probabilistic statements about the population if the sample is selected in a certain fashion. Of particular importance is the case of a simple random sample, usually called a random sample, which is defined in Definition 2 below for any population which has a density. That is, we assume that each element in our population has some numerical value associated with it and the distribution of these numerical values is given by a density. For such a population we define a random sample.

Definition 2 Random sample Let the random variables $X_1, X_2, \ldots, X_n$ have a joint density $f_{X_1, \ldots, X_n}(\cdot, \ldots, \cdot)$ that factors as follows:

$$f_{X_1, X_2, \ldots, X_n}(x_1, x_2, \ldots, x_n) = f(x_1)f(x_2) \cdot \cdots \cdot f(x_n),$$

where $f(\cdot)$ is the (common) density of each X_i. Then $X_1, X_2, \ldots, X_n$ is defined to be a *random sample* of size n from a population with density $f(\cdot)$. ////

In the example in the previous subsection the 10 million seeds in the storage bin formed the population from which we propose to sample. Each seed is an element of the population and will produce a white or red flower; so, strictly speaking, there is not a numerical value associated with each element of the population. However, if we, say, associate the number 1 with white and the number 0 with red, then there is a numerical value associated with each element of the population, and we can discuss whether or not a particular sample is random. The random variable X_i is then 1 or 0 depending on whether the ith seed sampled produces a white or red flower, $i = 1, \ldots, n$. Now if the sampling of seeds is performed in such a way that the random variables $X_1, \ldots, X_n$ are independent and have the same density, then, according to Definition 2, the sample is called random.

An important part of the definition of a random sample is the meaning of the random variables $X_1, \ldots, X_n$. The random variable X_i is a representation for the numerical value that the ith item (or element) sampled will assume. After the sample is observed, the actual values of $X_1, \ldots, X_n$ are known, and as usual, we denote these observed values by $x_1, \ldots, x_n$. Sometimes the observations $x_1, \ldots, x_n$ are called a random sample if $x_1, \ldots, x_n$ are the values of $X_1, \ldots, X_n$, where $X_1, \ldots, X_n$ is a random sample.

Often it is not possible to select a random sample from the target population, but a random sample can be selected from some related population. To distinguish the two populations, we define sampled population.

Definition 3 Sampled population Let $X_1, X_2, \ldots, X_n$ be a random sample from a population with density $f(\cdot)$; then this population is called the *sampled population*. ////

Valid probability statements can be made about sampled populations on the basis of random samples, but statements about the target populations are not valid in a relative-frequency-probability sense unless the target population is also the sampled population. We shall give some examples to bring out the distinction between the sampled population and the target population.

EXAMPLE 1 Suppose that a sociologist desires to study the religious habits of 20-year-old males in the United States. He draws a sample from the 20-year-old males of a large city to make his study. In this case the target population is the 20-year-old males in the United States, and the sampled population is the 20-year-old males in the city which he sampled. He can draw valid relative-frequency-probabilistic conclusions about his sampled population, but he must use his personal judgment to extrapolate to the target population, and the reliability of the extrapolation cannot be measured in relative-frequency-probability terms. ////

EXAMPLE 2 A wheat researcher is studying the yield of a certain variety of wheat in the state of Colorado. He has at his disposal five farms scattered throughout the state on which he can plant the wheat and observe the yield. The sampled population consists of the yields on these five farms, whereas the target population consists of the yields of wheat on every farm in the state. ////

This book will be concerned with the problem of selecting (drawing) a sample from a sampled population with density $f(\cdot)$, and on the basis of these sample observations probability statements will be made about $f(\cdot)$, or *inferences* about $f(\cdot)$ will be made.

Remark We shall sometimes use the statement "population $f(\cdot)$" to mean "a population with density $f(\cdot)$." When we use the word "population" without an adjective "sampled" or "target," we shall always mean sampled population. ////

2.3 Distribution of Sample

Definition 4 Distribution of sample Let $X_1, X_2, \ldots, X_n$ denote a sample of size n. The *distribution of the sample* $X_1, \ldots, X_n$ is defined to be the joint distribution of $X_1, \ldots, X_n$. ////

Suppose that a random variable X has a density $f(\cdot)$ in some population, and suppose a sample of two values of X, say x_1 and x_2, is drawn at random.

x_1 is called the first observation, and x_2 the second observation. The pair of numbers (x_1, x_2) determines a point in a plane, and the collection of all such pairs of numbers that might have been drawn forms a bivariate population. We are interested in the distribution (bivariate) of this bivariate population in terms of the original density $f(\cdot)$. The pair of numbers (x_1, x_2) is a value of the joint random variable (X_1, X_2), and X_1, X_2 is a random sample (of size 2) from $f(\cdot)$. By definition of random sample, the joint distribution of X_1 and X_2, which we call the distribution of our random sample of size 2, is given by $f_{X_1, X_2}(x_1, x_2) = f(x_1)f(x_2)$.

As a simple example, suppose that X can have only two values, 0 and 1, with probabilities $q = 1 - p$ and p, respectively. That is, X is a discrete random variable which has the Bernoulli distribution

$$f(x) = p^x q^{1-x} I_{\{0,1\}}(x). \tag{1}$$

The joint density for a random sample of two values from $f(\cdot)$ is

$$f_{X_1, X_2}(x_1, x_2) = f(x_1)f(x_2) = p^{x_1 + x_2} q^{2 - x_1 - x_2} I_{\{0,1\}}(x_1) I_{\{0,1\}}(x_2). \tag{2}$$

It is to be observed that this (bivariate) density is not what we obtain as the distribution of the number of successes, say Y, in drawing two elements from a Bernoulli population. The density of Y is given by

$$f_Y(y) = \binom{2}{y} p^y q^{2-y} \qquad \text{for } y = 0, 1, 2.$$

The single random variable Y equals $X_1 + X_2$.

It should be noted that $f_{X_1, X_2}(x_1, x_2)$ gives us the distribution of the sample in *the order drawn*. For instance, in Eq. (2), $f_{X_1, X_2}(0, 1) = pq$ refers to the probability of drawing first a 0 and then a 1.

Our comments for a random sample of size 2 generalize to a random sample of size n, and we have the following remark.

> **Remark** If $X_1, X_2, \ldots, X_n$ is a random sample of size n from $f(\cdot)$, then the distribution of the random sample $X_1, \ldots, X_n$, defined as the joint distribution of $X_1, \ldots, X_n$, is given by $f_{X_1, \ldots, X_n}(x_1, \ldots, x_n) = f(x_1) \cdots f(x_n)$. ////

Note that again this gives the distribution of the sample in the order drawn. Also, note that if $X_1, \ldots, X_n$ is a random sample, then $X_1, \ldots, X_n$ are stochastically independent.

We might further note that our definition of random sampling has automatically ruled out sampling from a finite population without replacement since, then, the results of the drawings are not independent.

2.4 Statistic and Sample Moments

One of the central problems in statistics is the following: It is desired to study a population which has a density $f(\cdot\,;\theta)$, where the form of the density is known but it contains an unknown parameter θ (if θ is known, then the density function is completely specified). The procedure is to take a random sample $X_1, X_2, \ldots, X_n$ of size n from this density and let the value of some function, say $t(x_1, x_2, \ldots, x_n)$, represent or estimate the unknown parameter θ. The problem is to determine which function will be the best one to estimate θ. This problem will be formulated in more detail in the next chapter. In this section we shall examine certain functions, namely, the sample moments, of a random sample. First, however, we shall define what we shall mean by a statistic.

> **Definition 5 Statistic** A *statistic* is a function of observable random variables, which is itself an observable random variable, which does not contain any unknown parameters. ////

The qualification imposed by the word "observable" is required because of the way we intend to use a statistic. ("Observable" means that we can observe the *values* of the random variables.) We intend to use a statistic to make inferences about the density of the random variables, and if the random variables were not observable, they would be of no use in making inferences.

For example, if the observable random variable X has the density $\phi_{\mu,\sigma^2}(x)$, where μ and σ^2 are unknown, then $X - \mu$ is not a statistic; neither is X/σ (since they are not functions of the observable random variable X only—they contain unknown parameters), but X, $X + 3$, and $X^2 + \log X^2$ are statistics.

In the formulation above, one of the central problems in statistics is to find a suitable statistic (function of the random variables $X_1, X_2, \ldots, X_n$) to represent θ.

EXAMPLE 3 If $X_1, \ldots, X_n$ is a random sample from a density $f(\cdot\,;\theta)$, then (provided $X_1, \ldots, X_n$ are observable)

$$\bar{X}_n = \frac{1}{n}\sum_{i=1}^{n} X_i$$

is a statistic, and

$$\tfrac{1}{2}\{\min[X_1, \ldots, X_n] + \max[X_1, \ldots, X_n]\}$$

is also a statistic. If $f(x;\theta) = \phi_{\theta,1}(x)$ and θ is unknown, $\bar{X}_n - \theta$ is not a statistic since it depends on θ, which is unknown. ////

Next we shall define and discuss some important statistics, the sample moments.

Definition 6 Sample moments Let $X_1, X_2, \ldots, X_n$ be a random sample from the density $f(\cdot)$. Then the *rth sample moment about* 0, denoted by M'_r, is defined to be

$$M'_r = \frac{1}{n} \sum_{i=1}^{n} X_i^r. \tag{3}$$

In particular, if $r = 1$, we get the *sample mean*, which is usually denoted by $\bar{X}$ or $\bar{X}_n$; that is,

$$\bar{X}_n = \frac{1}{n} \sum_{i=1}^{n} X_i. \tag{4}$$

Also, the *rth sample moment about* $\bar{X}_n$, denoted by M_r, is defined to be

$$M_r = \frac{1}{n} \sum_{i=1}^{n} (X_i - \bar{X}_n)^r. \tag{5}$$

////

Remark Note that sample moments are examples of statistics. ////

We will consider in detail some properties of the sample mean in Sec. 3 below.

In Chap. II we defined the rth moment of a random variable X, or the rth moment of its corresponding density $f_X(\cdot)$, to be $\mathscr{E}[X^r] = \mu'_r$. We could say that $\mathscr{E}[X^r]$ is the *rth population moment* of the population with density $f(x) = f_X(x)$. We shall now show that the *sample moments* reflect the *population moments* in the sense that the expected value of a sample moment (about 0) equals the corresponding population moment. Also, the variance of a sample moment will be shown to be $(1/n)$ times some function of population moments. The implication is that for a given population the values that the sample moment assume will tend to be more concentrated about the corresponding population moment for large sample size n than for small sample size. Thus a sample moment can be used to estimate its corresponding population moment (provided the population moment exists).

Theorem 1 Let $X_1, X_2, \ldots, X_n$ be a random sample from a population with a density $f(\cdot)$. The expected value of the rth sample moment (about 0) is equal to the rth population moment; that is,

$$\mathscr{E}[M'_r] = \mu'_r \qquad (\text{if } \mu'_r \text{ exists}). \tag{6}$$

Also,

$$\text{var}[M'_r] = \frac{1}{n}\{\mathscr{E}[X^{2r}] - (\mathscr{E}[X^r])^2\} = \frac{1}{n}[\mu'_{2r} - (\mu'_r)^2] \qquad (\text{if } \mu'_{2r} \text{ exists}). \tag{7}$$

PROOF

$$\mathscr{E}[M'_r] = \mathscr{E}\left[\frac{1}{n}\sum_{i=1}^{n} X_i^r\right] = \frac{1}{n}\mathscr{E}\left[\sum_{i=1}^{n} X_i^r\right]$$
$$= \frac{1}{n}\sum_{i=1}^{n}\mathscr{E}[X_i^r] = \frac{1}{n}\sum_{i=1}^{n}\mu'_r = \mu'_r.$$
$$\text{var}[M'_r] = \text{var}\left[\frac{1}{n}\sum_{i=1}^{n} X_i^r\right]$$
$$= \left(\frac{1}{n}\right)^2 \text{var}\left[\sum_{i=1}^{n} X_i^r\right] = \left(\frac{1}{n}\right)^2 \sum_{i=1}^{n}\text{var}[X_i^r]$$
$$= \left(\frac{1}{n}\right)^2 \sum_{i=1}^{n}\{\mathscr{E}[X_i^{2r}] - (\mathscr{E}[X_i^r])^2\} = \frac{1}{n}\{\mathscr{E}[X^{2r}] - (\mathscr{E}[X^r])^2\}$$
$$= \frac{1}{n}[\mu'_{2r} - (\mu'_r)^2].$$

In particular, if $r = 1$, we get the following corollary. ////

Corollary Let $X_1, X_2, \ldots, X_n$ be a random sample from a density $f(\cdot)$, and let $\bar{X}_n = \frac{1}{n}\sum_{i=1}^{n} X_i$ be the sample mean; then

$$\mathscr{E}[\bar{X}_n] = \mu \qquad \text{and} \qquad \text{var}[\bar{X}_n] = \frac{1}{n}\sigma^2, \tag{8}$$

where μ and σ^2 are, respectively, the mean and variance of $f(\cdot)$. ////

As we mentioned earlier, properties of the sample mean will be studied in detail in the next section.

Theorem 1 gives the mean and variance in terms of population moments of the rth sample moment about 0; a similar, though more complicated, result can be derived for the mean and variance of the rth sample moment about the sample mean. We will be content with looking only at the particular case $r = 2$, that is, at $M_2 = (1/n)\sum_{i=1}^{n}(X_i - \bar{X})^2$. M_2 is sometimes called the sample variance; although we will take Definition 7 as our definition of the sample variance.

Definition 7 Sample variance Let $X_1, X_2, \ldots, X_n$ be a random sample from a density $f(\cdot)$; then

$$S_n^2 = S^2 = \frac{1}{n-1}\sum_{i=1}^{n}(X_i - \bar{X})^2 \qquad \text{for} \quad n > 1 \tag{9}$$

is defined to be the *sample variance.* ////

The reason for taking S^2 rather than M_2 as our definition of the sample variance (both measure *dispersion* in the sample) is that the expected value of S^2 equals the population variance.

The proof of the following remark is left as an exercise.

Remark

$$S_n^2 = S^2 = \frac{1}{2n(n-1)}\sum_{i=1}^{n}\sum_{j=1}^{n}(X_i - X_j)^2. \qquad ////$$

Theorem 2 Let $X_1, X_2, \ldots, X_n$ be a random sample from a density $f(\cdot)$, and let

$$S^2 = \frac{1}{n-1}\sum_{i=1}^{n}(X_i - \bar{X})^2.$$

Then

$$\mathscr{E}[S^2] = \sigma^2 \qquad \text{and} \qquad \text{var}\,[S^2] = \frac{1}{n}\left(\mu_4 - \frac{n-3}{n-1}\sigma^4\right) \qquad \text{for } n > 1. \tag{10}$$

PROOF (Only the first part will be proved.) Recall that $\sigma^2 = \mathscr{E}[(X-\mu)^2]$ and $\mu_r = \mathscr{E}[(X-\mu)^r]$. We commence by noting and proving an identity that is quite useful.

$$\sum_{i=1}^{n}(X_i - \mu)^2 = \sum_{i=1}^{n}(X_i - \bar{X})^2 + n(\bar{X} - \mu)^2 \tag{11}$$

since

$$\begin{aligned}
\sum(X_i - \mu)^2 &= \sum(X_i - \bar{X} + \bar{X} - \mu)^2 = \sum[(X_i - \bar{X}) + (\bar{X} - \mu)]^2 \\
&= \sum[(X_i - \bar{X})^2 + 2(X_i - \bar{X})(\bar{X} - \mu) + (\bar{X} - \mu)^2] \\
&= \sum(X_i - \bar{X})^2 + 2(\bar{X} - \mu)\sum(X_i - \bar{X}) + n(\bar{X} - \mu)^2 \\
&= \sum(X_i - \bar{X})^2 + n(\bar{X} - \mu)^2.
\end{aligned}$$

Using the identity of Eq. (11), we obtain

$$\begin{aligned}
\mathscr{E}[S^2] &= \mathscr{E}\left[\frac{1}{n-1}\sum (X_i - \bar{X})^2\right] \\
&= \frac{1}{n-1}\mathscr{E}\left[\sum_{i=1}^{n}(X_i - \mu)^2 - n(\bar{X} - \mu)^2\right] \\
&= \frac{1}{n-1}\left\{\sum_{i=1}^{n}\mathscr{E}[(X_i - \mu)^2] - n\mathscr{E}[(\bar{X} - \mu)^2]\right\} \\
&= \frac{1}{n-1}\left\{\sum_{i=1}^{n}\sigma^2 - n\,\text{var}[\bar{X}]\right\} \\
&= \frac{1}{n-1}\left(n\sigma^2 - n\frac{\sigma^2}{n}\right) = \sigma^2.
\end{aligned}$$

Although the derivation of the formula for the variance of S^2 can be accomplished by utilizing the above identity [Eq. (11)] and

$$\begin{aligned}
\bar{X} - \mu &= \frac{1}{n}\sum X_i - \frac{1}{n}n\mu \\
&= \frac{1}{n}\sum X_i - \frac{1}{n}\sum \mu = \frac{1}{n}\sum (X_i - \mu),
\end{aligned}$$

such derivation is lengthy and is omitted here only to be relegated to the Problems. ////

Sample moments are examples of statistics that can be used to estimate their population counterparts; for example, M_r' estimates μ_r', $\bar{X}$ estimates μ, and S^2 estimates σ^2. In each case, we are taking some function of the sample, which we can observe, and using the value of that function of the sample to estimate the unknown population parameter.

3 SAMPLE MEAN

The first sample moment is the *sample mean* defined to be

$$\bar{X} = \bar{X}_n = \frac{1}{n}\sum_{i=1}^{n} X_i,$$

where $X_1, X_2, \ldots, X_n$ is a random sample from a density $f(\cdot)$. $\bar{X}$ is a function of the random variables $X_1, \ldots, X_n$, and hence theoretically the distribution of $\bar{X}$ can be found. In general, one would suspect that the distribution of $\bar{X}$

depends on the density $f(\cdot)$ from which the random sample was selected, and indeed it does. Two characteristics of the distribution of $\bar{X}$, its mean and variance, do not depend on the density $f(\cdot)$ per se but depend only on two characteristics of the density $f(\cdot)$. This idea is reviewed in the following subsection, while succeeding subsections consider other results involving the sample mean. The exact distribution of $\bar{X}$ will be given for certain specific densities $f(\cdot)$.

It might be helpful in reading this section to think of the sample mean $\bar{X}$ as an estimate of the mean μ of the density $f(\cdot)$ from which the sample was selected. We might think that one purpose in taking the sample is to estimate μ with $\bar{X}$.

3.1 Mean and Variance

Theorem 3 Let $X_1, X_2, \ldots, X_n$ be a random sample from a density $f(\cdot)$, which has mean μ and finite variance σ^2, and let $\bar{X} = (1/n) \sum_{i=1}^{n} X_i$. Then

$$\mathscr{E}[\bar{X}] = \mu_{\bar{X}} = \mu \qquad \text{and} \qquad \operatorname{var}[\bar{X}] = \sigma_{\bar{X}}^2 = \frac{1}{n}\sigma^2. \tag{12}$$

////

Theorem 3 is just a restatement of the corollary of Theorem 1. In light of using a value of $\bar{X}$ to estimate μ, let us note what Theorem 3 says. $\mathscr{E}[\bar{X}] = \mu$ says that on the average $\bar{X}$ is equal to the parameter μ being estimated or that the distribution of $\bar{X}$ is *centered* about μ. $\operatorname{var}[\bar{X}] = (1/n)\sigma^2$ says that the *spread* of the values of $\bar{X}$ about μ is small for a large sample size as compared to a small sample size. For instance, the variance of the distribution of $\bar{X}$ for a sample of size 20 is one-half the variance of the distribution of $\bar{X}$ for a sample of size 10. So for a large sample size the values of $\bar{X}$ (which are used to estimate μ) tend to be more concentrated about μ than for a small sample size. This notion is further exemplified by the law of large numbers considered in the next subsection.

3.2 Law of Large Numbers

Let $f(\cdot\,; \theta)$ be the density of a random variable X. We have discussed the fact that one way to get some information about the density function $f(\cdot\,; \theta)$ is to observe a random sample and make an inference from the sample to the population. If θ were known, the density functions would be completely specified, and no inference from the sample to the population would be necessary. Therefore, it seems that we would like to have the random sample tell us something about the unknown parameter θ. This problem will be discussed in detail in the next chapter. In this subsection we shall discuss a related particular problem.

Let $\mathscr{E}[X]$ be denoted by μ in the density $f(\cdot)$. The problem is to estimate μ. In a loose sense, $\mathscr{E}[X]$ is the average of an infinite number of values of the random variable X. In any real-world problem we can observe only a finite number of values of the random variable X. A very crucial question then is: Using only a finite number of values of X (a random sample of size n, say), can any reliable inferences be made about $\mathscr{E}[X]$, the average of an infinite number of values of X? The answer is "yes"; reliable inferences about $\mathscr{E}[X]$ can be made by using only a finite sample, and we shall demonstrate this by proving what is called the *weak law of large numbers*. In words, the law states the following: A positive integer n can be determined such that if a random sample of size n or larger is taken from a population with the density $f(\cdot)$ (with $\mathscr{E}[X] = \mu$), the probability can be made to be as close to 1 as desired that the sample mean $\bar{X}$ will deviate from μ by less than any arbitrarily specified small quantity. More precisely, the weak law of large numbers states that for any two chosen small numbers ε and δ, where $\varepsilon > 0$ and $0 < \delta < 1$, there exists an integer n such that if a random sample of size n or larger is obtained from $f(\cdot)$ and the sample mean, denoted by $\bar{X}_n$, computed, then the probability is greater than $1 - \delta$ (i.e., as close to 1 as desired) that $\bar{X}_n$ deviates from μ by less than ε (i.e., is arbitrarily close to μ). In symbols this is written: For any $\varepsilon > 0$ and $0 < \delta < 1$ there exists an integer n such that for all integers $m \geq n$

$$P[|\bar{X}_m - \mu| < \varepsilon] \geq 1 - \delta.$$

The weak law of large numbers is proved using the Chebyshev inequality given in Chap. II.

Theorem 4 Weak law of large numbers Let $f(\cdot)$ be a density with mean μ and finite variance σ^2, and let $\bar{X}_n$ be the sample mean of a random sample of size n from $f(\cdot)$. Let ε and δ be any two specified numbers satisfying $\varepsilon > 0$ and $0 < \delta < 1$. If n is any integer greater than $\sigma^2/\varepsilon^2 \delta$, then

$$P[-\varepsilon < \bar{X}_n - \mu < \varepsilon] \geq 1 - \delta. \tag{13}$$

PROOF Theorem 5 in Subsec. 4.4 of Chap. II stated that $P[g(X) \geq k] \leq \mathscr{E}[g(X)]/k$ for every $k > 0$, random variable X, and nonnegative function $g(\cdot)$. Equivalently, $P[g(X) < k] \geq 1 - \mathscr{E}[g(X)]/k$.

Let $g(X) = (\bar{X}_n - \mu)^2$ and $k = \varepsilon^2$; then

$$P[-\varepsilon < \bar{X}_n - \mu < \varepsilon] = P[|\bar{X}_n - \mu| < \varepsilon]$$

$$= P[|\bar{X}_n - \mu|^2 < \varepsilon^2] \geq 1 - \frac{\mathscr{E}[(\bar{X}_n - \mu)^2]}{\varepsilon^2}$$

$$= 1 - \frac{(1/n)\sigma^2}{\varepsilon^2} \geq 1 - \delta$$

for $\delta > \sigma^2/n\varepsilon^2$ or $n > \sigma^2/\varepsilon^2\,\delta$. ////

Below are two examples to illustrate how the weak law of large numbers can be used.

EXAMPLE 4 Suppose that some distribution with an unknown mean has variance equal to 1. How large a sample must be taken in order that the probability will be at least .95 that the sample mean $\bar{X}_n$ will lie within .5 of the population mean? We have $\sigma^2 = 1$, $\varepsilon = .5$, and $\delta = .05$; therefore

$$n > \frac{\sigma^2}{\delta\varepsilon^2} = \frac{1}{.05(.5)^2} = 80.$$ ////

EXAMPLE 5 How large a sample must be taken in order that you are 99 percent certain that $\bar{X}_n$ is within $.5\sigma$ of μ? We have $\varepsilon = .5\sigma$ and $\delta = .01$. Thus

$$n > \frac{\sigma^2}{\delta\varepsilon^2} = \frac{\sigma^2}{.01(.5)^2\sigma^2} = \frac{1}{.01(.5)^2} = 400.$$ ////

We have shown that by use of a random sample inductive inferences to populations can be made and the reliability of the inferences can be measured in terms of probability. For instance, in Example 4 above, the probability that the sample mean will be within one-half unit of the unknown population mean is at least .95 if a sample of size greater than 80 is taken.

3.3 Central-limit Theorem

Although we have already stated the central-limit theorem in our study of distribution theory in Chap. V, we will repeat it here in our study of the sample mean $\bar{X}$ because it gives the asymptotic distribution of $\bar{X}$. At the outset of this section we indicated that we were interested in the distribution of $\bar{X}$. The central-limit theorem, which is one of the most important theorems in all of probability and statistics, tells us approximately how $\bar{X}$ is distributed.

Theorem 5 Central-limit theorem Let $f(\cdot)$ be a density with mean μ and finite variance σ^2. Let $\bar{X}_n$ be the sample mean of a random sample of size n from $f(\cdot)$. Let the random variable Z_n be defined by

$$Z_n = \frac{\bar{X}_n - \mathscr{E}[\bar{X}_n]}{\sqrt{\text{var}\,[\bar{X}_n]}} = \frac{\bar{X}_n - \mu}{\sigma/\sqrt{n}}. \tag{14}$$

Then, the distribution of Z_n approaches the standard normal distribution as n approaches infinity. ////

Theorem 5 tells us that the limiting distribution of Z_n (which is $\bar{X}_n$ standardized) is a standard normal distribution, or it tells us that $\bar{X}_n$ itself is approximately, or asymptotically, distributed as a normal distribution with mean μ and variance σ^2/n.

The astonishing thing about Theorem 5 is the fact that nothing is said about the form of the original density function. Whatever the distribution function, provided only that it has a finite variance, the sample mean will have approximately the normal distribution for large samples. The condition that the variance be finite is not a critical restriction so far as applied statistics is concerned because in almost any practical situation the range of the random variable will be finite, in which case the variance must necessarily be finite.

The importance of Theorem 5, as far as practical applications are concerned, is the fact that the mean $\bar{X}_n$ of a random sample from any distribution with finite variance σ^2 and mean μ is approximately distributed as a normal random variable with mean μ and variance σ^2/n.

We shall not be able to prove Theorem 5 because it requires rather advanced mathematical techniques. However, in order to make the theorem plausible, we shall outline a proof for the more restricted situation in which the distribution has a moment generating function. The argument will be essentially a matter of showing that the moment generating function for the sample mean approaches the moment generating function for the normal distribution.

Recall that the moment generating function of a standard normal distribution is given by $e^{\frac{1}{2}t^2}$. (See Subsec. 3.2 of Chap. III.) Let $m(t) = e^{\frac{1}{2}t^2}$. Let $m_{Z_n}(t)$ denote the moment generating function of Z_n. It is our purpose to show that $m_{Z_n}(t)$ must approach $m(t)$ when n, the sample size, becomes large.

Now

$$m_{Z_n}(t) = \mathscr{E}[e^{tZ_n}] = \mathscr{E}[\exp tZ_n] = \mathscr{E}\left[\exp\left(t\,\frac{\bar{X} - \mu}{\sigma/\sqrt{n}}\right)\right] = \mathscr{E}\left[\exp\left(\frac{t}{n}\sum\frac{X_i - \mu}{\sigma/\sqrt{n}}\right)\right]$$

$$= \mathscr{E}\left[\prod_{i=1}^{n}\exp\left(\frac{t}{n}\cdot\frac{X_i - \mu}{\sigma/\sqrt{n}}\right)\right] = \prod_{i=1}^{n}\mathscr{E}\left[\exp\left(\frac{t}{\sqrt{n}}\cdot\frac{X_i - \mu}{\sigma}\right)\right]$$

using the independence of $X_1, \ldots, X_n$. Now if we let $Y_i = (X_i - \mu)/\sigma$, then $m_{Y_i}(t)$, the moment generating function of Y_i, is independent of i since all Y_i have the same distribution. Let $m_Y(t)$ denote $m_{Y_i}(t)$; then

$$\prod_{i=1}^{n}\mathscr{E}\left[\exp\left(\frac{t}{\sqrt{n}}\cdot\frac{X_i - \mu}{\sigma}\right)\right] = \prod_{i=1}^{n}\mathscr{E}\left[\exp\left(\frac{t}{\sqrt{n}}\cdot Y_i\right)\right]$$

$$= \prod_{i=1}^{n} m_{Y_i}\left(\frac{t}{\sqrt{n}}\right) = \prod_{i=1}^{n} m_Y\left(\frac{t}{\sqrt{n}}\right) = \left[m_Y\left(\frac{t}{\sqrt{n}}\right)\right]^n.$$

Hence,

$$m_{Z_n}(t) = \left[m_Y\left(\frac{t}{\sqrt{n}}\right)\right]^n. \tag{15}$$

The rth derivative of $m_Y(t/\sqrt{n})$ evaluated at $t = 0$ gives us the rth moment about the mean of the density $f(\cdot)$ divided by $(\sigma\sqrt{n})^r$, so we may write

$$m_Y\left(\frac{t}{\sqrt{n}}\right) = 1 + \frac{\mu_1}{\sigma}\frac{t}{\sqrt{n}} + \frac{1}{2!}\frac{\mu_2}{\sigma^2}\left(\frac{t}{\sqrt{n}}\right)^2 + \frac{1}{3!}\frac{\mu_3}{\sigma^3}\left(\frac{t}{\sqrt{n}}\right)^3 + \cdots, \tag{16}$$

and since $\mu_1 = 0$ and $\mu_2 = \sigma^2$, this may be written

$$m_Y\left(\frac{t}{\sqrt{n}}\right) = 1 + \frac{1}{n}\left(\frac{1}{2}t^2 + \frac{1}{3!\sqrt{n}}\frac{\mu_3}{\sigma^3}t^3 + \frac{1}{4!n}\frac{\mu_4}{\sigma^4}t^4 + \cdots\right). \tag{17}$$

Now $\lim_{n\to\infty}(1 + u/n)^n = e^{\frac{1}{2}t^2}$, where u represents the expression within the parentheses in Eq. (17). We have $\lim_{n\to\infty} m_{Z_n}(t) = e^{\frac{1}{2}t^2}$, so that in the limit, Z_n has the same moment generating function as a standard normal and, by a theorem similar to Theorem 7 in Chap. II, has the same distribution.

The degree of approximation depends, of course, on the sample size and on the particular density $f(\cdot)$. The approach to normality is illustrated in Fig. 2 for the particular function defined by $f(x) = e^{-x} I_{(0,\infty)}(x)$. The solid curves give the actual distributions, while the dashed curves give the normal approximations. Figure 2*a* gives the original distribution which corresponds to samples of 1; Fig. 2*b* shows the distribution of sample means for $n = 3$; Fig. 2*c*

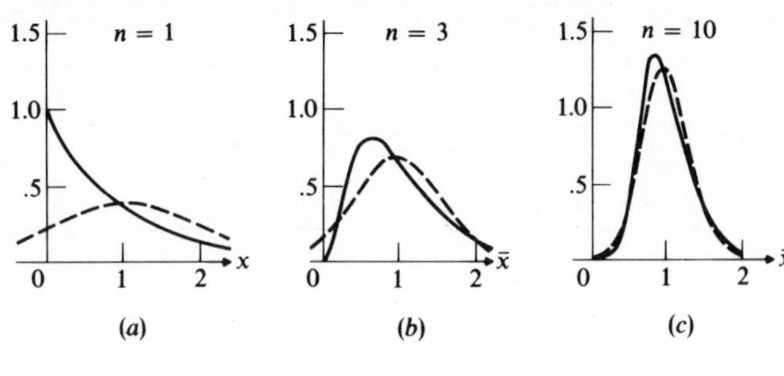

FIGURE 2

gives the distribution of sample means for $n = 10$. The curves rather exaggerate the approach to normality because they cannot show what happens on the tails of the distribution. Ordinarily distributions of sample means approach normality fairly rapidly with the sample size in the region of the mean, but more slowly at points distant from the mean; usually the greater the distance of a point from the mean, the more slowly the normal approximation approaches the actual distribution.

In the following subsections we will give the exact distribution of the sample mean for some specific densities $f(\cdot)$.

3.4 Bernoulli and Poisson Distributions

If $X_1, X_2, \ldots, X_n$ is a random sample from a Bernoulli distribution, we can find the exact distribution of $\bar{X}_n$. (We know that $\bar{X}_n$ is approximately normally distributed.) The density from which we are sampling is

$$f(x) = p^x(1-p)^{1-x}I_{\{0,1\}}(x).$$

We know (see Example 9 of Chap. V) that $\sum_1^n X_i$ has a binomial distribution; that is,

$$P\left[\sum_{i=1}^n X_i = k\right] = \binom{n}{k}p^k q^{n-k}I_{\{0,1,\cdots,n\}}(k);$$

hence, the distribution of $\bar{X}_n$ is given by

$$P\left[\bar{X}_n = \frac{k}{n}\right] = \binom{n}{k}p^k q^{n-k} \qquad \text{for } k = 0, 1, \ldots, n. \tag{18}$$

So $\bar{X}_n$, the sample mean of a random sample from a Bernoulli density, takes on the values 0, $1/n$, $2/n$, ..., 1 with respective binomial probabilities

$$\binom{n}{0}p^0q^n, \quad \binom{n}{1}p^1q^{n-1}, \quad \binom{n}{2}p^2q^{n-2}, \ldots, \quad \binom{n}{n}p^nq^0.$$

If $X_1, \ldots, X_n$ is a random sample from a Poisson distribution with mean λ, then $\sum X_i$ also has a Poisson distribution with parameter $n\lambda$ (see Example 10 of Chap. V), and hence

$$P\left[\bar{X}_n = \frac{k}{n}\right] = P\left[\sum_{i=1}^{n} X_i = k\right] = \frac{e^{-n\lambda}(n\lambda)^k}{k!} \quad \text{for} \quad k = 0, 1, 2, \ldots, \tag{19}$$

which gives the exact distribution of the sample mean for a sample from a Poisson density.

3.5 Exponential Distribution

Let $X_1, X_2, \ldots, X_n$ be a random sample from the exponential density

$$f(x) = \theta e^{-\theta x} I_{(0, \infty)}(x).$$

According to Example 11 of Chap. V, $\sum_1^n X_i$ has a gamma distribution with parameters n and θ; that is,

$$f_{\Sigma X_i}(z) = \frac{1}{\Gamma(n)} z^{n-1}\theta^n e^{-\theta z} I_{(0, \infty)}(z),$$

or

$$P[\sum X_i \le y] = \int_0^y \frac{1}{\Gamma(n)} z^{n-1}\theta^n e^{-\theta z}\, dz \qquad \text{for } y > 0,$$

and so

$$P\left[\bar{X}_n \le \frac{y}{n}\right] = \int_0^y \frac{1}{\Gamma(n)} z^{n-1}\theta^n e^{-\theta z}\, dz \qquad \text{for} \quad y > 0.$$

Or,

$$P[\bar{X}_n \le x] = \int_0^{nx} \frac{1}{\Gamma(n)} z^{n-1}\theta^n e^{-\theta z}\, dz$$
$$= \int_0^x \frac{1}{\Gamma(n)} (nu)^{n-1}\theta^n e^{-n\theta u} n\, du;$$

that is, $\bar{X}_n$ has a gamma distribution with parameters n and $n\theta$.

3.6 Uniform Distribution

Let $X_1, \ldots, X_n$ be a random sample from a uniform distribution on the interval (0, 1]. The exact density of $\bar{X}_n$ is given by

$$f_{\bar{X}_n}(x) = \sum_{k=0}^{n-1} \frac{n}{(n-1)!}\left[(nx)^{n-1} - \binom{n}{1}(nx-1)^{n-1} + \binom{n}{2}(nx-2)^{n-1} - \cdots + (-1)^k\binom{n}{k}(nx-k)^{n-1}\right] I_{(k/n,\,(k+1)/n]}(x). \tag{20}$$

The derivation of the above (using mathematical induction and the convolution formula) is rather tedious and is omitted. Instead let us look at the particular cases $n = 1, 2, 3$.

$$f_{\bar{X}_1}(x) = f_X(x) = I_{(0,1]}(x),$$

$$f_{\bar{X}_2}(x) = 2(2x)I_{(0,\frac{1}{2}]}(x) + 2[2x - 2(2x-1)]I_{(\frac{1}{2},1]}(x) = \begin{cases} 4x & \text{for } 0 < x \le \frac{1}{2} \\ 4(1-x) & \text{for } \frac{1}{2} < x \le 1, \end{cases}$$

and

$$f_{\bar{X}_3}(x) = \frac{3}{2}(3x)^2 I_{(0,\frac{1}{3}]}(x) + \frac{3}{2}\left[(3x)^2 - \binom{3}{1}(3x-1)^2\right] I_{(\frac{1}{3},\frac{2}{3}]}(x) + \frac{3}{2}\left[(3x)^2 - \binom{3}{1}(3x-1)^2 + \binom{3}{2}(3x-2)^2\right] I_{(\frac{2}{3},1]}(x)$$

$$= \begin{cases} \frac{27}{2}x^2 & \text{for } 0 < x \le \frac{1}{3} \\ 27[\frac{1}{12} - (x-\frac{1}{2})^2] & \text{for } \frac{1}{3} < x \le \frac{2}{3} \\ \frac{27}{2}(1-x)^2 & \text{for } \frac{2}{3} < x \le 1. \end{cases}$$

$f_{\bar{X}_1}(x)$, $f_{\bar{X}_2}(x)$, and $f_{\bar{X}_3}(x)$ are sketched in Fig. 3, and an approach to normality can be observed. (In fact, the inflection points of $f_{\bar{X}_3}(x)$ and of the normal approximation occur at the same points!)

We have given the distribution of the sample mean from a uniform distribution on the interval (0, 1]; the distribution of the sample mean from a uniform distribution over an arbitrary interval $(a, b]$ can be found by transformation.

3.7 Cauchy Distribution

Let $X_1, \ldots, X_n$ be a random sample from the Cauchy density

$$f(x) = \frac{1}{\pi\beta\{1 + [(x-\alpha)/\beta]^2\}};$$

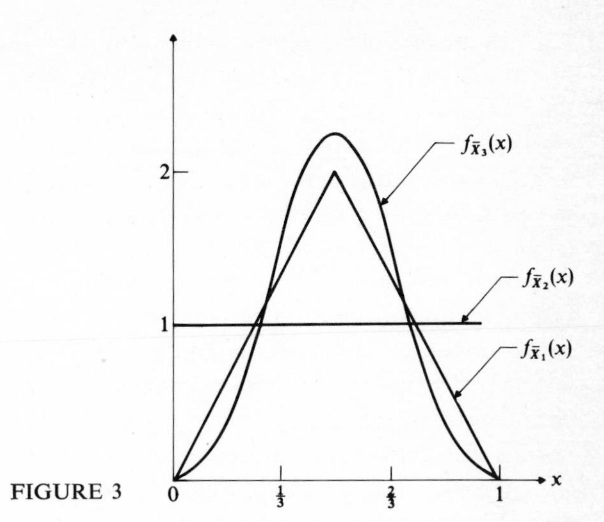

FIGURE 3

then $\bar{X}_n$ has this same Cauchy distribution for any n. That is, the sample mean has the same distribution as one of its components. We are unable to easily verify this result. The moment-generating-function technique fails us since the moment generating function of a Cauchy distribution does not exist. Mathematical induction in conjunction with the convolution formula produces integrations that are apt to be difficult for a nonadvanced calculus student to perform. The result, however, is easily obtained using complex-variable analysis. In fact, if we had defined the characteristic function of a random variable, which is a generalization of a moment generating function, then the above result would follow immediately from the fact that the product of the characteristic functions of independent and identically distributed random variables is the characteristic function of their sum. A major advantage of characteristic functions over moment generating functions is that they always exist.

4 SAMPLING FROM THE NORMAL DISTRIBUTIONS

4.1 Role of the Normal Distribution in Statistics

It will be found in the ensuing chapters that the normal distribution plays a very predominant role in statistics. Of course, the central-limit theorem alone ensures that this will be the case, but there are other almost equally important reasons.

In the first place, many populations encountered in the course of research in many fields seem to have a normal distribution to a good degree of approximation. It has often been argued that this phenomenon is quite reasonable in view of the central-limit theorem. We may consider the firing of a shot at a target as an illustration. The course of the projectile is affected by a great many factors, all admittedly with small effect. The net deviation is the net effect of all these factors. Suppose that the effect of each factor is an observation from some population; then the total effect is essentially the mean of a set of observations from a set of populations. Being of the nature of means, the actual observed deviations might therefore be expected to be approximately normally distributed. We do not intend to imply here that most distributions encountered in practice are normal, for such is not the case at all, but nearly normal distributions are encountered quite frequently.

Another consideration which favors the normal distribution is the fact that sampling distributions based on a parent normal distribution are fairly manageable analytically. In making inferences about populations from samples, it is necessary to have the distributions for various functions of the sample observations. The mathematical problem of obtaining these distributions is often easier for samples from a normal population than from any other, and the remaining subsections of this section will be devoted to the problem of finding the distributions of several different functions of a random sample from a normally distributed population.

In applying statistical methods based on the normal distribution, the experimenter must know, at least approximately, the general form of the distribution function which his data follow. If it is normal, he may use the methods directly; if it is not, he may sometimes transform his data so that the transformed observations follow a normal distribution. When the experimenter does not know the form of his population distribution, then he may use other more general but usually less powerful methods of analysis called *nonparametric* methods. Some of these methods will be presented in the final chapter of this book.

4.2 Sample Mean

One of the simplest of all the possible functions of a random sample is the sample mean, and for a random sample from a normal distribution the distribution (exact) of the sample mean is also normal. This result first appeared as a special case of Example 12 in Chap. V. It is repeated here.

Theorem 6 Let $\bar{X}_n$ denote the sample mean of a random sample of size n from a normal distribution with mean μ and variance σ^2. Then $\bar{X}_n$ has a normal distribution with mean μ and variance σ^2/n.

PROOF To prove this theorem we shall use the moment-generating-function technique.

$$\begin{aligned} m_{\bar{X}_n}(t) &= \mathscr{E}[\exp t\bar{X}_n] = \mathscr{E}\left[\exp \frac{t\sum X_i}{n}\right] \\ &= \mathscr{E}\left[\prod_{i=1}^{n} \exp \frac{tX_i}{n}\right] = \prod_{i=1}^{n} \mathscr{E}\left[\exp \frac{tX_i}{n}\right] \\ &= \prod_{i=1}^{n} m_{X_i}\left(\frac{t}{n}\right) = \prod_{i=1}^{n} \exp\left[\frac{\mu t}{n} + \frac{1}{2}\left(\frac{\sigma t}{n}\right)^2\right] \\ &= \exp\left[\mu t + \frac{\frac{1}{2}(\sigma t)^2}{n}\right], \end{aligned}$$

which is the moment generating function of a normal distribution with mean μ and variance σ^2/n. ////

Since we have the exact distribution of $\bar{X}_n$, in considering estimating μ with $\bar{X}_n$, we will be able to calculate, for instance, the (exact) probability that our "estimator" $\bar{X}_n$ is within any fixed amount of the unknown parameter μ.

4.3 The Chi-Square Distribution

The normal distribution has two unknown parameters μ and σ^2. In the previous subsection we found the distribution of $\bar{X}_n$, which "estimates" the unknown μ. In this subsection, we seek the distribution of

$$\mathcal{S}^2 = \frac{1}{n-1} \sum_{i=1}^{n} (X_i - \bar{X})^2,$$

which "estimates" the unknown σ^2. A density function which plays a central role in the derivation of the distribution of $\mathcal{S}^2$ is the chi-square distribution.

Definition 8 Chi-square distribution If X is a random variable with density

$$f_X(x) = \frac{1}{\Gamma(k/2)} \left(\frac{1}{2}\right)^{k/2} x^{k/2-1} e^{-\frac{1}{2}x} I_{(0,\infty)}(x), \tag{21}$$

then X is defined to have a *chi-square distribution with k degrees of freedom*; or the density given in Eq. (21) is called a *chi-square density with k degrees of freedom*, where the parameter k, called the *degrees of freedom*, is a positive integer. ////

Remark We note that a chi-square density is a particular case of a gamma density with gamma parameters r and λ equal, respectively, to $k/2$ and $\frac{1}{2}$. Hence, if a random variable X has a chi-square distribution,

$$\mathscr{E}[X] = \frac{k/2}{\frac{1}{2}} = k,$$

$$\operatorname{var}[X] = \frac{k/2}{(1/2)^2} = 2k, \tag{22}$$

and

$$m_X(t) = \left[\frac{\frac{1}{2}}{\frac{1}{2} - t}\right]^{k/2} = \left[\frac{1}{1 - 2t}\right]^{k/2}, \qquad t < 1/2. \tag{23}$$

////

Theorem 7 If the random variables X_i, $i = 1, 2, \ldots, k$, are normally and independently distributed with means μ_i and variances σ_i^2, then

$$U = \sum_{i=1}^{k} \left(\frac{X_i - \mu_i}{\sigma_i}\right)^2$$

has a chi-square distribution with k degrees of freedom.

PROOF Write $Z_i = (X_i - \mu_i)/\sigma_i$; then Z_i has a standard normal distribution. Now

$$m_U(t) = \mathscr{E}[\exp tU] = \mathscr{E}[\exp(t \sum Z_i^2)]$$
$$= \mathscr{E}\left[\prod_{i=1}^{n} \exp tZ_i^2\right] = \prod_{i=1}^{k} \mathscr{E}[\exp tZ_i^2].$$

But

$$\mathscr{E}[\exp tZ_i^2] = \int_{-\infty}^{\infty} e^{tz^2}\left(\frac{1}{\sqrt{2\pi}}\right)e^{-\frac{1}{2}z^2}\,dz$$
$$= \int_{-\infty}^{\infty} \frac{1}{\sqrt{2\pi}} e^{-\frac{1}{2}(1-2t)z^2}\,dz$$
$$= \frac{1}{\sqrt{1-2t}} \int_{-\infty}^{\infty} \frac{\sqrt{1-2t}}{\sqrt{2\pi}} e^{-\frac{1}{2}(1-2t)z^2}\,dz$$
$$= \frac{1}{\sqrt{1-2t}} \qquad \text{for} \quad t < \frac{1}{2},$$

the latter integral being unity since it represents the area under a normal curve with variance $1/(1-2t)$. Hence,

$$\prod_{i=1}^{k} \mathscr{E}[\exp tZ_i^2] = \prod_{i=1}^{k} \frac{1}{\sqrt{1-2t}} = \left(\frac{1}{1-2t}\right)^{k/2} \qquad \text{for} \quad t < \frac{1}{2},$$

the moment generating function of a chi-square distribution with k degrees of freedom. ////

Corollary If $X_1, \ldots, X_n$ is a random sample from a normal distribution with mean μ and variance σ^2, then $U = \sum_{i=1}^{n} (X_i - \mu)^2/\sigma^2$ has a chi-square distribution with n degrees of freedom. ////

We might note that if either μ or σ^2 is unknown, the U in the above corollary is not a statistic. On the other hand, if μ is known and σ^2 is unknown, we could estimate σ^2 with $(1/n)\sum_{i=1}^{n}(X_i - \mu)^2$ $\left\{\text{note that } \mathscr{E}\left[(1/n)\sum_{i=1}^{n}(X_i - \mu)^2\right] = (1/n)\sum_{i=1}^{n}\mathscr{E}[(X_i-\mu)^2] = (1/n)\sum_{i=1}^{n}\sigma^2 = \sigma^2\right\}$, and find the distribution of $(1/n)\sum_{i=1}^{n}(X_i - \mu)^2$ by using the corollary.

Remark In words, Theorem 7 says, "the sum of the squares of independent standard normal random variables has a chi-square distribution with degrees of freedom equal to the number of terms in the sum." ////

Theorem 8 If $Z_1, Z_2, \ldots, Z_n$ is a random sample from a standard normal distribution, then:

(i) $\bar{Z}$ has a normal distribution with mean 0 and variance $1/n$.

(ii) $\bar{Z}$ and $\sum_{i=1}^{n}(Z_i - \bar{Z})^2$ are independent.

(iii) $\sum_{i=1}^{n}(Z_i - \bar{Z})^2$ has a chi-square distribution with $n-1$ degrees of freedom.

PROOF (Our proof will be incomplete.) (i) is a special case of Theorem 6. We will prove (ii) for the case $n = 2$. If $n = 2$,

$$\bar{Z} = \frac{Z_1 + Z_2}{2}$$

and

$$\sum (Z_i - \bar{Z})^2 = \left(Z_1 - \frac{Z_1 + Z_2}{2}\right)^2 + \left(Z_2 - \frac{Z_1 + Z_2}{2}\right)^2$$
$$= \frac{(Z_1 - Z_2)^2}{4} + \frac{(Z_2 - Z_1)^2}{4}$$
$$= \frac{(Z_2 - Z_1)^2}{2};$$

so $\bar{Z}$ is a function of $Z_1 + Z_2$, and $\sum (Z_i - \bar{Z})^2$ is a function of $Z_2 - Z_1$; so to prove $\bar{Z}$ and $\sum (Z_i - \bar{Z})^2$ are independent, it suffices to show that $Z_1 + Z_2$ and $Z_2 - Z_1$ are independent. Now

$$m_{Z_1+Z_2}(t_1) = \mathscr{E}[e^{t_1(Z_1+Z_2)}] = \mathscr{E}[e^{t_1Z_1}e^{t_1Z_2}] = \mathscr{E}[e^{t_1Z_1}]\mathscr{E}[e^{t_1Z_2}]$$
$$= \exp \tfrac{1}{2}t_1^2 \exp \tfrac{1}{2}t_1^2 = \exp t_1^2,$$

and, similarly,

$$m_{Z_2-Z_1}(t_2) = \exp t_2^2.$$

Also,

$$m_{Z_1+Z_2, Z_2-Z_1}(t_1, t_2) = \mathscr{E}[e^{t_1(Z_1+Z_2)+t_2(Z_2-Z_1)}]$$
$$= \mathscr{E}[e^{(t_1-t_2)Z_1}e^{(t_1+t_2)Z_2}] = \mathscr{E}[e^{(t_1-t_2)Z_1}]\mathscr{E}[e^{(t_1+t_2)Z_2}]$$
$$= e^{\frac{1}{2}(t_1-t_2)^2}e^{\frac{1}{2}(t_1+t_2)^2} = \exp t_1^2 \exp t_2^2$$
$$= m_{Z_1+Z_2}(t_1)m_{Z_2-Z_1}(t_2);$$

and since the joint moment generating function factors into the product of the marginal moment generating functions, $Z_1 + Z_2$ and $Z_2 - Z_1$ are independent.

To prove (iii), we accept the independence of $\bar{Z}$ and $\sum_1^n (Z_i - \bar{Z})^2$ for arbitrary n. Let us note that $\sum Z_i^2 = \sum (Z_i - \bar{Z} + \bar{Z})^2 = \sum (Z_i - \bar{Z})^2 + 2\bar{Z} \sum (Z_i - \bar{Z}) + \sum \bar{Z}^2 = \sum (Z_i - \bar{Z})^2 + n\bar{Z}^2$; also $\sum (Z_i - \bar{Z})^2$ and $n\bar{Z}^2$ are independent; hence

$$m_{\Sigma Z_i^2}(t) = m_{\Sigma (Z_i - \bar{Z})^2}(t)m_{n\bar{Z}^2}(t).$$

So,

$$m_{\Sigma (Z_i - \bar{Z})^2}(t) = \frac{m_{\Sigma Z_i^2}(t)}{m_{n\bar{Z}^2}(t)} = \frac{(1/(1-2t))^{n/2}}{(1/(1-2t))^{\frac{1}{2}}} = \left(\frac{1}{1-2t}\right)^{(n-1)/2}, \qquad t < 1/2$$

noting that $\sqrt{n}\bar{Z}$ has a standard normal distribution implying that $n\bar{Z}^2$ has a chi-square distribution with one degree of freedom. We have shown that the moment generating function of $\sum (Z_i - \bar{Z})^2$ is that of a chi-square distribution with $n - 1$ degrees of freedom, which completes the proof. ////

Theorem 8 was stated for a random sample from a standard normal distribution, whereas if we wish to make inferences about μ and σ^2, our sample is from a normal distribution with mean μ and variance σ^2. Let $X_1, \ldots, X_n$ denote the sample from the normal distribution with mean μ and variance σ^2; then the Z_i of Theorem 8 could be taken equal to $(X_i - \mu)/\sigma$.

(i) of Theorem 8 becomes:

(i′) $\bar{Z} = (1/n) \sum (X_i - \mu)/\sigma = (\bar{X} - \mu)/\sigma$ has a normal distribution with mean 0 and variance $1/n$.

(ii) of Theorem 8 becomes:

(ii′) $\bar{Z} = (\bar{X} - \mu)/\sigma$ and $\sum (Z_i - \bar{Z})^2 = \sum [(X_i - \mu)/\sigma - (\bar{X} - \mu)/\sigma]^2 = \sum [(X_i - \bar{X})^2/\sigma^2]$ are independent, which implies $\bar{X}$ and $\sum (X_i - \bar{X})^2$ are independent.

(iii) of Theorem 8 becomes:

(iii′) $\sum (Z_i - \bar{Z})^2 = \sum [(X_i - \bar{X})^2/\sigma^2]$ has a chi-square distribution with $n - 1$ degrees of freedom.

Corollary If $S^2 = [1/(n-1)] \sum_{i=1}^{n} (X_i - \bar{X})^2$ is the sample variance of a random sample from a normal distribution with mean μ and variance σ^2, then

$$U = \frac{(n-1)S^2}{\sigma^2} \tag{24}$$

has a chi-square distribution with $n - 1$ degrees of freedom.

PROOF This is just (iii′). ////

Remark Since S^2 is a linear function of U in Eq. (24), the density of S^2 can be obtained from the density of U. It is

$$f_{S^2}(y) = \left(\frac{n-1}{2\sigma^2}\right)^{(n-1)/2} \frac{1}{\Gamma[(n-1)/2]} y^{(n-3)/2} e^{-(n-1)y/2\sigma^2} I_{(0, \infty)}(y). \tag{25}$$

////

Remark The phrase "degrees of freedom" can refer to the number of independent squares in the sum. For example, the sum of Theorem 7 has k independent squares, but the sum in (iii) of Theorem 8 has only $n - 1$ independent terms since the relation $\sum (Z_i - \bar{Z}) = 0$ enables one to compute any one of the deviations $Z_i - \bar{Z}$, given the other $n - 1$ of them. ////

All the results of this section apply only to normal populations. In fact, it can be proved that for no other distributions (i) are the sample mean and sample variance independently distributed or (ii) is the sample mean exactly normally distributed.

4.4 The F Distribution

A distribution, the F distribution, which we shall later find to be of considerable practical interest, is the distribution of the ratio of two independent chi-square random variables divided by their respective degrees of freedom. We suppose that U and V are independently distributed with chi-square distributions with m and n degrees of freedom, respectively. Their joint density is then [see Eq. (21)]

$$f_{U,V}(u, v) = \frac{1}{\Gamma(m/2)\Gamma(n/2)2^{(m+n)/2}} u^{(m-2)/2}v^{(n-2)/2}e^{-\frac{1}{2}(u+v)}I_{(0,\infty)}(u)I_{(0,\infty)}(v). \tag{26}$$

We shall find the distribution of the quantity

$$X = \frac{U/m}{V/n}, \tag{27}$$

which is sometimes referred to as the *variance ratio*. To find the distribution of X, we make the transformation $X = (U/m)/(V/n)$ and $Y = V$, obtain the joint distribution of X and Y, and then get the marginal distribution of X by integrating out the y variable. The Jacobian of the transformation is $(m/n)y$; so

$$f_{X,Y}(x, y) = \frac{m}{n} y \frac{1}{\Gamma(m/2)\Gamma(n/2)2^{(m+n)/2}} \left(\frac{m}{n} xy\right)^{(m-2)/2} y^{(n-2)/2}e^{-\frac{1}{2}[(m/n)xy+y]},$$

and

$$\begin{aligned} f_X(x) &= \int_0^\infty f_{X,Y}(x, y)\, dy \\ &= \frac{1}{\Gamma(m/2)\Gamma(n/2)2^{(m+n)/2}} \left(\frac{m}{n}\right)^{m/2} x^{(m-2)/2} \int_0^\infty y^{(m+n-2)/2}e^{-\frac{1}{2}[(m/n)x+1]y}\, dy \\ &= \frac{\Gamma[(m+n)/2]}{\Gamma(m/2)\Gamma(n/2)} \left(\frac{m}{n}\right)^{m/2} \frac{x^{(m-2)/2}}{[1 + (m/n)x]^{(m+n)/2}} I_{(0,\infty)}(x). \end{aligned} \tag{28}$$

Definition 9 ***F* distribution** If X is a random variable having density given by Eq. (28), then X is defined to be an *F-distributed random variable with degrees of freedom m and n.* ////

The order in which the degrees of freedom are given is important since the density of the F distribution is not symmetrical in m and n. The number of degrees of freedom of the numerator of the ratio m/n that appears in Eq. (28) is always quoted first. Or if the F-distributed random variable is a ratio of two independent chi-square-distributed random variables divided by their respective degrees of freedom, as in the derivation above, then the degrees of freedom of the chi-square random variable that appears in the numerator are always quoted first.

We have proved the following theorem.

Theorem 9 Let U be a chi-square random variable with m degrees of freedom; let V be a chi-square random variable with n degrees of freedom, and let U and V be independent. Then the random variable

$$X = \frac{U/m}{V/n}$$

is distributed as an F distribution with m and n degrees of freedom. The density of X is given in Eq. (28). ////

The following corollary shows how the result of Theorem 9 can be useful in sampling.

Corollary If $X_1, \ldots, X_{m+1}$ is a random sample of size $m + 1$ from a normal population with mean μ_X and variance σ^2, if $Y_1, \ldots, Y_{n+1}$ is a random sample of size $n + 1$ from a normal population with mean μ_Y and variance σ^2, and if the two samples are independent, then it follows that $(1/\sigma^2) \sum_1^{m+1} (X_i - \bar{X})^2$ is chi-square distributed with m degrees of freedom, and $(1/\sigma^2) \sum_1^{n+1} (Y_j - \bar{Y})^2$ is chi-square-distributed with n degrees of freedom; so that the statistic

$$\frac{\sum (X_i - \bar{X})^2/m}{\sum (Y_j - \bar{Y})^2/n}$$

has an F distribution with m and n degrees of freedom. ////

We close this subsection with several further remarks about the F distribution.

Remark If X is an F-distributed random variable with m and n degrees of freedom, then

$$\mathscr{E}[X] = \frac{n}{n-2} \qquad \text{for } n > 2$$

and

$$\text{var}\,[X] = \frac{2n^2(m+n-2)}{m(n-2)^2(n-4)} \qquad \text{for } n > 4. \tag{29}$$

PROOF At first it might be surprising that the mean depends only on the degrees of freedom of the denominator. Write X as in Eq. (27); that is,

$$X = \frac{U/m}{V/n};$$

then

$$\mathscr{E}[X] = \mathscr{E}\left[\frac{U/m}{V/n}\right] = \frac{n}{m}\mathscr{E}[U]\mathscr{E}\left[\frac{1}{V}\right].$$

But $\mathscr{E}[U] = m$ by Eq. (22), and

$$\begin{aligned}\mathscr{E}\left[\frac{1}{V}\right] &= \frac{1}{\Gamma(n/2)}\left(\frac{1}{2}\right)^{n/2}\int_0^\infty \frac{1}{v}\cdot v^{(n-2)/2}e^{-\frac{1}{2}v}\,dv \\ &= \frac{1}{\Gamma(n/2)}\left(\frac{1}{2}\right)^{n/2}\int_0^\infty v^{(n-4)/2}e^{-\frac{1}{2}v}\,dv \\ &= \frac{\Gamma[(n-2)/2]}{\Gamma(n/2)}\left(\frac{1}{2}\right)^{n/2}\left(\frac{1}{2}\right)^{-(n-2)/2} = \frac{1}{n-2};\end{aligned}$$

and so

$$\mathscr{E}[X] = \left(\frac{n}{m}\right)\mathscr{E}[U]\mathscr{E}\left[\frac{1}{V}\right] = \frac{n}{m}\frac{m}{n-2} = \frac{n}{n-2}.$$

The variance formula is similarly derived. ////

Remark If X has an F distribution with m and n degrees of freedom, then $1/X$ has an F distribution with n and m degrees of freedom. This result allows one to table the F distribution for the upper tail only. For example, if the quantile $\xi_{.95}$ is given for an F distribution with m and n degrees of freedom, then the quantile $\xi'_{.05}$ for an F distribution with n and m degrees of freedom is given by $1/\xi_{.95}$. In general, if X has an F distribution with m and n degrees of freedom and Y has an F distribution with n and m degrees of freedom, then the pth quantile point of X, ξ_p, is the reciprocal of the $(1-p)$th quantile point of Y, ξ'_{1-p}, as the following shows:

$$p = P[X \le \xi_p] = P\left[\frac{1}{X} \ge \frac{1}{\xi_p}\right] = P\left[Y \ge \frac{1}{\xi_p}\right] = 1 - P\left[Y \le \frac{1}{\xi_p}\right],$$

but

$$1 - p = P[Y \le \xi'_{1-p}];$$

so

$$\xi'_{1-p} = \frac{1}{\xi_p}. \qquad ////$$

Remark If X is an F-distributed random variable with m and n degrees of freedom, then

$$W = \frac{mX/n}{1 + mX/n}$$

has a beta density with parameters $a = m/2$ and $b = n/2$. ////

4.5 Student's t Distribution

Another distribution of considerable practical importance is that of the ratio of a standard normally distributed random variable to the square root of an independently distributed chi-square random variable divided by its degrees of freedom. That is, if Z has a standard normal distribution, if U has a chi-square distribution with k degrees of freedom, and if Z and U are independent, we seek the distribution of

$$X = \frac{Z}{\sqrt{U/k}}.$$

The joint density of Z and U is given by

$$f_{Z,U}(z, u) = \frac{1}{\sqrt{2\pi}} \frac{1}{\Gamma(k/2)} \left(\frac{1}{2}\right)^{k/2} u^{(k/2)-1} e^{-\frac{1}{2}u} e^{-\frac{1}{2}z^2} I_{(0,\infty)}(u). \tag{30}$$

If we make the transformation $X = Z/\sqrt{U/k}$ and $Y = U$, the Jacobian is $\sqrt{y/k}$, and so

$$f_{X,Y}(x, y) = \sqrt{\frac{y}{k}} \frac{1}{\sqrt{2\pi}} \frac{1}{\Gamma(k/2)} \left(\frac{1}{2}\right)^{k/2} y^{(k/2)-1} e^{-\frac{1}{2}y} e^{-\frac{1}{2}x^2y/k} I_{(0,\infty)}(y)$$

$$f_X(x) = \int_{-\infty}^{\infty} f_{X,Y}(x, y)\, dy$$

$$= \frac{1}{\sqrt{2k\pi}} \frac{1}{\Gamma(k/2)} \left(\frac{1}{2}\right)^{k/2} \int_0^{\infty} y^{k/2-1+\frac{1}{2}} e^{-\frac{1}{2}(1+x^2/k)y}\, dy$$

$$= \frac{\Gamma[(k+1)/2]}{\Gamma(k/2)} \frac{1}{\sqrt{k\pi}} \frac{1}{(1+x^2/k)^{(k+1)/2}}. \tag{31}$$

Definition 10 Student's *t* distribution If X is a random variable having density given by Eq. (31), then X is defined to have a *Student's t distribution*, or the density given in Eq. (31) is called a *Student's t distribution with k degrees of freedom.* ////

We have derived the following result.

Theorem 10 If Z has a standard normal distribution, if U has a chi-square distribution with k degrees of freedom, and if Z and U are independent, then $Z/\sqrt{U/k}$ has a Student's t distribution with k degrees of freedom. ////

The following corollary shows how the result of Theorem 10 is applicable to sampling from a normal population.

Corollary If $X_1, \ldots, X_n$ is a random sample from a normal distribution with mean μ and variance σ^2, then $Z = (\bar{X} - \mu)/(\sigma/\sqrt{n})$ has a standard normal distribution and $U = (1/\sigma^2) \sum (X_i - \bar{X})^2$ has a chi-square distribution with $n - 1$ degrees of freedom. Furthermore, Z and U are independent (see Theorem 8); hence

$$\frac{(\bar{X} - \mu)/(\sigma/\sqrt{n})}{\sqrt{(1/\sigma^2) \sum (X_i - \bar{X})^2/(n-1)}} = \frac{\sqrt{n(n-1)}(\bar{X} - \mu)}{\sqrt{\sum (X_i - \bar{X})^2}}$$

has a Student's t distribution with $n - 1$ degrees of freedom. ////

We might note that for one degree of freedom the Student's t distribution reduces to a Cauchy distribution; and as the number of degrees of freedom increases, the Student's t distribution approaches the standard normal distribution. Also, the square of a Student's t-distributed random variable with k degrees of freedom has an F distribution with 1 and k degrees of freedom.

Remark If X is a random variable having a Student's t distribution with k degrees of freedom, then

$$\mathscr{E}[X] = 0 \quad \text{if } k > 1 \quad \text{and} \quad \text{var}[X] = k/(k-2) \quad \text{if } k > 2. \tag{32}$$

PROOF The first two moments of X can be found by writing $X = Z/\sqrt{U/k}$ as in Theorem 10 and using the independence of Z and U. The actual derivation is left as an exercise. ////

This completes Sec. 4 on sampling from the normal distribution. Note that we considered the distribution of functions of only two different statistics, namely, the sample mean and sample variance. In the next chapter we will find that these two statistics are the only ones of interest in sampling from a normal distribution; they will turn out to be sufficient statistics.

5 ORDER STATISTICS

5.1 Definition and Distributions

In Subsec. 2.4 we defined what we meant by statistic and then gave the sample moments as examples of easy-to-understand statistics. In this section the concept of *order statistics* will be defined, and some of their properties will be investigated. Order statistics, like sample moments, play an important role in statistical inference. Order statistics are to population quantiles as sample moments are to population moments.

Definition 11 Order statistics Let $X_1, X_2, \ldots, X_n$ denote a random sample of size n from a cumulative distribution function $F(\cdot)$. Then $Y_1 \leq Y_2 \leq \cdots \leq Y_n$, where Y_i are the X_i arranged in order of increasing magnitudes and are defined to be the *order statistics* corresponding to the random sample $X_1, \ldots, X_n$. ////

We note that the Y_i are statistics (they are functions of the random sample $X_1, X_2, \ldots, X_n$) and are in order. Unlike the random sample itself, the order statistics are clearly not independent, for if $Y_j \geq y$, then $Y_{j+1} \geq y$.

We seek the distribution, both marginal and joint, of the order statistics. We have already found the marginal distributions of $Y_1 = \min [X_1, \ldots, X_n]$ and $Y_n = \max [X_1, \ldots, X_n]$ in Chap. V. Now we will find the marginal cumulative distribution of an arbitrary order statistic.

Theorem 11 Let $Y_1 \leq Y_2 \leq \cdots \leq Y_n$ represent the order statistics from a cumulative distribution function $F(\cdot)$. The marginal cumulative distribution function of Y_α, $\alpha = 1, 2, \ldots, n$, is given by

$$F_{Y_\alpha}(y) = \sum_{j=\alpha}^{n} \binom{n}{j} [F(y)]^j [1 - F(y)]^{n-j} \tag{33}$$

PROOF For fixed y, let

$$Z_i = I_{(-\infty, y]}(X_i);$$

then

$$\sum_{i=1}^{n} Z_i = \text{the number of } X_i \leq y.$$

Note that $\sum_{i=1}^{n} Z_i$ has a binomial distribution with parameters n and $F(y)$. Now

$$F_{Y_\alpha}(y) = P[Y_\alpha \leq y] = P[\sum Z_i \geq \alpha] = \sum_{j=\alpha}^{n} \binom{n}{j} [F(y)]^j [1 - F(y)]^{n-j}.$$

The key step in the proof is the equivalence of the two events $\{Y_\alpha \leq y\}$ and $\{\sum Z_i \geq \alpha\}$. If the αth order statistic is less than or equal to y, then surely the number of X_i less than or equal to y is greater than or equal to α, and conversely. ////

Corollary $F_{Y_n}(y) = \sum_{j=n}^{n} \binom{n}{j} [F(y)]^j [1 - F(y)]^{n-j} = [F(y)]^n,$

and

$$F_{Y_1}(y) = \sum_{j=1}^{n} \binom{n}{j} [F(y)]^j [1 - F(y)]^{n-j} = 1 - [1 - F(y)]^n.$$ ////

Theorem 11 gives the marginal distribution of an individual order statistic in terms of the cumulative distribution function $F(\cdot)$. For the remainder of this subsection, we will assume that our random sample $X_1, \ldots, X_n$ came from a probability density function $f(\cdot)$; that is, we assume that the random variables X_i are continuous. We seek the density of Y_α, which, of course, could be obtained from Eq. (33) by differentiation of $F_{Y_\alpha}(y)$. Note that

$$\begin{aligned}
&f_{Y_\alpha}(y) \\
&= \lim_{\Delta y \to 0} \frac{F_{Y_\alpha}(y + \Delta y) - F_{Y_\alpha}(y)}{\Delta y} = \lim_{\Delta y \to 0} \frac{P[y < Y_\alpha \le y + \Delta y]}{\Delta y} \\
&= \lim_{\Delta y \to 0} \frac{P[(\alpha - 1) \text{ of the } X_i \le y; \text{ one } X_i \text{ in } (y, y + \Delta y]; (n - \alpha) \text{ of the } X_i > y + \Delta y]}{\Delta y} \\
&= \lim_{\Delta y \to 0} \left\{ \frac{n!}{(\alpha - 1)!1!(n - \alpha)!} \frac{[F(y)]^{\alpha - 1}[F(y + \Delta y) - F(y)][1 - F(y + \Delta y)]^{n - \alpha}}{\Delta y} \right\} \\
&= \frac{n!}{(\alpha - 1)!(n - \alpha)!} [F(y)]^{\alpha - 1}[1 - F(y)]^{n - \alpha} f(y).
\end{aligned}$$

We have made sensible use of the multinomial distribution. Similarly, we can derive the joint density of Y_α and Y_β for $1 \le \alpha < \beta \le n$.

$$\begin{aligned}
f_{Y_\alpha, Y_\beta}(x, y)\, \Delta x\, \Delta y &\approx P[x < Y_\alpha \le x + \Delta x; y < Y_\beta \le y + \Delta y] \\
&\approx P[(\alpha - 1) \text{ of the } X_i \le x; \text{ one } X_i \text{ in } (x, x + \Delta x]; \\
&\quad (\beta - \alpha - 1) \text{ of the } X_i \text{ in } (x + \Delta x, y]; \\
&\quad \text{one } X_i \text{ in } (y, y + \Delta y]; (n - \beta) \text{ of the } X_i > y + \Delta y] \\
&\approx \frac{n!}{(\alpha - 1)!\,1!\,(\beta - \alpha - 1)!\,1!\,(n - \beta)!} \\
&\times [F(x)]^{\alpha - 1}[F(y) - F(x + \Delta x)]^{\beta - \alpha - 1}[1 - F(y + \Delta y)]^{n - \beta} f(x)\, \Delta x\, f(y)\, \Delta y;
\end{aligned}$$

hence

$$f_{Y_\alpha, Y_\beta}(x, y) = \frac{n!}{(\alpha - 1)!\,(\beta - \alpha - 1)!\,(n - \beta)!} [F(x)]^{\alpha - 1}[F(y) - F(x)]^{\beta - \alpha - 1}[1 - F(y)]^{n - \beta} f(x) f(y)$$

for $x < y$, and

$$f_{Y_\alpha, Y_\beta}(x, y) = 0 \qquad \text{for } x \ge y.$$

In general, $f_{Y_1, \ldots, Y_n}(y_1, \ldots, y_n)$

$$= \lim_{\Delta y_i \to 0} \frac{1}{\prod_{i=1}^{n} \Delta y_i} P[y_1 < Y_1 \le y_1 + \Delta y_1; \ldots; y_n < Y_n \le y_n + \Delta y_n]$$

$$= \lim_{\Delta y_i \to 0} \frac{1}{\prod_{i=1}^{n} \Delta y_i} P[\text{one } X_i \text{ in } (y_1, y_1 + \Delta y_1]; \ldots; \text{one } X_i \text{ in } (y_n, y_n + \Delta y_n]]$$

$$= \lim_{\Delta y_i \to 0} \frac{n!}{\prod_{i=1}^{n} \Delta y_i} [F(y_1 + \Delta y_1) - F(y_1)] \cdot \cdots \cdot [F(y_n + \Delta y_n) - F(y_n)]$$

$$= n! f(y_1) \cdot \cdots \cdot f(y_n) \qquad \text{for} \quad y_1 < y_2 < \cdots < y_n,$$

and $f_{Y_1, \ldots, Y_n}(y_1, \ldots, y_n) = 0$, otherwise.

We have derived the following theorem.

Theorem 12 Let $X_1, X_2, \ldots, X_n$ be a random sample from the probability density function $f(\cdot)$ with cumulative distribution function $F(\cdot)$. Let $Y_1 \le Y_2 \le \cdots \le Y_n$ denote the corresponding order statistics; then

$$f_{Y_\alpha}(y) = \frac{n!}{(\alpha - 1)!(n - \alpha)!} [F(y)]^{\alpha - 1}[1 - F(y)]^{n - \alpha} f(y); \tag{34}$$

$$f_{Y_\alpha, Y_\beta}(x, y) = \frac{n!}{(\alpha - 1)!(\beta - \alpha - 1)!(n - \beta)!} \times [F(x)]^{\alpha - 1}[F(y) - F(x)]^{\beta - \alpha - 1} \times [1 - F(y)]^{n - \beta} f(x) f(y) I_{(x, \infty)}(y); \tag{35}$$

$$f_{Y_1, \ldots, Y_n}(y_1, \ldots, y_n) = \begin{cases} n! f(y_1) \cdot \cdots \cdot f(y_n) & \text{for } y_1 < y_2 < \cdots < y_n \\ 0 & \text{otherwise.} \end{cases} \tag{36}$$

////

Any set of marginal densities can be obtained from the joint density $f_{Y_1, \ldots, Y_n}(y_1, \ldots, y_n)$ by simply integrating out the unwanted variables.

5.2 Distribution of Functions of Order Statistics

In the previous subsection we derived the joint and marginal distributions of the order statistics themselves. In this subsection we will find the distribution of certain functions of the order statistics. One possible function of the order statistics is their arithmetic mean, equal to

$$\frac{1}{n} \sum_{j=1}^{n} Y_j.$$

Note, however, that $(1/n)\sum_{j=1}^{n} Y_j = (1/n)\sum_{i=1}^{n} X_i$, the sample mean, which was the subject of Sec. 3 of this chapter. We define now some other functions of the order statistics.

Definition 12 Sample median, sample range, sample midrange Let $Y_1 \leq \cdots \leq Y_n$ denote the order statistics of a random sample $X_1, \ldots, X_n$ from a density $f(\cdot)$. The *sample median* is defined to be the middle order statistic if n is odd and the average of the middle two order statistics if n is even. The *sample range* is defined to be $Y_n - Y_1$, and the *sample mid-range* is defined to be $(Y_1 + Y_n)/2$. ////

If the sample size is odd, then the distribution of the sample median is given by Eq. (34); for example, if $n = 2k + 1$, where k is some positive integer, then Y_{k+1} is the sample median whose distribution is given by Eq. (34). If the sample size is even, say $n = 2k$, then the sample median is $(Y_k + Y_{k+1})/2$, the distribution of which can be obtained by a transformation starting with the joint density of Y_k and Y_{k+1}, which is given by Eq. (35).

We derive now the joint distribution of the sample range and midrange, from which the marginals can be obtained.

By Eq. (35), we have

$$f_{Y_1, Y_n}(x, y) = n(n-1)[F(y) - F(x)]^{n-2} f(x) f(y) \qquad \text{for } x < y. \tag{37}$$

Make the transformation $R = Y_n - Y_1$ and $T = (Y_1 + Y_n)/2$ or $r = y - x$ and $t = (x + y)/2$. Now $x = t - r/2$, and $y = t + r/2$; hence

$$J = \begin{vmatrix} \dfrac{\partial x}{\partial r} & \dfrac{\partial x}{\partial t} \\ \dfrac{\partial y}{\partial r} & \dfrac{\partial y}{\partial t} \end{vmatrix} = \begin{vmatrix} -\frac{1}{2} & 1 \\ \frac{1}{2} & 1 \end{vmatrix} = -1,$$

and we obtain Theorem 13.

Theorem 13 If R is the sample range and T the sample midrange from a probability density function, then their joint distribution is given by

$$f_{R,T}(r, t) =$$
$$n(n-1)[F(t + r/2) - F(t - r/2)]^{n-2} f(t - r/2) f(t + r/2) \qquad \text{for } r > 0, \tag{38}$$

and the marginal distributions are given by

$$f_R(r) = \int_{-\infty}^{\infty} f_{R,T}(r, t)\, dt \qquad \text{and} \qquad f_T(t) = \int_0^{\infty} f_{R,T}(r, t)\, dr. \tag{39}$$

////

EXAMPLE 6 Let $X_1, \ldots, X_n$ be a random sample from a uniform distribution on $(\mu - \sqrt{3}\sigma, \mu + \sqrt{3}\sigma)$. Here μ is the mean, and σ^2 is the variance of the sampled population.

$$f(x) = \frac{1}{2\sqrt{3}\sigma} I_{(\mu - \sqrt{3}\sigma,\, \mu + \sqrt{3}\sigma)}(x),$$

and

$$F(x) = \left(\frac{1}{2\sqrt{3}\sigma} x - \frac{\mu - \sqrt{3}\sigma}{2\sqrt{3}\sigma}\right) I_{(\mu - \sqrt{3}\sigma,\, \mu + \sqrt{3}\sigma]}(x) + I_{(\mu + \sqrt{3}\sigma,\, \infty)}(x).$$

$$f_{R,T}(r, t) = \frac{n(n-1)}{(2\sqrt{3}\sigma)^n} r^{n-2} I_{(\mu - \sqrt{3}\sigma + r/2,\, \mu + \sqrt{3}\sigma - r/2)}(t) I_{(0,\, 2\sqrt{3}\sigma)}(r). \tag{40}$$

$$f_R(r) = \int f_{R,T}(r, t)\, dt = \frac{n(n-1)}{(2\sqrt{3}\sigma)^n} r^{n-2} (2\sqrt{3}\sigma - r) I_{(0,\, 2\sqrt{3}\sigma)}(r). \tag{41}$$

We note that $f_R(r)$ is independent of the parameter μ.

$$f_T(t) = \int f_{R,T}(r, t)\, dr$$

$$= \frac{n(n-1)}{(2\sqrt{3}\sigma)^n} \int_0^{\min[2t - 2(\mu - \sqrt{3}\sigma),\, 2(\mu + \sqrt{3}\sigma) - 2t]} r^{n-2}\, dr \cdot I_{(\mu - \sqrt{3}\sigma,\, \mu + \sqrt{3}\sigma)}(t),$$

which simplifies to

$$f_T(t) = \frac{n}{2\sqrt{3}\sigma} \left(\frac{t - \mu}{\sqrt{3}\sigma} + 1\right)^{n-1} I_{(-1,\, 0)}\left(\frac{t - \mu}{\sqrt{3}\sigma}\right) + \frac{n}{2\sqrt{3}\sigma} \left(1 - \frac{t - \mu}{\sqrt{3}\sigma}\right)^{n-1} I_{[0,\, 1)}\left(\frac{t - \mu}{\sqrt{3}\sigma}\right). \tag{42}$$

From Eq. (41), we can derive $\mathscr{E}[R] = 2\sqrt{3}\sigma(n-1)/(n+1)$. ////

Certain functions of the order statistics are again statistics and may be used to make statistical inferences. For example, both the sample median and the midrange can be used to estimate μ, the mean of the population. For the uniform density given in the above example, the variances of the sample mean, the sample median, and the sample midrange are compared in Problem 33.

5.3 Asymptotic Distributions

In Subsec. 3.3, we discussed the asymptotic distribution of the sample mean $\bar{X}_n$. We saw that $\bar{X}_n$ was asymptotically normally distributed with mean μ and variance σ^2/n. We now consider the question: Is there an asymptotic distribution for the sample median? We will state (without proof) a more general result.

Since for asymptotic results the sample size n increases, we let $Y_1^{(n)} \leq Y_2^{(n)} \leq \cdots \leq Y_n^{(n)}$ denote the order statistics for a sample of size n. The superscript denotes the sample size. We will give the asymptotic distribution of that order statistic which is approximately the (np)th order statistic for a sample of size n for any $0 < p < 1$. We say "approximately" the (np)th order statistic because np may not be an integer. Define p_n to be such that np_n is an integer and p_n is approximately equal to p; then $Y_{np_n}^{(n)}$ is the (np_n)th order statistic for a sample of size n. (If $X_1, \ldots, X_n$ are independent for each positive integer n, we will say $X_1, \ldots, X_n, \ldots$ are independent.)

Theorem 14 Let $X_1, \ldots, X_n, \ldots$ be independent identically distributed random variables with common probability density $f(\cdot)$ and cumulative distribution function $F(\cdot)$. Assume that $F(x)$ is strictly monotone for $0 < F(x) < 1$. Let ξ_p be the unique solution in x of $F(x) = p$ for some $0 < p < 1$. (ξ_p is the pth quantile.) Let p_n be such that np_n is an integer and $n|p_n - p|$ is bounded. Finally, let $Y_{np_n}^{(n)}$ denote the (np_n)th order statistic for a random sample of size n. Then $Y_{np_n}^{(n)}$ is asymptotically distributed as a normal distribution with mean ξ_p and variance $p(1-p)/n[f(\xi_p)]^2$. ////

EXAMPLE 7 Let $p = \frac{1}{2}$; then ξ_p is the population median, and Theorem 14 states that the sample median is asymptotically distributed as a normal distribution with mean the population median and variance $1/4n[f(\xi_{1/2})]^2$. In particular, if $f(\cdot)$ is a normal density with mean μ and variance σ^2, then the sample median is asymptotically normally distributed with mean μ and variance $1/4n[f(\mu)]^2 = \pi\sigma^2/2n$. Recall that the sample mean is normally distributed with mean μ and variance σ^2/n. ////

In Theorem 14 above we considered a certain kind of asymptotic distribution of order statistics. We will now consider yet another kind. In the above we looked at the asymptotic distribution of that order statistic which was approximately the (np)th order statistic for a sample of size n. Such an order statistic had (approximately) $100p$ percent of the n observations to its left. That is, its *relative* position remained unchanged as n, the sample size, increased; it always had (approximately) the same percentage of the n observations to its left. We will now consider the asymptotic distribution of that order statistic whose *absolute* position remains unchanged. That is, we consider the asymptotic distribution of, say, $Y_k^{(n)}$ for fixed k and increasing n. $Y_k^{(n)}$ is the kth smallest order statistic for a sample size $n \geq k$, and k remains fixed. In order to make the presentation somewhat simpler, we will take $k = 1$, in which case $Y_1^{(n)}$ is the smallest of the n observations. We note that we could just as well consider the kth largest order statistic, namely $Y_{n-k+1}^{(n)}$, which for $k = 1$ specializes to $Y_n^{(n)}$, the largest order statistic for a sample of size n. Either $Y_1^{(n)}$ or $Y_n^{(n)}$ is often referred to as an *extreme-value statistic.*

Practical applications of extreme-value statistics are many. The old adage that a chain is no stronger than its weakest link provides a simple example. If X_i denotes the "strength" of the ith link of a chain with n similar links, then $Y_1^{(n)} = \min [X_1, \ldots, X_n]$ is the "strength" of the chain. Also, in measuring the results of certain physical phenomena such as floods, droughts, earthquakes, winds, temperatures, etc., it can be seen that under certain circumstances one is more interested in extreme values than in average values. For instance, it is the extreme earthquake or flood, and not the average earthquake or flood, that is more damaging. We can see that results, whether exact or asymptotic, for extreme-value statistics can be just as important as results for averages.

For the most part we will concentrate on finding the asymptotic distribution of $Y_n^{(n)}$. One might wonder why we should be interested in an asymptotic distribution of $Y_n^{(n)}$ when the exact distribution, which is given by $F_{Y_n^{(n)}}(y) = [F(y)]^n$, where $F(\cdot)$ is the c.d.f. sampled from, is known. The hope is that we will find an asymptotic distribution which does not depend on the sampled c.d.f. $F(\cdot)$. We recall that the central-limit theorem gave an asymptotic distribution for $\bar{X}_n$ which did not depend on the sampled distribution even though the exact distribution of $\bar{X}_n$ could be found.

In searching for the asymptotic distribution of $Y_n^{(n)}$, let us pattern our development after what was done in deriving the asymptotic distribution of $\bar{X}_n$. According to the law of large numbers, $\bar{X}_n$ has a *degenerate* limiting distribution; that is, the limiting c.d.f. of $\bar{X}_n$ is the cumulative distribution that assigns all its mass to the point μ. Such a limiting distribution is not useful if one intends to use the limiting distribution to approximate probabilities of events since it assigns each event a probability of either 0 or 1. To circumvent such difficulty, we first "centered" the values of $\bar{X}_n$ by subtracting μ, and then we "inflated" the values of $\bar{X}_n - \mu$ by multiplying them by $\sqrt{n}/\sigma$, and, consequently, we were able to get a *nondegenerate* limiting distribution; that is, according to the central-limit theorem, $\sqrt{n}(\bar{X}_n - \mu)/\sigma$ had a standard normal distribution as its limiting distribution. A general procedure, when one is looking for a limiting distribution of, say, Z_n, is to first "center" the Z_n by subtracting a constant, say a_n, and to then "scale" $Z_n - a_n$ by dividing by another constant, say b_n. In the case of the central-limit theorem, $Z_n = \bar{X}_n$, $a_n \equiv \mu$, and $b_n = \sigma/\sqrt{n}$. In the case of Theorem 14 above, $Z_n = Y_{np_n}^{(n)}$, $a_n \equiv \xi_p$, and $b_n = \sqrt{p(1-p)/n[f(\xi_p)]^2}$. For both of these two cases the sequence of constants $\{a_n\}$ did not depend on n. In the case at hand, namely when $Z_n = Y_n^{(n)}$, the sequence of constants $\{a_n\}$ is likely to depend on n since $Y_n^{(n)}$ tends to increase with n. Let us look at a couple of examples.

EXAMPLE 8 Consider sampling from the logistic distribution; that is, $F(x) = (1 + e^{-x})^{-1}$. Find the limiting distribution of $(Y_n^{(n)} - a_n)/b_n$. There are two problems: First, what should we take the sequences of constants $\{a_n\}$ and $\{b_n\}$ to be? And, second, what is the limiting distribution of $(Y_n^{(n)} - a_n)/b_n$ for the selected constants $\{a_n\}$ and $\{b_n\}$? It seems reasonable that the "centering" constants $\{a_n\}$ should be close to $\mathscr{E}[Y_n^{(n)}]$; so we seek an approximation to $\mathscr{E}[Y_n^{(n)}]$. Now $F(X_1), \ldots, F(X_n)$ is a random sample from the uniform distribution over (0, 1); hence $F(Y_n^{(n)})$ is the largest of a sample of size n from a uniform distribution over (0, 1). That $\mathscr{E}[F(Y_n^{(n)})] = n/(n+1)$ can then be routinely derived. Now $F(\mathscr{E}[Y_n^{(n)}]) \approx \mathscr{E}[F(Y_n^{(n)})]$ or

$$\begin{aligned}
F(\mathscr{E}[Y_n^{(n)}]) &= \{1 + \exp(-\mathscr{E}[Y_n^{(n)}])\}^{-1} \\
&= 1 - \{1 + \exp(\mathscr{E}[Y_n^{(n)}])\}^{-1} \\
&\approx 1 - (n+1)^{-1} \\
&= \frac{n}{n+1} \\
&= \mathscr{E}[F(Y_n^{(n)})],
\end{aligned}$$

which implies that

$$n \approx \exp(\mathscr{E}[Y_n^{(n)}])$$

or that

$$\mathscr{E}[Y_n^{(n)}] \approx \log_e n.$$

Finally, since $\mathscr{E}[Y_n^{(n)}] \approx \log n$ (from here on we use $\log n$ for $\log_e n$), a reasonable choice for the sequence of "centering" constants $\{a_n\}$ seems to be the sequence $\{\log n\}$. We are seeking

$$\begin{aligned}
\lim_{n\to\infty} P\left[\frac{Y_n^{(n)} - a_n}{b_n} \le y\right] &= \lim_{n\to\infty} P\left[\frac{Y_n^{(n)} - \log n}{b_n} \le y\right] \\
&= \lim_{n\to\infty} P[Y_n^{(n)} \le b_n y + \log n] \\
&= \lim_{n\to\infty} [F(b_n y + \log n)]^n \\
&= \lim_{n\to\infty} (1 + e^{-b_n y - \log n})^{-n} \\
&= \lim_{n\to\infty} (1 + (1/n)e^{-b_n y})^{-n} \\
&= \exp(-e^{-y}) \qquad \text{for} \quad b_n \equiv 1.
\end{aligned}$$

Hence, if $\{a_n\}$ and $\{b_n\}$ are selected so that $\{a_n\} = \{\log n\}$ and $\{b_n\} = \{1\}$, respectively, then the limiting distribution of $(Y_n^{(n)} - a_n)/b_n = Y_n^{(n)} - \log n$ is $\exp(-e^{-y})$. ////

EXAMPLE 9 Consider sampling from the exponential distribution so that $F(x) = (1 - e^{-\lambda x})I_{(0, \infty)}(x)$. Again, let us find the limiting distribution of $(Y_n^{(n)} - a_n)/b_n$. As in Example 8, $\mathscr{E}[F(Y_n^{(n)})] = n/(n+1) = 1 - 1/(n+1)$. Now

$$F(\mathscr{E}[Y_n^{(n)}]) = 1 - \exp\{-\lambda\mathscr{E}[Y_n^{(n)}]\},$$

and

$$F(\mathscr{E}[Y_n^{(n)}]) \approx \mathscr{E}[F(Y_n^{(n)})];$$

so

$$\frac{1}{n+1} \approx \exp\{-\lambda\mathscr{E}[Y_n^{(n)}]\},$$

or

$$\mathscr{E}[Y_n^{(n)}] \approx \frac{1}{\lambda}\log(n+1) \approx \frac{1}{\lambda}\log n.$$

Hence, it seems reasonable to use $a_n = (1/\lambda)\log n$.

$$\begin{aligned}
\lim_{n\to\infty} P\left[\frac{Y_n - a_n}{b_n} \le y\right] &= \lim_{n\to\infty} P\left[Y_n - \frac{1}{\lambda}\log n \le b_n y\right] \\
&= \lim_{n\to\infty}\left[F\left(b_n y + \frac{1}{\lambda}\log n\right)\right]^n \\
&= \lim_{n\to\infty}(1 - e^{-\lambda b_n y - \log n})^n \\
&= \lim_{n\to\infty}\left(1 - \frac{1}{n}e^{-\lambda b_n y}\right)^n \\
&= \exp(-e^{-y}) \qquad \text{for } b_n \equiv \frac{1}{\lambda}.
\end{aligned}$$

Hence the limiting distribution of $(Y_n^{(n)} - a_n)/b_n = [Y_n^{(n)} - (1/\lambda)\log n]/(1/\lambda)$ is $\exp(-e^{-y})$. We note that we obtained the same limiting distribution here as in Example 8. Here we were sampling from an exponential distribution, and there we were sampling from a logistic distribution.

////

In each of the above two examples we were able to obtain the limiting distribution of $(Y_n^{(n)} - a_n)/b_n$ by using the exact distribution of $Y_n^{(n)}$ and ordinary algebraic manipulation. There are some rather powerful theoretical results concerning extreme-value statistics that tell us, among other things, what limiting distributions we can expect. We can only sketch such results here. The interested reader is referred to Refs. 13, 30, and 35.

Theorem 15 Let $X_1, \ldots, X_n, \ldots$ be independent and identically distributed random variables with c.d.f. $F(\cdot)$. If $(Y_n^{(n)} - a_n)/b_n$ has a limiting distribution, then that limiting distribution must be one of the following three types:

$$G_1(y; \gamma) = e^{-y^{-\gamma}} I_{(0,\infty)}(y), \qquad \text{where } \gamma > 0.$$

$$G_2(y; \gamma) = e^{-|y|^{\gamma}} I_{(-\infty,0)}(y) + I_{[0,\infty)}(y), \qquad \text{where } \gamma > 0.$$

$$G_3(y) = \exp(-e^{-y}). \qquad ////$$

Theorem 15 states what types of limiting distributions can be expected. The following theorem gives conditions on the sampled $F(\cdot)$ that enable us to determine which of the three types of limiting distributions correspond to the sampled $F(\cdot)$.

Theorem 16 Let $X_1, \ldots, X_n, \ldots$ be independent and identically distributed random variables with c.d.f. $F(\cdot)$. Assume that $(Y_n^{(n)} - a_n)/b_n$ has a limiting distribution. The limiting distribution is:

(i) $G_1(\cdot\,; \gamma)$ if and only if

$$\lim_{x \to \infty} \frac{1 - F(x)}{1 - F(\tau x)} = \tau^{\gamma} \qquad \text{for every } \tau > 0.$$

(ii) $G_2(\cdot\,; \gamma)$ if and only if there exists an x_0 such that

$$F(x_0) = 1 \qquad \text{and} \qquad F(x_0 - \varepsilon) < 1 \qquad \text{for every } \varepsilon > 0.$$

and

$$\lim_{0 < x \to 0} \frac{1 - F(x_0 - \tau x)}{1 - F(x_0 - x)} = \tau^{\gamma} \qquad \text{for every } \tau > 0.$$

(iii) $G_3(\cdot)$ if and only if

$$\lim_{n \to \infty} n[1 - F(\beta_n x + \alpha_n)] = e^{-x} \qquad \text{for each } x,$$

where

$$\alpha_n = \inf\left\{z: \frac{n-1}{n} \le F(z)\right\}$$

and

$$\beta_n = \inf\{z: 1 - (ne)^{-1} \le F(\alpha_n + z)\}. \qquad ////$$

Note that if $F(\cdot)$ is strictly monotone and continuous, then α_n is given by $F(\alpha_n) = (n-1)/n$, or $\alpha_n = F^{-1}(\{n-1\}/n)$; and β_n is given by $F(\alpha_n + \beta_n) = 1 - (ne)^{-1}$, or $\beta_n = F^{-1}(1 - \{ne\}^{-1}) - \alpha_n = F^{-1}(1 - \{ne\}^{-1}) - F^{-1}(\{n-1\}/n)$.

EXAMPLE 10 Take $F(x) = (1 - e^{-\lambda x})I_{(0, \infty)}(x)$ as in Example 9. α_n is such that $F(\alpha_n) = (n-1)/n$ or $1 - e^{-\lambda\alpha_n} = 1 - 1/n$, which implies that $\alpha_n = (1/\lambda)\log n$. β_n is such that $F(\alpha_n + \beta_n) = 1 - (ne)^{-1}$ or $1 - e^{-\lambda(1/\lambda)\log n - \lambda\beta_n} = 1 - (ne)^{-1}$, or $\beta_n = 1/\lambda$.

$$\lim_{n\to\infty} n[1 - F(\beta_n x + \alpha_n)] = \lim_{n\to\infty} n(e^{-\lambda(\beta_n x + \alpha_n)}) = e^{-x} \qquad \text{for each } x,$$

so, as we saw in Example 9, the exponential distribution has $G_3(\cdot)$ as its corresponding limiting extreme-value distribution. ////

EXAMPLE 11 Take $F(x) = F(x; \gamma) = [1 - (1-x)^\gamma]I_{(0,1)}(x) + I_{[1,\infty)}(x)$. Note that for $x_0 = 1$, $F(x_0) = 1$ and $F(x_0 - \varepsilon) < 1$ for every $\varepsilon > 0$. Also,

$$\lim_{0<x\to 0} \frac{1 - F(x_0 - \tau x)}{1 - F(x_0 - x)} = \lim_{0<x\to 0} \frac{(1 - x_0 + \tau x)^\gamma}{(1 - x_0 + x)^\gamma} = \tau^\gamma;$$

so $F(\cdot\,; \gamma)$ has $G_2(\cdot\,; \gamma)$ as its limiting extreme-value distribution. ////

EXAMPLE 12 Take $F(x) = F(x; \gamma)$ the c.d.f. of a t distribution with γ degrees of freedom.

$$\lim_{x\to\infty} \frac{1 - F(x)}{1 - F(\tau x)} = \lim_{x\to\infty} \frac{f(x)}{\tau f(\tau x)} = \lim_{x\to\infty} \frac{[1 + (\tau x)^2/\gamma]^{(\gamma+1)/2}}{\tau(1 + x^2/\gamma)^{(\gamma+1)/2}} = \tau^\gamma;$$

so the t distribution with γ degrees of freedom has $G_1(\cdot\,; \gamma)$ as its limiting extreme-value distribution. ////

Theorem 16 gives conditions on the sampled c.d.f. $F(\cdot)$ that enable us to determine the proper limiting extreme-value distribution for $(Y_n^{(n)} - a_n)/b_n$. The theorem does not tell us what the constants $\{a_n\}$ and $\{b_n\}$ should be. If, however, the conditions for the third type are satisfied, then we have

$$n[1 - F(\beta_n x + \alpha_n)] \to e^{-x}, \text{ as } n \to \infty,$$

and

$$P\left[\frac{Y_n^{(n)} - a_n}{b_n} \le x\right] \to \exp(-e^{-x}).$$

Now, $P[(Y_n^{(n)} - a_n)/b_n \le x] = [F(b_n x + a_n)]^n$; hence

$$[F(b_n x + a_n)]^n \to \exp(-e^{-x}),$$

or

$$n \log F(b_n x + a_n) \to -e^{-x},$$

or

$$n[1 - F(b_n x + a_n)] \to e^{-x};$$

and we see that a_n can be taken equal to α_n and $b_n = \beta_n$. Thus, for the third type the constants $\{a_n\}$ and $\{b_n\}$ are actually determined by the condition for that type. We shall see below that for certain practical applications it is possible to estimate $\{a_n\}$ and $\{b_n\}$.

Since the types $G_1(\cdot\,; \gamma)$ and $G_2(\cdot\,; \gamma)$ both contain a parameter, it can be surmised that the third type $G_3(\cdot)$ is more convenient than the other two in applications. Also, $G_3(y) = \exp(-e^{-y})$ is the correct limiting extreme-value distribution for a number of families of distributions. We saw that it was correct for the logistic and exponential distributions in Examples 8 and 9; it is also correct for the gamma and normal distributions. What is often done in practice is to assume that the sampled distribution $F(\cdot)$ is such that $\exp(-e^{-y})$ is the proper limiting extreme-value distribution; one can do this without assuming exactly which parametric family the sampled distribution $F(\cdot)$ belongs to. One then knows that $P[(Y_n^{(n)} - a_n)/b_n \le y] \to \exp(-e^{-y})$ for every y as $n \to \infty$. Hence,

$$P\left[\frac{Y_n^{(n)} - a_n}{b_n} \le y\right] \approx \exp(-e^{-y})$$

for large fixed n. Or

$$P[Y_n^{(n)} \le a_n + b_n y] \approx \exp(-e^{-y}),$$

or

$$P[Y_n^{(n)} \le z] \approx \exp(-e^{-(z - a_n)/b_n}).$$

It is true that a_n and b_n are given in terms of the $(1 - 1/n)$th quantile and the $(1 - 1/ne)$th quantile of the sampled distribution; however, for certain applications they can be estimated, in which case we would have an approximate distribution for $Y_n^{(n)}$, a distribution that is valid for a variety of different distributions that could be sampled from. (One might note that in applications of the central-limit theorem, which states that $\bar{X}_n$ is approximately distributed as $N(\mu, \sigma^2/n)$, often μ and σ^2 are unknown and consequently they also have to be estimated.) The preceding indicates how powerful the asymptotic extreme-value theory can be. We have merely introduced the subject. For instance, we stated some results for the asymptotic distribution of $Y_n^{(n)}$; one could state similar results for $Y_1^{(n)}$, $Y_k^{(n)}$, or $Y_{n-k+1}^{(n)}$. The interested reader is referred to Refs. 13, 30, and 35.

5.4 Sample Cumulative Distribution Function

We have repeatedly stated in this chapter that our purpose in sampling from some distribution was to make inferences about the sampled distribution, or population, which was assumed to be at least partly unknown. One question that might be posed is: Why not estimate the unknown distribution itself? The answer is that we can estimate the unknown cumulative distribution function using the *sample*, or *empirical*, *cumulative distribution function*, which is a function of the order statistics.

Definition 13 Sample cumulative distribution function Let $X_1, X_2, \ldots, X_n$ denote a random sample from a cumulative distribution function $F(\cdot)$, and let $Y_1 \leq Y_2 \leq \cdots \leq Y_n$ denote the corresponding order statistics. The *sample cumulative distribution function*, denoted by $F_n(x)$, is defined by $F_n(x) = (1/n) \times$ (number of Y_j less than or equal to x) or, equivalently, by $F_n(x) = (1/n) \times$ (number of X_i less than or equal to x). ////

For fixed x, $F_n(x)$ is a statistic since it is a function of the sample. (The dependence of $F_n(x)$ on the sample may not be clear from the notation itself.) We shall see that $F_n(x)$ has the same distribution as that of the sample mean of a Bernoulli distribution.

Theorem 17 Let $F_n(x)$ denote the sample cumulative distribution function of a random sample of size n from $F(\cdot)$; then

$$P\left[F_n(x) = \frac{k}{n}\right] = \binom{n}{k} [F(x)]^k[1 - F(x)]^{n-k}, \qquad k = 0, 1, \ldots, n. \tag{43}$$

PROOF Let $Z_i = I_{(-\infty, x]}(X_i)$; then Z_i has a Bernoulli distribution with parameter $F(x)$. Hence, $\sum_1^n Z_i$, which is the number of X_i less than or equal to x, has a binomial distribution with parameters n and $F(x)$. But $F_n(x) = (1/n) \sum_1^n Z_i$. The result follows. ////

Much more could be said about the sample cumulative distribution function, but we will wait until Chap. XI on nonparametric statistics to do so.

PROBLEMS

1 (*a*) Give an example where the target population and the sampled population are the same.

(*b*) Give an example where the target population and the sampled population are not the same.

2 (*a*) A company manufactures transistors in three different plants A, B, and C whose manufacturing methods are very similar. It is decided to inspect those transistors that are manufactured in plant A since plant A is the largest plant and statisticians are available there. In order to inspect a week's production, 100 transistors will be selected at random and tested for defects. Define the sampled population and target population.

(*b*) In part (*a*) above, it is decided to use the results in plant A to draw conclusions about plants B and C. Define the target population.

3 (*a*) What is the probability that the two observations of a random sample of two from a population with a rectangular distribution over the unit interval will not differ by more than $\frac{1}{2}$?

(*b*) What is the probability that the mean of a sample of two observations from a rectangular distribution over the unit interval will be between $\frac{1}{4}$ and $\frac{3}{4}$?

4 (*a*) Balls are drawn with replacement from an urn containing one white and two black balls. Let $X = 0$ for a white ball and $X = 1$ for a black ball. For samples $X_1, X_2, \ldots, X_9$ of size 9, what is the joint distribution of the observations? The distribution of the sum of the observations?

(*b*) Referring to part (*a*) above, find the expected values of the sample mean and sample variance.

5 Let $X_1, \ldots, X_n$ be a random sample from a distribution which has a finite fourth moment. Define $\mu = \mathscr{E}[X_1]$, $\sigma^2 = \text{var}[X_1]$, $\mu_3 = \mathscr{E}[(X_1 - \mu)^3]$, $\mu_4 = \mathscr{E}[(X_1 - \mu)^4]$, $\bar{X} = (1/n) \sum_1^n X_i$, and $S^2 = [1/(n-1)] \sum_1^n (X_i - \bar{X})^2$.

(*a*) Does $S^2 = [1/2n(n-1)] \sum_{i=1}^n \sum_{j=1}^n (X_i - X_j)^2$?

*(b) Find var $[\mathcal{S}^2]$.

*(c) Find cov $[\bar{X}, \mathcal{S}^2]$, and note that cov $[\bar{X}, \mathcal{S}^2] = 0$ if $\mu_3 = 0$.

Possible HINT: $\Sigma(X_i - \mu)^2 = \Sigma(X_i - \bar{X})^2 + (1/n)\Sigma\Sigma(X_i - \mu)(X_j - \mu)$.

6 *(a) For a random sample of size 2 from a population with a finite $(2r)$th moment, find $\mathscr{E}[M_r]$ and var $[M_r]$, where $M_r = (1/n) \sum_{i=1}^{n} (X_i - \bar{X}_n)^r$.

(b) For a random sample of size n from a population with mean μ and rth central moment μ_r, show that

$$\mathscr{E}\left[\frac{1}{n}\sum_{i=1}^{n}(X_i - \mu)^r\right] = \mu_r.$$

7 (a) Use the Chebyshev inequality to find how many times a coin must be tossed in order that the probability will be at least .90 that $\bar{X}$ will lie between .4 and .6. (Assume that the coin is true.)

(b) How could one determine the number of tosses required in part (a) more accurately, i.e., make the probability very nearly equal to .90? What is the number of tosses?

8 If a population has $\sigma = 2$ and $\bar{X}$ is the mean of samples of size 100, find limits between which $\bar{X} - \mu$ will lie with probability .90. Use both the Chebyshev inequality and the central-limit theorem. Why do the two results differ?

9 Suppose that $\bar{X}_1$ and $\bar{X}_2$ are means of two samples of size n from a population with variance σ^2. Determine n so that the probability will be about .01 that the two sample means will differ by more than σ. (Consider $Y = \bar{X}_1 - \bar{X}_2$.)

10 Suppose that light bulbs made by a standard process have an average life of 2000 hours with a standard deviation of 250 hours, and suppose that it is considered worthwhile to replace the process if the mean life can be increased by at least 10 percent. An engineer wishes to test a proposed new process, and he is willing to assume that the standard deviation of the distribution of lives is about the same as for the standard process. How large a sample should he examine if he wishes the probability to be about .01 that he will fail to adopt the new process if in fact it produces bulbs with a mean life of 2250 hours?

11 A research worker wishes to estimate the mean of a population using a sample large enough that the probability will be .95 that the sample mean will not differ from the population mean by more than 25 percent of the standard deviation. How large a sample should he take?

12 A polling agency wishes to take a sample of voters in a given state large enough that the probability is only .01 that they will find the proportion favoring a certain candidate to be less than 50 percent when in fact it is 52 percent. How large a sample should be taken?

13 A standard drug is known to be effective in about 80 percent of the cases in which it is used to treat infections. A new drug has been found effective in 85 of the first 100 cases tried. Is the superiority of the new drug well established? (If

the new drug were as equally effective as the old, what would be the probability of obtaining 85 or more successes in a sample of 100?)

14 Find the third moment about the mean of the sample mean for samples of size n from a Bernoulli population. Show that it approaches 0 as n becomes large (as it must if the normal approximation is to be valid).

15 (*a*) A bowl contains five chips numbered from 1 to 5. A sample of two drawn without replacement from this finite population is said to be random if all possible pairs of the five chips have an equal chance to be drawn. What is the expected value of the sample mean? What is the variance of the sample mean?

(*b*) Suppose that the two chips of part (*a*) were drawn with replacement; what would be the variance of the sample mean? Why might one guess that this variance would be larger than the one obtained before?

*(*c*) Generalize part (*a*) by considering N chips and samples of size n. Show that the variance of the sample mean is

$$\frac{\sigma^2}{n}\frac{N-n}{N-1},$$

where σ^2 is the population variance; that is

$$\sigma^2 = \frac{1}{N}\sum_{i=1}^{N}\left(i - \frac{N+1}{2}\right)^2.$$

16 If X_1, X_2, X_3 are independent random variables and each has a uniform distribution over (0, 1), derive the distribution of $(X_1 + X_2)/2$ and $(X_1 + X_2 + X_3)/3$.

17 If $X_1, \ldots, X_n$ is a random sample from $N(\mu, \sigma^2)$, find the mean and variance of

$$S = \sqrt{\frac{\sum_{1}^{n}(X_i - \bar{X})^2}{n-1}}.$$

18 On the F distribution:

(*a*) Derive the variance of the F distribution. [See part (*d*).]

(*b*) If X has an F distribution with m and n degrees of freedom, argue that $1/X$ has an F distribution with n and m degrees of freedom.

(*c*) If X has an F distribution with m and n degrees of freedom, show that

$$W = \frac{mX/n}{1 + mX/n}$$

has a beta distribution.

(*d*) Use the result of part (*c*) and the beta function to find the mean and variance of the F distribution. [Find the first two moments of $mX/n = W/(1 - W)$].

19 On the t distribution:

(*a*) Find the mean and variance of Student's t distribution. (Be careful about existence.)

(*b*) Show that the density of a t distributed random variable approaches the standard normal density as the degrees of freedom increase. (Assume that the "constant" part of the density does what it has to do.)

(*c*) If X is t-distributed, show that X^2 is F-distributed.

(*d*) If X is t-distributed with k degrees of freedom, show that $1/(1 + X^2/k)$ has a beta distribution.

20 Let X_1, X_2 be a random sample from $N(0, 1)$. Using the results of Sec. 4 of Chap. VI, answer the following:

(*a*) What is the distribution of $(X_2 - X_1)/\sqrt{2}$?

(*b*) What is the distribution of $(X_1 + X_2)^2/(X_2 - X_1)^2$?

(*c*) What is the distribution of $(X_2 + X_1)/\sqrt{(X_1 - X_2)^2}$?

(*d*) What is the distribution of $1/Z$ if $Z = X_1^2/X_2^2$?

21 Let $X_1, \ldots, X_n$ be a random sample from $N(0, 1)$. Define

$$\bar{X}_k = \frac{1}{k} \sum_1^k X_i \quad \text{and} \quad \bar{X}_{n-k} = \frac{1}{n-k} \sum_{k+1}^n X_i.$$

Using the results of Sec. 4, answer the following:

(*a*) What is the distribution of $\frac{1}{2}(\bar{X}_k + \bar{X}_{n-k})$?

(*b*) What is the distribution of $k\bar{X}_k^2 + (n-k)\bar{X}_{n-k}^2$?

(*c*) What is the distribution of X_1^2/X_2^2?

(*d*) What is the distribution of X_1/X_n?

22 Let $X_1, \ldots, X_n$ be a random sample from $N(\mu, \sigma^2)$. Define

$$\bar{X}_k = \frac{1}{k} \sum_1^k X_i,$$

$$\bar{X}_{n-k} = \frac{1}{n-k} \sum_{k+1}^n X_i,$$

$$\bar{X} = \frac{1}{n} \sum_1^n X_i.$$

$$S_k^2 = \frac{1}{k-1} \sum_1^k (X_i - \bar{X}_k)^2,$$

$$S_{n-k}^2 = \frac{1}{n-k-1} \sum_{k+1}^n (X_i - \bar{X}_{n-k})^2$$

and

$$S^2 = \frac{1}{n-1} \sum_1^n (X_i - \bar{X})^2.$$

Using the results of Sec. 4, answer the following:

(a) What is the distribution of $\sigma^{-2}[(k-1)\mathcal{S}_k^2 + (n-k-1)\mathcal{S}_{n-k}^2]$?

(b) What is the distribution of $(\frac{1}{2})(\bar{X}_k + \bar{X}_{n-k})$?

(c) What is the distribution of $\sigma^{-2}(X_i - \mu)^2$?

(d) What is the distribution of $\mathcal{S}_k^2/\mathcal{S}_{n-k}^2$?

(e) What is the distribution of $(\bar{X} - \mu)/(\mathcal{S}/\sqrt{n})$?

23 Let Z_1, Z_2 be a random sample of size 2 from $N(0, 1)$ and X_1, X_2 a random sample of size 2 from $N(1, 1)$. Suppose the Z_i's are independent of the X_j's. Use the results of Sec. 4 to answer the following:

(a) What is the distribution of $\bar{X} + \bar{Z}$?

(b) What is the distribution of $(Z_1 + Z_2)/\sqrt{[(X_2 - X_1)^2 + (Z_2 - Z_1)^2]/2}$?

(c) What is the distribution of $[(X_1 - X_2)^2 + (Z_1 - Z_2)^2 + (Z_1 + Z_2)^2]/2$?

(d) What is the distribution of $(X_2 + X_1 - 2)^2/(X_2 - X_1)^2$?

24 Let X_i be a random variable distributed $N(i, i^2)$, $i = 1, 2, 3$. Assume that the random variables X_1, X_2, and X_3 are independent. Using only the three random variables X_1, X_2, and X_3:

(a) Give an example of a statistic that has a chi-square distribution with three degrees of freedom.

(b) Give an example of a statistic that has an F distribution with one and two degrees of freedom.

(c) Give an example of a statistic that has t distribution with two degrees of freedom.

25 Let X_1, X_2 be a random sample of size 2 from the density

$$f(x) = \tfrac{1}{2}e^{-\frac{1}{2}x}I_{(0,\infty)}(x).$$

Use results on the chi-square and F distributions to give the distribution of X_1/X_2.

26 Let U_1, U_2 be a random sample of size 2 from a uniform distribution over the interval (0, 1). Let Y_1 and Y_2 be the corresponding order statistics.

(a) For $0 < y_2 < 1$, what is $f_{Y_1 \mid Y_2 = y_2}(y_1 \mid y_2)$, the conditional density of Y_1 given $Y_2 = y_2$?

(b) What is the distribution of $Y_2 - Y_1$?

27 If $X_1, X_2, \ldots, X_n$ are independently and normally distributed with the same mean but different variances $\sigma_1^2, \sigma_2^2, \ldots, \sigma_n^2$ and assuming that $U = \Sigma(X_i/\sigma_i^2)/\Sigma(1/\sigma_j^2)$ and $V = \Sigma(X_i - U)^2/\sigma_i^2$ are independently distributed, show that U is normal and V has the chi-square distribution with $n - 1$ degrees of freedom.

28 For three samples from normal populations (with variances σ_1^2, σ_2^2, and σ_3^2), the sample sizes being n_1, n_2, and n_3, find the joint density of

$$U = \frac{\mathcal{S}_1^2}{\mathcal{S}_3^2} \quad \text{and} \quad V = \frac{\mathcal{S}_2^2}{\mathcal{S}_3^2},$$

where the $\mathcal{S}_1^2$, $\mathcal{S}_2^2$, and $\mathcal{S}_3^2$ are the sample variances. (Assume that the samples are independent.)

29 Let a sample of size n_1 from a normal population (with variance σ_1^2) have sample variance S_1^2, and let a second sample of size n_2 from a second normal population (with mean μ_2 and variance σ_2^2) have mean $\bar{X}$ and sample variance S_2^2. Find the joint density of

$$U = \frac{\sqrt{n_2}(\bar{X} - \mu_2)}{S_2} \quad \text{and} \quad V = \frac{S_1^2}{S_2^2}.$$

(Assume that the samples are independent.)

30 For a random sample of size 2 from a normal density with mean 0 and variance 1, find the distribution of the range.

31 (*a*) What is the probability that the larger of two random observations from any continuous distribution will exceed the median?

(*b*) Generalize the result of part (*a*) to samples of size n.

32 Considering random samples of size n from a population with density $f(x)$, what is the expected value of the area under $f(x)$ to the left of the smallest sample observation?

***33** Consider a random sample $X_1, \ldots, X_n$ from the uniform distribution over the interval $(\mu - \sqrt{3}\sigma, \ \mu + \sqrt{3}\sigma)$. Let $Y_1 \leq \cdots \leq Y_n$ denote the corresponding order statistics.

(*a*) Find the mean and variance of $Y_n - Y_1$.

(*b*) Find the mean and variance of $(Y_1 + Y_n)/2$.

(*c*) Find the mean and variance of Y_{k+1} if $n = 2k + 1$, $k = 0, 1, \ldots$.

(*d*) Compare the variances of $\bar{X}_n$, Y_{k+1}, $(Y_1 + Y_n)/2$.

HINT: It might be easier to solve the problem for $U_1, \ldots, U_n$, a random sample from the uniform distribution over either (0, 1) or (−1, 1), and then make an appropriate transformation.

34 Let $X_1, \ldots, X_n$ be a random sample from the density

$$f(x; \alpha, \beta) = \frac{1}{2\beta} \exp\left[-|(x - \alpha)/\beta|\right],$$

where $-\infty < \alpha < \infty$ and $\beta > 0$. Compare the *asymptotic* distributions of the sample mean and the sample median. In particular, compare the asymptotic variances.

***35** Let $X_1, \ldots, X_n$ be a random sample from the cumulative distribution function $F(x) = \{1 - \exp[-x/(1 - x)]\}I_{(0,1)}(x) + I_{[1,\infty)}(x)$. What is the limiting distribution of $(Y_n^{(n)} - a_n)/b_n$, where $a_n = \log n/(1 + \log n)$ and $b_n^{-1} = (\log n)(1 + \log n)$? What is the asymptotic distribution of $Y_n^{(n)}$?

36 Let $X_1, \ldots, X_n$ be a random sample from $f(x; \theta) = \theta e^{-\theta x} I_{(0,\infty)}(x)$, $\theta > 0$.

(*a*) Compare the asymptotic distribution of $\bar{X}_n$ with the asymptotic distribution of the sample median.

(*b*) For your choice of $\{a_n\}$ and $\{b_n\}$, find a limiting distribution of $(Y_n^{(n)} - a_n)/b_n$.

(*c*) For your choice of $\{a_n\}$ and $\{b_n\}$, find a limiting distribution of $(Y_1^{(n)} - a_n)/b_n$.

VII

PARAMETRIC POINT ESTIMATION

1 INTRODUCTION AND SUMMARY

Chapter VI commenced with some general comments about *inference*. There, it was indicated that a sample from the distribution of a population is useful in making inferences about the population. Two important problems in *statistical inference* are *estimation* and *tests of hypotheses*. One type of estimation, namely *point estimation*, is to be the subject of this chapter.

The problem of estimation, as it shall be considered herein, is loosely defined as follows: Assume that some characteristic of the elements in a population can be represented by a random variable X whose density is $f_X(\cdot\,; \theta) = f(\cdot\,; \theta)$, where the *form* of the density is assumed known except that it contains an unknown parameter θ (if θ were known, the density function would be completely specified, and there would be no need to make inferences about it). Further assume that the values $x_1, x_2, \ldots, x_n$ of a random sample $X_1, X_2, \ldots, X_n$ from $f(\cdot\,; \theta)$ can be observed. On the basis of the observed sample values $x_1, x_2, \ldots, x_n$ it is desired to estimate the value of the unknown parameter θ or the value of some function, say $\tau(\theta)$, of the unknown parameter. This estimation can be made in two ways. The first, called *point estimation*, is to let the value of some statistic, say $\ell(X_1, \ldots, X_n)$, represent, or estimate, the unknown $\tau(\theta)$; such a statistic $\ell(X_1, \ldots, X_n)$ is called a *point estimator*. The second, called *interval estimation*, is to define two statistics, say $\ell_1(X_1, \ldots, X_n)$ and $\ell_2(X_1, \ldots, X_n)$, where $\ell_1(X_1, \ldots, X_n) < \ell_2(X_1, \ldots, X_n)$, so that $(\ell_1(X_1, \ldots, X_n), \ell_2(X_1, \ldots, X_n))$ constitutes an interval for which the probability can be determined that it contains the unknown $\tau(\theta)$. For example, if $f(\cdot\,; \theta)$ is the normal density, that is,

$$f(x; \theta) = f(x; \mu, \sigma) = \phi_{\mu, \sigma^2}(x) = \frac{1}{\sqrt{2\pi}\sigma} \exp\left[-\frac{1}{2}\left(\frac{x - \mu}{\sigma}\right)^2\right],$$

where the parameter θ is (μ, σ), and if it is desired to estimate the mean, that is, $\tau(\theta) = \mu$, then the statistic $\bar{X} = (1/n) \sum_1^n X_i$ is a possible point estimator of $\tau(\theta) = \mu$, and $(\bar{X} - 2\sqrt{S^2/n}, \bar{X} + 2\sqrt{S^2/n})$ is a possible interval estimator of $\tau(\theta) = \mu$. {Recall that $S^2 = [1/(n-1)] \sum_1^n (X_i - \bar{X})^2$.} Point estimation will be discussed in this chapter and interval estimation in the next.

Point estimation admits two problems: the first, to devise some means of obtaining a statistic to use as an estimator; the second, to select criteria and techniques to define and find a "best" estimator among many possible estimators. Several methods of finding point estimators are introduced in Sec. 2. One of these, and probably the most important, is the *method of maximum likelihood*. In Sec. 3 several "optimum" properties that an estimator or sequence of estimators may possess are defined. These include closeness, bias and variance, efficiency, and consistency. The loss and risk functions, essential elements in *decision theory*, are defined as possible tools in assessing the *goodness* of estimators.

Section 4 is devoted to *sufficiency*, an important and useful concept in the study of mathematical statistics that will also be utilized in succeeding chapters. Unbiased estimation is considered in Sec. 5. The Cramér-Rao lower bound for the variance of unbiased estimators is given, as well as the Rao-Blackwell theorem concerning sufficient statistics. A brief look at invariant estimators is presented in Sec. 6. A Bayes estimation is considered in Sec. 7. A Bayes estimator is given as the mean of the posterior or from the decision-theoretical viewpoint as an estimator having smallest average risk. Some results in the simultaneous estimation of several parameters are given in Sec. 8. Included is the notion of *ellipsoid of concentration* of a vector of point estimators and the Lehmann-Scheffé theorem. Section 9 is devoted to a brief discussion of some optimum properties of maximum-likelihood estimators.

Frequent use of some of the distribution-theoretical results for statistics, which were derived in earlier chapters, especially Chaps. V and VI, will be noted throughout this chapter. After all, estimators are statistics, and to study properties of estimators, it is desirable to look at their distributions.

2 METHODS OF FINDING ESTIMATORS

Assume that $X_1, \ldots, X_n$ is a random sample from a density $f(\cdot\,; \theta)$, where the form of the density is known but the parameter θ is unknown. Further assume that θ is a vector of real numbers, say $\theta = (\theta_1, \ldots, \theta_k)$. (Often k will be unity.) We sometimes say that $\theta_1, \ldots, \theta_k$ are k parameters. We will let $\underline{\overline{\Theta}}$, called the *parameter space*, denote the set of possible values that the parameter θ can assume. The object is to find statistics, functions of the observations $X_1, \ldots, X_n$, to be used as estimators of the θ_j, $j = 1, \ldots, k$. Or, more generally, our object is to find statistics to be used as estimators of certain functions, say $\tau_1(\theta), \ldots, \tau_r(\theta)$, of $\theta = (\theta_1, \ldots, \theta_k)$. A variety of methods of finding such estimators has been proposed on more or less intuitive grounds. Several such methods will be presented, along with examples, in this section. Another method, that of the *method of least squares* will be discussed in Chap. X.

An estimator can be defined as in Definition 1.

> **Definition 1 Estimator** Any statistic (known function of observable random variables that is itself a random variable) whose values are used to estimate $\tau(\theta)$, where $\tau(\cdot)$ is some function of the parameter θ, is defined to be an *estimator* of $\tau(\theta)$. ////

An estimator is always a statistic which is both a random variable and a function. For instance, suppose $X_1, \ldots, X_n$ is a random sample from a density $f(\cdot\,; \theta)$ and it is desired to estimate $\tau(\theta)$, where $\tau(\cdot)$ is some function of θ. Let $\ell(X_1, \ldots, X_n)$ be an estimator of $\tau(\theta)$. The estimator $\ell(X_1, \ldots, X_n)$ can be thought of in two related ways: first, as the random variable, say T, where $T = \ell(X_1, \ldots, X_n)$, and, second, as the function $\ell(\cdot, \ldots, \cdot)$. Naturally, one needs to specify the function $\ell(\cdot, \ldots, \cdot)$ before the random variable $T = \ell(X_1, \ldots, X_n)$ is defined. In all we have three types of tees: the capital Latin T, which represents the random variable $\ell(X_1, \ldots, X_n)$, the small script ℓ, which represents the function $\ell(\cdot, \ldots, \cdot)$, and the small Latin t, which represents a value of T; that is, $t = \ell(x_1, \ldots, x_n)$. Let us adopt the convention of calling the statistic (or random variable) that is used as an estimator an "estimator" and calling a value that the statistic takes on an "estimate." Thus the word "estimator" stands for the function, and the word "estimate" stands for a value of that function; for example, $\bar{X}_n = \frac{1}{n} \sum_{i=1}^{n} X_i$ is an estimator of a mean μ, and $\bar{x}_n$ is an estimate of μ. Here T is $\bar{X}_n$, t is $\bar{x}_n$, and $\ell(\cdot, \ldots, \cdot)$ is the function defined by summing the arguments and then dividing by n.

Notation in estimation that has widespread usage is the following: $\hat{\theta}$ is used to denote an estimate of θ, and, more generally, $(\hat{\theta}_1, \ldots, \hat{\theta}_k)$ is a vector that estimates the vector $(\theta_1, \ldots, \theta_k)$, where $\hat{\theta}_j$ estimates θ_j, $j = 1, \ldots, k$. If $\hat{\theta}$ is an estimate of θ, then $\hat{\Theta}$ is the corresponding estimator of θ; and if the discussion requires that the function that defines both $\hat{\theta}$ and $\hat{\Theta}$ be specified, then it can be denoted by a small script theta, that is, $\hat{\Theta} = \hat{\vartheta}(X_1, \ldots, X_n)$.

When we speak of estimating θ, we are speaking of estimating the fixed yet unknown value that θ has. That is, we assume that the random sample $X_1, \ldots, X_n$ came from the density $f(\cdot\,; \theta)$, where θ is unknown but fixed. Our object is, after looking at the values of the random sample, to estimate the fixed unknown θ. And when we speak of estimating $\tau(\theta)$, we are speaking of estimating the value $\tau(\theta)$ that the known function $\tau(\cdot)$ assumes for the unknown but fixed θ.

2.1 Methods of Moments

Let $f(\cdot\,; \theta_1, \ldots, \theta_k)$ be a density of a random variable X which has k parameters $\theta_1, \ldots, \theta_k$. As before let μ'_r denote the rth moment about 0; that is, $\mu'_r = \mathscr{E}[X^r]$. In general μ'_r will be a known function of the k parameters $\theta_1, \ldots, \theta_k$. Denote this by writing $\mu'_r = \mu'_r(\theta_1, \ldots, \theta_k)$. Let $X_1, \ldots, X_n$ be a random sample from the density $f(\cdot\,; \theta_1, \ldots, \theta_k)$, and, as before, let M'_j be the jth sample moment; that is,

$$M'_j = \frac{1}{n} \sum_{i=1}^{n} X_i^j.$$

Form the k equations

$$M'_j = \mu'_j(\theta_1, \ldots, \theta_k), \qquad j = 1, \ldots, k, \tag{1}$$

in the k variables $\theta_1, \ldots, \theta_k$, and let $\hat{\Theta}_1, \ldots, \hat{\Theta}_k$ be their solution (we assume that there is a unique solution). We say that the estimator $(\hat{\Theta}_1, \ldots, \hat{\Theta}_k)$, where $\hat{\theta}_j$ estimates θ_j, is the estimator of $(\theta_1, \ldots, \theta_k)$ obtained by the *method of moments*. The estimators were obtained by replacing population moments by sample moments. Some examples follow.

EXAMPLE 1 Let $X_1, \ldots, X_n$ be a random sample from a normal distribution with mean μ and variance σ^2. Let $(\theta_1, \theta_2) = (\mu, \sigma)$. Estimate the parameters μ and σ by the method of moments. Recall that $\sigma^2 = \mu_2' - (\mu_1')^2$ and $\mu = \mu_1'$. The method-of-moments equations become

$$M_1' = \mu_1' = \mu_1'(\mu, \sigma) = \mu$$
$$M_2' = \mu_2' = \mu_2'(\mu, \sigma) = \sigma^2 + \mu^2,$$

and their solution is the following: The method-of-moments estimator of μ is $M_1' = \bar{X}$, and the method-of-moments estimator of σ is $\sqrt{M_2' - \bar{X}^2} = \sqrt{(1/n)\sum X_i^2 - \bar{X}^2} = \sqrt{\sum (X_i - \bar{X})^2/n}$. Note that the method-of-moments estimator of σ given above is not $\sqrt{S^2}$. ////

EXAMPLE 2 Let $X_1, \ldots, X_n$ be a random sample from a Poisson distribution with parameter λ. Estimate λ. There is only one parameter, hence only one equation, which is

$$M_1' = \mu_1' = \mu_1'(\lambda) = \lambda.$$

Hence the method-of-moments estimator of λ is $M_1' = \bar{X}$, which says estimate the population mean λ with the sample mean $\bar{x}$. ////

EXAMPLE 3 Let $X_1, \ldots, X_n$ be a random sample from the negative exponential density $f(x; \theta) = \theta e^{-\theta x} I_{(0, \infty)}(x)$. Estimate θ. The method-of-moments equation is

$$M_1' = \mu_1' = \mu_1'(\theta) = \frac{1}{\theta};$$

hence the method-of-moments estimator of θ is $1/M_1' = 1/\bar{X}$. ////

EXAMPLE 4 Let $X_1, \ldots, X_n$ be a random sample from a uniform distribution on $(\mu - \sqrt{3}\sigma, \mu + \sqrt{3}\sigma)$. Here the unknown parameters are two, namely μ and σ, which are the population mean and standard deviation. The method-of-moments equations are

$$M_1' = \mu_1' = \mu_1'(\mu, \sigma) = \mu$$

and

$$M_2' = \mu_2' = \mu_2'(\mu, \sigma) = \sigma^2 + \mu^2;$$

hence the method-of-moments estimators are $\bar{X}$ for μ and

$$\sqrt{\frac{1}{n}\sum X_i^2 - \bar{X}^2} = \sqrt{\frac{1}{n}\sum (X_i - \bar{X})^2} \qquad \text{for } \sigma.$$

We shall see later that there are better estimators of μ and σ for this distribution. ////

Method-of-moments estimators are not uniquely defined. The method-of-moments equations given in Eq. (1) are obtained by using the first k raw moments. Central moments (rather than raw moments) could also be used to obtain equations whose solution would also produce estimators that would be labeled method-of-moments estimators. Also, moments other than the first k could be used to obtain estimators that would be labeled method-of-moments estimators.

If, instead of estimating $(\theta_1, \ldots, \theta_k)$, method-of-moments estimators of, say, $\tau_1(\theta_1, \ldots, \theta_k), \ldots, \tau_r(\theta_1, \ldots, \theta_k)$ are desired, they can be obtained in several ways. One way would be to first find method-of-moments estimates, say $\hat{\theta}_1, \ldots, \hat{\theta}_k$, of $\theta_1, \ldots, \theta_k$ and then use $\tau_j(\hat{\theta}_1, \ldots, \hat{\theta}_k)$ as an estimate of $\tau_j(\theta_1, \ldots, \theta_k)$ for $j = 1, \ldots, r$. Another way would be to form the equations

$$M_j' = \mu_j'(\tau_1, \ldots, \tau_r), \qquad j = 1, \ldots, r$$

and solve them for $\tau_1, \ldots, \tau_r$. Estimators obtained using either way are called method-of-moments estimators and may not be the same in both cases.

2.2 Maximum Likelihood

To introduce the *method of maximum likelihood*, consider a very simple estimation problem. Suppose that an urn contains a number of black and a number of white balls, and suppose that it is known that the ratio of the numbers is 3/1 but that it is not known whether the black or the white balls are more numerous. That is, the probability of drawing a black ball is either $\frac{1}{4}$ or $\frac{3}{4}$. If n balls are drawn with replacement from the urn, the distribution of X, the number of black balls, is given by the binomial distribution

$$f(x; p) = \binom{n}{x} p^x q^{n-x} \qquad \text{for } x = 0, 1, 2, \ldots, n,$$

where $q = 1 - p$ and p is the probability of drawing a black ball. Here $p = \frac{1}{4}$, or $p = \frac{3}{4}$.

We shall draw a sample of three balls, that is, $n = 3$, with replacement and attempt to estimate the unknown parameter p of the distribution. The estimation problem is particularly simple in this case because we have only to choose between the two numbers .25 and .75. Let us anticipate the results of the drawing of the sample. The possible outcomes and their probabilities are given below:

Outcome: x	0	1	2	3
$f(x; \frac{3}{4})$	$\frac{1}{64}$	$\frac{9}{64}$	$\frac{27}{64}$	$\frac{27}{64}$
$f(x; \frac{1}{4})$	$\frac{27}{64}$	$\frac{27}{64}$	$\frac{9}{64}$	$\frac{1}{64}$

In the present example, if we found $x = 0$ in a sample of 3, the estimate .25 for p would be preferred over .75 because the probability $\frac{27}{64}$ is greater than $\frac{1}{64}$, i.e., because a sample with $x = 0$ is more likely (in the sense of having larger probability) to arise from a population with $p = \frac{1}{4}$ than from one with $p = \frac{3}{4}$. And in general we should estimate p by .25 when $x = 0$ or 1 and by .75 when $x = 2$ or 3. The estimator may be defined as

$$\hat{p} = \hat{p}(x) = \begin{cases} .25 & \text{for } x = 0, 1 \\ .75 & \text{for } x = 2, 3. \end{cases}$$

The estimator thus selects for every possible x the value of p, say $\hat{p}$, such that

$$f(x; \hat{p}) > f(x; p'),$$

where p' is the alternative value of p.

More generally, if several alternative values of p were possible, we might reasonably proceed in the same manner. Thus if we found $x = 6$ in a sample of 25 from a binomial population, we should substitute all possible values of p in the expression

$$f(6; p) = \binom{25}{6} p^6(1 - p)^{19} \qquad \text{for } 0 \leq p \leq 1 \tag{2}$$

and choose as our estimate that value of p which maximized $f(6; p)$. For the given possible values of p we should find our estimate to be $\frac{6}{25}$. The position of its maximum value can be found by putting the derivative of the function defined in Eq. (2) with respect to p equal to 0 and solving the resulting equation for p. Thus,

$$\frac{d}{dp} f(6; p) = \binom{25}{6} p^5(1 - p)^{18}[6(1 - p) - 19p],$$

and on putting this equal to 0 and solving for p, we find that $p = 0, 1, \frac{6}{25}$ are the roots. The first two roots give a minimum, and so our estimate is therefore $\hat{p} = \frac{6}{25}$. This estimate has the property that

$$f(6; \hat{p}) > f(6; p'),$$

where p' is any other value of p in the interval $0 \leq p \leq 1$.

In order to define maximum-likelihood estimators, we shall first define the likelihood function.

Definition 2 Likelihood function The *likelihood function* of n random variables $X_1, X_2, \ldots, X_n$ is defined to be the joint density of the n random variables, say $f_{X_1, \ldots, X_n}(x_1, \ldots, x_n; \theta)$, which is considered to be a function of θ. In particular, if $X_1, \ldots, X_n$ is a random sample from the density $f(x; \theta)$, then the likelihood function is $f(x_1; \theta)f(x_2; \theta) \cdot \cdots \cdot f(x_n; \theta)$. ////

Notation To remind ourselves to think of the likelihood function as a function of θ, we shall use the notation $L(\theta; x_1, \ldots, x_n)$ or $L(\cdot\,; x_1, \ldots, x_n)$ for the likelihood function. ////

The likelihood function $L(\theta; x_1, \ldots, x_n)$ gives the *likelihood* that the random variables assume a particular value $x_1, x_2, \ldots, x_n$. The *likelihood* is the value of a density function; so for discrete random variables it is a probability. Suppose for a moment that θ is known; denote the value by θ_0. The particular value of the random variables which is "most likely to occur" is that value $x_1', x_2', \ldots, x_n'$ such that $f_{X_1, \ldots, X_n}(x_1, \ldots, x_n; \theta_0)$ is a maximum. For example, for simplicity let us assume that $n = 1$ and X_1 has the normal density with mean 6 and variance 1. Then the value of the random variable which is most likely to occur is $X_1 = 6$. By "most likely to occur" we mean the value x_1' of X_1 such that $\phi_{6,1}(x_1') > \phi_{6,1}(x_1)$. Now let us suppose that the joint density of n random variables is $f_{X_1, \ldots, X_n}(x_1, \ldots, x_n; \theta)$, where θ is unknown. Let the particular values which are observed be represented by $x_1', x_2', \ldots, x_n'$. We want to know from which density is this particular set of values most likely to have come. We want to know from which density (what value of θ) is the likelihood largest that the set $x_1', \ldots, x_n'$ was obtained. In other words, we want to find the value of θ in $\underline{\Theta}$, denoted by $\hat{\theta}$, which maximizes the likelihood function $L(\theta; x_1', \ldots, x_n')$. The value $\hat{\theta}$ which maximizes the likelihood function is, in general, a function of $x_1, \ldots, x_n$, say $\hat{\theta} = \hat{\vartheta}(x_1, x_2, \ldots, x_n)$. When this is the case, the random variable $\hat{\Theta} = \hat{\vartheta}(X_1, X_2, \ldots, X_n)$ is called the *maximum-likelihood estimator* of θ. (We are assuming throughout that the maximum of the likelihood function exists.) We shall now formalize the definition of a maximum-likelihood estimator.

Definition 3 Maximum-likelihood estimator Let

$$L(\theta) = L(\theta; x_1, \ldots, x_n)$$

be the likelihood function for the random variables $X_1, X_2, \ldots, X_n$. If $\hat{\theta}$ [where $\hat{\theta} = \hat{\vartheta}(x_1, x_2, \ldots, x_n)$ is a function of the observations $x_1, \ldots, x_n$] is the value of θ in $\underline{\Theta}$ which maximizes $L(\theta)$, then $\hat{\Theta} = \hat{\vartheta}(X_1, X_2, \ldots, X_n)$ is the *maximum-likelihood estimator* of θ. $\hat{\theta} = \hat{\vartheta}(x_1, \ldots, x_n)$ is the maximum-likelihood estimate of θ for the sample $x_1, \ldots, x_n$. ////

The most important cases which we shall consider are those in which $X_1, X_2, \ldots, X_n$ is a *random sample* from some density $f(x; \theta)$, so that the likelihood function is

$$L(\theta) = f(x_1; \theta) f(x_2; \theta) \cdots f(x_n; \theta).$$

Many likelihood functions satisfy regularity conditions; so the maximum-likelihood estimator is the solution of the equation

$$\frac{dL(\theta)}{d\theta} = 0.$$

Also $L(\theta)$ and $\log L(\theta)$ have their maxima at the same value of θ, and it is sometimes easier to find the maximum of the logarithm of the likelihood.

If the likelihood function contains k parameters, that is, if

$$L(\theta_1, \theta_2, \ldots, \theta_k) = \prod_{i=1}^{n} f(x_i; \theta_1, \theta_2, \ldots, \theta_k),$$

then the maximum-likelihood estimators of the parameters $\theta_1, \theta_2, \ldots, \theta_k$ are the random variables $\hat{\Theta}_1 = \hat{\vartheta}_1(X_1, \ldots, X_n)$, $\hat{\Theta}_2 = \hat{\vartheta}_2(X_1, \ldots, X_n)$, $\ldots$, $\hat{\Theta}_k = \hat{\vartheta}_k(X_1, \ldots, X_n)$, where $\hat{\theta}_1, \hat{\theta}_2, \ldots, \hat{\theta}_k$ are the values in $\underline{\Theta}$ which maximize $L(\theta_1, \theta_2, \ldots, \theta_k)$.

If certain regularity conditions are satisfied, the point where the likelihood is a maximum is a solution of the k equations

$$\frac{\partial L(\theta_1, \ldots, \theta_k)}{\partial \theta_1} = 0$$

$$\frac{\partial L(\theta_1, \ldots, \theta_k)}{\partial \theta_2} = 0$$

$$\vdots$$

$$\frac{\partial L(\theta_1, \ldots, \theta_k)}{\partial \theta_k} = 0.$$

In this case it may also be easier to work with the logarithm of the likelihood.

We shall illustrate these definitions with some examples.

EXAMPLE 5 Suppose that a random sample of size n is drawn from the Bernoulli distribution

$$f(x; p) = p^x q^{1-x} I_{\{0,1\}}(x), \qquad 0 \le p \le 1 \text{ and } q = 1 - p.$$

The sample values $x_1, x_2, \ldots, x_n$ will be a sequence of 0s and 1s, and the likelihood function is

$$L(p) = \prod_{i=1}^{n} p^{x_i} q^{1-x_i} = p^{\sum x_i} q^{n - \sum x_i},$$

and if we let

$$y = \sum x_i,$$

we obtain

$$^*\log L(p) = y \log p + (n - y) \log q$$

and

$$\frac{d \log L(p)}{dp} = \frac{y}{p} - \frac{n-y}{q},$$

remembering that $q = 1 - p$. On putting this last expression equal to 0 and solving for p, we find the estimate

$$\hat{p} = \frac{y}{n} = \frac{1}{n} \sum x_i = \bar{x}, \tag{3}$$

which is intuitively what the estimate for this parameter should be. It is also a method-of-moments estimate. For $n = 3$, let us sketch the likelihood function. Note that the likelihood function depends on the x_i's only through $\sum x_i$; thus the likelihood function can be represented by the following four curves:

$$\begin{aligned} L_0 &= L(p; \textstyle\sum x_i = 0) = (1 - p)^3 \\ L_1 &= L(p; \textstyle\sum x_i = 1) = p(1 - p)^2 \\ L_2 &= L(p; \textstyle\sum x_i = 2) = p^2(1 - p) \\ L_3 &= L(p; \textstyle\sum x_i = 3) = p^3, \end{aligned}$$

which are sketched in Fig. 1. Note that the point where the maximum of each of the curves takes place for $0 \le p \le 1$ is the same as that given in Eq. (3) when $n = 3$. ////

* Recall that log x means $\log_e x$.

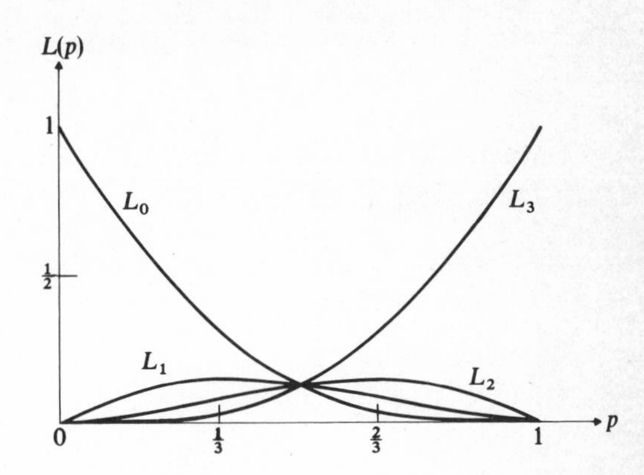

FIGURE 1

EXAMPLE 6 A random sample of size n from the normal distribution has the density

$$\prod_{i=1}^{n} \frac{1}{\sqrt{2\pi}\sigma} e^{-(1/2\sigma^2)(x_i-\mu)^2} = \left(\frac{1}{2\pi\sigma^2}\right)^{n/2} \exp\left[-\frac{1}{2\sigma^2}\sum(x_i-\mu)^2\right].$$

The logarithm of the likelihood function is

$$L^* = -\frac{n}{2}\log 2\pi - \frac{n}{2}\log\sigma^2 - \frac{1}{2\sigma^2}\sum(x_i-\mu)^2,$$

where $\sigma > 0$ and $-\infty < \mu < \infty$.

To find the location of its maximum, we compute

$$\frac{\partial L^*}{\partial\mu} = \frac{1}{\sigma^2}\sum(x_i-\mu)$$

and

$$\frac{\partial L^*}{\partial\sigma^2} = -\frac{n}{2}\frac{1}{\sigma^2} + \frac{1}{2\sigma^4}\sum(x_i-\mu)^2,$$

and on putting these derivatives equal to 0 and solving the resulting equations for μ and σ^2, we find the estimates

$$\hat{\mu} = \frac{1}{n}\sum x_i = \bar{x} \tag{4}$$

$$\hat{\sigma}^2 = \frac{1}{n}\sum(x_i-\bar{x})^2, \tag{5}$$

which turn out to be the sample moments corresponding to μ and σ^2.

////

EXAMPLE 7 Let the random variable X have a uniform density given by

$$f(x; \theta) = I_{[\theta - \frac{1}{2}, \theta + \frac{1}{2}]}(x),$$

where $-\infty < \theta < \infty$; that is, $\underline{\Theta}$ = real line. The likelihood function for a sample of size n is

$$L(\theta; x_1, \ldots, x_n) = \prod_{i=1}^{n} f(x_i; \theta) = \prod_{i=1}^{n} I_{[\theta - \frac{1}{2}, \theta + \frac{1}{2}]}(x_i)$$

$$= I_{[y_n - \frac{1}{2}, y_1 + \frac{1}{2}]}(\theta), \tag{6}$$

where y_1 is the smallest of the observations and y_n is the largest. The last equality in Eq. (6) follows since $\prod_{i=1}^{n} I_{[\theta - \frac{1}{2}, \theta + \frac{1}{2}]}(x_i)$ is unity if and only if all $x_1, \ldots, x_n$ are in the interval $[\theta - \frac{1}{2}, \theta + \frac{1}{2}]$, which is true if and only if $\theta - \frac{1}{2} \le y_1$ and $y_n \le \theta + \frac{1}{2}$, which is true if and only if $y_n - \frac{1}{2} \le \theta \le y_1 + \frac{1}{2}$. We see that the likelihood function is either 1 (for $y_n - \frac{1}{2} \le \theta \le y_1 + \frac{1}{2}$) or 0 (otherwise); hence any statistic with value $\hat{\theta}$ satisfying $y_n - \frac{1}{2} \le \hat{\theta} \le y_1 + \frac{1}{2}$ is a maximum-likelihood estimate. Examples are $y_n - \frac{1}{2}$, $y_1 + \frac{1}{2}$, and $\frac{1}{2}(y_1 + y_n)$. This latter is the midpoint between $y_n - \frac{1}{2}$ and $y_1 + \frac{1}{2}$, or the midpoint between y_1 and y_n, the smallest and largest observations. ////

EXAMPLE 8 Let the random variable X have a uniform distribution with density given by

$$f(x; \theta) = f(x; \mu, \sigma) = \frac{1}{2\sqrt{3}\,\sigma} I_{[\mu - \sqrt{3}\sigma, \mu + \sqrt{3}\sigma]}(x),$$

where $-\infty < \mu < \infty$ and $\sigma > 0$. (Recall Example 4.) Here the likelihood function for a sample of size n is

$$L(\mu, \sigma; x_1, \ldots, x_n) = \left(\frac{1}{2\sqrt{3}\,\sigma}\right)^n \prod_{i=1}^{n} I_{[\mu - \sqrt{3}\sigma, \mu + \sqrt{3}\sigma]}(x_i)$$

$$= \left(\frac{1}{2\sqrt{3}\,\sigma}\right)^n I_{[\mu - \sqrt{3}\sigma, y_n]}(y_1) I_{[y_1, \mu + \sqrt{3}\sigma]}(y_n)$$

$$= \left(\frac{1}{2\sqrt{3}\,\sigma}\right)^n I_{[(\mu - y_1)/\sqrt{3}, \infty)}(\sigma) I_{[(y_n - \mu)/\sqrt{3}, \infty)}(\sigma) I_{[y_1, \infty)}(y_n),$$

where y_1 is the smallest of the observations and y_n is the largest. The likelihood function is $(2\sqrt{3}\,\sigma)^{-n}$ in the shaded area of Fig. 2 and 0 elsewhere. $(2\sqrt{3}\,\sigma)^{-n}$ within the shaded area is clearly a maximum when σ

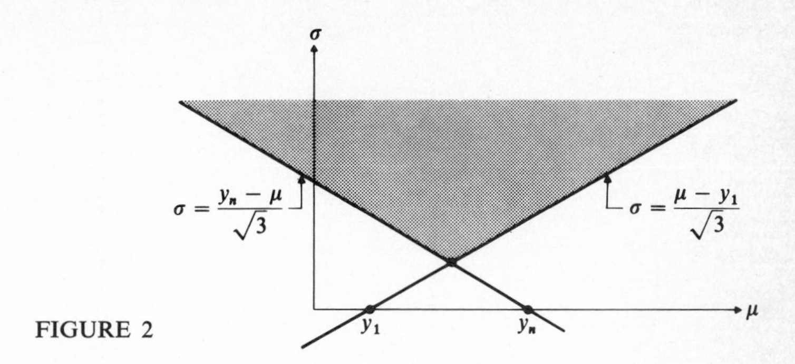

FIGURE 2

is smallest, which is at the intersection of the lines $\mu - \sqrt{3}\,\sigma = y_1$ and $\mu + \sqrt{3}\,\sigma = y_n$. Hence the maximum-likelihood estimates of μ and σ are

$$\hat{\mu} = \frac{1}{2}(y_1 + y_n) \tag{7}$$

and

$$\hat{\sigma} = \frac{1}{2\sqrt{3}}(y_n - y_1), \tag{8}$$

which are quite different from the method-of-moments estimates given in Example 4. ////

The above four examples are sufficient to illustrate the application of the method of maximum likelihood. The last two show that one must not always rely on the differentiation process to locate the maximum.

The function $L(\theta)$ may, for example, be represented by the curve in Fig. 3, where the actual maximum is at $\hat{\theta}$, but the derivative set equal to 0 would locate θ' as the maximum. One must also remember that the equation $\partial L/\partial\theta = 0$ locates minima as well as maxima, and hence one must avoid using a root of the equation which actually locates a minimum.

We shall see in later sections (especially Sec. 9 of this chapter) that the maximum-likelihood estimator has some desirable optimum properties other than the intuitively appealing property that it maximizes the likelihood function. In addition, the maximum-likelihood estimators possess a property which is sometimes called the *invariance property of maximum-likelihood estimators*. A little reflection on the meaning of a single-valued inverse will convince one of the validity of the following theorem.

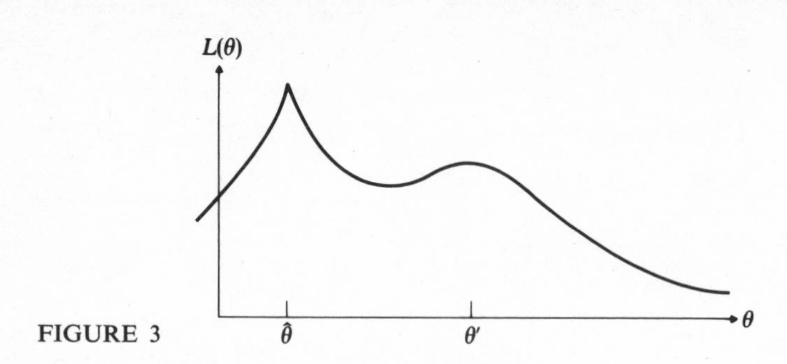

FIGURE 3

Theorem 1 Invariance property of maximum-likelihood estimators Let $\hat{\Theta} = \hat{\vartheta}(X_1, X_2, \ldots, X_n)$ be the maximum-likelihood estimator of θ in the density $f(x; \theta)$, where θ is assumed unidimensional. If $\tau(\cdot)$ is a function with a single-valued inverse, then the maximum-likelihood estimator of $\tau(\theta)$ is $\tau(\hat{\Theta})$. ////

For example, in the normal density with μ_0 known the maximum-likelihood estimator of σ^2 is

$$\frac{1}{n}\sum_{i=1}^{n}(X_i - \mu_0)^2.$$

By the invariance property of maximum-likelihood estimators, the maximum-likelihood estimator of σ is

$$\sqrt{\frac{1}{n}\sum_{i=1}^{n}(X_i - \mu_0)^2}.$$

Similarly, the maximum-likelihood estimator of, say, $\log \sigma^2$ is

$$\log\left[\frac{1}{n}\sum_{i=1}^{n}(X_i - \mu_0)^2\right].$$

The invariance property of maximum-likelihood estimators that is exhibited in Theorem 1 above can and should be extended. Following Zehna [43], we extend in two directions: First, θ will be taken as k-dimensional rather than unidimensional, and, second, the assumption that $\tau(\cdot)$ has a single-valued inverse will be removed. It can be noted that such extension is necessary by considering two simple examples. As a first example, suppose an estimate of the variance, namely $\theta(1 - \theta)$, of a Bernoulli distribution is desired. Example 5 gives the maximum-likelihood estimate of θ to be $\bar{x}$, but since $\theta(1 - \theta)$ is not a one-to-one function of θ, Theorem 1 does not give the maximum-likelihood estimator of $\theta(1-\theta)$. Theorem 2 below will give such an estimate, and it will be $\bar{x}(1-\bar{x})$. As a second example, consider sampling from a normal distribution where both μ and σ^2 are unknown, and suppose an estimate of $\mathscr{E}[X^2] = \mu^2 + \sigma^2$ is desired. Example 6 gives the maximum-likelihood estimates of μ and σ^2, but $\mu^2 + \sigma^2$ is not a one-to-one function of μ and σ^2, and so the maximum-likelihood estimate of $\mu^2 + \sigma^2$ is not known. Such an estimate will be obtainable from Theorem 2 below. It will be $\bar{x}^2 + (1/n) \sum (x_i - \bar{x})^2$.

Let $\theta = (\theta_1, \ldots, \theta_k)$ be a k-dimensional parameter, and, as before, let $\underline{\Theta}$ denote the parameter space. Suppose that the maximum-likelihood estimate of $\tau(\theta) = (\tau_1(\theta), \ldots, \tau_r(\theta))$, where $1 \le r \le k$, is desired. Let T denote the range space of the transformation $\tau(\cdot) = (\tau_1(\cdot), \ldots, \tau_r(\cdot))$. T is an r-dimensional space. Define $M(\tau; x_1, \ldots, x_n) = \sup_{\{\theta : \tau(\theta) = \tau\}} L(\theta; x_1, \ldots, x_n)$. $M(\cdot\,; x_1, \ldots, x_n)$ is called the *likelihood function induced by* $\tau(\cdot)$.* When estimating θ we maximized the likelihood function $L(\theta; x_1, \ldots, x_n)$ as a function of θ for fixed $x_1, \ldots, x_n$; when estimating $\tau = \tau(\theta)$ we will maximize the likelihood function induced by $\tau(\cdot)$, namely $M(\tau; x_1, \ldots, x_n)$, as a function of τ for fixed $x_1, \ldots, x_n$. Thus, the maximum-likelihood estimate of $\tau = \tau(\theta)$, denoted by $\hat{\tau}$, is any value that maximizes the induced likelihood function for fixed $x_1, \ldots, x_n$; that is, $\hat{\tau}$ is such that $M(\hat{\tau}; x_1, \ldots, x_n) \ge M(\tau; x_1, \ldots, x_n)$ for all $\tau \in T$. The *invariance property* of maximum-likelihood estimation is given in the following theorem.

Theorem 2 Let $\hat{\Theta} = (\hat{\Theta}_1, \ldots, \hat{\Theta}_k)$, where $\hat{\Theta}_j = \hat{\vartheta}_j(X_1, \ldots, X_n)$, be a maximum-likelihood estimator of $\theta = (\theta_1, \ldots, \theta_k)$ in the density $f(\cdot\,; \theta_1, \ldots, \theta_k)$. If $\tau(\theta) = (\tau_1(\theta), \ldots, \tau_r(\theta))$ for $1 \le r \le k$ is a transformation of the parameter space $\underline{\Theta}$, then a maximum-likelihood estimator of $\tau(\theta) = (\tau_1(\theta), \ldots, \tau_r(\theta))$ is $\tau(\hat{\Theta}) = (\tau_1(\hat{\Theta}), \ldots, \tau_r(\hat{\Theta}))$. [Note that $\tau_j(\theta) = \tau_j(\theta_1, \ldots, \theta_k)$; so the maximum-likelihood estimator of $\tau_j(\theta_1, \ldots, \theta_k)$ is $\tau_j(\hat{\Theta}_1, \ldots, \hat{\Theta}_k)$, $j = 1, \ldots, r$.]

PROOF Let $\hat{\theta} = (\hat{\theta}_1, \ldots, \hat{\theta}_k)$ be a maximum-likelihood estimate of $\theta = (\theta_1, \ldots, \theta_k)$. It suffices to show that $M(\tau(\hat{\theta}); x_1, \ldots, x_n) \ge M(\tau; x_1, \ldots, x_n)$ for any $\tau \in T$, which follows immediately from the inequality $M(\tau; x_1, \ldots, x_n) = \sup_{\{\theta : \tau(\theta) = \tau\}} L(\theta; x_1, \ldots, x_n) \le \sup_{\theta \in \underline{\Theta}} L(\theta; x_1, \ldots, x_n)$ $= L(\hat{\theta}; x_1, \ldots, x_n) = \sup_{\{\theta : \tau(\theta) = \tau(\hat{\theta})\}} L(\theta; x_1, \ldots, x_n) = M(\tau(\hat{\theta}); x_1, \ldots, x_n)$. ////

*The notation "sup" is used here, and elsewhere in this book, as it is usually used in mathematics. For those readers who are not acquainted with this notation, not much is lost if "sup" is replaced by "max," where max is an abbreviation for maximum.

It is precisely this property of invariance enjoyed by maximum-likelihood estimators that allowed us in our discussion of maximum-likelihood estimation to consider estimating $(\theta_1, \ldots, \theta_k)$ rather than the more general $\tau_1(\theta_1, \ldots, \theta_k)$, $\ldots$, $\tau_r(\theta_1, \ldots, \theta_k)$.

EXAMPLE 9 In the normal density, let $\theta = (\theta_1, \theta_2) = (\mu, \sigma^2)$. Suppose $\tau(\theta) = \mu + z_q \sigma$, where z_q is given by $\phi(z_q) = q$. $\tau(\theta)$ is the qth quantile. According to Theorem 2, the maximum-likelihood estimator of $\tau(\theta)$ is $\bar{X} + z_q\sqrt{(1/n)\sum (X_i - \bar{X})^2}$. ////

2.3 Other Methods

There are several other methods of obtaining point estimators of parameters. Among these are (i) the method of *least squares*, to be discussed in Chap. X, (ii) the *Bayes* method, to be discussed later in this chapter, (iii) the *minimum-chi-square* method, and (iv) the *minimum-distance* method. In this subsection we will briefly consider the last two. Neither will be used again in this book.

Minimum-chi-square method Let $X_1, \ldots, X_n$ be a random sample from a density given by $f_X(x; \theta)$, and let $\mathscr{S}_1, \ldots, \mathscr{S}_k$ be a partition of the range of X. The probability that an observation falls in cell $\mathscr{S}_j$, $j = 1, \ldots, k$, denoted by $p_j(\theta)$, can be found. For instance, if $f_X(x; \theta)$ is the density function of a continuous random variable, then $p_j(\theta) = P[X \text{ falls in cell } \mathscr{S}_j] = \int_{\mathscr{S}_j} f_X(x; \theta)\, dx$. Note that $\sum_{j=1}^{k} p_j(\theta) = 1$. Let the random variable N_j denote the number of X_i's in the sample which falls in cell $\mathscr{S}_j$, $j = 1, \ldots, k$; then $\sum_{j=1}^{k} N_j = n$, the sample size. Form the following summation:

$$\chi^2 = \sum_{j=1}^{k} \frac{[n_j - np_j(\theta)]^2}{np_j(\theta)},$$

where n_j is a value of N_j. The numerator of the jth term in the sum is the square of the difference between the observed and the expected number of observations falling in cell $\mathscr{S}_j$. The *minimum-chi-square* estimate of θ is that $\hat{\theta}$ which minimizes χ^2. It is that θ among all possible θ's which makes the expected number of observation in cell $\mathscr{S}_j$ "nearest" the observed number. The minimum-chi-square estimator depends on the partition $\mathscr{S}_1, \ldots, \mathscr{S}_k$ selected.

EXAMPLE 10 Let $X_1, \ldots, X_n$ be a random sample from a Bernoulli distribution; that is, $f_X(x; \theta) = \theta^x(1-\theta)^{1-x}$ for $x = 0, 1$. Take $N_j =$ the number of observations equal to j for $j = 0, 1$. Here the range of the observation X is partitioned into the two sets consisting of the numbers 0 and 1 respectively.

$$\chi^2 = \sum_{j=0}^{1} \frac{[n_j - np_j(\theta)]^2}{np_j(\theta)} = \frac{[n_0 - n(1-\theta)]^2}{n(1-\theta)} + \frac{(n_1 - n\theta)^2}{n\theta}$$

$$= \frac{[n - n_1 - n(1-\theta)]^2}{n(1-\theta)} + \frac{(n_1 - n\theta)^2}{n\theta} = \frac{(n_1 - n\theta)^2}{n} \frac{1}{\theta(1-\theta)}.$$

The minimum of χ^2 as a function of θ can be found by inspection by noting that $\chi^2 = 0$ for $\theta = n_1/n$. Hence $\hat{\theta} = n_1/n$. For this example there was only one choice for the partition $\mathscr{S}_1, \ldots, \mathscr{S}_k$. The estimator found is the same as what would be obtained by either the method of moments or maximum likelihood. ////

Often it is difficult to locate that $\hat{\theta}$ which minimizes χ^2; hence, the denominator $np_j(\theta)$ is sometimes changed to n_j (if $n_j = 0$, unity is used) forming a *modified* $\chi^2 = \sum_{j=1}^{k} \{[n_j - np_j(\theta)]^2/n_j\}$. The *modified minimum-chi-square estimate* of θ is then that $\hat{\theta}$ which minimizes the modified χ^2.

Minimum-distance method Let $X_1, \ldots, X_n$ be a random sample from the distribution given by the cumulative distribution function $F_X(x; \theta) = F(x; \theta)$, and let $d(F, G)$ be a distance function that measures how "far apart" two cumulative distribution functions F and G are. An example of a distance function is $d(F, G) = \sup_x |F(x) - G(x)|$, which is the largest vertical distance between F and G. See Fig. 4.

The *minimum-distance estimate* of θ is that $\hat{\theta}$ among all possible θ for which $d(F(x; \theta), F_n(x))$ is minimized, where $F_n(x)$ is the sample cumulative distribution function. Thus, $\hat{\theta}$ is chosen so that $F(x; \hat{\theta})$ will be "closest" to $F_n(x)$, which is desirable since we saw in Subsec. 5.4 of Chap. VI that for a fixed argument x the sample cumulative distribution function has the same distribution as the mean of a binomial distribution; hence, by the law of large numbers $F_n(x)$ "converges" to $F(x)$. The minimum-distance estimator might be intuitively appealing, but it is almost always difficult to find since locating $\hat{\theta}$ which minimizes $d(F(x; \theta), F_n(x))$ is seldom easy. The following example is an exception.

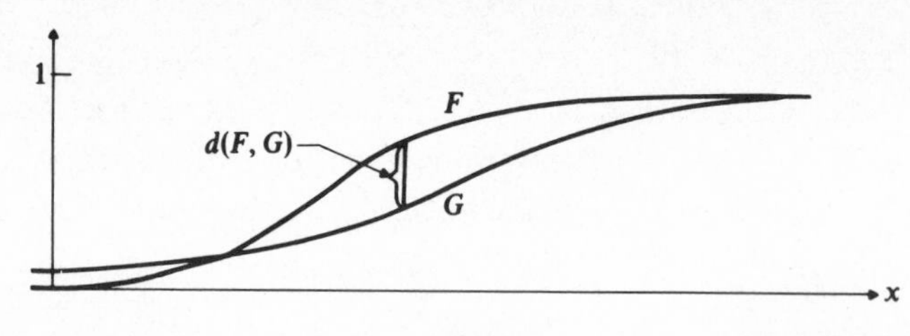

FIGURE 4

EXAMPLE 11 Again let $X_1, \ldots, X_n$ be a random sample from a Bernoulli distribution; then

$$F(x; \theta) = (1 - \theta)I_{[0,1)}(x) + I_{[1,\infty)}(x),$$

where $0 \le \theta \le 1$. Let n_j = the number of observations equal to j; $j = 0, 1$. Then

$$F_n(x) = \frac{n_0}{n} I_{[0,1)}(x) + I_{[1,\infty)}(x).$$

Now if the distance function $d(F, G) = \sup_x |F(x) - G(x)|$ is used, then $d(F(x; \theta), F_n(x))$ is minimized if $1 - \theta$ is taken equal to n_0/n or $\theta = n_1/n = \sum x_i/n$. Hence $\hat{\theta} = \bar{x}$. ////

For a more thorough discussion of the minimum-chi-square method, see Cramér [11] or Rao [17]. The minimum-distance method is discussed in Wolfowitz [42].

3 PROPERTIES OF POINT ESTIMATORS

We presented several methods of obtaining point estimators in the preceding section. All the methods were arrived at on a more or less intuitive basis. The question that now arises is: Are some of many possible estimators better, in some sense, than others? In this section we will define certain properties, which an estimator may or may not possess, that will help us in deciding whether one estimator is better than another.

3.1 Closeness

If we have a random sample $X_1, \ldots, X_n$ from a density, say $f(x; \theta)$, which is known except for θ, then a point estimator of $\tau(\theta)$ is a statistic, say $\ell(X_1, \ldots, X_n)$, whose value is used as an estimate of $\tau(\theta)$. We will assume here that $\tau(\theta)$ is a real-valued (not a vector) function of the unknown parameter θ. [Often $\tau(\theta)$ will be θ itself.] Ideally, we would like the value of $\ell(X_1, \ldots, X_n)$ to be the unknown $\tau(\theta)$, but this is not possible except in trivial cases, one of which follows.

EXAMPLE 12 Assume that one can sample from a density given by

$$f(x; \theta) = I_{(\theta - \frac{1}{2}, \theta + \frac{1}{2})}(x),$$

where θ is known to be an integer. That is, $\underline{\overline{\Theta}}$, the parameter space, consists of all integers. Consider estimating θ on the basis of a single observation x_1. If $\ell(x_1)$ is assigned as its value the integer nearest x_1, then the statistic or estimator $\ell(X_1)$ will always correctly estimate θ. In a sense, the problem posed in this example is really not statistical since one knows the value of θ after taking one observation. ////

Not being able to achieve the ultimate of always correctly estimating the unknown $\tau(\theta)$, we look for an estimator $\ell(X_1, \ldots, X_n)$ that is "close" to $\tau(\theta)$. There are several ways of defining "close." $T = \ell(X_1, \ldots, X_n)$ is a statistic and hence has a distribution, or rather a family of distributions, depending on what θ is. The distribution of T tells how the values t of T are distributed, and we would like to have the values of T distributed near $\tau(\theta)$; that is, we would like to select $\ell(\cdot, \ldots, \cdot)$ so that the values of $T = \ell(X_1, \ldots, X_n)$ are concentrated near $\tau(\theta)$. We saw that the mean and variance of a distribution were, respectively, measures of location and spread. So what we might require of an estimator is that it have its mean near or equal to $\tau(\theta)$ and have small variance. These two notions are explored in Subsec. 3.2 below and then again in Sec. 5.

Rather than resorting to characteristics of a distribution, such as its mean and variance, one can define what "concentration" might mean in terms of the distribution itself. Two such definitions follow.

Definition 4 More concentrated and most concentrated Let $T = \ell(X_1, \ldots, X_n)$ and $T' = \ell'(X_1, \ldots, X_n)$ be two estimators of $\tau(\theta)$. T' is called a *more concentrated* estimator of $\tau(\theta)$ than T if and only if $P_\theta[\tau(\theta) - \lambda < T' \leq \tau(\theta) + \lambda] \geq P_\theta[\tau(\theta) - \lambda < T \leq \tau(\theta) + \lambda]$ for all $\lambda > 0$ and for each θ in $\underline{\overline{\Theta}}$. An estimator $T^* = \ell^*(X_1, \ldots, X_n)$ is called *most concentrated* if it is more concentrated than any other estimator. ////

Remark The subscript θ on the probability symbol $P_\theta[\cdot]$ is there to emphasize that, in general, such probability depends on θ. For instance, in $P_\theta[\tau(\theta) - \lambda < T \le \tau(\theta) + \lambda]$, the event $\{\tau(\theta) - \lambda < T \le \tau(\theta) + \lambda\}$ is described in terms of the random variable T, and, in general, the distribution of T is indexed by θ. ////

We see from the definition that the property of most concentrated is highly desirable (Pitman [41], in defense of his calling a most concentrated estimator best, stated that such an estimator is "undeniably best"); unfortunately, most concentrated estimators seldom exist. There are just too many possible estimators for any one of them to be most concentrated. What is then sometimes done is to restrict the totality of possible estimators under consideration by requiring that each estimator possess some other desirable property and to look for a best or most concentrated estimator in this restricted class. We will not pursue the problem of finding most concentrated estimators, even within some restricted class, in this book.

Another criterion for comparing estimators is the following one.

Definition 5 Pitman-closer and Pitman-closest Let $T = \ell(X_1, \ldots, X_n)$ and $T' = \ell'(X_1, \ldots, X_n)$ be two estimators of $\tau(\theta)$. T' is called a *Pitman-closer* estimator of $\tau(\theta)$ than T if and only if

$$P_\theta[|T' - \tau(\theta)| < |T - \tau(\theta)|] \ge \tfrac{1}{2} \qquad \text{for each } \theta \text{ in } \underline{\overline{\Theta}}.$$

An estimator T^* is called *Pitman-closest* if it is Pitman-closer than any other estimator. ////

The property of Pitman-closest is, like the property of most concentrated, desirable, yet rarely will there exist a Pitman-closest estimator. Both Pitman-closer and more concentrated are intuitively attractive properties to be used to compare estimators, yet they are not always useful. Given two estimators T and T', one does not have to be more concentrated or Pitman-closer than the other. What often happens is that one, say T, is Pitman-closer or more concentrated for some θ in $\underline{\overline{\Theta}}$, and the other T' is Pitman-closer or more concentrated for other θ in $\underline{\overline{\Theta}}$; and since θ is unknown, we cannot say which estimator is preferred. Since Pitman-closest estimators rarely exist for applied problems, we will not devote further study to the notion in this book; instead, we will consider other ways of measuring the closeness of an estimator to $\tau(\theta)$.

Competing estimators can be compared by defining a *measure* of the closeness of an estimate to the unknown $\tau(\theta)$. An estimator $T' = \ell'(X_1, \ldots, X_n)$ of $\tau(\theta)$ will be judged better than an estimator $T = \ell(X_1, \ldots, X_n)$ if the *measure* of the closeness of T' to $\tau(\theta)$ indicates that T' is closer to $\tau(\theta)$ than T. Such concepts of closeness will be discussed in Subsecs. 3.2 and 3.4.

In the above we were assuming that n, the sample size, was fixed. Still another meaning can be affixed to "closeness" if one thinks in terms of increasing sample size. It seems that a good estimator should do better when it is based on a large sample than when it is based on a small sample. *Consistency* and *asymptotic efficiency* are two properties that are defined in terms of increasing sample size; they are considered in Subsec. 3.3. Properties of point estimators that are defined for a fixed sample size are sometimes referred to as *small-sample* properties, whereas properties that are defined for increasing sample size are sometimes referred to as *large-sample* properties.

3.2 Mean-squared Error

A useful, though perhaps crude, measure of goodness or closeness of an estimator $\ell(X_1, \ldots, X_n)$ of $\tau(\theta)$ is what is called the *mean-squared error* of the estimator.

Definition 6 Mean-squared error Let $T = \ell(X_1, \ldots, X_n)$ be an estimator of $\tau(\theta)$. $\mathscr{E}_\theta[[T - \tau(\theta)]^2]$ is defined to be the *mean-squared error* of the estimator $T = \ell(X_1, \ldots, X_n)$. ////

Notation Let $\text{MSE}_\ell(\theta)$ denote the mean-squared error of the estimator $T = \ell(X_1, \ldots, X_n)$ of $\tau(\theta)$. ////

Remark The subscript θ on the expectation symbol $\mathscr{E}_\theta$ indicates from which density in the family under consideration the sample came. That is,

$$\begin{aligned}&\mathscr{E}_\theta[[T - \tau(\theta)]^2]\\ &\quad= \mathscr{E}_\theta[[\ell(X_1, \ldots, X_n) - \tau(\theta)]^2]\\ &\quad= \int \cdots \int [\ell(x_1, \ldots, x_n) - \tau(\theta)]^2 f(x_1; \theta) \cdots f(x_n; \theta)\, dx_1 \cdots dx_n,\end{aligned}$$

where $f(x; \theta)$ is the probability density function from which the random sample was selected. ////

The name "mean-squared error" can be justified if one first thinks of the difference $t - \tau(\theta)$, where t is a value of T used to estimate $\tau(\theta)$, as the error made in estimating $\tau(\theta)$, and then interprets the "mean" in "mean-squared error" as expected or average. To support the contention that the mean-squared error of an estimator is a measure of goodness, one merely notes that $\mathscr{E}_\theta[[T - \tau(\theta)]^2]$ is a measure of the spread of T values about $\tau(\theta)$, just as the variance of a random variable is a measure of its spread about its mean. If we

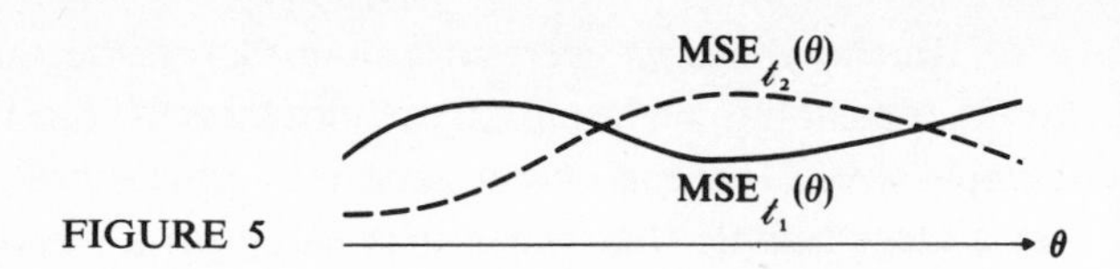

FIGURE 5

were to compare estimators by looking at their respective mean-squared errors, naturally we would prefer one with small or smallest mean-squared error. We could define as best that estimator with smallest mean-squared error, but such estimators rarely exist. In general, the mean-squared error of an estimator depends on θ.

For any two estimators $T_1 = t_1(X_1, \ldots, X_n)$ and $T_2 = t_2(X_1, \ldots, X_n)$ of $\tau(\theta)$, their respective mean-squared errors $\text{MSE}_{t_1}(\theta)$ and $\text{MSE}_{t_2}(\theta)$ as functions of θ are likely to cross; so for some θ, t_1 has smaller MSE, and for others t_2 has smaller MSE. We would then have no basis for preferring one of the estimators over the other. See Fig. 5.

The following example shows that except in very rare cases an estimator with smallest mean-squared error will not exist.

EXAMPLE 13 Let $X_1, \ldots, X_n$ be a random sample from the density $f(x; \theta)$, where θ is a real number, and consider estimating θ itself; that is, $\tau(\theta) = \theta$. We seek an estimator, say $T^* = t^*(X_1, \ldots, X_n)$, such that $\text{MSE}_{t^*}(\theta) \leq \text{MSE}_t(\theta)$ for every θ and for any other estimator $T = t(X_1, \ldots, X_n)$ of θ. Consider the family of estimators $T_{\theta_0} = t_{\theta_0}(X_1, \ldots, X_n) \equiv \theta_0$ indexed by θ_0 for $\theta_0 \in \underline{\overline{\Theta}}$. For each θ_0 belonging to $\underline{\overline{\Theta}}$, the estimator T_{θ_0} ignores the observations and estimates θ to be θ_0. Note that

$$\begin{aligned}\text{MSE}_{t_{\theta_0}}(\theta_0) &= \mathscr{E}_\theta[[t_{\theta_0}(X_1, \ldots, X_n) - \theta]^2] \\ &= \mathscr{E}_\theta[(\theta_0 - \theta)^2] = (\theta_0 - \theta)^2;\end{aligned}$$

so $\text{MSE}_{t_{\theta_0}}(\theta_0) = 0$; that is, the mean-squared error of t_{θ_0} evaluated at $\theta = \theta_0$ is 0. Hence, if there is to exist an estimator $T^* = t^*(X_1, \ldots, X_n)$ satisfying $\text{MSE}_{t^*}(\theta) \leq \text{MSE}_t(\theta)$ for every θ and for any estimator t, $\text{MSE}_{t^*}(\theta) \equiv 0$. [For any θ_0, $\text{MSE}_{t^*}(\theta_0) = 0$ since $\text{MSE}_{t^*}(\theta_0) \leq \text{MSE}_{t_{\theta_0}}(\theta_0) = 0$.] In order for an estimator t^* to have its mean-squared error identically 0, it must always estimate θ correctly, which means that from the sample you must be able to identify the true parameter value. ////

One reason for being unable to find an estimator with uniformly smallest mean-squared error is that the class of all possible estimators is too large—it includes some estimators that are extremely prejudiced in favor of particular θ. For instance, in the example above $t_{\theta_0}(X_1, \ldots, X_n)$ is highly partial to θ_0 since it always estimates θ to be θ_0. One could restrict the totality of estimators by considering only estimators that satisfy some other property. One such property is that of *unbiasedness.*

Definition 7 Unbiased An estimator $T = t(X_1, \ldots, X_n)$ is defined to be an *unbiased* estimator of $\tau(\theta)$ if and only if

$$\mathscr{E}_\theta[T] = \mathscr{E}_\theta[t(X_1, \ldots, X_n)] = \tau(\theta) \qquad \text{for all } \theta \in \underline{\Theta}. \qquad ////$$

An estimator is unbiased if the mean of its distribution equals $\tau(\theta)$, the function of the parameter being estimated. Consider again the estimator $t_{\theta_0}(X_1, \ldots, X_n) \equiv \theta_0$ of the above example; $\mathscr{E}_\theta[t_{\theta_0}(X_1, \ldots, X_n)] = \mathscr{E}_\theta[\theta_0] = \theta_0 \neq \theta$; so $t_{\theta_0}(X_1, \ldots, X_n)$ is not an unbiased estimator of θ. If we restricted the totality of estimators under consideration by considering only unbiased estimators, we could hope to find an estimator with uniformly smallest mean-squared error within the restricted class, that is, within the class of unbiased estimators. The problem of finding an unbiased estimator with uniformly smallest mean-squared error among all unbiased estimators is dealt with in Sec. 5 below.

Remark

$$\text{MSE}_t(\theta) = \text{var}\,[T] + \{\tau(\theta) - \mathscr{E}_\theta[T]\}^2. \tag{9}$$

So if T is an unbiased estimator of $\tau(\theta)$, then $\text{MSE}_t(\theta) = \text{var}\,[T]$.

PROOF

$$\begin{aligned}\text{MSE}_t(\theta) &= \mathscr{E}_\theta[[T - \tau(\theta)]^2] = \mathscr{E}_\theta[((T - \mathscr{E}_\theta[T]) - \{\tau(\theta) - \mathscr{E}_\theta[T]\})^2] \\ &= \mathscr{E}_\theta[(T - \mathscr{E}_\theta[T])^2] - 2\{\tau(\theta) - \mathscr{E}_\theta[T]\}\mathscr{E}_\theta[T - \mathscr{E}_\theta[T]] \\ &\quad + \mathscr{E}_\theta[\{\tau(\theta) - \mathscr{E}_\theta[T]\}^2] = \text{var}\,[T] + \{\tau(\theta) - \mathscr{E}_\theta[T]\}^2. \qquad ////\end{aligned}$$

The term $\tau(\theta) - \mathscr{E}_\theta[T]$ is called the *bias* of the estimator T and can be either positive, negative, or zero. The remark shows that the mean-squared error is the sum of two nonnegative quantities; it also shows how the mean-squared error, variance, and bias of an estimator are related.

EXAMPLE 14 Let $X_1, \ldots, X_n$ be a random sample from $f(x; \theta) = \phi_{\mu, \sigma^2}(x)$. Recall that the maximum-likelihood estimators of μ and σ^2 are, respectively, $\bar{X}$ and $(1/n)\sum (X_i - \bar{X})^2$. (See Example 6.) Now $\mathscr{E}_\theta[\bar{X}] = \mu$; so

$\bar{X}$ is an unbiased estimator of μ, and hence the mean-squared error of $\bar{X} = \mathscr{E}_\theta[(\bar{X} - \mu)^2] = \text{var}[\bar{X}] = \sigma^2/n$. We know that $\mathscr{E}_\theta[S^2] = \sigma^2$; so

$$\begin{aligned}\mathscr{E}_\theta[(1/n)\sum(X_i - \bar{X})^2] &= [(n-1)/n]\mathscr{E}_\theta[[1/(n-1)]\sum(X_i - \bar{X})^2] \\ &= [(n-1)/n]\mathscr{E}_\theta[S^2] = [(n-1)/n]\sigma^2.\end{aligned}$$

Hence the maximum-likelihood estimator of σ^2 is not unbiased. The mean-squared error of $(1/n)\sum(X_i - \bar{X})^2$ is

$$\begin{aligned}&\mathscr{E}_\theta[[(1/n)\sum(X_i - \bar{X})^2 - \sigma^2]^2] \\ &\quad = \text{var}[(1/n)\sum(X_i - \bar{X})^2] + \{\sigma^2 - \mathscr{E}_\theta[(1/n)\sum(X_i - \bar{X})^2]\}^2 \\ &\quad = \frac{(n-1)^2}{n^2}\text{var}[S^2] + \left(\sigma^2 - \frac{n-1}{n}\sigma^2\right)^2 \\ &\quad = \frac{(n-1)^2}{n^2}\frac{1}{n}\left(\mu_4 - \frac{n-3}{n-1}\sigma^4\right) + \frac{\sigma^4}{n^2},\end{aligned}$$

using Eq. (10) of Theorem 2 in Chap. VI. ////

Remark For the most part, in the remainder of this book we will take the mean-squared error of an estimator as our standard in assessing the goodness of an estimator. ////

3.3 Consistency and BAN

In the previous subsection we defined the mean-squared error of an estimator and the property of unbiasedness. Both concepts were defined for a fixed sample size. In this subsection we will define two concepts that are defined for increasing sample size. In our notation for an estimator of $\tau(\theta)$, let us use $T_n = t_n(X_1, \ldots, X_n)$, where the subscript n of t indicates sample size. Actually we will be considering a sequence of estimators, say $T_1 = t_1(X_1)$, $T_2 = t_2(X_1, X_2)$ $T_3 = t_3(X_1, X_2, X_3), \ldots, T_n = t_n(X_1, \ldots, X_n), \ldots$. An obvious example is $T_n = t_n(X_1, \ldots, X_n) = \bar{X}_n = (1/n)\sum_{i=1}^{n} X_i$. Ordinarily the functions t_n in the sequence will be the same *kind* of function for each n.

When considering a sequence of estimators, it seems that a good sequence of estimators should be one for which the values of the estimators tend to get closer to the quantity being estimated as the sample size increases. The following definitions formalize this intuitively desirable notion of limiting closeness.

Definition 8 Mean-squared-error consistency Let $T_1, T_2, \ldots, T_n \ldots$ be a sequence of estimators of $\tau(\theta)$, where $T_n = t_n(X_1, \ldots, X_n)$ is based on a sample of size n. This sequence of estimators is defined to be a *mean-squared-error consistent* sequence of estimators of $\tau(\theta)$, if and only if $\lim_{n\to\infty} \mathscr{E}_\theta[[T_n - \tau(\theta)]^2] = 0$ for all θ in $\underline{\Theta}$. ////

Remark Mean-squared-error consistency implies that both the bias and the variance of T_n approach 0 since $\mathscr{E}_\theta[[T_n - \tau(\theta)]^2] = \text{var}[T_n] + \{\tau(\theta) - \mathscr{E}_\theta[T_n]\}^2$. ////

EXAMPLE 15 In sampling from any density having mean μ and variance σ^2, let $\overline{X}_n = (1/n)\sum_{i=1}^{n} X_i$ be a sequence of estimators of μ and $S_n^2 = [1/(n-1)]\sum_{i=1}^{n}(X_i - \overline{X}_n)$ be a sequence of estimators of σ^2. $\mathscr{E}[(\overline{X}_n - \mu)^2] = \text{var}[\overline{X}_n] = \sigma^2/n \to 0$ as $n \to \infty$; hence the sequence $\{\overline{X}_n\}$ is a mean-squared-error consistent sequence of estimators of μ.

$$\mathscr{E}[(S_n^2 - \sigma^2)^2] = \text{var}[S_n^2] = \frac{1}{n}\left(\mu_4 - \frac{n-3}{n-1}\sigma^4\right) \to 0$$

as $n \to \infty$, using Eq. (10) of Chap. VI; hence the sequence $\{S_n^2\}$ is a mean-squared-error consistent sequence of estimators of σ^2. Note that if $T_n = (1/n)\sum(X_i - \overline{X})^2$, then the sequence $\{T_n\}$ is also a mean-squared-error consistent sequence of estimators of σ^2. ////

There is another weaker notion of consistency given in the following definition.

Definition 9 Simple consistency Let $T_1, T_2, \ldots, T_n, \ldots$ be a sequence of estimators of $\tau(\theta)$, where $T_n = t_n(X_1, \ldots, X_n)$. The sequence $\{T_n\}$ is defined to be a *simple* (or *weakly*) *consistent* sequence of estimators of $\tau(\theta)$ if for every $\varepsilon > 0$ the following is satisfied:

$$\lim_{n\to\infty} P_\theta[\tau(\theta) - \varepsilon < T_n < \tau(\theta) + \varepsilon] = 1 \qquad \text{for every } \theta \text{ in } \underline{\Theta}. \qquad ////$$

Remark If an estimator is a mean-squared-error consistent estimator, it is also a simple consistent estimator, but not necessarily vice versa.

PROOF

$$P_\theta[\tau(\theta) - \varepsilon < T_n < \tau(\theta) + \varepsilon] = P[|T_n - \tau(\theta)| < \varepsilon]$$
$$= P_\theta[[T_n - \tau(\theta)]^2 < \varepsilon^2] \ge 1 - \frac{\mathscr{E}_\theta[[T_n - \tau(\theta)]^2]}{\varepsilon^2}$$

by the Chebyshev inequality. As n approaches infinity, $\mathscr{E}_\theta[[T_n - \tau(\theta)]^2]$ approaches 0; hence $\lim_{n\to\infty} P_\theta[\tau(\theta) - \varepsilon < T_n < \tau(\theta) + \varepsilon] = 1$. ////

We close this subsection with one further large-sample definition.

Definition 10 Best asymptotically normal estimators (BAN estimators) A sequence of estimators $T_1^*, \ldots, T_n^*, \ldots$ of $\tau(\theta)$ is defined to be *best asymptotically normal* (BAN) if and only if the following four conditions are satisfied:

(i) The distribution of $\sqrt{n}[T_n^* - \tau(\theta)]$ approaches the normal distribution with mean 0 and variance $\sigma^{*^2}(\theta)$ as n approaches infinity.

(ii) For every $\varepsilon > 0$,

$$\lim_{n\to\infty} P_\theta[|T_n^* - \tau(\theta)| > \varepsilon] = 0 \qquad \text{for each } \theta \text{ in } \underline{\Theta}.$$

(iii) Let $\{T_n\}$ be any other sequence of simple consistent estimators for which the distribution of $\sqrt{n}[T_n - \tau(\theta)]$ approaches the normal distribution with mean 0 and variance $\sigma^2(\theta)$.

(iv) $\sigma^2(\theta)$ is not less than $\sigma^{*^2}(\theta)$ for all θ in any open interval. ////

Remark The abbreviation BAN is sometimes replaced by CANE, standing for *consistent asymptotically normal efficient.* ////

The usefulness of this definition derives partially from theorems proving the existence of BAN estimators and from the fact that ordinarily reasonable estimators are asymptotically normally distributed.

It can be shown that for samples drawn from a normal density with mean μ and variance σ^2 the sequence $T_n^* = (1/n)\sum_{i=1}^n X_i = \bar{X}_n$ for $n = 1, 2, \ldots$ is a BAN estimator of μ. In fact, the limiting distribution of $\sqrt{n}(\bar{X}_n - \mu)$ is normal with mean 0 and variance σ^2, and no other estimator can have smaller limiting variance in any interval of μ values. However, there are many other estimators for this problem which are also BAN estimators of μ, that is, estimators with the same normal distribution in the limit. For example,

$$T_n' = \frac{1}{n+1}\sum_{i=1}^n X_i, \qquad n = 1, 2, \ldots,$$

is a BAN estimator of μ. BAN estimators are necessarily weakly consistent by (ii) of the definition.

3.4 Loss and Risk Functions

In Subsec. 3.2 we used *mean-squared error* of an estimator as a measure of the closeness of the estimator to $\tau(\theta)$. Other measures are possible, for example,

$$\mathscr{E}_\theta[|T - \tau(\theta)|],$$

called the *mean absolute deviation.* In order to exhibit and consider still other measures of closeness, we will borrow and rely on the language of decision theory. On the basis of an observed random sample from some density function, the statistician has to *decide* what to estimate $\tau(\theta)$ to be. One might then call the value of some estimator $T = t(X_1, \ldots, X_n)$ a *decision* and call the estimator itself a *decision function* since it tells us what *decision* to make. Now the estimate t of $\tau(\theta)$ might be in error; if so, some measure of the severity of the error seems appropriate. The word "loss" is used in place of "error," and "loss function" is used as a measure of the "error." A formal definition follows.

Definition 11 Loss function Consider estimating $\tau(\theta)$. Let t denote an estimate of $\tau(\theta)$. The *loss function*, denoted by $\ell(t; \theta)$, is defined to be a real-valued function satisfying (i) $\ell(t; \theta) \geq 0$ for all possible estimates t and all θ in $\bar{\Theta}$ and (ii) $\ell(t; \theta) = 0$ for $t = \tau(\theta)$. $\ell(t; \theta)$ equals the *loss* incurred if one estimates $\tau(\theta)$ to be t when θ is the true parameter value. ////

In a given estimation problem one would have to define an appropriate loss function for the particular problem under study. It is a measure of the error and presumably would be greater for large error than for small error. We would want the loss to be small; or, stated another way, we want the error in estimation to be small, or we want the estimate to be close to what it is estimating.

EXAMPLE 16 Several possible loss functions are:

(i) $\ell_1(t; \theta) = [t - \tau(\theta)]^2$.

(ii) $\ell_2(t; \theta) = |t - \tau(\theta)|$.

(iii) $\ell_3(t; \theta) = \begin{cases} A & \text{if } |t - \tau(\theta)| > \varepsilon \\ 0 & \text{if } |t - \tau(\theta)| \leq \varepsilon, \text{ where } A > 0. \end{cases}$

(iv) $\ell_4(t; \theta) = \rho(\theta)|t - \tau(\theta)|^r \quad$ for $\rho(\theta) \geq 0$ and $r > 0$.

ℓ_1 is called the *squared-error* loss function, and ℓ_2 is called the *absolute-error* loss function. Note that both ℓ_1 and ℓ_2 increase as the error $t - \tau(\theta)$ increases in magnitude. ℓ_3 says that you lose nothing if the estimate t is within ε units of $\tau(\theta)$ and otherwise you lose amount A. ℓ_4 is a general loss function that includes both ℓ_1 and ℓ_2 as special cases. ////

We assume now that an appropriate loss function has been defined for our estimation problem, and we think of the loss function as a measure of error or loss. Our object is to select an estimator $T = \ell(X_1, \ldots, X_n)$ that makes this error or loss small. (Admittedly, we are not considering a very important, substantive problem by assuming that a suitable loss function is given. In general, selection of an appropriate loss function is not trivial.) The loss function in its first argument depends on the estimate t, and t is a value of the estimator T; that is, $t = \ell(x_1, \ldots, x_n)$. Thus, our loss depends on the sample $X_1, \ldots, X_n$. We cannot hope to make the loss small for every possible sample, but we can try to make the loss small on the average. Hence, if we alter our objective of picking that estimator that makes the loss small to picking that estimator that makes the *average* loss small, we can remove the dependence of the loss on the sample $X_1, \ldots, X_n$. This notion is embodied in the following definition.

Definition 12 Risk function For a given loss function $\ell(\cdot\,;\,\cdot)$, the *risk function*, denoted by $\mathscr{R}_t(\theta)$, of an estimator $T = \ell(X_1, \ldots, X_n)$ is defined to be

$$\mathscr{R}_t(\theta) = \mathscr{E}_\theta[\ell(T;\theta)]. \qquad (10)$$

////

The risk function is the *average loss*. The expectation in Eq. (10) can be taken in two ways. For example, if the density $f(x;\theta)$ from which we sampled is a probability density function, then

$$\mathscr{E}_\theta[\ell(T;\theta)] = \mathscr{E}_\theta[\ell(\ell(X_1, \ldots, X_n);\theta)]$$
$$= \int \cdots \int \ell(\ell(x_1, \ldots, x_n);\theta) \prod_{i=1}^{n} f(x_i;\theta)\,dx_i.$$

Or we can consider the random variable T and the density of T. We get

$$\mathscr{E}_\theta[\ell(T;\theta)] = \int \ell(\ ;\theta) f_T(t)\,dt,$$

where $f_T(t)$ is the density of the estimator T. In either case, the expectation averages out the values of $x_1, \ldots, x_n$.

EXAMPLE 17 Consider the same loss functions given in Example 16. The corresponding risks are given by:

(i) $\mathscr{E}_\theta[[T - \tau(\theta)]^2]$, our familiar mean-squared error.
(ii) $\mathscr{E}_\theta[|T - \tau(\theta)|]$, the mean absolute error.
(iii) $A \cdot P_\theta[|T - \tau(\theta)| > \varepsilon]$.
(iv) $\rho(\theta)\mathscr{E}_\theta[|T - \tau(\theta)|^r]$. ////

Our object now is to select an estimator that makes the average loss (risk) small and ideally select an estimator that has the smallest risk. To help meet this objective, we use the concept of admissible estimators.

Definition 13 Admissible estimator For two estimators $T_1 = \ell_1(X_1, \ldots, X_n)$ and $T_2 = \ell_2(X_1, \ldots, X_n)$, estimator ℓ_1 is defined to be a *better* estimator than ℓ_2 if and only if

$$\mathscr{R}_{\ell_1}(\theta) \le \mathscr{R}_{\ell_2}(\theta) \qquad \text{for all } \theta \text{ in } \underline{\overline{\Theta}}$$

and

$$\mathscr{R}_{\ell_1}(\theta) < \mathscr{R}_{\ell_2}(\theta) \qquad \text{for at least one } \theta \text{ in } \underline{\overline{\Theta}}.$$

An estimator $T = \ell(X_1, \ldots, X_n)$ is defined to be *admissible* if and only if there is no better estimator. ////

In general, given two estimators ℓ_1 and ℓ_2 neither is better than the other; that is, their respective risk functions as functions of θ, cross. We observed this same phenomenon when we studied the mean-squared error. Here, as there, there will not, in general, exist an estimator with uniformly smallest risk. The problem is the dependence of the risk function on θ. What we might do is average out θ, just as we average out the dependence on $x_1, \ldots, x_n$ when going from the loss function to the risk function. The question then is: Just how should θ be averaged out? We will consider just this problem in Sec. 7 on the Bayes estimators. Another way of removing the dependence of the risk function on θ is to replace the risk function by its maximum value and compare estimators by looking at their respective maximum risks, naturally preferring that estimator with smallest maximum risk. Such an estimator is said to be *minimax.*

Definition 14 Minimax An estimator ℓ^* is defined to be a *minimax* estimator if and only if $\sup_\theta \mathscr{R}_{\ell^*}(\theta) \le \sup_\theta \mathscr{R}_{\ell}(\theta)$ for every estimator ℓ. ////

Minimax estimators will be discussed in Sec. 7.

4 SUFFICIENCY

Prior to continuing our pursuit of finding best estimators, we introduce the concept of *sufficiency* of statistics. In many of the estimation problems that we will encounter, we will be able to summarize the information in the sample

$x_1, \ldots, x_n$. That is, we will be able to find some function of the sample that tells us just as much about θ as the sample itself. Such a function would be sufficient for estimation purposes and accordingly is called a *sufficient statistic.*

Sufficient statistics are of interest in themselves, as well as being useful in statistical inference problems such as estimation or testing of hypotheses. Because the concept of sufficiency is widely applicable, possibly the notion should have been isolated in a chapter by itself rather than buried in this chapter on estimation.

4.1 Sufficient Statistics

Let $X_1, \ldots, X_n$ be a random sample from some density, say $f(\cdot\,; \theta)$. We defined a statistic to be a function of the sample; that is, a statistic is a function with domain the range of values that $(X_1, \ldots, X_n)$ can take on and counterdomain the real numbers. A statistic $T = \ell(X_1, \ldots, X_n)$ is also a random variable; it condenses the n random variables $X_1, X_2, \ldots, X_n$ into a single random variable. Such condensing is appealing since we would rather work with unidimensional quantities than n-dimensional quantities. We shall be interested in seeing if we lost any "information" by this condensing process. The condensing can also be viewed another way. Let $\mathfrak{X}$ denote the range of values that $(X_1, \ldots, X_n)$ can assume. For example, if we sample from a Bernoulli distribution, then $\mathfrak{X}$ is a collection of all n-dimensional vectors with components either 0 or 1; or if we sample from a normal distribution, then $\mathfrak{X}$ is an n-dimensional euclidean space. Now a statistic induces or defines a *partition* of $\mathfrak{X}$. (Recall that a partition of $\mathfrak{X}$ is a collection of mutually disjoint subsets of $\mathfrak{X}$ whose union is $\mathfrak{X}$.) Let $\ell(\cdot, \ldots, \cdot)$ be the function corresponding to the statistic $T = \ell(X_1, \ldots, X_n)$. The partition induced by $\ell(\cdot, \ldots, \cdot)$ is brought about as follows: Let t_0 denote any value of the function $\ell(\cdot, \ldots, \cdot)$; that subset of $\mathfrak{X}$ consisting of all those points $(x_1, \ldots, x_n)$ for which $\ell(x_1, \ldots, x_n) = t_0$ is one subset in the collection of subsets which the partition comprises; the other subsets are similarly formed by considering other values of $\ell(\cdot, \ldots, \cdot)$. For example, if a sample of size 3 is selected from a Bernoulli distribution, then $\mathfrak{X}$ consists of eight points (0, 0, 0), (0, 0, 1), (0, 1, 0), (1, 0, 0), (0, 1, 1), (1, 0, 1), (1, 1, 0), (1, 1, 1). Let $\ell(x_1, x_2, x_3) = x_1 + x_2 + x_3$; then $\ell(\cdot, \cdot, \cdot)$ takes on the values 0, 1, 2, and 3. The partition of $\mathfrak{X}$ induced by $\ell(\cdot, \cdot, \cdot)$, consists of the four subsets {(0, 0, 0)}, {(0, 0, 1), (0, 1, 0), (1, 0, 0)}, {(0, 1, 1), (1, 0, 1), (1, 1, 0)}, and {(1, 1, 1)} corresponding, respectively, to the four values 0, 1, 2, and 3 of $\ell(\cdot, \cdot, \cdot)$. A statistic then is really a condensation of $\mathfrak{X}$. In the above example, if we use the statistic $\ell(\cdot, \cdot, \cdot)$, we have only four different values to worry about instead of the eight different points of $\mathfrak{X}$.

Several different statistics can induce the same partition. In fact, if $t(\cdot, \ldots, \cdot)$ is a statistic, then any one-to-one function of t has the same partition as t. In the example above $t'(x_1, x_2, x_3) = 6(x_1 + x_2 + x_3)^2$, or even $t''(x_1, x_2, x_3) = x_1^2 + x_2^2 + x_3^2$, induces the same partition as $t(x_1, x_2, x_3) = x_1 + x_2 + x_3$. One of the reasons for using statistics is that they do condense $\mathfrak{X}$ and if such is our only reason for using a statistic, then any two statistics with the same partition are of the same utility. The important aspect of a statistic is the partition of $\mathfrak{X}$ that it induces, not the values that it assumes.

A sufficient statistic is a particular kind of statistic. It is a statistic that condenses $\mathfrak{X}$ in such a way that no "information about θ" is lost. The only information about the parameter θ in the density $f(\cdot\,; \theta)$ from which we sampled is contained in the sample $X_1, \ldots, X_n$; so, when we say that a statistic loses no information, we mean that it contains all the information about θ that is contained in the sample. We emphasize that the type of information of which we are speaking is that information about θ contained in the sample given that we know the form of the density; that is, we know the function $f(\cdot\,; \cdot)$ in $f(\cdot\,; \theta)$, and the parameter θ is the only unknown. We are not speaking of information in the sample that might be useful in checking the validity of our assumption that the density does indeed have form $f(\cdot\,; \cdot)$.

Now we shall formalize the definition of a sufficient statistic; in fact, we shall give two definitions, namely, Definitions 15 and 16. It can be argued that the two definitions are equivalent, but we will not do it.

Definition 15 Sufficient statistic Let $X_1, \ldots, X_n$ be a random sample from the density $f(\cdot\,; \theta)$, where θ may be a vector. A statistic $S = s(X_1, \ldots, X_n)$ is defined to be a *sufficient statistic* if and only if the conditional distribution of $X_1, \ldots, X_n$ given $S = s$ does not depend on θ for any value s of S. ////

Note that we use $S = s(X_1, \ldots, X_n)$, instead of $T = t(X_1, \ldots, X_n)$, to denote a sufficient statistic. Some care is required in interpreting the conditional distribution of $X_1, \ldots, X_n$ given $S = s$, as Example 19 and the paragraph preceding it demonstrate.

The definition says that a statistic $S = s(X_1, \ldots, X_n)$ is sufficient if the conditional distribution of the sample given the value of the statistic does not depend on θ. The idea is that if you know the value of the sufficient statistic, then the sample values themselves are not needed and can tell you nothing more about θ, and this is true since the distribution of the sample given the sufficient statistic does not depend on θ. One cannot hope to learn anything about θ by sampling from a distribution that does not depend on θ.

EXAMPLE 18 Let X_1, X_2, X_3 be a sample of size 3 from the Bernoulli distribution. Consider the two statistics $S = \mathscr{s}(X_1, X_2, X_3) = X_1 + X_2 + X_3$ and $T = \mathscr{t}(X_1, X_2, X_3) = X_1X_2 + X_3$. We will show that $\mathscr{s}(\cdot, \cdot, \cdot)$ is sufficient and $\mathscr{t}(\cdot, \cdot, \cdot)$ is not. This first column of Fig. 6 is $\mathfrak{X}$.

	Values of S	Values of T	$f_{X_1,X_2,X_3\|S}$	$f_{X_1,X_2,X_3\|T}$
(0, 0, 0)	0	0	1	$\dfrac{1-p}{1+p}$
(0, 0, 1)	1	1	$\dfrac{1}{3}$	$\dfrac{1-p}{1+2p}$
(0, 1, 0)	1	0	$\dfrac{1}{3}$	$\dfrac{p}{1+p}$
(1, 0, 0)	1	0	$\dfrac{1}{3}$	$\dfrac{p}{1+p}$
(0, 1, 1)	2	1	$\dfrac{1}{3}$	$\dfrac{p}{1+2p}$
(1, 0, 1)	2	1	$\dfrac{1}{3}$	$\dfrac{p}{1+2p}$
(1, 1, 0)	2	1	$\dfrac{1}{3}$	$\dfrac{p}{1+2p}$
(1, 1, 1)	3	2	1	1

FIGURE 6

The conditional densities given in the last two columns are routinely calculated. For instance,

$$
\begin{aligned}
f_{X_1, X_2, X_3|S=1}(0, 1, 0|1) &= P[X_1 = 0; X_2 = 1; X_3 = 0|S = 1] \\
&= \frac{P[X_1 = 0; X_2 = 1; X_3 = 0; S = 1]}{P[S = 1]} \\
&= \frac{(1-p)p(1-p)}{\binom{3}{1}p(1-p)^2} = \frac{1}{3};
\end{aligned}
$$

and

$$f_{X_1, X_2, X_3|T=0}(0, 1, 0|0) = \frac{P[X_1 = 0; X_2 = 1; X_3 = 0; T = 0]}{P[T = 0]}$$

$$= \frac{(1-p)^2 p}{(1-p)^3 + 2(1-p)^2 p} = \frac{p}{1 - p + 2p} = \frac{p}{1+p}.$$

The conditional distribution of the sample given the values of S is independent of p; so S is a sufficient statistic; however, the conditional distribution of the sample given the values of T depends on p; so T is not sufficient. We might note that the statistic T provides a greater condensation of $\mathfrak{X}$ than does S. A question that might be asked is: Is there a statistic which provides greater condensation of $\mathfrak{X}$ than does S which is sufficient as well? The answer is "no" and can be verified by trying all possible partitions of $\mathfrak{X}$ consisting of three or fewer subsets. ////

In the case of sampling from a probability density function, the meaning of the term "the conditional distribution of $X_1, \ldots, X_n$ given $S = s$" that appears in Definition 15 may not be obvious since then $P[S = s] = 0$. We can give two interpretations. The first deals with the joint cumulative distribution function and uses Eq. (9) of Subsec. 3.3 in Chap. IV; that is, to show that $S = \mathscr{s}(X_1, \ldots, X_n)$ is sufficient, one shows that $P[X_1 \le x_1; \ldots; X_n \le x_n | S = s]$ is independent of θ, where $P[X_1 \le x_1; \ldots; X_n \le x_n | S = s]$ is defined as in Eq. (9) of Chap. IV. The second interpretation is obtained if a one-to-one transformation of $X_1, X_2, \ldots, X_n$ to, say, $S, Y_2, \ldots, Y_n$ is made, and then it is demonstrated that the density of $Y_2, \ldots, Y_n$ given $S = s$ is independent of θ. If the distribution of $Y_2, \ldots, Y_n$ given $S = s$ is independent of θ, then the distribution of $S, Y_2, \ldots, Y_n$ given $S = s$ is independent of θ, and hence the distribution of $X_1, X_2, \ldots, X_n$ given $S = s$ is independent of θ. These two interpretations are illustrated in the following example.

EXAMPLE 19 Let $X_1, \ldots, X_n$ be a random sample from $f(\cdot; \theta) = \phi_{\theta,1}(\cdot)$; that is, $X_1, \ldots, X_n$ is a random sample from a normal distribution with mean θ and variance unity. In order to expedite calculations, we take $n = 2$. Let us argue that $S = \mathscr{s}(X_1, X_2) = X_1 + X_2$ is sufficient using the second interpretation above. The transformation of (X_1, X_2) to (S, Y_2), where $S = X_1 + X_2$ and $Y_2 = X_2 - X_1$, is one-to-one; so it suffices to show that $f_{Y_2|S}(y_2|s)$ is independent of θ. Now

$$f_{Y_2|S}(y_2|s) = \frac{f_{Y_2,S}(y_2, s)}{f_S(s)} = \frac{f_{Y_2}(y_2) f_S(s)}{f_S(s)} = f_{Y_2}(y_2)$$

(using the independence of $X_1 + X_2$ and $X_2 - X_1$ that was proved in Theorem 8 of Chap. VI), but

$$f_{Y_2}(y_2) = \frac{1}{\sqrt{2\pi}\sqrt{2}} e^{-\frac{1}{2}(y_2^2/2)}$$

since

$$Y_2 \sim N(0, 2),$$

which is independent of θ.

The necessary calculations for the first interpretation above are less simple. We must show that $P[X_1 \le x_1;\ X_2 \le x_2 | S = s]$ is independent of θ. According to Eq. (9) of Chap. IV,

$$P[X_1 \le x_1;\ X_2 \le x_2 | S = s] = \lim_{h \to 0} P[X_1 \le x_1;\ X_2 \le x_2 | s - h < S < s + h].$$

Without loss of generality, assume that $x_1 \le x_2$. We have the following three cases to consider: (i) $s \le x_1$, (ii) $x_1 < s \le x_2$, and (iii) $x_2 < s$. $P[X_1 \le x_1;\ X_2 \le x_2 | S = s]$ is clearly 0 (and hence independent of θ) for case (iii). Let us consider (i). [Case (ii) is similar.]

$$\begin{aligned} P[X_1 \le x_1;\ X_2 \le x_2 | S = s] &= \lim_{h \to 0} P[X_1 \le x_1;\ X_2 \le x_2 | s - h < S < s + h] \\ &= \frac{\lim_{h \to 0} \frac{1}{2h} P[X_1 \le x_1;\ X_2 \le x_2; s - h < S < s + h]}{\lim_{h \to 0} \frac{1}{2h} P[s - h < S < s + h]} \\ &= \frac{\lim_{h \to 0} \frac{1}{2h} P[X_1 \le x_1;\ X_2 \le x_2;\ s - h < S < s + h]}{f_S(s)}. \end{aligned}$$

Note that (see Fig. 7)

$$\begin{aligned} &\lim_{h \to 0} \frac{1}{2h} \int_{s+h-x_2}^{x_1} \int_{s-h-u}^{s+h-u} f_{X_1}(u) f_{X_2}(v)\, dv\, du \\ &\le \lim_{h \to 0} \frac{1}{2h} P[X_1 \le x_1;\ X_2 \le x_2;\ s - h < X_1 + X_2 < s + h] \\ &\le \lim_{h \to 0} \frac{1}{2h} \int_{s-h-x_2}^{x_1} \int_{s-h-u}^{s+h-u} f_{X_1}(u) f_{X_2}(v)\, dv\, du, \end{aligned}$$

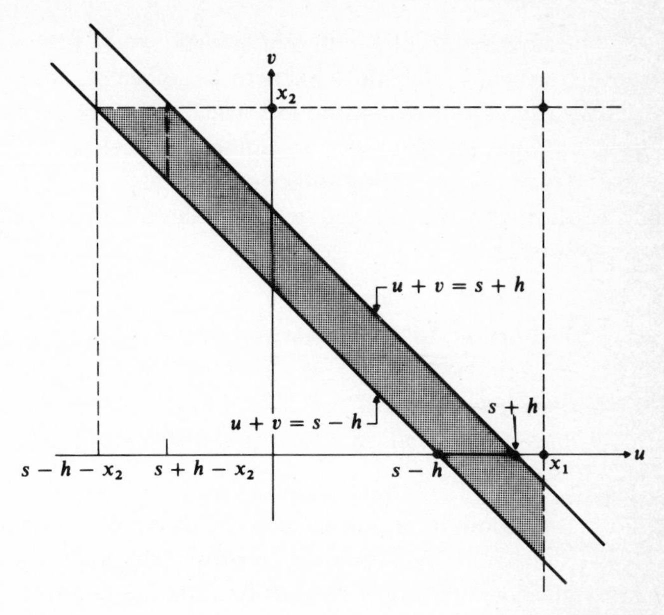

FIGURE 7

and hence

$$\lim_{h \to 0} \frac{1}{2h} P[X_1 \le x_1; X_2 \le x_2; s - h < X_1 + X_2 < s + h]$$

$$= \int_{s-x_2}^{x_1} f_{X_1}(u) f_{X_2}(s - u)\, du.$$

Finally, then,

$$P[X_1 \le x_1; X_2 \le x_2 | S = s]$$

$$= \frac{\int_{s-x_2}^{x_1} f_{X_1}(u) f_{X_2}(s - u)\, du}{f_S(s)}$$

$$= \frac{\int_{s-x_2}^{x_1} (1/2\pi) e^{-\frac{1}{2}[u^2 - 2u\theta + \theta^2 + (s-u)^2 - 2(s-u)\theta + \theta^2]}\, du}{(1/\sqrt{4\pi}) e^{-\frac{1}{2}[(s-2\theta)/\sqrt{2}]^2}}$$

$$= \frac{(1/\sqrt{2\pi}) \int_{s-x_2}^{x_1} e^{-\frac{1}{2}[u^2 + (s-u)^2]}\, du}{(1/\sqrt{2}) e^{-\frac{1}{2}(s^2/2)}},$$

which is independent of θ. ////

Definition 15 of a sufficient statistic is not very workable. First, it does not tell us which statistic is likely to be sufficient, and, second, it requires us to derive a conditional distribution which may not be easy, especially for continuous random variables. In Subsec. 4.2 below, we will present a criterion that may aid us in finding sufficient statistics.

Although we will not so argue, the following definition is equivalent to Definition 15.

Definition 16 Sufficient statistic Let $X_1, \ldots, X_n$ be a random sample from the density $f(\cdot\,; \theta)$. A statistic $S = \mathscr{s}(X_1, \ldots, X_n)$ is defined to be a *sufficient statistic* if and only if the conditional distribution of T given S does not depend on θ for *any* statistic $T = \mathscr{t}(X_1, \ldots, X_n)$. ////

Definition 16 is particularly useful in showing that a particular statistic is *not* sufficient. For instance, to prove that a statistic $T' = \mathscr{t}'(X_1, \ldots, X_n)$ is not sufficient, one needs only to find another statistic $T = \mathscr{t}(X_1, \ldots, X_n)$ for which the conditional distribution of T given T' depends on θ.

For some problems, no single sufficient statistic exists. However, there will always exist jointly sufficient statistics.

Definition 17 Jointly sufficient statistics Let $X_1, \ldots, X_n$ be a random sample from the density $f(\cdot\,; \theta)$. The statistics $S_1, \ldots, S_r$ are defined to be *jointly sufficient* if and only if the conditional distribution of $X_1, \ldots, X_n$ given $S_1 = s_1, \ldots, S_r = s_r$ does not depend on θ. ////

The sample $X_1, \ldots, X_n$ itself is always jointly sufficient since the conditional distribution of the sample given the sample does not depend on θ. Also, the order statistics $Y_1, \ldots, Y_n$ are jointly sufficient for random sampling. If the order statistics are given, say, by $(Y_1 = y_1, \ldots, Y_n = y_n)$, then the only values that can be taken on by $(X_1, \ldots, X_n)$ are the permutations of $y_1, \ldots, y_n$. Since the sampling is random, each of the $n!$ permutations is equally likely. So, given the values of the order statistics the probability that the sample equals a particular permutation of these given values of the order statistics is $1/n!$, which is independent of θ. (Sufficiency of the order statistics also follows from Theorem 5 below.)

If we recall that the important aspect of a statistic or set of statistics is the partition of $\mathfrak{X}$ that it induces, and not the values that it takes on, then the validity of the following theorem is evident.

Theorem 3 If $S_1 = \mathscr{s}_1(X_1, \ldots, X_n), \ldots, S_r = \mathscr{s}_r(X_1, \ldots, X_n)$ is a set of jointly sufficient statistics, then any set of one-to-one functions, or transformations, of $S_1, \ldots, S_r$ is also jointly sufficient. ////

For example, if $\sum X_i$ and $\sum X_i^2$ are jointly sufficient, then $\bar{X}$ and $\sum (X_i - \bar{X})^2 = \sum X_i^2 - n\bar{X}^2$ are also jointly sufficient. Note, however, that $\bar{X}^2$ and $\sum (X_i - \bar{X})^2$ may not be jointly sufficient since they are not one-to-one functions of $\sum X_i$ and $\sum X_i^2$.

We note again that the parameter θ that appears in any of the above three definitions of sufficient statistics can be a vector.

4.2 Factorization Criterion

The concept of sufficiency of statistics was defined in Definitions 15 to 17 above. In many cases, a relatively easy criterion for examining a statistic or set of statistics for sufficiency has been developed. This is given in the next two theorems, the proofs of which are omitted.

Theorem 4 Factorization theorem (single sufficient statistic) Let X_1, $X_2, \ldots, X_n$ be a random sample of size n from the density $f(\cdot\,; \theta)$, where the parameter θ may be a vector. A statistic $S = \mathscr{s}(X_1, \ldots, X_n)$ is sufficient if and only if the joint density of $X_1, \ldots, X_n$, which is $\prod_{i=1}^{n} f(x_i; \theta)$, factors as

$$\begin{aligned} f_{X_1, \ldots, X_n}(x_1, \ldots, x_n; \theta) &= g(\mathscr{s}(x_1, \ldots, x_n); \theta)h(x_1, \ldots, x_n) \\ &= g(s; \theta)h(x_1, \ldots, x_n), \end{aligned} \tag{11}$$

where the function $h(x_1, \ldots, x_n)$ is nonnegative and does not involve the parameter θ and the function $g(\mathscr{s}(x_1, \ldots, x_n); \theta)$ is nonnegative and depends on $x_1, \ldots, x_n$ only through the function $\mathscr{s}(\cdot, \ldots, \cdot)$. ////

Theorem 5 Factorization theorem (jointly sufficient statistics) Let X_1, $X_2, \ldots, X_n$ be a random sample of size n from the density $f(\cdot\,; \theta)$, where the parameter θ may be a vector. A set of statistics $S_1 = \mathscr{s}_1(X_1, \ldots, X_n)$, $\ldots, S_r = \mathscr{s}_r(X_1, \ldots, X_n)$ is jointly sufficient if and only if the joint density of $X_1, \ldots, X_n$ can be factored as

$$\begin{aligned} &f_{X_1, \ldots, X_n}(x_1, \ldots, x_n; \theta) \\ &\quad = g(\mathscr{s}_1(x_1, \ldots, x_n), \ldots, \mathscr{s}_r(x_1, \ldots, x_n); \theta)h(x_1, \ldots, x_n) \qquad (12) \\ &\quad = g(s_1, \ldots, s_r; \theta)h(x_1, \ldots, x_n), \end{aligned}$$

where the function $h(x_1, \ldots, x_n)$ is nonnegative and does not involve the parameter θ and the function $g(s_1, \ldots, s_r; \theta)$ is nonnegative and depends on $x_1, \ldots, x_n$ only through the functions $\mathscr{s}_1(\cdot, \ldots, \cdot), \ldots, \mathscr{s}_r(\cdot, \ldots, \cdot)$. ////

Note that, according to Theorem 3, there are many possible sets of sufficient statistics. The above two theorems give us a relatively easy method for judging whether a certain statistic is sufficient or a set of statistics is jointly sufficient. However, the method is not the complete answer since a particular statistic may be sufficient yet the user may not be clever enough to factor the joint density as in Eq. (11) or (12). The theorems may also be useful in *discovering* sufficient statistics.

Actually, the result of either of the above factorization theorems is intuitively evident if one notes the following: If the joint density factors as indicated in, say, Eq. (12), then the likelihood function is proportional to $g(s_1, \ldots, s_r; \theta)$, which depends on the observations $x_1, \ldots, x_n$ only through $\mathscr{s}_1, \ldots, \mathscr{s}_r$ [the likelihood function is viewed as a function of θ, so $h(x_1, \ldots, x_n)$ is just a proportionality constant], which means that the information about θ that the likelihood function contains is embodied in the statistics $\mathscr{s}_1(\cdot, \ldots, \cdot), \ldots, \mathscr{s}_r(\cdot, \ldots, \cdot)$.

Before giving several examples, we remark that the function $h(\cdot, \ldots, \cdot)$ appearing in either Eq. (11) or (12) may be constant.

EXAMPLE 20 Let $X_1, \ldots, X_n$ be a random sample from the Bernoulli density with parameter θ; that is,

$$f(x; \theta) = \theta^x(1 - \theta)^{1-x} I_{\{0,1\}}(x) \qquad \text{and} \qquad 0 \le \theta \le 1.$$

Then

$$\prod_{i=1}^{n} f(x_i; \theta) = \prod_{i=1}^{n} \theta^{x_i}(1 - \theta)^{1-x_i} I_{\{0,1\}}(x_i)$$

$$= \theta^{\Sigma x_i}(1 - \theta)^{n - \Sigma x_i} \prod_{i=1}^{n} I_{\{0,1\}}(x_i).$$

If we take $\theta^{\Sigma x_i}(1 - \theta)^{n - \Sigma x_i}$ as $g(\mathscr{s}(x_1, \ldots, x_n); \theta)$ and $\prod_{i=1}^{n} I_{\{0,1\}}(x_i)$ as $h(x_1, \ldots, x_n)$ and set $\mathscr{s}(x_1, \ldots, x_n) = \sum x_i$, then the joint density of $X_1, \ldots, X_n$ factors as in Eq. (11), indicating that $S = \mathscr{s}(X_1, \ldots, X_n) = \sum X_i$ is a sufficient statistic. ////

EXAMPLE 21 Let $X_1, \ldots, X_n$ be a random sample from the normal density with mean μ and variance unity. Here the parameter is denoted by μ instead of θ. The joint density is given by

$$
\begin{aligned}
f_{X_1,\ldots,X_n}(x_1, \ldots, x_n; \mu) &= \prod_{i=1}^{n} \phi_{\mu,1}(x_i) \\
&= \prod_{i=1}^{n} \frac{1}{\sqrt{2\pi}} \exp\left[-\frac{1}{2}(x_i - \mu)^2\right] \\
&= \frac{1}{(2\pi)^{n/2}} \exp\left[-\frac{1}{2}\sum_{i=1}^{n}(x_i - \mu)^2\right] \\
&= \frac{1}{(2\pi)^{n/2}} \exp\left[-\frac{1}{2}\left(\sum x_i^2 - 2\mu\sum x_i + n\mu^2\right)\right] \\
&= \frac{1}{(2\pi)^{n/2}} \exp\left(\mu\sum x_i - \frac{n}{2}\mu^2\right) \exp\left(-\frac{1}{2}\sum x_i^2\right).
\end{aligned}
$$

If we take $h(x_1, \ldots, x_n) = [1/(2\pi)^{n/2}] \exp(-\frac{1}{2}\sum x_i^2)$ and $g(\mathscr{s}(x_1, \ldots, x_n); \mu) = \exp[\mu \sum x_i - (n/2)\mu^2]$, then the joint density has been factored as in Eq. (11) with $\mathscr{s}(x_1, \ldots, x_n) = \sum x_i$; hence $\sum X_i$ is a sufficient statistic. (Recall that $\overline{X}_n$ is also sufficient since any one-to-one function of a sufficient statistic is also sufficient.) ////

EXAMPLE 22 Let $X_1, \ldots, X_n$ be a random sample from the normal density $\phi_{\mu,\sigma^2}(\cdot)$. Here the parameter θ is a vector of two components; that is, $\theta = (\mu, \sigma)$. The joint density of $X_1, \ldots, X_n$ is given by

$$
\begin{aligned}
\prod_{i=1}^{n} \phi_{\mu,\sigma^2}(x_i) &= \prod_{i=1}^{n} \frac{1}{\sqrt{2\pi}\sigma} \exp\left[-\frac{1}{2}\left(\frac{x_i - \mu}{\sigma}\right)^2\right] \\
&= \frac{1}{(2\pi)^{n/2}} \sigma^{-n} \exp\left[-\frac{1}{2}\sum\left(\frac{x_i - \mu}{\sigma}\right)^2\right] \\
&= \frac{1}{(2\pi)^{n/2}} \sigma^{-n} \exp\left[-\frac{1}{2\sigma^2}\left(\sum x_i^2 - 2\mu\sum x_i + n\mu^2\right)\right];
\end{aligned}
$$

so the joint density itself depends on the observations $x_1, \ldots, x_n$ only through the statistics $\mathscr{s}_1(x_1, \ldots, x_n) = \sum x_i$ and $\mathscr{s}_2(x_1, \ldots, x_n) = \sum x_i^2$; that is, the joint density is factored as in Eq. (12) with $h(x_1, \ldots, x_n) = 1$. Hence, $\sum X_i$ and $\sum X_i^2$ are jointly sufficient. It can be shown that $\overline{X}_n$ and $S^2 = [1/(n-1)] \sum (X_i - \overline{X})^2$ are one-to-one functions of $\sum X_i$ and $\sum X_i^2$; so $\overline{X}_n$ and S^2 are also jointly sufficient. ////

EXAMPLE 23 Let $X_1, \ldots, X_n$ be a random sample from a uniform distribution over the interval $[\theta_1, \theta_2]$. The joint density of $X_1, \ldots, X_n$ is given by

$$f_{X_1, \ldots, X_n}(x_1, \ldots, x_n; \theta_1, \theta_2) = \prod_{i=1}^{n} \frac{1}{\theta_2 - \theta_1} I_{[\theta_1, \theta_2]}(x_i)$$

$$= \frac{1}{(\theta_2 - \theta_1)^n} \prod_{i=1}^{n} I_{[\theta_1, \theta_2]}(x_i)$$

$$= \frac{1}{(\theta_2 - \theta_1)^n} I_{[\theta_1, y_n]}(y_1) I_{[y_1, \theta_2]}(y_n),$$

where

$$y_1 = \min [x_1, \ldots, x_n] \qquad \text{and} \qquad y_n = \max [x_1, \ldots, x_n].$$

The joint density itself depends on $x_1, \ldots, x_n$ only through y_1 and y_n; hence it factors as in Eq. (12) with $h(x_1, \ldots, x_n) = 1$. The statistics Y_1 and Y_n are jointly sufficient. Note that if we take $\theta_1 = \theta$ and $\theta_2 = \theta + 1$, then Y_1 and Y_n are still jointly sufficient. However, if we take $\theta_1 = 0$ and $\theta_2 = \theta$, then our factorization can be expressed as

$$f_{X_1, \ldots, X_n}(x_1, \ldots, x_n; \theta) = \frac{1}{\theta^n} \prod_{i=1}^{n} I_{[0, \theta]}(x_i)$$

$$= \frac{1}{\theta^n} I_{[0, \theta]}(y_n) I_{[0, y_n]}(y_1).$$

Taking $g(\mathscr{s}(x_1, \ldots, x_n); \theta) = (1/\theta^n) I_{[0, \theta]}(y_n)$ and $h(x_1, \ldots, x_n) = I_{[0, y_n]}(y_1)$, we see that Y_n alone is sufficient. ////

The factorization criterion of Eqs. (11) and (12) is primarily useful in showing that a statistic or set of statistics is sufficient. It is not useful in proving that a statistic or set of statistics is *not* sufficient. The fact that we cannot factor the joint density does not mean that it cannot be factored; it could be that we are just not able to find a correct factorization.

If we go back and look through our examples on maximum-likelihood estimators (see Examples 5 to 8), we will see that all the maximum-likelihood estimators that appear there depend on the sample $X_1, \ldots, X_n$ through sufficient statistics. This is not something that is characteristic of the relatively simple examples we had given but something that is true in general.

Theorem 6 A maximum-likelihood estimator or set of maximum-likelihood estimators depends on the sample through any set of jointly sufficient statistics.

PROOF If $S_1 = \mathscr{s}_1(X_1, \ldots, X_n), \ldots, S_k = \mathscr{s}_k(X_1, \ldots, X_n)$ are jointly sufficient, then the likelihood function can be written as

$$L(\theta; x_1, \ldots, x_n) = \prod_{i=1}^{n} f(x_i; \theta) = g(\mathscr{s}_1(x_1, \ldots, x_n), \ldots, \mathscr{s}_k(x_1, \ldots, x_n); \theta)h(x_1, \ldots, x_n).$$

As a function of θ, $L(\theta; x_1, \ldots, x_n)$ will have its maximum at the same place that $g(s_1, \ldots, s_k; \theta)$ has its maximum, but the place where g attains its maximum can depend on $x_1, \ldots, x_n$ only through $s_1, \ldots, s_k$ since g does. ////

We might note that method-of-moment estimators may not be functions of sufficient statistics. See Examples 4 and 23.

4.3 Minimal Sufficient Statistics

When we introduced the concept of sufficiency, we said that our objective was to condense the data without losing any information about the parameter. We have seen that there is more than one set of sufficient statistics. For example, in sampling from a normal distribution with both the mean and variance unknown, we have noted three sets of jointly sufficient statistics, namely, the sample $X_1, \ldots, X_n$ itself, the order statistics $Y_1, \ldots, Y_n$, and $\bar{X}$ and S^2. We naturally prefer the jointly sufficient set $\bar{X}$ and S^2 since they condense the data more than either of the other two. (Note that the order statistics do condense the data.) The question that we might ask is: Does there exist a set of sufficient statistics that condenses the data more than $\bar{X}$ and S^2? The answer is that there does not, but we will not develop the necessary tools to establish this answer. The notion that we are alluding to is that of a minimum set of sufficient statistics, which we label *minimal sufficient statistics.*

We noted earlier that corresponding to any statistic is the partition of $\mathfrak{X}$ that it induces. The same is true of a set of statistics; a set of statistics induces a partition of $\mathfrak{X}$. Loosely speaking, the condensation of the data that a statistic or set of statistics exhibits can be measured by the number of subsets in the partition induced by that statistic or set of statistics. If a set of statistics has fewer subsets in its induced partition than does the induced partition of another set of statistics, then we say that the first statistic condenses the data more than the latter. Still loosely speaking, a minimal sufficient set of statistics is then a sufficient set of statistics that has fewer subsets in its partition than the induced partition of any other set of sufficient statistics. So a set of sufficient statistics is minimal if no other set of sufficient statistics condenses the data more. A formal definition is the following.

Definition 18 Minimal sufficient statistic A set of jointly sufficient statistics is defined to be *minimal sufficient* if and only if it is a function of every other set of sufficient statistics. ////

Like many definitions, Definition 18 is of little use in finding minimal sufficient statistics. A technique for finding minimal sufficient statistics has been devised by Lehmann and Scheffé [19], but we will not present it. If the joint density is properly factored, the factorization criterion will give us minimal sufficient statistics. All the sets of sufficient statistics found in Examples 20 to 23 are minimal.

4.4 Exponential Family

Many of the parametric families of densities that we have considered are members of what is called the *exponential class*, or *exponential family*, not to be confused with the negative exponential family of densities which is a special case.

Definition 19 Exponential family of densities A one-parameter family (θ is unidimensional) of densities $f(\cdot\,;\theta)$ that can be expressed as

$$f(x;\theta) = a(\theta)b(x)\exp\left[c(\theta)d(x)\right] \tag{13}$$

for $-\infty < x < \infty$, for all $\theta \in \underline{\overline{\Theta}}$, and for a suitable choice of functions $a(\cdot)$, $b(\cdot)$, $c(\cdot)$, and $d(\cdot)$ is defined to belong to the *exponential family* or *exponential class*. ////

EXAMPLE 24 If $f(x;\theta) = \theta e^{-\theta x} I_{(0,\infty)}(x)$, then $f(x;\theta)$ belongs to the exponential family for $a(\theta) = \theta$, $b(x) = I_{(0,\infty)}(x)$, $c(\theta) = -\theta$, and $d(x) = x$ in Eq. (13). ////

EXAMPLE 25 If $f(x;\theta) = f(x;\lambda)$ is the Poisson density, then

$$\begin{aligned} f(x;\lambda) &= \frac{e^{-\lambda}\lambda^x}{x!} I_{\{0,1,\ldots\}}(x) \\ &= e^{-\lambda}\left(\frac{1}{x!} I_{\{0,1,\ldots\}}(x)\right)\exp\left(x\log\lambda\right) \end{aligned}$$

In Eq. (13), we can take $a(\lambda) = e^{-\lambda}$, $b(x) = (1/x!)I_{\{0, 1, \ldots\}}(x)$, $c(\lambda) = \log \lambda$, and $d(x) = x$; so $f(x; \lambda)$ belongs to the exponential family. ////

Remark If $f(x; \theta) = a(\theta)b(x) \exp [c(\theta)d(x)]$, then

$$\prod_{i=1}^{n} f(x_i; \theta) = a^n(\theta)\left[\prod_{i=1}^{n} b(x_i)\right] \exp \left[c(\theta)\sum_{i=1}^{n} d(x_i)\right],$$

and hence by the factorization criterion $\sum d(X_i)$ is a sufficient statistic. ////

The above remark shows that, under random sampling, if a density belongs to the one-parameter exponential family, then there is a sufficient statistic. In fact, it can be shown that the sufficient statistic so obtained is minimal.

The one-parameter exponential family can be generalized to the k-parameter exponential family.

Definition 20 ***k*-parameter exponential family** A family of densities $f(\cdot\,; \theta_1, \ldots, \theta_k)$ that can be expressed as

$$f(x; \theta_1, \ldots, \theta_k) = a(\theta_1, \ldots, \theta_k)b(x) \exp \sum_{j=1}^{k} c_j(\theta_1, \ldots, \theta_k)d_j(x) \tag{14}$$

for a suitable choice of functions $a(\cdot, \ldots, \cdot)$, $b(\cdot)$, $c_j(\cdot, \ldots, \cdot)$, and $d_j(\cdot)$, $j = 1, \ldots, k$, is defined to belong to the exponential family. ////

In Definition 20, note that the number of terms in the sum of the exponent is k, which is also the dimension of the parameter.

EXAMPLE 26 If $f(x; \theta_1, \theta_2) = \phi_{\mu, \sigma^2}(x)$, where $(\theta_1, \theta_2) = (\mu, \sigma)$, then $f(x; \theta_1, \theta_2)$ belongs to the exponential family.

$$\begin{aligned} f(x; \theta_1, \theta_2) &= \frac{1}{\sqrt{2\pi}\sigma} \exp \left[-\frac{1}{2}\left(\frac{x-\mu}{\sigma}\right)^2\right] \\ &= \frac{1}{\sqrt{2\pi}\sigma} \exp \left(-\frac{1}{2} \cdot \frac{\mu^2}{\sigma^2}\right) \exp \left(-\frac{1}{2\sigma^2}x^2 + \frac{\mu}{\sigma^2}x\right). \end{aligned}$$

Take $a(\mu, \sigma) = (1/\sqrt{2\pi}\sigma) \exp(-\frac{1}{2} \cdot \mu^2/\sigma^2)$, $b(x) = 1$, $c_1(\mu, \sigma) = -1/2\sigma^2$, $c_2(\mu, \sigma) = \mu/\sigma^2$, $d_1(x) = x^2$, and $d_2(x) = x$ to show that $\phi_{\mu, \sigma^2}(x)$ can be expressed as in Eq. (14). ////

EXAMPLE 27 If

$$f(x; \theta_1, \theta_2) = \frac{1}{B(\theta_1, \theta_2)} x^{\theta_1 - 1}(1 - x)^{\theta_2 - 1} I_{(0,1)}(x),$$

then

$$f(x; \theta_1, \theta_2) = \frac{1}{B(\theta_1, \theta_2)} I_{(0,1)}(x) \exp[(\theta_1 - 1)\log x + (\theta_2 - 1)\log(1 - x)];$$

so $f(x; \theta_1, \theta_2)$ belongs to the exponential family with $a(\theta_1, \theta_2) = 1/B(\theta_1, \theta_2)$, $b(x) = I_{(0,1)}(x)$, $c_1(\theta) = \theta_1 - 1$, $c_2(\theta) = \theta_2 - 1$, $d_1(x) = \log x$, and $d_2(x) = \log(1 - x)$. ////

Remark If $f(x; \theta_1, \ldots, \theta_k) = a(\theta_1, \ldots, \theta_k)b(x) \exp \sum_{j=1}^{k} c_j(\theta_1, \ldots, \theta_k)d_j(x)$, then, under random sampling,

$$\prod_{i=1}^{n} f(x_i; \theta_1, \ldots, \theta_k)$$
$$= a^n(\theta_1, \ldots, \theta_k)\left[\prod_{i=1}^{n} b(x_i)\right] \exp\left[\sum_{j=1}^{k} c_j(\theta_1, \ldots, \theta_k) \sum_{i=1}^{n} d_j(x_i)\right],$$

and so by the factorization criterion $\sum_{i=1}^{n} d_1(X_i), \ldots, \sum_{i=1}^{n} d_k(X_i)$ is a set of jointly sufficient statistics. $\sum d_1(X_i), \ldots, \sum d_k(X_i)$ are in fact minimal sufficient statistics. ////

EXAMPLE 28 From Example 27, we see that $\sum_{i=1}^{n} \log X_i$ and $\sum_{i=1}^{n} \log(1 - X_i)$ are jointly minimal sufficient when sampling from a beta density. ////

Our main use of the exponential family will not be in finding sufficient statistics, but it will be in showing that the sufficient statistics are *complete*, a concept that is useful in obtaining "best" estimators. This concept will be defined in Sec. 5.

Lest one get the impression that all parametric families belong to the exponential family, we remark that a family of uniform densities does not belong to the exponential family. In fact, any family of densities for which the range of the values where the density is nonnegative depends on the parameter θ does not belong to the exponential class.

5 UNBIASED ESTIMATION

Since estimators with uniformly minimum mean-squared error rarely exist, a reasonable procedure is to restrict the class of estimating functions and look for estimators with uniformly minimum mean-squared error within the restricted class. One way of restricting the class of estimating functions would be to consider only unbiased estimators and then among the class of unbiased estimators search for an estimator with minimum mean-squared error. Consideration of unbiased estimators and the problem of finding one with uniformly minimum mean-squared error are to be the subjects of this section.

According to Eq. (9) the mean-squared error of an estimator T of $\tau(\theta)$ can be written as

$$\mathscr{E}_\theta[[T - \tau(\theta)]^2] = \operatorname{var}_\theta[T] + \{\tau(\theta) - \mathscr{E}_\theta[T]\}^2,$$

and if T is an unbiased estimator of $\tau(\theta)$, then $\mathscr{E}_\theta[T] = \tau(\theta)$, and so $\mathscr{E}_\theta[[T - \tau(\theta)]^2] = \operatorname{var}_\theta[T]$. Hence, seeking an estimator with uniformly minimum mean-squared error among unbiased estimators is tantamount to seeking an estimator with uniformly minimum variance among unbiased estimators.

Definition 21 Uniformly minimum-variance unbiased estimator (UMVUE) Let $X_1, \ldots, X_n$ be a random sample from $f(\cdot\,; \theta)$. An estimator $T^* = t^*(X_1, \ldots, X_n)$ of $\tau(\theta)$ is defined to be a *uniformly minimum-variance unbiased estimator* of $\tau(\theta)$ if and only if (i) $\mathscr{E}_\theta[T^*] = \tau(\theta)$, that is, T^* is unbiased, and (ii) $\operatorname{var}_\theta[T^*] \leq \operatorname{var}_\theta[T]$ for any other estimator $T = t(X_1, \ldots, X_n)$ of $\tau(\theta)$ which satisfies $\mathscr{E}_\theta[T] = \tau(\theta)$. ////

In Subsec. 5.1 below we will derive a lower bound for the variance of unbiased estimators and show how it can sometimes be useful in finding an UMVUE. In Subsec. 5.2 we will introduce the concept of *completeness* and show how it in conjunction with sufficiency can sometimes be used to find an UMVUE.

5.1 Lower Bound for Variance

Let $X_1, \ldots, X_n$ be a random sample from $f(\cdot\,; \theta)$, where θ belongs to $\overline{\Theta}$. Assume that $\overline{\Theta}$ is a subset of the real line. Let $T = t(X_1, \ldots, X_n)$ be an unbiased estimator of $\tau(\theta)$. We will consider the case where $f(\cdot\,; \theta)$ is a probability density function; the development for discrete density functions is analogous. We make the following assumptions, called *regularity conditions:*

(i) $\dfrac{\partial}{\partial\theta} \log f(x;\theta)$ exists for all x and all θ.

(ii) $$\frac{\partial}{\partial\theta}\int\cdots\int\prod_{i=1}^{n} f(x_i;\theta)\,dx_1\cdots dx_n = \int\cdots\int\frac{\partial}{\partial\theta}\prod_{i=1}^{n} f(x_i;\theta)\,dx_1\cdots dx_n.$$

(iii) $$\frac{\partial}{\partial\theta}\int\cdots\int t(x_1,\ldots,x_n)\prod_{i=1}^{n} f(x_i;\theta)\,dx_1\cdots dx_n = \int\cdots\int t(x_1,\ldots,x_n)\frac{\partial}{\partial\theta}\prod_{i=1}^{n} f(x_i;\theta)\,dx_1\cdots dx_n.$$

(iv) $0 < \mathscr{E}_\theta\left[\left[\frac{\partial}{\partial\theta}\log f(X;\theta)\right]^2\right] < \infty$ for all θ in $\underline{\Theta}$.

Theorem 7 Cramér-Rao inequality Under assumptions (i) to (iv) above

$$\text{var}_\theta\,[T] \geq \frac{[\tau'(\theta)]^2}{n\mathscr{E}_\theta\left[\left[\frac{\partial}{\partial\theta}\log f(X;\theta)\right]^2\right]} \tag{15}$$

where $T = t(X_1, \ldots, X_n)$ is an unbiased estimator of $\tau(\theta)$. Equality prevails in Eq. (15) if and only if there exists a function, say $K(\theta, n)$, such that

$$\sum_1^n \frac{\partial}{\partial\theta}\log f(x_i;\theta) = K(\theta, n)[t(x_1,\ldots,x_n) - \tau(\theta)]. \tag{16}$$

Equation (15) is called the *Cramér-Rao inequality*, and the right-hand side is called the *Cramér-Rao lower bound* for the variance of unbiased estimators of $\tau(\theta)$.

PROOF

$$\begin{aligned}
\tau'(\theta) &= \frac{\partial}{\partial\theta}\tau(\theta) = \frac{\partial}{\partial\theta}\int\cdots\int t(x_1,\ldots,x_n)\prod_{i=1}^{n} f(x_i;\theta)\,dx_1\cdots dx_n \\
&= \int\cdots\int t(x_1,\ldots,x_n)\frac{\partial}{\partial\theta}\left[\prod_{i=1}^{n} f(x_i;\theta)\right]dx_1\cdots dx_n \\
&\quad - \tau(\theta)\frac{\partial}{\partial\theta}\int\cdots\int\prod_{i=1}^{n}[f(x_i;\theta)\,dx_i] \\
&= \int\cdots\int t(x_1,\ldots,x_n)\frac{\partial}{\partial\theta}\left[\prod_{i=1}^{n} f(x_i;\theta)\right]dx_1\cdots dx_n \\
&\quad - \tau(\theta)\int\cdots\int\frac{\partial}{\partial\theta}\left[\prod_{i=1}^{n} f(x_i;\theta)\right]dx_1\cdots dx_n
\end{aligned}$$

$$= \int \cdots \int [t(x_1, \ldots, x_n) - \tau(\theta)] \frac{\partial}{\partial \theta} \left[\prod_{i=1}^{n} f(x_i; \theta \right] dx_1 \cdots dx_n$$

$$= \int \cdots \int [t(x_1, \ldots, x_n) - \tau(\theta)] \left[\frac{\partial}{\partial \theta} \log \prod_{i=1}^{n} f(x_i; \theta) \right]$$

$$\times \prod_{i=1}^{n} f(x_i; \theta)\, dx_1 \cdots dx_n$$

$$= \mathscr{E}_\theta \left[[t(X_1, \ldots, X_n) - \tau(\theta)] \left[\frac{\partial}{\partial \theta} \log \prod_{i=1}^{n} f(X_i; \theta) \right] \right].$$

Now by the Cauchy-Schwarz inequality

$$[\tau'(\theta)]^2 \le \mathscr{E}_\theta[[t(X_1, \ldots, X_n) - \tau(\theta)]^2] \mathscr{E}_\theta \left[\left[\frac{\partial}{\partial \theta} \log \prod_{i=1}^{n} f(X_i; \theta) \right]^2 \right],$$

or

$$\operatorname{var}_\theta [T] \ge \frac{[\tau'(\theta)]^2}{\mathscr{E}_\theta \left[\left(\frac{\partial}{\partial \theta} \log \prod_{i=1}^{n} f(X_i; \theta) \right)^2 \right]};$$

but

$$\mathscr{E}_\theta \left[\left[\frac{\partial}{\partial \theta} \log \prod_{i=1}^{n} f(X_i; \theta) \right]^2 \right] = \mathscr{E}_\theta \left[\left[\sum_{i=1}^{n} \frac{\partial}{\partial \theta} \log f(X_i; \theta) \right]^2 \right]$$

$$= \sum_i \sum_j \mathscr{E}_\theta \left[\left[\frac{\partial}{\partial \theta} \log f(X_i; \theta) \right] \left[\frac{\partial}{\partial \theta} \log f(X_j; \theta) \right] \right]$$

$$= n \mathscr{E}_\theta \left[\left[\frac{\partial}{\partial \theta} \log f(X; \theta) \right]^2 \right],$$

using the independence of X_i and X_j and noting that

$$\mathscr{E}_\theta \left[\frac{\partial}{\partial \theta} \log f(X; \theta) \right] = \int \left[\frac{\partial}{\partial \theta} \log f(x; \theta) \right] f(x; \theta)\, dx$$

$$= \int \frac{\partial}{\partial \theta} f(x; \theta)\, dx = \frac{\partial}{\partial \theta} \int f(x; \theta)\, dx = \frac{\partial}{\partial \theta}(1) = 0.$$

The inequality in the Cauchy-Schwarz inequality becomes an equality if and only if one function is proportional to the other; in our case this requires that $\frac{\partial}{\partial \theta} \log \prod_{i=1}^{n} f(x_i; \theta)$ be proportional to $t(x_1, \ldots, x_n) - \tau(\theta)$ or that there exists $K = K(\theta, n)$ such that

$$\frac{\partial}{\partial \theta} \log \left[\prod_{i=1}^{n} f(x_i; \theta) \right] = K(\theta, n)[t(x_1, \ldots, x_n) - \tau(\theta)]. \qquad ////$$

The *regularity conditions*, which were stated for probability density functions, can be modified for discrete density functions, leaving the statement of the theorem unchanged.

The theorem has two uses: First, it gives a lower bound for the variance of unbiased estimators. An experimenter using an unbiased estimator whose variance was close to the Cramér-Rao lower bound would know that he was using a good unbiased estimator. Second, if an unbiased estimator whose variance coincides with the Cramér-Rao lower bound can be found, then this estimator is an UMVUE. Equation (16) aids in finding an estimator whose variance coincides with the Cramér-Rao lower bound. In fact, if there exists a $T^* = \ell^*(X_1, \ldots, X_n)$ such that

$$\sum_{1}^{n} \frac{\partial}{\partial \theta} \log f(x_i; \theta) = K(\theta, n)[\ell^*(x_1, \ldots, x_n) - \tau^*(\theta)]$$

for some functions $K(\theta, n)$ and $\tau^*(\theta)$, then T^* is an UMVUE of $\tau^*(\theta)$.

EXAMPLE 29 Let $X_1, \ldots, X_n$ be a random sample from $f(x; \theta) = \theta e^{-\theta x} I_{(0, \infty)}(x)$. Take $\tau(\theta) = \theta$. It can be shown that the regularity conditions are satisfied. $\tau'(\theta) = 1$; hence

$$\operatorname{var}_\theta [T] \geq \frac{1}{n\mathscr{E}_\theta\left[\left[\frac{\partial}{\partial \theta} \log f(X; \theta)\right]^2\right]}.$$

Note that $\frac{\partial}{\partial \theta} \log f(x; \theta) = \frac{\partial}{\partial \theta} (\log \theta - \theta x) = 1/\theta - x$, and so

$$\mathscr{E}_\theta\left[\left[\frac{\partial}{\partial \theta} \log f(X; \theta)\right]^2\right] = \mathscr{E}_\theta\left[\left(\frac{1}{\theta} - X\right)^2\right] = \operatorname{var}[X] = \frac{1}{\theta^2}.$$

Hence, the Cramér-Rao lower bound for the variance of unbiased estimators of θ is given by

$$\operatorname{var}_\theta [T] \geq \frac{1}{n(1/\theta^2)} = \frac{\theta^2}{n}.$$

Similarly the Cramér-Rao lower bound for the variance of unbiased estimators of $\tau(\theta) = 1/\theta$ is given by

$$\operatorname{var}_\theta [T] \geq \frac{[\tau'(\theta)]^2}{n(1/\theta^2)} = \frac{1}{n\theta^2}.$$

The left-hand side of Eq. (16) is

$$\sum_1^n \frac{\partial}{\partial \theta} \log f(x_i; \theta) = \sum_1^n \frac{\partial}{\partial \theta} (\log \theta - \theta x_i) = \sum_1^n \left(\frac{1}{\theta} - x_i\right) = -n\left(\bar{x}_n - \frac{1}{\theta}\right).$$

By taking $K(\theta, n) = -n$ and utilizing the result of Eq. (16), we see that $\bar{X}_n$ is an UMVUE of $1/\theta$ since its variance coincides with the Cramér-Rao lower bound. ////

EXAMPLE 30 Let $X_1, \ldots, X_n$ be a random sample from $f(x; \theta) = f(x; \lambda) = e^{-\lambda}\lambda^x/x!$ for $x = 0, 1, 2, \ldots$.

$$\begin{aligned}\frac{\partial}{\partial \lambda} \log f(x; \lambda) &= \frac{\partial}{\partial \lambda} \log \frac{e^{-\lambda}\lambda^x}{x!} \\ &= \frac{\partial}{\partial \lambda} (-\lambda + x \log \lambda - \log x!) = -1 + \frac{x}{\lambda}.\end{aligned}$$

Therefore

$$\begin{aligned}\mathscr{E}\left[\left[\frac{\partial}{\partial \lambda} \log f(X; \lambda)\right]^2\right] &= \mathscr{E}\left[\left(\frac{X}{\lambda} - 1\right)^2\right] = \frac{1}{\lambda^2} \mathscr{E}[(X - \lambda)^2] \\ &= \frac{1}{\lambda^2} \operatorname{var}[X] = \frac{1}{\lambda^2} \lambda = \frac{1}{\lambda},\end{aligned}$$

and so the denominator of the Cramér-Rao lower bound is n/λ. Now, if $\tau(\lambda) = e^{-\lambda} = P[X = 0]$, then the Cramér-Rao lower bound for the variance of unbiased estimators of $\tau(\lambda) = e^{-\lambda}$ is given by $\operatorname{var}[T] \geq \lambda e^{-2\lambda}/n$. Note that $T = (1/n) \sum_{i=1}^n I_{\{0\}}(X_i)$ is an unbiased estimator of $\tau(\lambda) = e^{-\lambda}$ since $\mathscr{E}[T] = (1/n) \sum_{i=1}^n \mathscr{E}[I_{\{0\}}(X_i)] = (1/n) \sum_{i=1}^n e^{-\lambda} = e^{-\lambda}$. $I_{\{0\}}(X_i) = 1$ if $X_i = 0$, and $I_{\{0\}}(X_i) = 0$ otherwise; so $\mathscr{E}[I_{\{0\}}(X_i)] = 1 \cdot P[X_i = 0] + 0 \cdot P[X_i \neq 0] = e^{-\lambda}$. T is the proportion of observations in the sample that are equal to 0. $\operatorname{var}[T] = (1/n)e^{-\lambda}(1 - e^{-\lambda})$, as compared to the Cramér-Rao lower bound, which is $(1/n)\lambda e^{-2\lambda}$. Note that

$$(1/n)e^{-\lambda}(1 - e^{-\lambda}) \geq (1/n)\lambda e^{-2\lambda},$$

as it should be. An UMVUE of $\tau(\lambda) = e^{-\lambda}$ is found in Example 34.

We note that $\sum (\partial/\partial\lambda) \log f(x_i; \lambda) = \sum (-1 + x_i/\lambda) = (n/\lambda)(\bar{x} - \lambda)$; hence, $\bar{X}$ is the UMVUE of λ by Eq. (16). ////

In general, the Cramér-Rao lower bound is not an attainable lower bound; that is, there often exists a lower bound for variance that is greater than the Cramér-Rao lower bound. We will see several such examples in Subsec. 5.2 below. We will see that an UMVUE can exist whose variance does not coincide with the Cramér-Rao lower bound.

We conclude this subsection with several remarks, the statements of which are not necessarily mathematically precise. All the same, the remarks are important and do relate some earlier concepts to the Cramér-Rao lower bound.

Remark Under certain assumptions involving the existence of second derivatives and the validity of interchanging the order of certain differentiations and integrations,

$$\mathscr{E}_\theta\left[\left[\frac{\partial}{d\theta}\log f(X;\theta)\right]^2\right] = -\mathscr{E}_\theta\left[\frac{\partial^2}{\partial\theta^2}\log f(X;\theta)\right]. \qquad ////$$

This remark is computationally useful if the first expectation is more difficult to obtain than the second. The proof is left as an exercise.

Remark If the maximum-likelihood estimate of θ, say $\hat{\theta} = \hat{\vartheta}(x_1, \ldots, x_n)$, is given by a solution to the equation

$$\frac{\partial}{\partial\theta}\log L(\theta; x_1, \ldots, x_n) \equiv \frac{\partial}{\partial\theta}\log\prod_{i=1}^{n} f(x_i;\theta) = 0,$$

and if $T^* = t^*(X_1, \ldots, X_n)$ is an unbiased estimator of $\tau^*(\theta)$ whose variance coincides with the Cramér-Rao lower bound, then $t^*(x_1, \ldots, x_n) = \tau^*(\hat{\vartheta}(x_1, \ldots, x_n))$.

PROOF

$$0 = \frac{\partial}{\partial\theta}\log\prod_{i=1}^{n} f(x_i;\theta)\bigg|_{\theta=\hat{\theta}} = K(\theta, n)[t^*(x_1, \ldots, x_n) - \tau^*(\theta)]\bigg|_{\theta=\hat{\theta}}$$

by Eq. (16) and the definition of $\hat{\theta}$. ////

This remark tells us that under the conditions of the remark a maximum-likelihood estimator is an UMVUE!

Remark If $T^* = t^*(X_1, \ldots, X_n)$ is an unbiased estimator of some $\tau^*(\theta)$ whose variance coincides with the Cramér-Rao lower bound, then $f(\cdot\,;\theta)$ is a member of the exponential class; and, conversely, if $f(\cdot\,;\theta)$ is a member of the exponential class, then there exists an unbiased estimator, say T^*, of some function, say $\tau^*(\theta)$, whose variance coincides with the Cramér-Rao lower bound. ////

We will omit the proof of this remark. It relates the Cramér-Rao lower bound to the exponential family; in fact, it tells us that we will be able to find an estimator whose variance coincides with the Cramér-Rao lower bound if and only if the density from which we are sampling is a member of the exponential class. Although the remark does not explicitly so state, the following is true: There is essentially only one function (one function and then any linear function of the one function) of the parameter for which there exists an unbiased estimator whose variance coincides with the Cramér-Rao lower bound. So, what this remark and the comments following it really tell is: The Cramér-Rao lower bound is of limited use in finding UMVUEs. It is useful only if we sample from a member of the one-parameter exponential family, and even then it is useful in finding the UMVUE of only one function of the parameter. Hence, it behooves us to search for other techniques for finding UMVUEs, and that is what we do in the next subsection.

5.2 Sufficiency and Completeness

In this subsection we will continue our search for UMVUEs. Our first result will show how sufficiency aids in this search. Loosely speaking, an unbiased estimator which is a function of sufficient statistics has smaller variance than an unbiased estimator which is not based on sufficient statistics. In fact, let $f(\cdot;\theta)$ be the density from which we can sample, and suppose that we want to estimate $\tau(\theta)$. Let us assume that $T = t(X_1, \ldots, X_n)$ is an unbiased estimator of $\tau(\theta)$ and that $S = s(X_1, \ldots, X_n)$ is a sufficient statistic. It can be shown that another unbiased estimator, denoted by T', can be derived from T such that (i) T' is a function of the sufficient statistic S and (ii) T' is an unbiased estimator of $\tau(\theta)$ with variance less than or equal to the variance of T. Therefore, in our search for UMVUEs we need to consider only unbiased estimators that are functions of sufficient statistics. We shall formalize these ideas in the following theorem.

Theorem 8 Rao-Blackwell Let $X_1, \ldots, X_n$ be a random sample from the density $f(\cdot;\theta)$, and let $S_1 = s_1(X_1, \ldots, X_n), \ldots, S_k = s_k(X_1, \ldots, X_n)$ be a set of jointly sufficient statistics. Let the statistic $T = t(X_1, \ldots, X_n)$ be an unbiased estimator of $\tau(\theta)$. Define T' by $T' = \mathscr{E}[T \mid S_1, \ldots, S_k]$. Then,

(i) T' is a statistic, and it is a function of the sufficient statistics $S_1, \ldots, S_k$. Write $T' = t'(S_1, \ldots, S_k)$.

(ii) $\mathscr{E}_\theta[T'] = \tau(\theta)$; that is, T' is an unbiased estimator of $\tau(\theta)$.

(iii) $\text{var}_\theta\,[T'] \le \text{var}_\theta\,[T]$ for every θ, and $\text{var}_\theta\,[T'] < \text{var}_\theta\,[T]$ for some θ unless T is equal to T' with probability 1.

PROOF (i) $S_1, \ldots, S_k$ are sufficient statistics; so the conditional distribution of any statistic, in particular the statistic T, given $S_1, \ldots, S_k$ is independent of θ; hence $T' = \mathscr{E}[T \mid S_1, \ldots, S_k]$ is independent of θ, and so T' is a statistic which is obviously a function of $S_1, \ldots, S_k$. (ii) $\mathscr{E}_\theta[T'] = \mathscr{E}_\theta[\mathscr{E}[T \mid S_1, \ldots, S_k]] = \mathscr{E}_\theta[T] = \tau(\theta)$ [using Eq. (26) of Chap. IV]. (iii) We can write

$$\begin{aligned}\operatorname{var}_\theta[T] &= \mathscr{E}_\theta[(T - \mathscr{E}_\theta[T'])^2] = \mathscr{E}_\theta[(T - T' + T' - \mathscr{E}_\theta[T'])^2] \\ &= \mathscr{E}_\theta[(T - T')^2] + 2\mathscr{E}_\theta[(T - T')(T' - \mathscr{E}_\theta[T'])] + \operatorname{var}_\theta[T'].\end{aligned}$$

But

$$\mathscr{E}_\theta[(T - T')(T' - \mathscr{E}[T'])] = \mathscr{E}_\theta[\mathscr{E}[(T - T')(T' - \mathscr{E}[T']) \mid S_1, \ldots, S_k]],$$

and

$$\begin{aligned}&\mathscr{E}_\theta[(T - T')(T' - \mathscr{E}[T']) \mid S_1 = s_1; \ldots; S_k = s_k] \\ &\quad = \{t'(s_1, \ldots, s_k) - \mathscr{E}[T']\}\mathscr{E}[(T - T') \mid S_1 = s_1; \ldots; S_k = s_k] \\ &\quad = \{t'(s_1, \ldots, s_k) - \mathscr{E}[T']\}(\mathscr{E}[T \mid S_1 = s_1; \ldots; S_k = s_k] \\ &\qquad - \mathscr{E}[T' \mid S_1 = s_1; \ldots; S_k = s_k]) \\ &\quad = \{t'(s_1, \ldots, s_k) - \mathscr{E}[T']\}[t'(s_1, \ldots, s_k) - t'(s_1, \ldots, s_k)] \\ &\quad = 0,\end{aligned}$$

and therefore

$$\operatorname{var}_\theta[T] = \mathscr{E}_\theta[(T - T')^2] + \operatorname{var}_\theta[T'] \geq \operatorname{var}_\theta[T'].$$

Note that $\operatorname{var}_\theta[T] > \operatorname{var}_\theta[T']$ unless T equals T' with probability 1. ////

For many applications (particularly where the density involved has only one unknown parameter) there will exist a single sufficient statistic, say $S = s(X_1, \ldots, X_n)$, which would then be used in place of the jointly sufficient set of statistics $S_1, \ldots, S_k$. What the theorem says is that, given an unbiased estimator, another unbiased estimator that is a function of sufficient statistics can be derived and it will not have larger variance. To find the derived statistic, the calculation of a conditional expectation, which may or may not be easy, is required.

EXAMPLE 31 Let $X_1, \ldots, X_n$ be a random sample from the Bernoulli density $f(x; \theta) = \theta^x(1 - \theta)^{1-x}$ for $x = 0$ or 1. X_1 is an unbiased estimator of $\tau(\theta) = \theta$. We use X_1 as $T = t(X_1, \ldots, X_n)$ in the above theorem. $\sum X_i$ is a sufficient statistic; so we use $S = \sum X_i$ as our set (of one element) of sufficient statistics. According to the above theorem $T' = \mathscr{E}[T \mid S] = \mathscr{E}[X_1 \mid \sum X_i]$ is an unbiased estimator of θ with no larger variance than $T = X_1$. Let us evaluate $\mathscr{E}[T \mid S]$. We first find the conditional distribution of X_1 given $\sum X_i = s$. X_1 takes on at most the two values 0 and 1.

$$P[X_1 = 0 \mid \textstyle\sum X_i = s] = \frac{P\left[X_1 = 0; \sum_{i=1}^{n} X_i = s\right]}{P\left[\sum_{i=1}^{n} X_i = s\right]}$$

$$= \frac{P\left[X_1 = 0; \sum_{i=2}^{n} X_i = s\right]}{P\left[\sum_{i=1}^{n} X_i = s\right]} =$$

$$\frac{P[X_1 = 0] \cdot P\left[\sum_{i=2}^{n} X_i = s\right]}{P\left[\sum_{i=1}^{n} X_i = s\right]} = \frac{(1-\theta)\binom{n-1}{s}\theta^s(1-\theta)^{n-1-s}}{\binom{n}{s}\theta^s(1-\theta)^{n-s}} = \frac{n-s}{n}.$$

$$P\left[X_1 = 1 \,\middle|\, \sum_{i=1}^{n} X_i = s\right] = \frac{P\left[X_1 = 1; \sum_{i=1}^{n} X_i = s\right]}{P\left[\sum_{i=1}^{n} X_i = s\right]}$$

$$= \frac{P\left[X_1 = 1; \sum_{i=2}^{n} X_i = s - 1\right]}{P\left[\sum_{i=1}^{n} X_i = s\right]} =$$

$$\frac{P[X_1 = 1] \cdot P\left[\sum_{i=2}^{n} X_i = s - 1\right]}{P\left[\sum_{i=1}^{n} X_i = s\right]} = \frac{\theta \cdot \binom{n-1}{s-1}\theta^{s-1}(1-\theta)^{n-1-s+1}}{\binom{n}{s}\theta^s(1-\theta)^{n-s}} = \frac{s}{n}.$$

We note in passing that the conditional distribution of X_1 given $\sum X_i = s$ is independent of θ, as it should be. Also, we could have derived the conditional distribution with much less effort by asking: Given that you have observed s successes in n trials, what is the probability that the first trial resulted in a success? This probability is s/n. (See Example 28 in Chap. I.)

$$\mathscr{E}\left[X_1 \,\middle|\, \sum_{i=1}^{n} X_i = s\right] = 0 \cdot \frac{n-s}{n} + 1 \cdot \frac{s}{n} = \frac{s}{n};$$

hence,

$$T' = \frac{\sum_{i=1}^{n} X_i}{n}.$$

The variance of X_1 is $\theta(1-\theta)$, and the variance of T' is $\theta(1-\theta)/n$; so for $n > 1$ the variance of T' is actually smaller than the variance of $T = X_1$. ////

Before leaving Theorem 8, two comments are appropriate: First, if the unbiased estimator T is already a function of only $S_1, \ldots, S_k$, then the derived statistic T' will be identical to T, and hence no improvement in variance can be expected. Second, although the set of jointly sufficient statistics is an arbitrary set, in practice one would naturally use a *minimal* set of jointly sufficient statistics if such were available.

Theorem 8 tells us how to improve on an unbiased estimator by conditioning on sufficient statistics. For some estimation problems this unbiased estimator, obtained by conditioning on sufficient statistics, will be an UMVUE. To aid in identifying those estimation problems for which a derived estimator is an UMVUE, the concept of completeness of a family of densities is useful.

Definition 22 Complete family of densities Let $X_1, \ldots, X_n$ denote a random sample from the density $f(\cdot; \theta)$ with parameter space $\overline{\Theta}$, and let $T = t(X_1, \ldots, X_n)$ be a statistic. The family of densities of T is defined to be *complete* if and only if $\mathscr{E}_\theta[z(T)] \equiv 0$ for all $\theta \in \overline{\Theta}$ implies that $P_\theta[z(T) = 0] \equiv 1$ for all $\theta \in \overline{\Theta}$, where $z(T)$ is a statistic. Also, the statistic T is said to be *complete* if and only if its family of densities is complete. ////

Another way of stating that a statistic T is complete is the following: T is complete if and only if the *only* unbiased estimator of 0 that is a function of T is the statistic that is identically 0 with probability 1.

EXAMPLE 32 Let $X_1, \ldots, X_n$ be a random sample from the Bernoulli density. The statistic $T = X_1 - X_2$ is not complete since $\mathscr{E}_\theta[X_1 - X_2] = 0$ and $X_1 - X_2$ is not 0 with probability 1. Consider the statistic $T = \sum_1^n X_i$. Let $z(T)$ be any statistic that is a function of T for which $\mathscr{E}_\theta[z(T)] \equiv 0$

for all $\theta \in \bar{\Theta}$, that is, for $0 \le \theta \le 1$. To argue that T is complete, we must show that $z(t) = 0$ for $t = 0, 1, \ldots, n$. Now

$$\mathscr{E}_\theta[z(T)] = \sum_{t=0}^{n} z(t)\binom{n}{t}\theta^t(1-\theta)^{n-t} = (1-\theta)^n \sum_{t=0}^{n} z(t)\binom{n}{t}[\theta/(1-\theta)]^t;$$

hence, $\mathscr{E}_\theta[z(T)] \equiv 0$ for all $0 \le \theta \le 1$ implies that

$$\sum_{t=0}^{n} z(t)\binom{n}{t}\left(\frac{\theta}{1-\theta}\right)^t \equiv 0$$

or

$$\sum_{t=0}^{n} z(t)\binom{n}{t}\alpha^t \equiv 0$$

for all α, where $\alpha = \theta/(1-\theta)$. Now in order for a polynomial in α to be identically 0, each coefficient of α^t, $t = 0, \ldots, n$, must be 0; that is, $z(t)\binom{n}{t} = 0$ for $t = 0, \ldots, n$, but $\binom{n}{t} \ne 0$; so $z(t) = 0$ for $t = 0, \ldots, n$. ////

EXAMPLE 33 Let $X_1, \ldots, X_n$ be a random sample from the uniform distribution over the interval $(0, \theta)$, where $\bar{\Theta} = \{\theta\colon \theta > 0\}$. Show that the statistic Y_n is complete. We must show that if $\mathscr{E}_\theta[z(Y_n)] \equiv 0$ for all $\theta > 0$, then $P_\theta[z(Y_n) = 0] \equiv 1$ for all $\theta > 0$.

$$\mathscr{E}_\theta[z(Y_n)] = \int z(y) f_{Y_n}(y)\,dy = \int_0^\theta z(y)\theta^{-n} n y^{n-1}\,dy,$$

and $\mathscr{E}_\theta[z(Y_n)] \equiv 0$ for all $\theta > 0$ when and only when

$$\frac{n}{\theta^n}\int_0^\theta z(y) y^{n-1}\,dy \equiv 0 \qquad \text{for all } \theta > 0$$

or

$$\int_0^\theta z(y) y^{n-1}\,dy \equiv 0 \qquad \text{for all } \theta > 0.$$

Differentiating both sides of this last identity with respect to θ produces $z(\theta)\theta^{n-1} \equiv 0$, which in turn implies that $z(\theta) \equiv 0$ for $\theta > 0$. ////

In general, demonstrating completeness can require tricky analysis. The two above examples are exceptions. We state now, without proof, a theorem that gives us completeness of a statistic. It will be our main tool for arguing completeness.

Theorem 9 Let $X_1, \ldots, X_n$ be a random sample from $f(\cdot\,; \theta)$, $\theta \in \overline{\Theta}$, where $\overline{\Theta}$ is an interval (possibly infinite). If $f(x; \theta) = a(\theta)b(x) \exp [c(\theta)d(x)]$, that is, $f(\cdot\,; \theta)$ is a member of the one-parameter exponential family, then $\sum d(X_i)$ is a complete minimal sufficient statistic. ////

Theorem 9 shows once again the importance of the exponential family or exponential class. We are finally adequately prepared to state the theorem that is useful in finding UMVUEs.

Theorem 10 Lehmann-Scheffé Let $X_1, \ldots, X_n$ be a random sample from a density $f(\cdot\,; \theta)$. If $S = \mathscr{s}(X_1, \ldots, X_n)$ is a complete sufficient statistic and if $T^* = \mathscr{t}^*(S)$, a function of S, is an unbiased estimator of $\tau(\theta)$, then T^* is an UMVUE of $\tau(\theta)$.

PROOF Let T' be any unbiased estimator of $\tau(\theta)$ which is a function of S; that is, $T' = \mathscr{t}'(S)$. Then $\mathscr{E}_\theta[T^* - T'] \equiv 0$ for all $\theta \in \overline{\Theta}$, and $T^* - T'$ is a function of S; so by completeness of S, $P_\theta[\mathscr{t}^*(S) = \mathscr{t}'(S)] \equiv 1$ for all $\theta \in \overline{\Theta}$. Hence there is only one unbiased estimator of $\tau(\theta)$ that is a function of S. Now let T be any unbiased estimator of $\tau(\theta)$. T^* must be equal to $\mathscr{E}[T \mid S]$ since $\mathscr{E}[T \mid S]$ is an unbiased estimator of $\tau(\theta)$ depending on S. By Theorem 8, $\text{var}_\theta\,[T^*] \le \text{var}_\theta\,[T]$ for all $\theta \in \overline{\Theta}$; so T^* is an UMVUE. ////

Let us review what this important theorem says: First, if a complete sufficient statistic S exists and if there is an unbiased estimator for $\tau(\theta)$, then there is an UMVUE for $\tau(\theta)$; second, the UMVUE is the unique unbiased estimator of $\tau(\theta)$ which is a function of S.

To actually find that unbiased estimator of $\tau(\theta)$ which is a function of S, we have several ways of proceeding. First, simply guess the correct form of the function of S that defines the desired estimator. Second, guess or find *any* unbiased estimator of $\tau(\theta)$, and then calculate the conditional expectation of the unbiased estimator given the sufficient statistic. Third, solve for $\mathscr{t}^*(\cdot)$ in the equation $\mathscr{E}_\theta[\mathscr{t}^*(S)] \equiv \tau(\theta)$. Such an equation becomes the integral equation $\int \mathscr{t}^*(s)f_S(s)\,ds \equiv \tau(\theta)$ in the case of a continuous random variable S and becomes the summation $\sum \mathscr{t}^*(s)f_S(s) \equiv \tau(\theta)$ for S a discrete random variable. We will employ two of these methods in the following examples.

EXAMPLE 34 Let $X_1, \ldots, X_n$ be a random sample from the Poisson density

$$f(x; \lambda) = \frac{e^{-\lambda}\lambda^x}{x!} \qquad \text{for } x = 0, 1, \ldots.$$

We saw in Example 25 that $f(x; \lambda)$ belongs to the exponential family with $d(x) = x$. By Theorem 9, the statistic $\sum X_i$ is complete and sufficient. To find the UMVUE of λ itself, it suffices to guess a function of $\sum X_i$ whose expectation is λ. Noting that λ is the population mean, $(1/n) \sum X_i$ is the obvious choice; so $(1/n) \sum X_i$ is the UMVUE of λ.

Consider now estimating $\tau(\lambda) = e^{-\lambda} = P[X_i = 0]$. (Recall Example 30.) Let us derive the UMVUE of $e^{-\lambda}$ by calculating the conditional expectation of some unbiased estimator given the sufficient statistic. Any unbiased estimator will do as the preliminary estimator whose conditional expectation needs to be calculated; so we may as well choose one that would make the calculations easy. $I_{\{0\}}(X_1)$ is an unbiased estimator of $e^{-\lambda}$ and is relatively simple since it can assume only the two values 0 and 1. By Theorem 10, $\mathscr{E}[I_{\{0\}}(X_1) | \sum X_i]$ is the UMVUE of $e^{-\lambda}$. To find the desired conditional expectation, we first find the conditional distribution of X_1 given $\sum X_i$.

$$P\left[X_1 = 0 \Big| \sum_1^n X_i = s\right] = \frac{P\left[X_1 = 0; \sum_1^n X_i = s\right]}{P\left[\sum_1^n X_i = s\right]}$$

$$= \frac{P\left[X_1 = 0; \sum_2^n X_i = s\right]}{P\left[\sum_1^n X_i = s\right]} = \frac{P[X_1 = 0]P\left[\sum_2^n X_i = s\right]}{P\left[\sum_1^n X_i = s\right]}$$

$$= \frac{e^{-\lambda}e^{-(n-1)\lambda}[(n-1)\lambda]^s/s!}{e^{-n\lambda}(n\lambda)^s/s!} = \left(\frac{n-1}{n}\right)^s \qquad \text{for } n > 1.$$

Therefore,

$$\mathscr{E}[I_{\{0\}}(X_1) | \sum X_i = s] = P[X_1 = 0 | \sum X_i = s] = \left(\frac{n-1}{n}\right)^s;$$

hence

$$\left(\frac{n-1}{n}\right)^{\Sigma X_i}$$

is the UMVUE of $e^{-\lambda}$ for $n > 1$. For $n = 1$, $I_{\{0\}}(X_1)$ is an unbiased estimator which is a function of the complete sufficient statistic X_1, and hence $I_{\{0\}}(X_1)$ itself is the UMVUE of $e^{-\lambda}$. The reader may want to derive the mean and variance of

$$\left(\frac{n-1}{n}\right)^{\Sigma X_i}$$

and compare them with the mean and variance of the estimator $(1/n)\sum_{i=1}^{n} I_{\{0\}}(X_i)$ given in Example 30. ////

EXAMPLE 35 Let $X_1, \ldots, X_n$ be a random sample from $f(x;\ \theta) = \theta e^{-\theta x} I_{(0,\,\infty)}(x)$. Our object is to find the UMVUE of each of the following functions of the parameter θ: θ, $1/\theta$, and $e^{-K\theta} = P[X > K]$ for given K. Since $\theta e^{-\theta x} I_{(0,\,\infty)}(x)$ is a member of the exponential class (see Example 24), the statistic $S = \sum_1^n X_i$ is complete and sufficient.

$\bar{X}_n = (1/n)\sum_{i=1}^{n} X_i$, which is a function of the complete sufficient statistic $S = \sum_{i=1}^{n} X_i$, is an unbiased estimator of $1/\theta$; hence by Theorem 10, $\bar{X}_n$ is the UMVUE of $1/\theta$.

To find the UMVUE of θ, one might suspect that the estimator is of the form $c/\sum_1^n X_i$, where c is a constant which may depend on n. Now

$$\begin{aligned}
\mathscr{E}_\theta\left[\frac{c}{\sum X_i}\right] &= c\mathscr{E}_\theta\left[\frac{1}{S}\right] = c\int_0^\infty \frac{1}{s} f_S(s)\, ds = c\int_0^\infty \frac{1}{s}\frac{1}{\Gamma(n)}\theta^n s^{n-1} e^{-\theta s}\, ds \\
&= c\frac{1}{\Gamma(n)}\int_0^\infty \theta^n s^{n-2} e^{-\theta s}\, ds = \frac{c\theta}{\Gamma(n)}\int_0^\infty u^{n-2} e^{-u}\, du \\
&= \frac{c\theta}{\Gamma(n)}\cdot\Gamma(n-1) = \frac{c\theta}{n-1}
\end{aligned}$$

for $n > 1$. So $\mathscr{E}_\theta[c/\sum X_i] = \theta$ when $c = n - 1$; hence $(n-1)/\sum X_i$ is the UMVUE of θ for $n > 1$. The variance of $(n-1)/\sum X_i$ is given by $\theta^2/(n-2)$ for $n > 2$.

Although one might be able to guess which function of $S = \sum X_i$ is an unbiased estimator for $e^{-K\theta}$, let us derive the desired estimator by starting with the following simple unbiased estimator of $e^{-K\theta}$: $I_{(K,\,\infty)}(X_1)$. Note that $\mathscr{E}_\theta[I_{(K,\,\infty)}(X_1)] = 0 \cdot P[X_1 \le K] + 1 \cdot P[X_1 > K] = P[X_1 > K] = e^{-K\theta}$; so $I_{(K,\,\infty)}(X_1)$ is indeed an unbiased estimator of $e^{-K\theta}$, and therefore by Theorems 8 and 10 $\mathscr{E}_\theta[I_{(K,\,\infty)}(X_1) \mid S]$ is the UMVUE of $e^{-K\theta}$. Now, $\mathscr{E}_\theta[I_{(K,\,\infty)}(X_1) \mid S = s] = P[I_{(K,\,\infty)}(X_1) = 1 \mid S = s] = P[X_1 > K \mid S = s]$. In order to obtain $P[X_1 > K \mid S = s]$, we will first find the conditional distribution of X_1 given $S = s$.

$$f_{X_1|S=s}(x_1|s)\,\Delta x_1$$

$$= \frac{f_{X_1,S}(x_1,s)\,\Delta x_1\,\Delta s}{f_S(s)\,\Delta s}$$

$$\approx \frac{P\left[x_1 < X_1 < x_1 + \Delta x_1;\, s < \sum_1^n X_i < s + \Delta s\right]}{[1/\Gamma(n)]\theta^n s^{n-1} e^{-\theta s}\,\Delta s}$$

$$\approx \frac{P\left[x_1 < X_1 < x_1 + \Delta x_1;\, s - x_1 < \sum_2^n X_i < s - x_1 + \Delta s\right]}{[1/\Gamma(n)]\theta^n s^{n-1} e^{-\theta s}\,\Delta s}$$

$$= \frac{P[x_1 < X_1 < x_1 + \Delta x_1]P\left[s - x_1 < \sum_2^n X_i < s - x_1 + \Delta s\right]}{[1/\Gamma(n)]\theta^n s^{n-1} e^{-\theta s}\,\Delta s}$$

$$\approx \frac{\theta e^{-\theta x_1}[1/\Gamma(n-1)]\theta^{n-1}(s-x_1)^{n-2}e^{-\theta(s-x_1)}\,\Delta x_1\,\Delta s}{[1/\Gamma(n)]\theta^n s^{n-1} e^{-\theta s}\,\Delta s}$$

$$= \frac{\Gamma(n)}{\Gamma(n-1)} \cdot \frac{(s-x_1)^{n-2}}{s^{n-1}}\,\Delta x_1$$

for $x_1 < s$ and $n > 1$.

$$\mathscr{E}_\theta[I_{(K,\infty)}(X_1)|S=s] = P[X_1 > K|S=s] = \int_K^\infty f_{X_1|S=s}(x_1|s)\,dx_1$$

$$= \int_K^s \frac{\Gamma(n)}{\Gamma(n-1)} \frac{(s-x_1)^{n-2}}{s^{n-1}}\,dx_1$$

$$= \frac{n-1}{s^{n-1}} \int_K^s (s-x_1)^{n-2}\,dx_1$$

$$= \frac{n-1}{s^{n-1}} \int_{s-K}^0 y^{n-2}(-dy)$$

$$= \frac{n-1}{s^{n-1}} \frac{y^{n-1}}{n-1}\bigg|_0^{s-K}$$

$$= \left(\frac{s-K}{s}\right)^{n-1}$$

for $s > K$ and $n > 1$, where the substitution $y = s - x_1$ was made. Hence,

$$\left(\frac{\sum X_i - K}{\sum X_i}\right)^{n-1} \cdot I_{(K,\infty)}(\sum X_i)$$

is the UMVUE for $e^{-K\theta}$ for $n > 1$. (Actually the estimator is applicable for $n = 1$ as well.) It may be of interest and would serve as a check to verify directly that

$$\left(\frac{\sum X_i - K}{\sum X_i}\right)^{n-1} I_{(K,\infty)}(\sum X_i)$$

is unbiased.

$$\begin{aligned}
\mathscr{E}_\theta\left[\left(\frac{\sum X_i - K}{\sum X_i}\right)^{n-1} I_{(K,\infty)}(\sum X_i)\right] &= \int_K^\infty \left(\frac{s-K}{s}\right)^{n-1} f_S(s)\, ds \\
&= \int_K^\infty \left(\frac{s-K}{s}\right)^{n-1} \frac{1}{\Gamma(n)} \theta^n s^{n-1} e^{-\theta s}\, ds \\
&= \int_0^\infty u^{n-1} \frac{1}{\Gamma(n)} \theta^n e^{-\theta(u+K)}\, du = e^{-\theta K},
\end{aligned}$$

where the substitution $u = s - K$ was made. ////

In closing this section on unbiased estimation, we make several remarks.

Remark For some functions of the parameter there is no unbiased estimator. For example, in a sample of size 1 from a binomial density there is no unbiased estimator for $1/\theta$. Suppose there were; let $T = t(X)$ denote it. Then $\mathscr{E}_\theta[T] = \sum_{x=0}^{n} t(x) \binom{n}{x} \theta^x (1-\theta)^{n-x} \equiv 1/\theta$, which says that an nth-degree polynomial in θ is identical to $1/\theta$, which cannot be. ////

Remark We mentioned in Subsec. 5.1 that the Cramér-Rao lower bound is not necessarily the best lower bound. For example, the Cramér-Rao lower bound for the variance of unbiased estimators of θ in sampling from the negative exponential distribution is given by θ^2/n (see Example 29), and the variance of the UMVUE of θ is given by $\theta^2/(n-2)$ (see Example 35). $\theta^2/(n-2)$ is necessarily the best lower bound. ////

Remark For some estimation problems there is an unbiased estimator but no UMVUE. Consider the following example. ////

EXAMPLE 36 Let $X_1, \ldots, X_n$ be a random sample from the uniform density over the interval $(\theta, \theta + 1]$. We want to estimate θ. $\bar{X}_n - \frac{1}{2}$ and $(Y_1 + Y_n)/2 - \frac{1}{2}$ are unbiased estimators of θ, yet there is no UMVUE of θ. For fixed $0 \le p < 1$, consider the estimator $g(X_1 - p) + p$, where the function $g(y)$ is defined to be the greatest integer less than y. Now

$$\mathscr{E}[g(X_1 - p) + p]$$
$$= \int_{\theta}^{\theta+1} g(x_1 - p)\, dx_1 + p = \int_{\theta-p}^{\theta+1-p} g(y)\, dy + p.$$

For fixed θ and p, there exists an integer, say $N = N(\theta, p)$, satisfying $\theta - p < N < \theta + 1 - p$. Hence,

$$\mathscr{E}[g(X_1 - p) + p]$$
$$= \int_{\theta-p}^{\theta+1-p} g(y)\, dy + p = \int_{\theta-p}^{N} (N-1)\, dy + \int_{N}^{\theta+1-p} N\, dy + p = \theta.$$

So $g(X_1 - p) + p$ is an unbiased estimator of θ. Moreover, if $\theta + 1 - p$ is an integer, say J, then $g(x_1 - p) = J - 1$ for all x_1 satisfying $\theta - p < x_1 - p \le \theta + 1 - p$; so $g(x_1 - p) + p = J - 1 + p = \theta + 1 - p - 1 + p = \theta$ for all $\theta < x_1 \le \theta + 1$; that is, $g(X_1 - p) + p$ estimates θ with no error, and hence has zero variance for $\theta + 1 - p$ equal to any integer. So, we have an estimator, namely $g(X_1 - p) + p$, which has zero variance for $\theta =$ any integer $- 1 + p$. But $0 \le p < 1$ is arbitrary; so for any fixed θ, say θ_0, we can find an unbiased estimator of θ which has zero variance at θ_0. Hence, in order for an estimator to be the UMVUE of θ, it must have zero variance for all θ; that is, it must always estimate θ without error. Clearly, no such estimator exists. {The reader may wish to show that var $[g(X_1 - p) + p] = [N - (\theta - p)][(\theta + 1 - p) - N]$, where $N = N(\theta, p)$ is an integer satisfying $\theta - p < N \le \theta + 1 - p$.} ////

Remark It is sometimes possible to find an UMVUE even when a minimal sufficient statistic is not complete. See Prob. 11, p. 313, in Rao [17]. ////

6 LOCATION OR SCALE INVARIANCE

In the last section we employed the property of unbiasedness as a means of restricting the class of estimators with the hope of finding an estimator having minimum mean-squared error within the restricted class. In this section we will indicate how an alternative property, the property of *invariance*, can be used to restrict the class of estimators. Our discussion will be limited to only two types of invariance, namely, *location* invariance and *scale* invariance; a fuller discussion, which is beyond the scope of this book, can be found in Refs. [12] and [19].

6.1 Location Invariance

If the observations $X_1, \ldots, X_n$ represented measurements of some sort and the parameter being estimated was also measured in the same units, one might reasonably require that an estimator $\ell(\cdot, \ldots, \cdot)$ satisfy the property $\ell(x_1 + c, x_2 + c, \ldots, x_n + c) = \ell(x_1, \ldots, x_n) + c$ for every constant c. The idea is that if a constant c is added to each of the measurements $x_1, \ldots, x_n$, then the estimator evaluated at the adjusted measurements $x_1 + c, \ldots, x_n + c$ ought to adjust the estimated values $\ell(x_1, \ldots, x_n)$ by adding the same constant to it. For example, suppose that it is desired to estimate the average weight of a group of pigs when the only method available for weighing is for a person to stand on a scale holding a pig; so both the pig and person are weighed. If one person were to hold the pigs, the measurements (weights) $x_1 + c, \ldots, x_n + c$ would be obtained, where x_i is the weight of the ith pig and c is the person's weight. If, on the other hand, someone else were to hold the pigs, the measurements $x_1 + c', \ldots, x_n + c'$ would be obtained, where c' is the other person's weight. It seems reasonable that the estimate of the average weight of the group of pigs obtained should not depend on which person held the pigs; that is, the estimate should not vary with c, the weight of the pig holder. We define a *location-invariant* estimator accordingly.

Definition 23 Location invariant An estimator $T = \ell(X_1, \ldots, X_n)$ is defined to be *location-invariant* if and only if $\ell(x_1 + c, \ldots, x_n + c) = \ell(x_1, \ldots, x_n) + c$ for all values $x_1, \ldots, x_n$ and all c. ////

A number of the estimators that we have encountered are location-invariant, for example, $\bar{X}_n$ and $(Y_1 + Y_n)/2$, as the following shows:

$$\ell(x_1 + c, \ldots, x_n + c) = \frac{\sum (x_i + c)}{n} = \frac{\sum x_i}{n} + c = \ell(x_1, \ldots, x_n) + c$$

for $\ell(x_1, \ldots, x_n) = \bar{x}_n$; and

$$\begin{aligned} \ell(x_1 + c, \ldots, x_n + c) & \\ &= \frac{\min [x_1 + c, \ldots, x_n + c] + \max [x_1 + c, \ldots, x_n + c]}{2} \\ &= \frac{\min [x_1, \ldots, x_n] + c + \max [x_1, \ldots, x_n] + c}{2} \\ &= \frac{\min [x_1, \ldots, x_n] + \max [x_1, \ldots, x_n]}{2} + c \\ &= \ell(x_1, \ldots, x_n) + c \end{aligned}$$

for $t(x_1, \ldots, x_n) = (y_1 + y_n)/2$. On the other hand, quite a number of estimators are not location-invariant; for example, S^2 and $Y_n - Y_1$, as the following shows: Take $T = t(X_1, \ldots, X_n) = S^2 = \sum (X_i - \bar{X}_n)^2/(n-1)$; then $t(x_1 + c, \ldots, x_n + c) = \sum [x_i + c - \sum (x_i + c)/n]^2/(n-1) = t(x_1, \ldots, x_n)$, instead of $t(x_1, \ldots, x_n) + c$. Now take $T = t(X_1, \ldots, X_n) = Y_n - Y_1$; then $t(x_1 + c, \ldots, x_n + c) = \max [x_1 + c, \ldots, x_n + c] - \min [x_1 + c, \ldots, x_n + c] = \max [x_1, \ldots, x_n] + c - \min [x_1, \ldots, x_n] - c = t(x_1, \ldots, x_n)$, instead of $t(x_1, \ldots, x_n) + c$.

Our use of location invariance will be similar to our use of unbiasedness. We will restrict ourselves to looking at location-invariant estimators and seek an estimator within the class of location-invariant estimators that has uniformly smallest mean-squared error. The property of location invariance is intuitively appealing and turns out also to be practically appealing if the parameter we are estimating represents location.

Definition 24 Location parameter Let $\{f(\cdot\,; \theta), \theta \in \underline{\Theta}\}$ be a family of densities indexed by a parameter θ, where $\underline{\Theta}$ is the real line. The parameter θ is defined to be a *location parameter* if and only if the density $f(x; \theta)$ can be written as a function of $x - \theta$; that is, $f(x; \theta) = h(x - \theta)$ for some function $h(\cdot)$. Equivalently, θ is a location parameter for the density $f_X(x; \theta)$ of a random variable X if and only if the distribution of $X - \theta$ does not depend on θ. ////

We note that if θ is a location parameter for the family of densities $\{f(\cdot\,; \theta), \theta \in \underline{\Theta}\}$, then the function $h(\cdot)$ of the definition is a density function given by $h(\cdot) = f(\cdot\,; 0)$.

EXAMPLE 37 We will give examples of several different location parameters.

If $f(x; \theta) = \phi_{\theta,1}(x)$, then θ is a location parameter since

$$\phi_{\theta,1}(x) = \frac{1}{\sqrt{2\pi}} \exp\left[-\frac{1}{2}(x-\theta)^2\right] = \phi_{0,1}(x - \theta).$$

Or if X is distributed normally with mean θ and variance 1, then $X - \theta$ has a standard normal distribution; hence the distribution of $X - \theta$ is independent of θ.

If $f(x; \theta) = I_{(\theta - \frac{1}{2}, \theta + \frac{1}{2})}(x)$, then θ is a location parameter since $f(x; \theta) = I_{(\theta - \frac{1}{2}, \theta + \frac{1}{2})}(x) = I_{(-\frac{1}{2}, \frac{1}{2})}(x - \theta)$, a function of $x - \theta$.

If $f(x; \alpha) = 1/\pi[1 + (x - \alpha)^2]$, then α is a location parameter since $f(x; \alpha)$ is a function of $x - \alpha$. ////

We will now state, without proof, a theorem that gives within the class of location-invariant estimators the uniformly smallest mean-squared error estimator of a location parameter. The theorem is from Pitman [41].

Theorem 11 Let $X_1, \ldots, X_n$ denote a random sample from the density $f(\cdot\,;\,\theta)$, where θ is a location parameter and $\underline{\Theta}$ is the real line. The estimator

$$t(X_1, \ldots, X_n) = \frac{\int \theta \prod_{i=1}^{n} f(X_i;\,\theta)\,d\theta}{\int \prod_{i=1}^{n} f(X_i;\,\theta)\,d\theta} \tag{17}$$

is the estimator of θ which has uniformly smallest mean-squared error within the class of location-invariant estimators. ////

Definition 25 Pitman estimator for location The estimator given in Eq. (17) is defined to be the *Pitman estimator for location.* ////

According to the formula given in Eq. (17), determining the Pitman estimator requires evaluating the integrals given in the numerator and denominator; such evaluation may not be easy. Note that the integration is with respect to the parameter; so the resulting ratio will be a function of $X_1, \ldots, X_n$.

EXAMPLE 38 Let $X_1, \ldots, X_n$ be a random sample from a normal distribution with mean θ and variance unity. We saw in Example 37 that θ is a location parameter. Our object is to find the Pitman estimator of θ, which is given by Eq. (17). In the following series of equalities one should be forewarned that cancellations and insertions are being made simultaneously in the numerator and denominator.

$$\frac{\int \theta \prod_{i=1}^{n} \phi_{\theta,1}(X_i)\,d\theta}{\int \prod_{i=1}^{n} \phi_{\theta,1}(X_i)\,d\theta} = \frac{\int \theta(1/\sqrt{2\pi})^n \exp\left[-\frac{1}{2}\sum(X_i-\theta)^2\right]d\theta}{\int (1/\sqrt{2\pi})^n \exp\left[-\frac{1}{2}\sum(X_i-\theta)^2\right]d\theta}$$

$$= \frac{\int \theta \exp\left[-(n/2)\theta^2 + \theta\sum X_i\right]d\theta}{\int \exp\left[-(n/2)\theta^2 + \theta\sum X_i\right]d\theta}$$

$$= \frac{\int \theta \exp[-(n/2)(\theta - \bar{X}_n)^2] \, d\theta}{\int \exp[-(n/2)(\theta - \bar{X}_n)^2] \, d\theta}$$

$$= \frac{\int \theta[1/\sqrt{2\pi}(1/\sqrt{n})]\exp\{-\frac{1}{2}[(\theta - \bar{X}_n)/(1/\sqrt{n})]^2\} d\theta}{\int [1/\sqrt{2\pi}(1/\sqrt{n})] \exp\{-\frac{1}{2}[(\theta - \bar{X}_n)/(1/\sqrt{n})]^2\} \, d\theta}$$

$$= \bar{X}_n$$

by noting that the last denominator is just the integral of a normal density with mean $\bar{X}_n$ and variance $1/n$ and hence is unity, and the last numerator is the mean of this same normal density and hence is $\bar{X}_n$.

We note that, for this example, the Pitman estimator of θ, which is uniformly minimum mean-squared error among location-invariant estimators, is identical to the UMVUE of θ; that is, the estimator that is best among location-invariant estimators is also best among unbiased estimators. ////

EXAMPLE 39 Let $X_1, \ldots, X_n$ be a random sample from a uniform distribution over the interval $(\theta - \frac{1}{2}, \theta + \frac{1}{2})$. According to Example 37, θ is a location parameter. The Pitman estimator of θ is

$$\frac{\int \theta \prod_{i=1}^{n} I_{(\theta - \frac{1}{2}, \theta + \frac{1}{2})}(X_i) \, d\theta}{\int \prod_{i=1}^{n} I_{(\theta - \frac{1}{2}, \theta + \frac{1}{2})}(X_i) \, d\theta} = \frac{\int \theta \prod_{i=1}^{n} I_{(X_i - \frac{1}{2}, X_i + \frac{1}{2})}(\theta) \, d\theta}{\int \prod_{i=1}^{n} I_{(X_i - \frac{1}{2}, X_i + \frac{1}{2})}(\theta) \, d\theta} =$$

$$\frac{\int_{Y_n - \frac{1}{2}}^{Y_1 + \frac{1}{2}} \theta \, d\theta}{\int_{Y_n - \frac{1}{2}}^{Y_1 + \frac{1}{2}} d\theta} = \frac{1}{2} \cdot \frac{(Y_1 + \frac{1}{2})^2 - (Y_n - \frac{1}{2})^2}{(Y_1 + \frac{1}{2}) - (Y_n - \frac{1}{2})} = \frac{Y_1 + Y_n}{2}.$$

Recall that for this example there is no UMVUE of θ. (See Example 36.) ////

Remark A Pitman estimator for location is a function of sufficient statistics.

PROOF If $S_1 = \mathscr{s}_1(X_1, \ldots, X_n), \ldots, S_k = \mathscr{s}_k(X_1, \ldots, X_n)$ is a set of sufficient statistics, then by the factorization criterion $\prod_{i=1}^{n} f(x_i; \theta) = g(s_1, \ldots, s_k; \theta)h(x_1, \ldots, x_n)$; so the Pitman estimator can be written as

$$\frac{\int \theta \prod_{i=1}^{n} f(X_i;\theta)\,d\theta}{\int \prod_{i=1}^{n} f(X_i;\theta)\,d\theta} = \frac{\int \theta g(S_1, \ldots, S_k; \theta)h(X_1, \ldots, X_n)\,d\theta}{\int g(S_1, \ldots, S_k; \theta)h(X_1, \ldots, X_n)\,d\theta}$$

$$= \frac{\int \theta\, g(S_1, \ldots, S_k; \theta)\,d\theta}{\int g(S_1, \ldots, S_k; \theta)\,d\theta}$$

which is a function of $S_1, \ldots, S_k$. ////

6.2 Scale Invariance

For those experiments in which measurements can be made in different units, such as length being measured in either inches or centimeters, weight being measured in either pounds or kilograms, or volume being measured in either quarts or liters, one might reasonably require that his statistical procedure be independent of the measurement units employed. If the statistical procedure is that of point estimation, then one might require that the estimator that is to be used satisfy the property of *scale invariance* defined below. The idea is that an estimator will be scale-invariant if the estimator does not depend on the scale of the measurement.

Definition 26 Scale-invariant An estimator $T = \ell(X_1, \ldots, X_n)$ is defined to be *scale-invariant* if and only if $\ell(cx_1, \ldots, cx_n) = c\ell(x_1, \ldots, x_n)$ for all values $x_1, \ldots, x_n$ and all $c > 0$. ////

A number of the estimators that we have considered are scale-invariant, including $\bar{X}_n$, $\sqrt{S^2}$, $(Y_1 + Y_n)/2$, and $Y_n - Y_1$. Our discussion of scale-invariant estimators will be limited to problems concerning estimation of scale parameters defined below.

Definition 27 Scale parameter Let $\{f(\cdot\,; \theta), \theta > 0\}$ be a family of densities indexed by a real parameter θ. The parameter θ is defined to be a *scale parameter* if and only if the density $f(x; \theta)$ can be written as $(1/\theta)h(x/\theta)$ for some density $h(\cdot)$. Equivalently, θ is a scale parameter for the density $f_X(x; \theta)$ of a random variable X if and only if the distribution of X/θ is independent of θ. ////

Note that if θ is a scale parameter for the family of densities $\{f(\cdot\,; \theta), \theta > 0\}$, then the density $h(\cdot)$ of the definition is given by $h(x) = f(x; 1)$.

EXAMPLE 40 We give several examples of scale parameters. If $f(x; \lambda) = (1/\lambda)e^{-x/\lambda}I_{(0,\infty)}(x)$, then λ is a scale parameter since $e^{-y}I_{(0,\infty)}(y)$ is a density. Note that this parameterization of the negative exponential distribution is not the parameterization that we have used previously.

If

$$f(x; \theta) = \phi_{0,\sigma^2}(x) = \frac{1}{\sqrt{2\pi}\sigma} \exp\left[-\frac{1}{2}\left(\frac{x}{\sigma}\right)^2\right],$$

then σ is a scale parameter since $(1/\sqrt{2\pi}) \exp(-\frac{1}{2}y^2)$ is a density.

If $f(x; \theta) = (1/\theta)I_{(0,\theta)}(x) = (1/\theta)I_{(0,1)}(x/\theta)$, then θ is a scale parameter since $I_{(0,1)}(y)$ is a density.

If $f(x; \theta) = (1/\theta)I_{(\theta,2\theta)}(x) = (1/\theta)I_{(1,2)}(x/\theta)$, then θ is a scale parameter since $I_{(1,2)}(y)$ is a density. ////

Our sole result for scale invariance, a result that is comparable to the result of Theorem 11 on location invariance, requires a slightly different framework. Instead of measuring error with squared-error loss function we measure it with the loss function $\ell(t; \theta) = (t - \theta)^2/\theta^2 = (t/\theta - 1)^2$. If $|t - \theta|$ represents error, then $100|t - \theta|/\theta$ can be thought of as percent error, and then $(t - \theta)^2/\theta^2$ is proportional to percent error squared. We state the following theorem, also from Pitman [41], without proof.

Theorem 12 Let $X_1, \ldots, X_n$ be a random sample from the density $f(\cdot\,; \theta)$, where $\theta > 0$ is a scale parameter. Assume that $f(x; \theta) = 0$ for $x \le 0$; that is, the random variables X_i assume only positive values. Within the class of scale-invariant estimators, the estimator

$$\ell(X_1, \ldots, X_n) = \frac{\int_0^\infty (1/\theta^2) \prod_{i=1}^n f(X_i; \theta)\, d\theta}{\int_0^\infty (1/\theta^3) \prod_{i=1}^n f(X_i; \theta)\, d\theta} \tag{18}$$

has uniformly smallest risk for the loss function $\ell(t; \theta) = (t - \theta)^2/\theta^2$. ////

Definition 28 Pitman estimator for scale The estimator given in Eq. (18) is defined to be the *Pitman estimator for scale*. ////

Remark The Pitman estimator for scale is a function of sufficient statistics. ////

EXAMPLE 41 Let $X_1, \ldots, X_n$ be a random sample from a density $f(x;\theta) = (1/\theta)I_{(0,\theta)}(x)$. The Pitman estimator for the scale parameter θ is

$$\frac{\int_0^\infty (1/\theta^2)\prod_{i=1}^n (1/\theta)I_{(0,\theta)}(X_i)\,d\theta}{\int_0^\infty (1/\theta^3)\prod_{i=1}^n (1/\theta)I_{(0,\theta)}(X_i)\,d\theta} = \frac{\int_{Y_n}^\infty \theta^{-n-2}\,d\theta}{\int_{Y_n}^\infty \theta^{-n-3}\,d\theta}$$

$$= \frac{\{1/[(n+2)-1]\}Y_n^{-(n+2)+1}}{\{1/[(n+3)-1]\}Y_n^{-(n+3)+1}} = \frac{n+2}{n+1}Y_n.$$

We know that Y_n is a complete sufficient statistic and $\mathscr{E}[Y_n] = [n/(n+1)]\theta$; so by the Lehmann-Scheffé theorem $[(n+1)/n]Y_n$ is the UMVUE of θ. ////

EXAMPLE 42 Let $X_1, \ldots, X_n$ be a random sample from the density $f(x;\lambda) = (1/\lambda)\exp(-x/\lambda)I_{(0,\infty)}(x)$. The Pitman estimator for the scale parameter λ is

$$\frac{\int_0^\infty (1/\lambda^2)\prod_{i=1}^n f(X_i;\lambda)\,d\lambda}{\int_0^\infty (1/\lambda^3)\prod_{i=1}^n f(X_i;\lambda)\,d\lambda} = \frac{\int_0^\infty (1/\lambda^{n+2})\exp(-\sum X_i/\lambda)\,d\lambda}{\int_0^\infty (1/\lambda^{n+3})\exp(-\sum X_i/\lambda)\,d\lambda}$$

$$= \frac{\int_0^\infty (\alpha/\sum X_i)^{n+2}e^{-\alpha}(\sum X_i/\alpha^2)\,d\alpha}{\int_0^\infty (\alpha/\sum X_i)^{n+3}e^{-\alpha}(\sum X_i/\alpha^2)\,d\alpha}$$

$$= \sum X_i \frac{\int_0^\infty \alpha^n e^{-\alpha}\,d\alpha}{\int_0^\infty \alpha^{n+1}e^{-\alpha}\,d\alpha}$$

$$= \sum X_i \frac{\Gamma(n+1)}{\Gamma(n+2)} = \frac{\sum X_i}{n+1}.$$

(It can be shown that the UMVUE of λ is $\sum X_i/n$.)

Note that $\sum X_i/n$ is a scale-invariant estimator, and, hence, since $\sum X_i/(n+1)$ is the scale-invariant estimator having uniformly smallest risk for the loss function $(t-\theta)^2/\theta^2$, the risk of $\sum X_i/(n+1)$ is uniformly smaller than the risk of $\sum X_i/n$. Also, since here risk equals $1/\theta^2$ times the MSE, the MSE of $\sum X_i/(n+1)$ is uniformly smaller than the MSE of $\sum X_i/n$. ////

7 BAYES ESTIMATORS

In our considerations of point-estimation problems in the previous sections of this chapter, we have assumed that our random sample came from some density $f(\cdot\,;\theta)$, where the function $f(\cdot\,;\,\cdot)$ was assumed known. Moreover, we have assumed that θ was some *fixed*, though unknown to us, point. In some real-world situations which the density $f(\cdot\,;\theta)$ represents, there is often additional information about θ (the only assumption which we heretofore have made about θ is that it can take on values in $\underline{\Theta}$). For example, the experimenter may have evidence that θ itself acts as a random variable for which he may be able to postulate a realistic density function. For instance, suppose that a machine which stamps out parts for automobiles is to be examined to see what fraction θ of defectives is being made. On a certain day, 10 pieces of the machine's output are examined, with the observations denoted by $X_1, X_2, \ldots, X_{10}$, where $X_i = 1$ if the ith piece is defective and $X_i = 0$ if it is nondefective. These can be viewed as a random sample of size 10 from the Bernoulli density

$$f(x;\theta) = \theta^x(1-\theta)^{1-x}I_{\{0,1\}}(x) \qquad \text{for } 0 \le \theta \le 1,$$

which indicates that the probability that a given part is defective is equal to the unknown number θ. The joint density of the 10 random variables $X_1, X_2, \ldots, X_{10}$ is

$$\theta^{\Sigma x_i}(1-\theta)^{10-\Sigma x_i}\prod_{i=1}^{10} I_{\{0,1\}}(x_i) \qquad \text{for } 0 \le \theta \le 1.$$

The maximum-likelihood estimator of θ, as explained in previous sections, is $\hat{\Theta} = \bar{X}$. The method of moments gives the same estimator. Suppose, however, that the experimenter has some additional information about θ; suppose that he has observed that on various days the value of θ changes and it appears that the change can be represented as a random variable with the density

$$g_\Theta(\theta) = 6\theta(1-\theta)I_{[0,1]}(\theta).$$

An important question is: How can this additional information about θ be used to estimate θ_0, where θ_0 is the value that Θ was equal to on the day the sample was drawn?

To examine this problem, we will assume, in addition to the assumption that our random sample came from a density $f(\cdot\,;\theta)$, that the unknown parameter θ is the value of some random variable, say Θ. We will still be interested in estimating some function of θ, say $\tau(\theta)$. If Θ is a random variable, it has a distribution. We let $G(\cdot) = G_\Theta(\cdot)$ denote the cumulative distribution function of Θ and $g(\cdot) = g_\Theta(\cdot)$ denote the density function of Θ, and we assume these functions contain no unknown parameters. In order to emphasize that the distribution of Θ is over the parameter space, we have departed from our custom of using $F(\cdot)$ and $f(\cdot)$ to represent a cumulative distribution function and density function, respectively, and have used $G(\cdot)$ and $g(\cdot)$ instead.

If we assume that the distribution of Θ is known, we have additional information. So an important question is: How can this additional information be used in estimation? It is this question that we will address ourselves to in the following two subsections. In many problems it may be unrealistic to assume that θ is the value of a random variable; in other problems, even though it seems reasonable to assume that θ is the value of a random variable Θ the distribution of Θ may not be known, or even if it is known, it may contain other unknown parameters. However, in some problems the assumption that the distribution of Θ is known is realistic, and we shall examine this situation.

7.1 Posterior Distribution

Heretofore we have used the notation $f(x; \theta)$ to indicate the density of a random variable X for each θ in $\underline{\overline{\Theta}}$. Whenever we want to indicate that the parameter θ is the value of a random variable Θ, we shall write the density of X as $f(x \mid \theta)$ instead of $f(x; \theta)$. We should note that $f(x \mid \theta)$ is a conditional density; it is the density of X given $\Theta = \theta$. A more complete notation for $f(x \mid \theta)$ would be $f_{X \mid \Theta = \theta}(x \mid \theta)$.

Let $X_1, \ldots, X_n$ be a random sample of size n from the density $f(\cdot \mid \theta)$, where θ is the value of a random variable Θ. Assume that the density of Θ, $g_\Theta(\cdot)$, is known and contains no unknown parameters, and suppose that we want to estimate $\tau(\theta)$. How do we incorporate the additional information of known $g_\Theta(\cdot)$ into our estimation procedures? In the past, we thought of the likelihood function as a single expression that contained all our information; the likelihood function included the observed sample $x_1, \ldots, x_n$ as well as the form of the density $f(x; \theta)$ we sampled from in its expression. Now we need an expression that contains all the information that the likelihood function contains plus the added information of the known density $g_\Theta(\cdot)$. $g_\Theta(\cdot)$ is called the *prior distribution* of Θ. It summarizes what we know about θ *prior* to taking a random sample. What we seek is an expression that summarizes what we know about θ *after* we take a random sample. We seek the *posterior* distribution of Θ given $X_1 = x_1, \ldots, X_n = x_n$.

Definition 29 Prior and posterior distributions The density $g_\Theta(\cdot)$ is called the *prior distribution* of Θ. The conditional density of Θ given $X_1 = x_1, \ldots, X_n = x_n$, denoted by $f_{\Theta \mid X_1 = x_1, \ldots, X_n = x_n}(\theta \mid x_1, \ldots, x_n)$, is called the *posterior* distribution of Θ. ////

Remark

$$f_{\Theta|X_1=x_1, \ldots, X_n=x_n}(\theta | x_1, \ldots, x_n) = \frac{f_{X_1, \ldots, X_n|\Theta=\theta}(x_1, \ldots, x_n | \theta) g_\Theta(\theta)}{f_{X_1, \ldots, X_n}(x_1, \ldots, x_n)}$$

$$= \frac{\left[\prod_{i=1}^{n} f(x_i | \theta)\right] g_\Theta(\theta)}{\int \left[\prod_{i=1}^{n} f(x_i | \theta)\right] g_\Theta(\theta)\, d\theta} \tag{19}$$

for random sampling. [Recall that $f_{Y|X=x}(y|x) = f_{X,Y}(x, y)/f_X(x) = f_{X|Y=y}(x|y)f_Y(y)/f_X(x)$.] ////

The posterior distribution replaces the likelihood function as an expression that incorporates all information. If we want to estimate θ and parallel the development of the maximum-likelihood estimator of θ, we could take as our estimator of θ that θ which maximizes the posterior distribution, that is, estimate θ with the *mode* of the posterior distribution. However, unlike the likelihood function (as a function of θ), the posterior distribution is a distribution function; so we could just as well estimate θ with the *median* or *mean* of the posterior distribution. We will use the mean of the posterior distribution as our estimate of θ, and in general we could estimate $\tau(\theta)$ as the mean of $\tau(\Theta)$ given $X_1 = x_1, \ldots, X_n = x_n$; that is, take $\mathscr{E}[\tau(\Theta) | X_1 = x_1, \ldots, X_n = x_n]$ as our estimate of $\tau(\theta)$.

Definition 30 Posterior Bayes estimator Let $X_1, \ldots, X_n$ be a random sample from a density $f(x|\theta)$, where θ is a value of the random variable Θ with known density $g_\Theta(\cdot)$. The *posterior Bayes estimator* of $\tau(\theta)$ *with respect to the prior* $g_\Theta(\cdot)$ is defined to be

$$\mathscr{E}[\tau(\Theta) | X_1, \ldots, X_n]. \tag{20}$$

////

Remark

$$\mathscr{E}[\tau(\Theta) | X_1 = x_1, \ldots, X_n = x_n] = \int \tau(\theta) f_{\Theta|X_1=x_1, \ldots, X_n=x_n}(\theta | x_1, \ldots, x_n)\, d\theta$$

$$= \frac{\int \tau(\theta) \left[\prod_{i=1}^{n} f(x_i | \theta)\right] g_\Theta(\theta)\, d\theta}{\int \left[\prod_{i=1}^{n} f(x_i | \theta)\right] g_\Theta(\theta)\, d\theta}. \tag{21}$$

////

One might note the similarity between the posterior Bayes estimator of $\tau(\theta) = \theta$ and the Pitman estimator of a location parameter [see Eq. (17)].

EXAMPLE 43 Let $X_1, \ldots, X_n$ denote a random sample from the Bernoulli density $f(x|\theta) = \theta^x(1-\theta)^{1-x}$ for $x = 0, 1$. Assume that the prior distribution of Θ is given by $g_\Theta(\theta) = I_{(0,1)}(\theta)$; that is, Θ is uniformly distributed over the interval (0, 1). Consider estimating θ and $\tau(\theta) = \theta(1-\theta)$. Now

$$f_{\Theta|X_1=x_1,\ldots,X_n=x_n}(\theta|x_1, \ldots, x_n) = \frac{\theta^{\Sigma x_i}(1-\theta)^{n-\Sigma x_i}I_{(0,1)}(\theta)}{\int_0^1 \theta^{\Sigma x_i}(1-\theta)^{n-\Sigma x_i}\,d\theta};$$

so the posterior Bayes estimator of θ with respect to the uniform prior distribution is given by

$$\begin{aligned}
&\mathscr{E}[\Theta|X_1 = x_1, \ldots, X_n = x_n] \\
&= \int \theta f_{\Theta|X_1=x_1,\ldots,X_n=x_n}(\theta|x_1, \ldots, x_n)\,d\theta \\
&= \frac{\int_0^1 \theta\theta^{\Sigma x_i}(1-\theta)^{n-\Sigma x_i}\,d\theta}{\int_0^1 \theta^{\Sigma x_i}(1-\theta)^{n-\Sigma x_i}\,d\theta} = \frac{B(\sum x_i + 2, n - \sum x_i + 1)}{B(\sum x_i + 1, n - \sum x_i + 1)} \\
&= \frac{\Gamma(\sum x_i + 2)\Gamma(n - \sum x_i + 1)}{\Gamma(n+3)} \cdot \frac{\Gamma(n+2)}{\Gamma(\sum x_i + 1)\Gamma(n - \sum x_i + 1)} \\
&= \frac{\sum x_i + 1}{n+2}.
\end{aligned}$$

Hence the posterior Bayes estimator of θ with respect to the uniform prior distribution is given by $(\sum X_i + 1)/(n+2)$. Contrast this to the maximum-likelihood estimator of θ, which is $\sum X_i/n$. $\sum X_i/n$ is unbiased and an UMVUE, whereas the posterior Bayes estimator is not unbiased.

To obtain the posterior Bayes estimator of, say $\tau(\theta) = \theta(1-\theta)$, we calculate

$$\begin{aligned}
&\mathscr{E}[\tau(\Theta)|X_1 = x_1, \ldots, X_n = x_n] \\
&= \int \theta(1-\theta) f_{\Theta|X_1=x_1,\ldots,X_n=x_n}(\theta|x_1, \ldots, x_n)\,d\theta \\
&= \frac{\int_0^1 \theta(1-\theta)\theta^{\Sigma x_i}(1-\theta)^{n-\Sigma x_i}\,d\theta}{\int_0^1 \theta^{\Sigma x_i}(1-\theta)^{n-\Sigma x_i}\,d\theta} \\
&= \frac{\Gamma(\sum x_i + 2)\Gamma(n - \sum x_i + 2)}{\Gamma(n+4)} \cdot \frac{\Gamma(n+2)}{\Gamma(\sum x_i + 1)\Gamma(n - \sum x_i + 1)} \\
&= \frac{(\sum x_i + 1)(n - \sum x_i + 1)}{(n+3)(n+2)}.
\end{aligned}$$

So the posterior Bayes estimator of $\theta(1-\theta)$ with respect to a uniform prior distribution is $(\sum X_i + 1)(n - \sum X_i + 1)/(n+3)(n+2)$. ////

We noted in the above example that the posterior Bayes estimator that we obtained was not unbiased. The following remark states that in general a posterior Bayes estimator is not unbiased.

Remark Let $T_G^* = \ell_G^*(X_1, \ldots, X_n)$ denote the posterior Bayes estimator of $\tau(\theta)$ with respect to a prior distribution $G(\cdot)$. If both T_G^* and $\tau(\Theta)$ have finite variance, then either $\text{var}\,[T_G^* | \theta] = 0$, or T_G^* is not an unbiased estimator of $\tau(\theta)$. That is, either T_G^* estimates $\tau(\theta)$ correctly with probability 1, or T_G^* is not an unbiased estimator of $\tau(\theta)$.

PROOF Let us suppose that T_G^* is an unbiased estimator of $\tau(\theta)$; that is, $\mathscr{E}[T_G^* | \theta] = \tau(\theta)$. By definition we have $T_G^* = \ell_G^*(X_1, \ldots, X_n) = \mathscr{E}[\tau(\Theta) | X_1, \ldots, X_n]$. Now

$$\begin{aligned} \text{var}\,[T_G^*] &= \mathscr{E}[\text{var}\,[T_G^* | \Theta]] + \text{var}\,[\mathscr{E}[T_G^* | \Theta]] \\ &= \mathscr{E}[\text{var}\,[T_G^* | \Theta]] + \text{var}\,[\tau(\Theta)], \end{aligned}$$

and

$$\begin{aligned} \text{var}\,[\tau(\Theta)] &= \mathscr{E}[\text{var}\,[\tau(\Theta) | X_1, \ldots, X_n]] + \text{var}\,[\mathscr{E}[\tau(\Theta) | X_1, \ldots, X_n]] \\ &= \mathscr{E}[\text{var}\,[\tau(\Theta) | X_1, \ldots, X_n]] + \text{var}\,[T_G^*]; \end{aligned}$$

hence, $\mathscr{E}[\text{var}\,[T_G^* | \Theta]] + \mathscr{E}[\text{var}\,[\tau(\Theta) | X_1, \ldots, X_n]] = 0$. Since both $\mathscr{E}[\text{var}\,[T_G^* | \Theta]]$ and $\mathscr{E}[\text{var}\,[\tau(\Theta) | X_1, \ldots, X_n]]$ are nonnegative and their sum is 0, both are 0. In particular, $\mathscr{E}[\text{var}\,[T_G^* | \Theta]] = 0$, and since $\text{var}\,[T_G^* | \Theta]$ is non-negative and has zero expectation, $\text{var}\,[T_G^* | \theta] = 0$. ////

7.2 Loss-function Approach

In Subsec. 3.4 we introduced the concepts of *loss* and *risk*. These two concepts were used to assess goodness of estimators. In this section we discuss how the additional information of knowledge of a prior distribution of Θ can be used in conjunction with loss and risk to define or select an optimum estimator.

We commence with a review of the problem we hope to solve. Let $X_1, \ldots, X_n$ be a random sample from a density $f(x | \theta)$, θ belonging to $\underline{\Theta}$, where the function $f(\cdot | \theta)$ is assumed known except for θ. We assume that the unknown θ is the value of some random variable Θ and that the distribution of Θ is known and contains no unknown parameters. On the basis of the random sample $X_1, \ldots, X_n$ we hope to estimate $\tau(\theta)$, some function of θ. In addition, we assume that a loss function $\ell(t; \theta)$ has been specified, where $\ell(t; \theta)$ represents the loss incurred if we estimate $\tau(\theta)$ to be t when θ is the parameter of the density from which we sampled. For any estimator $T = \ell(X_1, \ldots, X_n)$, we noted in Subsec. 3.4 that $\mathscr{E}_\theta[\ell(T; \theta)]$ represented the average loss of that estimator, and we defined this average loss to be the risk, denoted by $\mathscr{R}_t(\theta)$, of the estimator $t(\cdot, \ldots, \cdot)$. We further noted that two estimators, say $T_1 = t_1(X_1, \ldots, X_n)$ and $T_2 = t_2(X_1, \ldots, X_n)$, could be compared by looking at their respective risks $\mathscr{R}_{t_1}(\theta)$ and $\mathscr{R}_{t_2}(\theta)$, preference being given to that estimator with smaller risk. In general, the risk functions as functions of θ of two estimators may cross, one risk function being smaller for some θ and the other smaller for other θ. Then, since θ is unknown, it is difficult to make a choice between the two estimators. The difficulty is caused by the dependence of the risk function on θ. Now, since we have assumed that θ is the value of some random variable Θ, the distribution of which is also assumed known, we have a natural way of removing the dependence of the risk function on θ, namely, by averaging out the θ, using the density of Θ as our weight function.

Definition 31 Bayes risk Let $X_1, \ldots, X_n$ be a random sample from a density $f(x \mid \theta)$, where θ is the value of a random variable Θ with cumulative distribution function $G(\cdot) = G_\Theta(\cdot)$ and corresponding density $g(\cdot) = g_\Theta(\cdot)$. In estimating $\tau(\theta)$, let $\ell(t; \theta)$ be the loss function. The risk of estimator $T = t(X_1, \ldots, X_n)$ is denoted by $\mathscr{R}_t(\theta)$. The *Bayes risk* of estimator $T = t(X_1, \ldots, X_n)$ with respect to the loss function $\ell(\cdot\,;\,\cdot)$ and prior cumulative distribution $G(\cdot)$, denoted by $r(t) = r_{\ell, G}(t)$, is defined to be

$$r(t) = r_{\ell, G}(t) = \int_{\bar{\Theta}} \mathscr{R}_t(\theta) g(\theta)\, d\theta. \tag{22}$$

////

The Bayes risk of an estimator is an average risk, the averaging being over the parameter space $\bar{\Theta}$ with respect to the prior density $g(\cdot)$. For given loss function $\ell(\cdot\,;\,\cdot)$ and prior density $g(\cdot)$ the Bayes risk of an estimator is a real number; so now two competing estimators can be readily compared by comparing their respective Bayes risks, still preferring that estimator with smaller Bayes risk. In fact, we can now define the "best" estimator of $\tau(\theta)$ to be that estimator with smallest Bayes risk.

Definition 32 Bayes estimator The *Bayes estimator* of $\tau(\theta)$, denoted by $T^*_{\ell, G} = t^*_{\ell, G}(X_1, \ldots, X_n)$, with respect to the loss function $\ell(\cdot\,;\,\cdot)$ and prior cumulative distribution $G(\cdot)$, is defined to be that estimator with smallest Bayes risk. Or the *Bayes estimator* of $\tau(\theta)$ is that estimator $t^*_{\ell, G}$ satisfying

$$r_{\ell, G}(t^*) = r_{\ell, G}(t^*_{\ell, G}) \le r_{\ell, G}(t)$$

for every other estimator $T = t(X_1, \ldots, X_n)$ of $\tau(\theta)$. ////

The posterior Bayes estimator of $\tau(\theta)$, defined in Definition 30, was defined without regard to a loss function, whereas the definition given above requires specification of a loss function.

The definition leaves the problem of actually finding the Bayes estimator, which may not be easy for an arbitrary loss function, unsolved. However, for squared-error loss, finding the Bayes estimator is relatively easy. We seek that estimator, say $t^*(\cdot, \ldots, \cdot)$, which minimizes the expression $\int_{\underline{\Theta}} \mathscr{R}_t(\theta)\, g(\theta)\, d\theta = \int_{\underline{\Theta}} \mathscr{E}_\theta[[t(X_1, \ldots, X_n) - \tau(\theta)]^2] g(\theta)\, d\theta$ as a function over possible estimators $t(\cdot, \ldots, \cdot)$. Now,

$$\int_{\underline{\Theta}} \mathscr{E}_\theta[[t(X_1, \ldots, X_n) - \tau(\theta)]^2] g(\theta)\, d\theta$$

$$= \int_{\underline{\Theta}} \left\{ \int_{\mathfrak{X}} [t(x_1, \ldots, x_n) - \tau(\theta)]^2 f_{X_1, \ldots, X_n}(x_1, \ldots, x_n \mid \theta) \prod_{i=1}^{n} dx_i \right\} g(\theta)\, d\theta$$

$$= \int_{\mathfrak{X}} \left\{ \int_{\underline{\Theta}} [\tau(\theta) - t(x_1, \ldots, x_n)]^2 \frac{f_{X_1, \ldots, X_n}(x_1, \ldots, x_n \mid \theta) g(\theta)\, d\theta}{f_{X_1, \ldots, X_n}(x_1, \ldots, x_n)} \right\}$$

$$\cdot f_{X_1, \ldots, X_n}(x_1, \ldots, x_n) \prod_{i=1}^{n} dx_i$$

$$= \int_{\mathfrak{X}} \left\{ \int_{\underline{\Theta}} [\tau(\theta) - t(x_1, \ldots, x_n)]^2 f_{\Theta \mid X_1 = x_1, \ldots, X_n = x_n}(\theta \mid x_1, \ldots, x_n)\, d\theta \right\}$$

$$\cdot f_{X_1, \ldots, X_n}(x_1, \ldots, x_n) \prod_{i=1}^{n} dx_i,$$

and since the integrand is nonnegative, the double integral can be minimized if the expression within the braces is minimized for each $x_1, \ldots, x_n$. But the expression within the braces is the conditional expectation of $[\tau(\Theta) - t(x_1, \ldots, x_n)]^2$ with respect to the posterior distribution of Θ given $X_1 = x_1, \ldots, X_n = x_n$, which is minimized as a function of $t(x_1, \ldots, x_n)$ for $t^*(x_1, \ldots, x_n)$ equal to the conditional expectation of $\tau(\Theta)$ with respect to the posterior distribution of Θ given $X_1 = x_1, \ldots, X_n = x_n$. {Recall that $\mathscr{E}[(Z - a)^2]$ is minimized as a function of a for $a^* = \mathscr{E}[Z]$.} Hence the Bayes estimator of $\tau(\theta)$ with respect to the squared-error loss function is given by

$$\mathscr{E}[\tau(\Theta) \mid X_1 = x_1, \ldots, X_n = x_n] = \frac{\int \tau(\theta) \left[\prod_{i=1}^{n} f(x_i \mid \theta) \right] g(\theta)\, d\theta}{\int \left[\prod_{i=1}^{n} f(x_i \mid \theta) \right] g(\theta)\, d\theta}, \tag{23}$$

which is identical to the estimator given in Eq. (21).

For a general loss function, we seek that estimator which minimizes $\int_{\underline{\Theta}} \mathscr{R}_t(\theta) g(\theta)\, d\theta$. Again,

$$\int_{\underline{\Theta}} \mathscr{R}_t(\theta) g(\theta)\, d\theta$$

$$= \int_{\underline{\Theta}} \left[\int_{\mathfrak{X}} \ell(t(x_1, \ldots, x_n); \theta) f_{X_1, \ldots, X_n}(x_1, \ldots, x_n \mid \theta) \prod_{i=1}^{n} dx_i \right] g(\theta)\, d\theta$$

$$= \int_{\mathfrak{X}} \left[\int_{\underline{\Theta}} \ell(t(x_1, \ldots, x_n); \theta) f_{\Theta \mid X_1 = x_1, \ldots, X_n = x_n}(\theta \mid x_1, \ldots, x_n)\, d\theta \right]$$

$$\cdot f_{X_1, \ldots, X_n}(x_1, \ldots, x_n) \prod_{i=1}^{n} dx_i,$$

and minimizing the double integral is equivalent to minimizing the expression within the brackets, which is sometimes called the *posterior risk*. So, in general, the Bayes estimator of $\tau(\theta)$ with respect to the loss function $\ell(\cdot\,;\,\cdot)$ and prior density $g(\cdot)$ is that estimator which minimizes the posterior risk, which is the expected loss with respect to the posterior distribution of Θ given the observations $x_1, \ldots, x_n$. We have the following theorem and corollaries.

Theorem 13 Let $X_1, \ldots, X_n$ be a random sample from the density $f(x \mid \theta)$, and let $g(\theta)$ be the density of Θ. Further let $\ell(t; \theta)$ be the loss function for estimating $\tau(\theta)$. The Bayes estimator of $\tau(\theta)$ is that estimator $t^*(\cdot, \ldots, \cdot)$ which minimizes

$$\int_{\underline{\Theta}} \ell(t(x_1, \ldots, x_n); \theta) f_{\Theta \mid X_1 = x_1, \ldots, X_n = x_n}(\theta \mid x_1, \ldots, x_n)\, d\theta \tag{24}$$

as a function of $t(\cdot, \ldots, \cdot)$. ////

Corollary Under the assumptions of Theorem 13, the Bayes estimator of $\tau(\theta)$ is given by

$$\mathscr{E}[\tau(\Theta) \mid X_1 = x_1, \ldots, X_n = x_n] = \frac{\int \tau(\theta) \left[\prod_{i=1}^{n} f(x_i \mid \theta) \right] g(\theta)\, d\theta}{\int \left[\prod_{i=1}^{n} f(x_i \mid \theta) \right] g(\theta)\, d\theta} \tag{25}$$

for a squared-error loss function. ////

Corollary Under the assumptions of Theorem 13, the Bayes estimator of θ is given by the median of the posterior distribution of Θ for a loss function equal to the absolute deviation. ////

The proofs of the theorem and first corollary preceded the statement of the theorem. The second corollary follows from the observation that

$$\int_{\bar{\Theta}} |\theta - \ell(x_1, \ldots, x_n)| f_{\Theta|X_1=x_1, \ldots, X_n=x_n}(\theta | x_1, \ldots, x_n)\, d\theta$$

is minimized as a function of $\ell(\cdot, \ldots, \cdot)$ for $\ell^*(\cdot, \ldots, \cdot)$ equal to the median of the posterior distribution of Θ. {Recall that $\mathscr{E}[|Z - a|]$ is minimized as a function of a for $a^* =$ median of Z.}

EXAMPLE 44 Let $X_1, \ldots, X_n$ be a random sample from the normal density with mean θ and variance 1. Consider estimating θ with a squared-error loss function. Assume that Θ has a normal density with mean μ_0 and variance 1. Write $\mu_0 = x_0$ when convenient. According to Eq. (25) the Bayes estimator is given as the mean of the posterior distribution of Θ.

$$f_{\Theta|X_1=x_1, \ldots, X_n=x_n}(\theta | x_1, \ldots, x_n)$$

$$= \frac{f_{X_1, \ldots, X_n|\Theta}(x_1, \ldots, x_n | \theta) g(\theta)}{f_{X_1, \ldots, X_n}(x_1, \ldots, x_n)} = \frac{\left[\prod_{i=1}^{n} f(x_i | \theta)\right] g(\theta)}{\int \left[\prod_{i=1}^{n} f(x_i | \theta)\right] g(\theta)\, d\theta}$$

$$= \frac{(1/\sqrt{2\pi})^n \exp\left[-\frac{1}{2}\sum_1^n (x_i - \theta)^2\right](1/\sqrt{2\pi}) \exp\left[-\frac{1}{2}(\theta - \mu_0)^2\right]}{\int_{-\infty}^{\infty} (1/\sqrt{2\pi})^n \exp\left[-\frac{1}{2}\sum_1^n (x_i - \theta)^2\right](1/\sqrt{2\pi}) \exp\left[-\frac{1}{2}(\theta - \mu_0)^2\right] d\theta}$$

$$= \frac{\exp\left[-\frac{1}{2}\sum_{i=0}^{n} (x_i - \theta)^2\right]}{\int_{-\infty}^{\infty} \exp\left[-\frac{1}{2}\sum_{i=0}^{n} (x_i - \theta)^2\right] d\theta}$$

$$= \frac{\exp\left\{-\frac{1}{2}\left[(n+1)\theta^2 - 2\theta \sum_{i=0}^{n} x_i + \sum_{i=0}^{n} x_i^2\right]\right\}}{\int_{-\infty}^{\infty} \exp\left\{-\frac{1}{2}\left[(n+1)\theta^2 - 2\theta \sum_{i=0}^{n} x_i + \sum_{i=0}^{n} x_i^2\right]\right\} d\theta}$$

$$= \frac{\exp\left(-[(n+1)/2]\left\{\theta^2 - 2\theta \sum_0^n x_i/(n+1) + \left[\sum_0^n x_i/(n+1)\right]^2\right\}\right)}{\int_{-\infty}^{\infty} \exp\left(-[(n+1)/2]\left\{\theta^2 - 2\theta \sum_0^n x_i/(n+1) + \left[\sum_0^n x_i/(n+1)\right]^2\right\}\right) d\theta}$$

$$= \frac{[1/\sqrt{2\pi/(n+1)}] \exp\left\{-[(n+1)/2]\left[\theta - \sum_0^n x_i/(n+1)\right]^2\right\}}{\int_{-\infty}^{\infty} [1/\sqrt{2\pi/(n+1)}] \exp\left\{-[(n+1)/2]\left[\theta - \sum x_i/(n+1)\right]^2\right\} d\theta}$$

$$= \frac{1}{\sqrt{2\pi/(n+1)}} \exp\left\{-\frac{n+1}{2}\left[\theta - \sum_0^n x_i/(n+1)\right]^2\right\};$$

the denominator is unity since it is the integral of a density. We have shown that the posterior distribution of Θ is normal with mean $\sum_0^n x_i/(n+1)$ and variance $1/(n+1)$; hence the Bayes estimator of θ with respect to squared-error loss is

$$\frac{x_0 + \sum_1^n X_i}{n+1} = \frac{\mu_0 + \sum_1^n X_i}{n+1}.$$

Since the posterior distribution of Θ is normal, its mean and median are the same; hence

$$\frac{\mu_0 + \sum_1^n X_i}{n+1}$$

is also the Bayes estimator with respect to a loss function equal to the absolute deviation. ////

EXAMPLE 45 Let $X_1, \ldots, X_n$ be a random sample from the density $f(x \mid \theta) = (1/\theta)I_{(0,\theta)}(x)$. Estimate θ with the loss function $\ell(t; \theta) = (t-\theta)^2/\theta^2$. Assume that Θ has a density given by $g(\theta) = I_{(0,1)}(\theta)$. Let y_n denote $\max[x_1, \ldots, x_n]$. Find the posterior distribution of Θ.

$$\begin{aligned} f_{\Theta \mid X_1 = x_1, \ldots, X_n = x_n}(\theta \mid x_1, \ldots, x_n) &= \frac{(1/\theta)^n \prod_{i=1}^n I_{(0,\theta)}(x_i) I_{(0,1)}(\theta)}{\int_0^1 (1/\theta)^n \prod_{i=1}^n I_{(0,\theta)}(x_i)\, d\theta} \\ &= \frac{(1/\theta)^n I_{(y_n, 1)}(\theta)}{\int_{y_n}^1 (1/\theta)^n I_{(y_n,1)}(\theta)\, d\theta} \\ &= \frac{(1/\theta)^n I_{(y_n,1)}(\theta)}{[1/(n-1)](1/y_n^{n-1} - 1)}. \end{aligned}$$

We seek that estimator which minimizes Eq. (24), or we seek that estimator $t(\cdot)$ which minimizes

$$\frac{\int \{[t(y_n) - \theta]^2/\theta^2\}(1/\theta^n) I_{(y_n,1)}(\theta)\, d\theta}{[1/(n-1)](1/y_n^{n-1} - 1)},$$

or that estimator which minimizes

$$\int_{y_n}^{1} [\ell(y_n) - \theta]^2 \frac{1}{\theta^{n+2}}\, d\theta$$

$$= [\ell(y_n)]^2 \int_{y_n}^{1} \frac{1}{\theta^{n+2}}\, d\theta - 2\ell(y_n) \int_{y_n}^{1} \frac{1}{\theta^{n+1}}\, d\theta + \int_{y_n}^{1} \frac{1}{\theta^n}\, d\theta. \tag{26}$$

Equation (26) is a quadratic equation in $\ell(\cdot)$; this quadratic equation assumes its minimum for

$$\ell^*(y_n) = \frac{\int_{y_n}^{1} (1/\theta^{n+1})\, d\theta}{\int_{y_n}^{1} (1/\theta^{n+2})\, d\theta} = \frac{[1/(-n)](1 - 1/y_n^n)}{[1/(-n-1)](1 - 1/y_n^{n+1})} = \frac{n+1}{n} y_n \frac{1 - y_n^n}{1 - y_n^{n+1}}. \qquad ////$$

We note that the Bayes estimators derived in Examples 43 to 45 are functions of sufficient statistics. It can be shown that this is generally true; that is, a Bayes estimator is always a function of minimal sufficient statistics. In fact, under quite general conditions it can be shown that the Bayes estimator corresponding to an arbitrary prior probability density function, which is positive for all θ belonging to $\bar{\Theta}$, is consistent and BAN. So, even if you do not know the correct prior distribution, a Bayes estimator has some desirable optimum properties. And if you do know the correct prior distribution and accept the criterion that a best estimator is one that minimizes average loss, then the Bayes estimator corresponding to the known prior distribution is optimum.

Even in those problems when the prior distribution is unknown, the concept of Bayes estimation can benefit us. It provides us with a technique of determining many estimators that we might not have otherwise considered. Each possible prior distribution has a corresponding estimator, whose merits can be judged by using our standard methods of comparison. Thus, we have yet another method of finding estimators to append to the methods given in Sec. 2.

Bayes estimation can also sometimes be useful as a tool in obtaining an estimator possessing some desirable property that does not depend on prior distribution information. The property of minimax is such a property, and in the next subsection we will see how Bayes estimation can sometimes be used to find a minimax estimator. Another such property is given below. Our objective has been to minimize risk, but since risk depended on the parameter, we were unable to find one estimator that had smaller risk than all others for all parameter values. Minimax circumvented such difficulty by replacing the risk function by its maximum value and then seeking that estimator which minimized such maximum value. Another way of getting around the difficulty arising from attempting to uniformly minimize risk is to replace the risk function by the area under the risk function and to seek that estimator which has the least area under its risk function. We note that if the parameter space $\bar{\Theta}$ is an interval, the estimator having the least area under its risk function is the Bayes estimator corresponding to a uniform prior distribution over the interval $\bar{\Theta}$. This is true because for a uniform prior distribution the Bayes risk is proportional to the area under the risk function, and hence minimizing the Bayes risk is equivalent to minimizing area.

7.3 Minimax Estimator

We defined a minimax estimator at the end of Subsec. 3.4 as an estimator whose maximum risk is less than or equal to the maximum risk of any other estimator. Such an estimator might be considered "conservative" since it protects against the worst that can happen; it seeks to minimize the maximum risk. The following theorem is sometimes useful in finding a minimax estimator.

Theorem 14 If $T^* = t^*(X_1, \ldots, X_n)$ is a Bayes estimator having constant risk, that is, $\mathcal{R}_{t^*}(\theta) \equiv$ constant, then T^* is a minimax estimator.

PROOF Let $g^*(\cdot)$ be the prior density corresponding to the Bayes estimator $t^*(\cdot, \ldots, \cdot)$.

$$\sup_{\theta \in \bar{\Theta}} \mathcal{R}_{t^*}(\theta) = \text{constant} = \mathcal{R}_{t^*}(\theta)$$

$$= \int_{\bar{\Theta}} \mathcal{R}_{t^*}(\theta) g^*(\theta)\, d\theta \le \int_{\bar{\Theta}} \mathcal{R}_{t}(\theta) g^*(\theta)\, d\theta \le \sup_{\theta \in \bar{\Theta}} \mathcal{R}_{t}(\theta)$$

for any other estimator $t(\cdot, \ldots, \cdot)$. ////

EXAMPLE 46 Find the minimax estimator of θ in sampling from the Bernoulli distribution using a squared-error loss function. We seek a Bayes estimator with constant risk. The family of beta distributions is a family of possible prior distributions. We hope that for one of the beta prior distributions the corresponding Bayes estimator will have constant risk. A Bayes estimator is given by

$$\frac{\int_0^1 \theta \theta^{\Sigma x_i}(1-\theta)^{n-\Sigma x_i}[1/B(a,b)]\theta^{a-1}(1-\theta)^{b-1}\, d\theta}{\int_0^1 \theta^{\Sigma x_i}(1-\theta)^{n-\Sigma x_i}[1/B(a,b)]\theta^{a-1}(1-\theta)^{b-1}\, d\theta}$$

$$= \frac{\int_0^1 \theta^{\Sigma x_i + a}(1-\theta)^{n-\Sigma x_i+b-1}\, d\theta}{\int_0^1 \theta^{\Sigma x_i + a-1}(1-\theta)^{n-\Sigma x_i+b-1}\, d\theta}$$

$$= \frac{B(\sum x_i + a + 1, n - \sum x_i + b)}{B(\sum x_i + a, n - \sum x_i + b)} = \frac{\sum x_i + a}{n + a + b}.$$

So the Bayes estimator with respect to a beta prior distribution having parameters a and b is given by

$$\frac{\sum X_i + a}{n + a + b}.$$

We now evaluate the risk of $(\sum X_i + a)/(n + a + b)$ with the hope that we will be able to select a and b so that the risk will be constant. Write $t^*_{A,B}(x_1, \ldots, x_n) = A \sum x_i + B = (\sum x_i + a)/(n + a + b)$; then $\mathscr{R}_{t^*_{A,B}}(\theta) = \mathscr{E}[(A \sum X_i + B - \theta)^2] = \mathscr{E}[[A(\sum X_i - n\theta) + B - \theta + nA\theta]^2] = A^2\mathscr{E}[(\sum X_i - n\theta)^2] + (B - \theta + nA\theta)^2 = nA^2\theta(1 - \theta) + (B - \theta + nA\theta)^2 = \theta^2[(nA - 1)^2 - nA^2] + \theta[nA^2 + 2(nA - 1)B] + B^2$, which is constant if $(nA - 1)^2 - nA^2 = 0$ and $nA^2 + 2(nA - 1)B = 0$. Now $(nA - 1)^2 - nA^2 = 0$ if $A = 1/\sqrt{n}(\sqrt{n} \pm 1)$, and $nA^2 + 2(nA - 1)B = 0$ if $B = -nA^2/2(nA - 1)$, which is $1/2(\sqrt{n} + 1)$ for $A = 1/\sqrt{n}(\sqrt{n} + 1)$. On solving for a and b, we obtain $a = b = \sqrt{n}/2$; so $(\sum X_i + \sqrt{n}/2)/(n + \sqrt{n})$ is a Bayes estimator with constant risk and, hence, minimax. ////

8 VECTOR OF PARAMETERS

In this section we present a brief introduction to the problem of simultaneous point estimation of several functions of a vector parameter. We will assume that a random sample $X_1, \ldots, X_n$ of size n from the density $f(x; \theta_1, \ldots, \theta_k)$ is available, where the parameter $\theta = (\theta_1, \ldots, \theta_k)$ and parameter space $\bar{\Theta}$ are k-dimensional. We want to simultaneously estimate $\tau_1(\theta), \ldots, \tau_r(\theta)$, where $\tau_j(\theta)$, $j = 1, \ldots, r$, is some function of $\theta = (\theta_1, \ldots, \theta_k)$. Often $k = r$, but this need not be the case. An important special case is the estimation of $\theta = (\theta_1, \ldots, \theta_k)$ itself; then $r = k$, and $\tau_1(\theta) = \theta_1, \ldots, \tau_k(\theta) = \theta_k$. Another important special case is the estimation of $\tau(\theta)$; then $r = 1$. A point estimator of $(\tau_1(\theta), \ldots, \tau_r(\theta))$ is a vector of statistics, say $(T_1, \ldots, T_r)$, where $T_j = t_j(X_1, \ldots, X_n)$ and T_j is an estimator of $\tau_j(\theta)$.

Our presentation of the method of moments and maximum-likelihood method as techniques for finding estimators included the possibility that the parameter be vector-valued. So we already have methods of determining estimators. What we need are some criteria for assessing the goodness of an estimator, say $(T_1, \ldots, T_r)$, and for comparing two estimators, say $(T_1, \ldots, T_r)$ and $(T_1', \ldots, T_r')$. As was the case in estimating a real-valued function $\tau(\theta)$, where we wanted the values of our estimator to be close to $\tau(\theta)$, we now want the values of the estimator $(T_1, \ldots, T_r)$ to be close to $(\tau_1(\theta), \ldots, \tau_r(\theta))$. We want the distribution of $(T_1, \ldots, T_r)$ to be concentrated around $(\tau_1(\theta), \ldots, \tau_r(\theta))$.

There are a number of ways of measuring the closeness of an estimator. For instance, in comparing two estimators the definitions of "more concentrated" and "closer," given in Subsec. 3.1, can be generalized to r dimensions. We will, however, restrict ourselves to consideration of unbiased estimators and define several ways of measuring the closeness of an unbiased estimator. No attempt will be made in this book to generalize to r dimensions the notions of loss and/or risk, invariance, Bayes estimation, and minimax. As far as optimum estimation is concerned, we will be content to consider only unbiased estimators and look for a best estimator within the restricted class of unbiased estimators.

Definition 33 Unbiased An estimator $(T_1, \ldots, T_r)$, where $T_j = t_j(X_1, \ldots, X_n)$, $j = 1, \ldots, r$, is defined to be an *unbiased* estimator of $(\tau_1(\theta), \ldots, \tau_r(\theta))$ if and only if $\mathscr{E}_\theta[T_j] = \tau_j(\theta)$ for $j = 1, \ldots, r$ and for all $\theta \in \underline{\overline{\Theta}}$. ////

In Sec. 5, where we considered unbiased estimation of a real-valued function $\tau(\theta)$, we employed the variance of an estimator as a measure of its closeness to $\tau(\theta)$. Here we seek a generalization of the notion of variance to r dimensions. Several such generalizations have been proposed; we will consider four of them, called (i) vector of variances, (ii) linear combination (with nonnegative coefficients) of variances, (iii) ellipsoid of concentration, and (iv) Wilks' generalized variance. The last two require some knowledge of matrices.

Possibly the simplest way of generalizing the concept of variance to r dimensions is to use the vector of variances of the unbiased estimators $T_1, \ldots, T_r$. That is, let the vector $(\text{var}_\theta [T_1], \ldots, \text{var}_\theta [T_r])$ be a measure of the closeness of the estimator $(T_1, \ldots, T_r)$ to $(\tau_1(\theta), \ldots, \tau_r(\theta))$. The disadvantage of such a definition is that our measure is vector-valued and consequently sometimes difficult to work with. One way of circumventing this disadvantage is to use a linear combination of variances, that is, measure the closeness of the estimator $(T_1, \ldots, T_r)$ to $(\tau_1(\theta), \ldots, \tau_r(\theta))$ with $\sum_{j=1}^{r} a_j \text{ var}_\theta [T_j]$ for suitably chosen $a_j \geq 0$. Both of these generalizations of variance embody only the variances of the T_j, $j = 1, \ldots, r$. The T_j are likely to be correlated; so one might justifiably think that our measure of closeness of $(T_1, \ldots, T_r)$ to $(\tau_1(\theta), \ldots, \tau_r(\theta))$ should incorporate the covariances of the T_j's.

Notation If $(T_1, \ldots, T_r)$ is an unbiased estimator of $(\tau_1(\theta), \ldots, \tau_r(\theta))$, let $\sigma_{ij}(\theta) = \text{cov}_\theta [T_i, T_j]$. The matrix whose ijth element is $\sigma_{ij}(\theta)$ is called the *covariance matrix* of the estimator $(T_1, \ldots, T_r)$. Let $\sigma^{ij}(\theta)$ denote the ijth element of the inverse of the covariance matrix. ////

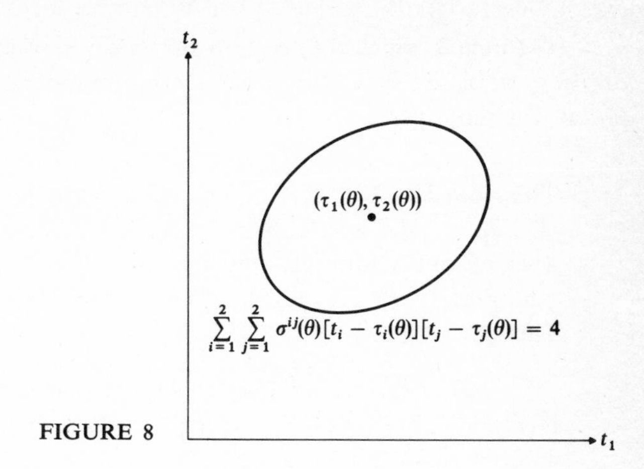

FIGURE 8

Definition 34 Ellipsoid of concentration Let $(T_1, \ldots, T_r)$ be an unbiased estimator of $(\tau_1(\theta), \ldots, \tau_r(\theta))$. Let $\sigma^{ij}(\theta)$ be the *ij*th element of the inverse of the covariance matrix of $(T_1, \ldots, T_r)$, where the *ij*th element of the covariance matrix is $\sigma_{ij}(\theta) = \text{cov}_\theta [T_i, T_j]$. The *ellipsoid of concentration* of $(T_1, \ldots, T_r)$ is defined as the interior and boundary of the ellipsoid

$$\sum_{i=1}^{r} \sum_{j=1}^{r} \sigma^{ij}(\theta)[t_i - \tau_i(\theta)][t_j - \tau_j(\theta)] = r + 2. \tag{27}$$

See Fig. 8 for $r = 2$. ////

Loosely speaking, the ellipsoid of concentration measures how concentrated the distribution of $(T_1, \ldots, T_r)$ is about $(\tau_1(\theta), \ldots, \tau_r(\theta))$. [In fact, if one considers the vector random variable, say $(U_1, \ldots, U_r)$, uniformly distributed over the ellipsoid of concentration, it can be proved that $(U_1, \ldots, U_r)$ and $(T_1, \ldots, T_r)$ have the same first- and second-order moments.] The distribution of an estimator $(T_1, \ldots, T_r)$ whose ellipsoid of concentration is contained within the ellipsoid of concentration of another estimator $(T_1', \ldots, T_r')$ is more highly concentrated about $(\tau_1(\theta), \ldots, \tau_r(\theta))$ than is the distribution of $(T_1', \ldots, T_r')$.

It is known that the determinant of the covariance matrix of an estimator is proportional to the square of the volume of the corresponding ellipsoid of concentration; hence another generalization of variance is as in Definition 35.

Definition 35 Wilks' generalized variance Let $(T_1, \ldots, T_r)$ be an unbiased estimator of $(\tau_1(\theta), \ldots, \tau_r(\theta))$. *Wilks' generalized variance* of $(T_1, \ldots, T_r)$ is defined to be the determinant of the covariance matrix of $(T_1, \ldots, T_r)$. ////

Theorem 8, which showed how sufficiency could be used to improve on an arbitrary unbiased estimator, generalizes to r dimensions. The generalization is stated without proof.

Theorem 15 Let $X_1, \ldots, X_n$ be a random sample from the density $f(x; \theta_1, \ldots, \theta_k)$, and let $S_1 = \mathscr{s}_1(X_1, \ldots, X_n), \ldots, S_m = \mathscr{s}_m(X_1, \ldots, X_n)$ be a set of jointly sufficient statistics. Let $(T_1, \ldots, T_r)$ be an unbiased estimator of $(\tau_1(\theta), \ldots, \tau_r(\theta))$. Define $T_j' = \mathscr{E}[T_j \mid S_1, \ldots, S_m], j = 1, \ldots, r$. Then,

(i) $(T_1', \ldots, T_r')$ is a statistic and an unbiased estimator of $(\tau_1(\theta), \ldots, \tau_r(\theta))$, and $T_j' = \mathscr{t}_j'(S_1, \ldots, S_m)$; that is, T_j' is a function of the sufficient statistics $S_1, \ldots, S_m, j = 1, \ldots, r$.

(ii) $\text{var}_\theta [T_j'] \leq \text{var}_\theta [T_j]$ for every $\theta \in \overline{\Theta}, j = 1, \ldots, r$.

(iii) The ellipsoid of concentration of $(T_1', \ldots, T_r')$ is contained in the ellipsoid of concentration of $(T_1, \ldots, T_r)$, for every $\theta \in \overline{\Theta}$. ////

We might note that (ii) implies that

$$\sum_{j=1}^{r} a_j \text{var}_\theta [T_j'] \leq \sum_{j=1}^{r} a_j \text{var}_\theta [T_j] \qquad \text{for } a_j \geq 0$$

and (iii) implies that Wilks' generalized variance of $(T_1', \ldots, T_r')$ is smaller than Wilks' generalized variance of $(T_1, \ldots, T_r)$.

Theorem 10 of Sec. 5 can also be generalized to r dimensions, but first the concept of completeness has to be generalized.

Definition 36 Joint completeness For $X_1, \ldots, X_n$, a random sample from the density $f(x; \theta_1, \ldots, \theta_k)$, let $(T_1, \ldots, T_m)$ be a set of statistics. $T_1, \ldots, T_m$ are defined to be *jointly complete* if and only if $\mathscr{E}_\theta[\mathscr{z}(T_1, \ldots, T_m)] \equiv 0$ for all $\theta \in \overline{\Theta}$ implies that $P_\theta[\mathscr{z}(T_1, \ldots, T_m) = 0] \equiv 1$ for all $\theta \in \overline{\Theta}$, where $\mathscr{z}(T_1, \ldots, T_m)$ is a statistic. ////

EXAMPLE 47 Let $X_1, \ldots, X_n$ be a random sample from

$$f(x; \theta_1, \theta_2) = \frac{1}{\theta_2 - \theta_1} I_{(\theta_1, \theta_2)}(x),$$

where $\theta_1 < \theta_2$. Write $\theta = (\theta_1, \theta_2)$. Let $Y_1 = \min [X_1, \ldots, X_n]$ and $Y_n = \max [X_1, \ldots, X_n]$. We want to show that Y_1 and Y_n are jointly complete. (We know that they are jointly sufficient.) Let $\not{z}(Y_1, Y_n)$ be an unbiased estimator of 0, that is,

$$\mathscr{E}_\theta[\not{z}(Y_1, Y_n)] \equiv 0 \qquad \text{for all } \theta \in \underline{\Theta}.$$

Now

$$\mathscr{E}_\theta[\not{z}(Y_1, Y_n)] = \iint \not{z}(y_1, y_n) f_{Y_1, Y_n}(y_1, y_n)\, dy_1\, dy_n$$

$$= \int_{\theta_1}^{\theta_2} \left[\int_{\theta_1}^{y_n} \not{z}(y_1, y_n)\, n(n-1) \left(\frac{y_n - \theta_1}{\theta_2 - \theta_1} - \frac{y_1 - \theta_1}{\theta_2 - \theta_1} \right)^{n-2} \right.$$

$$\left. \times \frac{1}{\theta_2 - \theta_1} \cdot \frac{1}{\theta_2 - \theta_1}\, dy_1 \right] dy_n,$$

which is identically 0 if and only if

$$\int_{\theta_1}^{\theta_2} \int_{\theta_1}^{y_n} \not{z}(y_1, y_n)(y_n - y_1)^{n-2}\, dy_1\, dy_n \equiv 0 \qquad \text{for } \theta_1 < \theta_2.$$

Differentiate both sides with respect to θ_2, and obtain

$$\int_{\theta_1}^{\theta_2} \not{z}(y_1, \theta_2)(\theta_2 - y_1)^{n-2}\, dy_1 \equiv 0 \qquad \text{for all } \theta_1 < \theta_2;$$

now differentiate both sides of the resulting identity with respect to θ_1, and obtain $-\not{z}(\theta_1, \theta_2)(\theta_2 - \theta_1)^{n-2} \equiv 0$ for all $\theta_1 < \theta_2$, and hence $\not{z}(\theta_1, \theta_2) = 0$ for $\theta_1 < \theta_2$; that is, $\not{z}(y_1, y_n) = 0$ for $y_1 < y_n$, where y_1 and y_n are the possible values of Y_1 and Y_n. We have shown that Y_1 and Y_n are jointly complete. ////

If the density $f(x; \theta_1, \ldots, \theta_k)$ is a member of the k-parameter exponential family, a set of jointly complete and sufficient statistics can be found using the following theorem. It is a k-dimensional analog of Theorem 9 and is stated without proof. The following theorem is not precisely stated; certain regularity conditions are omitted [16].

Theorem 16 Let $X_1, \ldots, X_n$ be a random sample from $f(x; \theta_1, \ldots, \theta_k)$. If $f(x; \theta_1, \ldots, \theta_k) = a(\theta_1, \ldots, \theta_k)b(x) \exp\left[\sum_{j=1}^{k} c_j(\theta_1, \ldots, \theta_k)\, d_j(x)\right]$, that is, $f(x; \theta_1, \ldots, \theta_k)$ is a member of the k-parameter exponential family, then $\left(\sum_{i=1}^{n} d_1(X_i), \ldots, \sum_{i=1}^{n} d_k(X_i)\right)$ is a minimal set of jointly complete and sufficient statistics. ////

EXAMPLE 48 Let $X_1, \ldots, X_n$ be a random sample from

$$f(x; \theta_1, \theta_2) = \phi_{\theta_1, \theta_2^2}(x) = \frac{1}{\sqrt{2\pi}\,\theta_2} \exp\left[-\tfrac{1}{2}\left(\frac{x-\theta_1}{\theta_2}\right)^2\right].$$

Now

$$f(x; \theta_1, \theta_2) = \frac{1}{\sqrt{2\pi}\,\theta_2} \exp\left[-\tfrac{1}{2}\left(\frac{\theta_1}{\theta_2}\right)^2\right] \exp\left(-\frac{x^2}{2\theta_2^2} + \frac{\theta_1 x}{\theta_2^2}\right);$$

so $\sum_{i=1}^{n} X_i$ and $\sum_{i=1}^{n} X_i^2$ are jointly complete and sufficient statistics by Theorem 16. ////

We will state without proof the vector analog of Theorem 10. In the same sense that an UMVUE was optimum, this following theorem gives an optimum estimator for a vector of functions of the parameter.

Theorem 17 Let $X_1, \ldots, X_n$ be a random sample from $f(x; \theta_1, \ldots, \theta_k)$. Write $\theta = (\theta_1, \ldots, \theta_k)$. If $S_1 = \mathscr{s}_1(X_1, \ldots, X_n), \ldots, S_m = \mathscr{s}_m(X_1, \ldots, X_n)$ is a set of jointly complete sufficient statistics and if there exists an unbiased estimator of $(\tau_1(\theta), \ldots, \tau_r(\theta))$, then there exists a unique unbiased estimator of $(\tau_1(\theta), \ldots, \tau_r(\theta))$, say $T_1^* = \mathscr{t}_1^*(S_1, \ldots, S_m), \ldots, T_r^* = \mathscr{t}_r^*(S_1, \ldots, S_m)$, where each $\mathscr{t}_j^*$ is a function of $S_1, \ldots, S_m$, which satisfies:

(i) $\text{var}_\theta [T_j^*] \le \text{var}_\theta [T_j]$ for every $\theta \in \overline{\Theta}$, $j = 1, \ldots, r$, for any unbiased estimator $(T_1, \ldots, T_r)$ of $(\tau_1(\theta), \ldots, \tau_r(\theta))$.

(ii) The ellipsoid of concentration of $(T_1^*, \ldots, T_r^*)$ is contained in the ellipsoid of concentration of $(T_1, \ldots, T_r)$, where $(T_1, \ldots, T_r)$ is any unbiased estimator of $(\tau_1(\theta), \ldots, \tau_r(\theta))$. ////

There are four different maximal subscripts, all of which are intended. n denotes the sample size, k denotes the dimension of the parameter θ, m is the number of real-valued statistics in our jointly complete and sufficient set, and r is the dimension of the vector of functions of the parameter that we are trying to estimate. In practice, it will turn out that usually $k = m$. The estimator $(T_1^*, \ldots, T_r^*)$ is optimal in the sense that among unbiased estimators it is the best estimator using any of the four generalizations of variance that have been proposed.

Just as was the case in using Theorem 10, we have two ways of finding $(T_1^*, \ldots, T_r^*)$. The first is to guess the correct form of the functions $\mathscr{t}_1^*, \ldots, \mathscr{t}_r^*$, which are functions of $S_1, \ldots, S_m$, that will make them unbiased estimators of

$\tau_1(\theta), \ldots, \tau_r(\theta)$. The second is to find any set of unbiased estimators of $\tau_1(\theta), \ldots, \tau_r(\theta)$ and then calculate the conditional expectation of these unbiased estimators given the set of jointly complete and sufficient statistics. We employ only the first method in the following examples.

EXAMPLE 49 Let $X_1, \ldots, X_n$ be a random sample from the density $f(x; \theta_1, \theta_2) = [1/(\theta_2 - \theta_1)]I_{(\theta_1, \theta_2)}(x)$. Suppose we want to jointly estimate the range and midrange, that is, $\tau_1(\theta) = \theta_2 - \theta_1$ and $\tau_2(\theta) = (\theta_1 + \theta_2)/2$. We know that $Y_1 = \min [X_1, \ldots, X_n]$ and $Y_n = \max [X_1, \ldots, X_n]$ are jointly sufficient (see Example 23); also, they are jointly complete (see Example 47). Hence, to find the unbiased estimator (T_1^*, T_2^*) which has uniformly smallest variance for each component among all unbiased estimators, it suffices to find the unbiased estimator that is a function of the jointly complete sufficient statistics. Since $\mathscr{E}[Y_1] = \theta_1 + (\theta_2 - \theta_1)/(n+1)$ and $\mathscr{E}[Y_n] = \theta_2 - (\theta_2 - \theta_1)/(n + 1)$, $([(n + 1)/(n - 1)](Y_n - Y_1), (Y_1 + Y_n)/2)$ is the unbiased estimator of $(\theta_2 - \theta_1, (\theta_1 + \theta_2)/2)$ that we are seeking. ////

EXAMPLE 50 Let $X_1, \ldots, X_n$ be a random sample from the normal density $f(x; \theta_1, \theta_2) = \phi_{\mu, \sigma^2}(x)$. By Examples 22 and 48, $\sum X_i$ and $\sum X_i^2$ are jointly complete and sufficient statistics. Hence, by Theorem 17, $(\sum X_i/n, \sum (X_i - \bar{X})^2/(n - 1))$ is an unbiased estimator of (μ, σ^2) whose corresponding ellipsoid of concentration is contained in the ellipsoid of concentration of any other unbiased estimator. [NOTE: $\sum (X_i - \bar{X})^2 = \sum X_i^2 - n\bar{X}^2$; so the estimator $\sum (X_i - \bar{X})^2/(n - 1)$ is a function of the jointly complete and sufficient statistics $\sum X_i$ and $\sum X_i^2$.]

For this same example, suppose we want to estimate that function of $\theta = (\mu, \sigma^2)$ satisfying the following integral equation:

$$\int_{\tau(\theta)}^{\infty} \phi_{\mu, \sigma^2}(x)\, dx = \alpha$$

for α fixed and known. $\tau(\theta)$ is that point which satisfies $P[X_i > \tau(\theta)] = \alpha$; that is, it is that point which has 100α percent of the mass of the population density to its right, or $\tau(\theta)$ is the $(1 - \alpha)$th quantile point. We have $1 - \alpha = \Phi([\tau(\theta) - \mu]/\sigma)$; so $\tau(\theta) = \mu + z_{1-\alpha}\sigma$, where $z_{1-\alpha}$ is given by $\Phi(z_{1-\alpha}) = 1 - \alpha$. Since α is known, $z_{1-\alpha}$ can be obtained from a table of the standard normal distribution. To find the UMVUE of $\tau(\theta)$, it suffices to find the unbiased estimator of $\mu + z_{1-\alpha}\sigma$ which is a function of

$\sum X_i$ and $\sum X_i^2$. We know that $\bar{X}$ is the UMVUE of μ, and it can be verified that

$$\frac{\Gamma[(n-1)/2]}{\Gamma(n/2)\sqrt{2}} \sqrt{\sum (X_i - \bar{X})^2} = T^*$$

say, is the UMVUE of σ; hence $\bar{X} + z_{1-\alpha} T^*$ is the UMVUE of $\tau(\theta)$. We have employed Theorem 17 for $r = 1$; our vector of functions of the parameter that we wanted to estimate was unidimensional. ////

9 OPTIMUM PROPERTIES OF MAXIMUM-LIKELIHOOD ESTIMATION

Several methods of finding point estimators were presented in Sec. 2 of this chapter. There, and in succeeding sections, we have particularly emphasized the method of maximum likelihood. In this section we will partially justify such emphasis by considering some optimum properties of maximum-likelihood estimators.

For simplicity of presentation, let us consider the maximum-likelihood estimation of the parameter θ, which is to be estimated on the basis of a random sample from a density $f(\cdot\,; \theta)$, where θ is assumed to be a real number. That is, let us consider the unidimensional-parameter case and estimate θ itself. Recall that for the observed sample $x_1, \ldots, x_n$ the maximum-likelihood estimate of θ is that value, say $\hat{\theta}$, of θ which maximizes the likelihood function $L(\theta; x_1, \ldots, x_n) = \prod_{i=1}^{n} f(x_i; \theta)$. Let $\hat{\Theta}_n = \hat{\vartheta}_n(X_1, \ldots, X_n)$ denote the maximum-likelihood estimator of θ based on a sample of size n. We defined and discussed in Sec. 3 of this chapter a number of properties that an estimator may or may not possess. Recall that some of these properties, such as unbiasedness and uniformly minimum variance, are referred to as small-sample properties, and others of these properties, such as consistency and best asymptotically normal, are referred to as large-sample properties. The use of the word "small" in "small-sample" is somewhat misleading since a small-sample property is really a property that is defined for a *fixed* sample size, which may be fixed to be either small or large. By a large-sample property, we mean a property that is defined in terms of the sample size increasing to infinity. Our main result of this section will be contained in Theorem 18 below and will concern optimum large-sample properties of maximum-likelihood estimation.

We have already observed some small-sample properties of maximum-likelihood estimation. For instance, we have noted two things: first, that some maximum-likelihood estimators are unbiased and others are not and, second, that some maximum-likelihood estimators are uniformly minimum-variance unbiased and others are not. For example, in the density $f(x; \theta) = \phi_{\theta,1}(x)$ the maximum-likelihood estimator of θ is $\bar{X}$, which is the uniformly minimum-variance unbiased estimator of θ, whereas in the density $f(x; \theta) = (1/\theta)I_{[0,\theta]}(x)$ the maximum-likelihood estimator of θ is $Y_n = \max [X_1, \ldots, X_n]$, which is biased. [We might note here that the Y_n in this last example can be corrected for bias by multiplying Y_n by $(n + 1)/n$ and that the estimator that is thus obtained is uniformly minimum variance unbiased.]

One property that it seems reasonable to expect of a sequence of estimators is that of *consistency*. Theorem 18 will show, in particular, that generally a sequence of maximum-likelihood estimators is consistent.

Theorem 18 If the density $f(x; \theta)$ satisfies certain regularity conditions and if $\hat{\Theta}_n = \hat{\vartheta}_n(X_1, \ldots, X_n)$ is the maximum-likelihood estimator of θ for a random sample of size n from $f(x; \theta)$, then:

(i) $\hat{\Theta}_n$ is asymptotically normally distributed with mean θ and variance $1/n\mathscr{E}_\theta\left[\left[\frac{\partial}{\partial\theta}\log f(X; \theta)\right]^2\right]$.

(ii) The sequence of maximum-likelihood estimators $\hat{\Theta}_1, \ldots, \hat{\Theta}_n, \ldots$ is best asymptotically normal (BAN). ////

We will not be able to prove Theorem 18. In fact, we have not precisely stated it, inasmuch as we have not delineated the regularity conditions. We do, however, want to emphasize what the theorem says. Loosely speaking, it says that for large sample size the maximum-likelihood estimator of θ is as good an estimator as there is. (Other estimators might be just as good but not better.)

We might point out one feature of the theorem, namely, that the asymptotic normal distribution of the maximum-likelihood estimator is not given in terms of the distribution of the maximum-likelihood estimator. It is given in terms of $f(\cdot\,; \theta)$, the density sampled. Also, the variance of the asymptotic normal distribution given in the theorem is the Cramér-Rao lower bound.

EXAMPLE 51 Let $X_1, \ldots, X_n$ be a random sample from the negative exponential distribution $f(x; \theta) = \theta e^{-\theta x}I_{[0,\infty)}(x)$. It can be routinely demonstrated that the maximum-likelihood estimator of θ is given by

$n/\sum_1^n X_i = 1/\bar{X}_n$. According to Theorem 18 above, the maximum-likelihood estimator has an asymptotic normal distribution with mean θ and variance equal to

$$\frac{1}{n\mathscr{E}_\theta\left[\left[\frac{\partial}{\partial\theta}\log f(X;\theta)\right]^2\right]} = \frac{\theta^2}{n}.$$

(See Example 29.) ////

We have ordinarily considered estimation of $\tau(\theta)$ some function of θ, rather than estimation of θ itself. For maximum-likelihood estimation, we noted (see Theorem 2) that the maximum-likelihood estimator of $\tau(\theta)$ was given by $\tau(\hat{\Theta})$, where $\hat{\Theta}$ was the maximum-likelihood estimator of θ. If we assume that $\tau(\cdot)$ is differentiable, then it can be shown that $\tau(\hat{\Theta})$ has an asymptotic normal distribution with mean $\tau(\theta)$ and variance

$$\frac{[\tau'(\theta)]^2}{n\mathscr{E}_\theta\left[\left[\frac{\partial}{\partial\theta}\log f(X;\theta)\right]^2\right]},$$

which is the Cramér-Rao lower bound. (See Theorem 7.)

Maximum-likelihood estimators possess similar optimum large-sample properties in the case of a k-dimensional parameter. For instance, it can be proved (again under regularity conditions) that the joint distribution of the maximum-likelihood estimators is asymptotically distributed as a multivariate normal distribution. Let us illustrate for the case when $k = 2$; that is, $\theta = (\theta_1, \theta_2)$. Recall that the bivariate normal distribution is specified by the five parameters μ_1, μ_2, σ_1^2, σ_2^2, and ρ. (See Sec. 5 of Chap. IV.) It turns out that under certain regularity conditions the joint distribution of the maximum-likelihood estimators $\hat{\Theta}_1$ and $\hat{\Theta}_2$ is asymptotically distributed as a bivariate normal distribution with parameters $\mu_1 = \theta_1$, $\mu_2 = \theta_2$,

$$\sigma_1^2 = \frac{-\mathscr{E}_\theta\left[\frac{\partial^2}{\partial\theta_2^2}\log f(X;\theta)\right]}{n\Delta},$$

$$\sigma_2^2 = \frac{-\mathscr{E}_\theta\left[\frac{\partial^2}{\partial\theta_1^2}\log f(X;\theta)\right]}{n\Delta},$$

and

$$\rho\sigma_1\sigma_2 = \frac{\mathscr{E}_\theta\left[\dfrac{\partial^2}{\partial\theta_2\,\partial\theta_1}\log f(X;\theta)\right]}{n\Delta},$$

where

$$\Delta = \mathscr{E}_\theta\left[\frac{\partial^2}{\partial\theta_1^2}\log f(X;\theta)\right]\mathscr{E}_\theta\left[\frac{\partial^2}{\partial\theta_2^2}\log f(X;\theta)\right] - \left(\mathscr{E}_\theta\left[\frac{\partial^2}{\partial\theta_2\,\partial\theta_1}\log f(X;\theta)\right]\right)^2.$$

EXAMPLE 52 Let $X_1, \ldots, X_n$ be a random sample from the density

$$f(x;\theta) = f(x;\theta_1,\theta_2) = \phi_{\theta_1,\theta_2}(x) = \frac{1}{\sqrt{2\pi\theta_2}}\,e^{-(1/2\theta_2)(x-\theta_1)^2}.$$

We have already derived, in Example 6, the maximum-likelihood estimators of θ_1 and θ_2; they are, respectively,

$$\hat{\Theta}_1 = \frac{1}{n}\sum_1^n X_i \qquad \text{and} \qquad \hat{\Theta}_2 = \frac{1}{n}\sum_1^n (X_i - \hat{\Theta}_1)^2.$$

According to the above, the asymptotic large-sample joint distribution of $\hat{\Theta}_1$ and $\hat{\Theta}_2$ is a bivariate normal distribution with means θ_1 and θ_2. Since $\log f(X;\theta) = -\frac{1}{2}\log 2\pi - \frac{1}{2}\log\theta_2 - (1/2\theta_2)(X-\theta_1)^2$, the required derivatives are

$$\frac{\partial^2}{\partial\theta_1^2}\log f(X;\theta) = -\frac{1}{\theta_2},$$

$$\frac{\partial^2}{\partial\theta_2\,\partial\theta_1}\log f(X;\theta) = -\frac{X-\theta_1}{\theta_2^2},$$

and

$$\frac{\partial^2}{\partial\theta_2^2}\log f(X;\theta) = \frac{1}{2\theta_2^2} - \frac{(X-\theta_1)^2}{\theta_2^3};$$

and because

$$\mathscr{E}[X] = \theta_1 \qquad \text{and} \qquad \mathscr{E}[(X-\theta_1)^2] = \theta_2,$$

$$-\mathscr{E}_\theta\left[\frac{\partial^2}{\partial\theta_1^2}\log f(X;\theta)\right] = \frac{1}{\theta_2},$$

$$-\mathscr{E}_\theta\left[\frac{\partial^2}{\partial\theta_2\,\partial\theta_1}\log f(X;\theta)\right] = 0,$$

and

$$-\mathscr{E}_\theta\left[\frac{\partial^2}{\partial\theta_2^2}\log f(X;\theta)\right]=\frac{1}{2\theta_2^2},$$

which gives $\Delta = 1/2\theta_2^3$. Finally, then, $\sigma_1^2 = \theta_2/n$, $\sigma_2^2 = 2\theta_2^2/n$, and $\rho = 0$. ////

PROBLEMS

1 An urn contains black and white balls. A sample of size n is drawn with replacement. What is the maximum-likelihood estimator of the ratio R of black to white balls in the urn? Suppose that one draws balls one by one with replacement until a black ball appears. Let X be the number of draws required (not counting the last draw). This operation is repeated n times to obtain a sample $X_1, X_2, \ldots, X_n$. What is the maximum-likelihood estimator of R on the basis of this sample?

2 Suppose that n cylindrical shafts made by a machine are selected at random from the production of the machine and their diameters and lengths measured. It is found that N_{11} have both measurements within the tolerance limits, N_{12} have satisfactory lengths but unsatisfactory diameters, N_{21} have satisfactory diameters but unsatisfactory lengths, and N_{22} are unsatisfactory as to both measurements. $\sum N_{ij} = n$. Each shaft may be regarded as a drawing from a multinomial population with density

$$p_{11}^{x_{11}}p_{12}^{x_{12}}p_{21}^{x_{21}}(1-p_{11}-p_{12}-p_{21})^{x_{22}} \qquad \text{for } x_{ij}=0, 1; \textstyle\sum x_{ij}=1$$

having three parameters. What are the maximum-likelihood estimates of the parameters if $N_{11}=90$, $N_{12}=6$, $N_{21}=3$, and $N_{22}=1$?

3 Referring to Prob. 2, suppose that there is no reason to believe that defective diameters can in any way be related to defective lengths. Then the distribution of the X_{ij} can be set up in terms of two parameters: p_1, the probability of a satisfactory length, and q_1, the probability of a satisfactory diameter. The density of the X_{ij} is then

$$(p_1q_1)^{x_{11}}[p_1(1-q_1)]^{x_{12}}[(1-p_1)q_1]^{x_{21}}[(1-p_1)(1-q_1)]^{x_{22}}$$
$$\text{for } x_{ij}=0, 1; \qquad \textstyle\sum x_{ij}=1.$$

What are the maximum-likelihood estimates for these parameters? Are the probabilities for the four classes different under this model from those obtained in the above problem?

4 A sample of size n_1 is to be drawn from a normal population with mean μ_1 and variance σ_1^2. A second sample of size n_2 is to be drawn from a normal population with mean μ_2 and variance σ_2^2. What is the maximum-likelihood estimator of $\theta = \mu_1 - \mu_2$? If we assume that the total sample size $n = n_1 + n_2$ is fixed, how should the n observations be divided between the two populations in order to minimize the variance of the maximum-likelihood estimator of θ?

5 A sample of size n is drawn from each of four normal populations, all of which have the same variance σ^2. The means of the four populations are $a+b+c$, $a+b-c$, $a-b+c$, and $a-b-c$. What are the maximum-likelihood estimators of a, b, c, and σ^2? (The sample observations may be denoted by X_{ij}, $i=1$, 2, 3, 4 and $j=1, 2, \ldots, n$.)

6 Observations $X_1, X_2, \ldots, X_n$ are drawn from normal populations with the same mean μ but with different variances $\sigma_1^2, \sigma_2^2, \ldots, \sigma_n^2$. Is it possible to estimate all the parameters? If we assume that the σ_i^2 are known, what is the maximum-likelihood estimator of μ?

7 The radius of a circle is measured with an error of measurement which is distributed $N(0, \sigma^2)$, σ^2 unknown. Given n independent measurements of the radius, find an unbiased estimator of the area of the circle.

8 Let X be a single observation from the Bernoulli density $f(x; \theta) = \theta^x(1-\theta)^{1-x}I_{\{0, 1\}}(x)$, where $0 < \theta < 1$. Let $t_1(X) = X$ and $t_2(X) = \frac{1}{2}$.

(*a*) Are both $t_1(X)$ and $t_2(X)$ unbiased? Is either?

(*b*) Compare the mean-squared error of $t_1(X)$ with that of $t_2(X)$.

9 Let X_1, X_2 be a random sample of size 2 from the Cauchy density

$$f(x; \theta) = \frac{1}{\pi[1 + (x-\theta)^2]}, \qquad -\infty < \theta < \infty.$$

Argue that $(X_1 + X_2)/2$ is a Pitman closer estimator of θ than X_1 is. [Note that $(X_1 + X_2)/2$ is not more concentrated than X_1 since they have identical distributions.]

10 Let θ denote some physical quantity, and let $X_1, \ldots, X_n$ denote n measurements of the physical quantity. If θ is estimated by $\hat{\Theta}$, then the residual of the ith measurement is defined by $X_i - \hat{\Theta}$, $i = 1, \ldots, n$. Show that there is only one estimator with the property that the residuals sum is 0, and find that estimator. Also, find that estimator which minimizes the sum of squared residuals.

11 Let $X_1, \ldots, X_n$ be a random sample from some density which has mean μ and variance σ^2.

(*a*) Show that $\sum_1^n a_i X_i$ is an unbiased estimator of μ for any set of known constants $a_1, \ldots, a_n$ satisfying $\sum_1^n a_i = 1$.

(*b*) If $\sum_1^n a_i = 1$, show that $\text{var}\,[\sum_1^n a_i X_i]$ is minimized for $a_i = 1/n$, $i = 1, \ldots, n$. [HINT: Prove that $\sum_1^n a_i^2 = \sum_1^n (a_i - 1/n)^2 + 1/n$ when $\sum_1^n a_i = 1$.]

12 Let $X_1, \ldots, X_n$ be a random sample from the discrete density function $f(x; \theta) = \theta^x(1-\theta)^{1-x}I_{\{0, 1\}}(x)$, where $0 \le \theta \le \frac{1}{2}$. Note that $\underline{\Theta} = \{\theta: 0 \le \theta \le \frac{1}{2}\}$.

(*a*) Find a method-of-moments estimator θ, and then find the mean and mean-squared error of your estimator.

(*b*) Find a maximum-likelihood estimator of θ, and then find the mean and mean-squared error of your estimator.

13 Let X_1, X_2 be a random sample of size 2 from a normal distribution with mean θ and variance 1. Consider the following three estimators of θ:

$$T_1 = t_1(X_1, X_2) = \tfrac{2}{3}X_1 + \tfrac{1}{3}X_2$$
$$T_2 = t_2(X_1, X_2) = \tfrac{1}{4}X_1 + \tfrac{3}{4}X_2$$
$$T_3 = t_3(X_1, X_2) = \tfrac{1}{2}X_1 + \tfrac{1}{2}X_2.$$

(*a*) For the loss function $\ell(t; \theta) = 3\theta^2(t - \theta)^2$, find $\mathscr{R}_{t_i}(\theta)$ for $i = 1, 2, 3$, and sketch it.

(*b*) Show that T_i is unbiased for $i = 1, 2, 3$.

14 Let $X_1, \ldots, X_n, \ldots$ be independent and identically distributed random variables from some distribution for which the first four central moments exist. We know that $\mathscr{E}[S^2] = \sigma^2$ and

$$\text{var}\,[S^2] = \frac{1}{n}\left(\mu_4 - \frac{n-3}{n-1}\,\sigma^4\right),$$

where

$$S^2 = \frac{1}{n-1}\sum (X_i - \bar{X})^2.$$

Is S^2 a mean-squared-error consistent estimator of σ^2?

15 In genetic investigations one frequently samples from a binomial distribution $f(x) = \binom{m}{x} p^x q^{m-x}$ except that observations of $x = 0$ are impossible; so, in fact, the sampling is from the conditional (truncated) distribution

$$\binom{m}{x} \frac{p^x q^{m-x}}{1 - q^m} I_{\{1, 2, \ldots, m\}}(x).$$

Find the maximum-likelihood estimator of p in the case $m = 2$ for samples of size n. Is the estimator unbiased?

16 Let X be a single observation from $N(0, \theta)$. ($\theta = \sigma^2$.)

(*a*) Is X a sufficient statistic?

(*b*) Is $|X|$ a sufficient statistic?

(*c*) Is X^2 an unbiased estimator of θ?

(*d*) What is a maximum-likelihood estimator of $\sqrt{\theta}$?

(*e*) What is a method-of-moments estimator of $\sqrt{\theta}$?

17 Let X have the density $f(x;\theta) = (\theta/2)^{|x|}(1 - \theta)^{1-|x|} I_{\{-1, 0, 1\}}(x)$, $0 \le \theta \le 1$. Define $t(x) = 2I_{\{1\}}(x)$.

(*a*) Is X a sufficient statistic? A complete statistic?

(*b*) Is $|X|$ a sufficient statistic? A complete statistic?

(*c*) What is a maximum-likelihood estimator of θ?

(*d*) Is $T = t(X)$ an unbiased estimator of θ?

(*e*) Does $f(x; \theta)$ belong to an exponential class?

(*f*) Find an estimator with uniformly smaller mean-squared error than that of $t(X)$, if such exists.

18 Let $X_1, X_2, \ldots, X_n$ be a random sample from the density

$$f(x; \theta) = \theta x^{-2} I_{[\theta, \infty)}(x)$$

where $\theta > 0$.

(*a*) Find a maximum-likelihood estimator of θ.

(*b*) Is $Y_1 = \min [X_1, \ldots, X_n]$ a sufficient statistic?

19 Let $X_1, \ldots, X_n$ be a random sample from $f(x; \theta) = \frac{1}{2}e^{-|x-\theta|}$, $-\infty < \theta < \infty$.

(*a*) Discuss sufficiency for this density.

(*b*) Obtain a method-of-moments estimator of θ.

(*c*) Find a maximum-likelihood estimator of θ.

(*d*) Does $f(x; \theta)$ belong to an exponential class?

20 Find a maximum-likelihood estimator for α in the density $f(x; \alpha) = (2/\alpha^2)(\alpha - x)I_{(0, \alpha)}(x)$ for samples of size 2. Is it a sufficient statistic? Estimate α by the method of moments. What is the maximum-likelihood estimator of the population mean?

21 Let $X_1, \ldots, X_n$ be a random sample from $f(x; \theta) = (1/\theta)I_{[0, \theta]}(x)$, where $\theta > 0$. Define $Y_n = \max [X_1, \ldots, X_n]$ and $Y_1 = \min [X_1, \ldots, X_n]$.

(*a*) Estimate θ by the method of moments. Call the estimator T_1. Find its mean and mean-squared error.

(*b*) Find the maximum-likelihood estimator of θ. Call the estimator T_2. Find its mean and mean-squared error.

(*c*) Among all estimators of the form aY_n, where a is a constant which may depend on n, find that estimator which has uniformly smallest mean-squared error. Call it T_3. Find its mean and mean-squared error.

(*d*) Find the UMVUE of θ. Call it T_4. Obtain its mean and mean-squared error.

(*e*) Let $T_5 = Y_1 + Y_n$. Find the mean and mean-squared error of T_5.

(*f*) What estimator of θ would you use and why?

(*g*) Find the maximum-likelihood estimator of the variance of the population.

22 Let $X_1, \ldots, X_n$ be a random sample from the Bernoulli distribution, say $P[X = 1] = \theta = 1 - P[X = 0]$.

(*a*) Find the Cramér-Rao lower bound for the variance of unbiased estimators of $\theta(1 - \theta)$.

(*b*) Find the UMVUE of $\theta(1 - \theta)$ if such exists.

23 Assuming r known, find the maximum-likelihood estimator for λ for a random sample of size n from a gamma distribution. Find a sufficient statistic if one exists. Is your maximum-likelihood estimator unbiased? Is there an UMVUE of λ?

24 Let $X_1, \ldots, X_n$ be a random sample from $\theta x^{\theta-1} I_{(0, 1)}(x)$, where $\theta > 0$.

(*a*) Find the maximum-likelihood estimator of $\mu = \theta/(1 + \theta)$.

(*b*) Find a sufficient statistic, and check completeness. Is $\sum X_i$ a sufficient statistic?

(*c*) Is there a function of θ for which there exists an unbiased estimator whose variance coincides with the Cramér-Rao lower bound?

*(*d*) Find the UMVUE of each of the following: θ, $1/\theta$, $\mu = \theta/(1 + \theta)$.

25 Let $X_1, \ldots, X_n$ be a random sample from the binomial distribution $\binom{m}{x} p^x(1-p)^{m-x}$, $x = 0, 1, \ldots, m$, where m is known and $0 \le p \le 1$.

(*a*) Estimate p by the method of moments and the method of maximum likelihood.

(*b*) Is there an UMVUE of p? If so, find it.

******26* Let $X_1, \ldots, X_n$ be a random sample from the discrete density function

$$f(x;\theta) = (1/\theta)I_{\{1,2,\ldots,\theta\}}(x),$$

where $\theta = 1, 2, \ldots$. That is, $\underline{\overline{\Theta}} = \{\theta\colon \theta = 1, 2, \ldots\}$ = the set of positive integers.

(*a*) Find a method-of-moments estimator of θ. Find its mean and mean-squared error.

(*b*) Find a maximum-likelihood estimator of θ. Find its mean and mean-squared error.

(*c*) Find a complete sufficient statistic.

(*d*) Let $T = Y_n$, the largest order statistic. Show that the UMVUE of θ is $[T^{n+1} - (T-1)^{n+1}]/[T^n - (T-1)^n]$.

27 Let X be a single observation from the density $[1/\mathrm{B}(\theta, \theta)]x^{\theta-1}(1-x)^{\theta-1}I_{(0,1)}(x)$. Is X a sufficient statistic? Is X complete?

28 An experimenter knows that the distribution of the lifetime of a certain component is negative exponentially distributed with mean $1/\theta$. On the basis of a random sample of size n of lifetimes he wants to estimate the *median* lifetime. Find both the maximum-likelihood and uniformly minimum-variance unbiased estimator of the median.

29 Let $X_1, \ldots, X_n$ be a random sample from $N(\theta, 1)$.

(*a*) Find the Cramér-Rao lower bound for the variance of unbiased estimators of θ, θ^2, and $P[X > 0]$.

(*b*) Is there an unbiased estimator of θ^2 for $n = 1$? If so, find it.

(*c*) Is there an unbiased estimator of $P[X > 0]$? If so, find it.

(*d*) What is the maximum-likelihood estimator of $P[X > 0]$?

(*e*) Is there an UMVUE of θ^2? If so, find it.

(*f*) Is there an UMVUE of $P[X > 0]$? If so, find it.

30 For a random sample from the Poisson distribution, find an unbiased estimator of $\tau(\lambda) = (1+\lambda)e^{-\lambda}$. Find a maximum-likelihood estimator of $\tau(\lambda)$. Find the UMVUE of $\tau(\lambda)$.

31 Let $X_1, \ldots, X_n$ be a random sample from the density

$$f(x;\theta) = \frac{2x}{\theta^2} I_{(0,\theta)}(x)$$

where $\theta > 0$.

(*a*) Find a maximum-likelihood estimator of θ.

(*b*) Is $Y_n = \max[X_1, \ldots, X_n]$ a sufficient statistic? Is Y_n complete?

(*c*) Is there an UMVUE of θ? If so, find it.

32 Let $X_1, \ldots, X_n$ be a random sample from the density

$$f(x; \theta) = \theta(1 + x)^{-(1+\theta)} I_{(0, \infty)}(x) \qquad \text{for} \quad \theta > 0.$$

(*a*) Estimate θ by the method of moments assuming $\theta > 1$.
(*b*) Find the maximum-likelihood estimator of $1/\theta$.
(*c*) Find a complete and sufficient statistic if one exists.
(*d*) Find the Cramér-Rao lower bound for unbiased estimators of $1/\theta$.
(*e*) Find the UMVUE of $1/\theta$ if such exists.
(*f*) Find the UMVUE of θ if such exists.

33 Let $X_1, \ldots, X_n$ be a random sample from

$$f(x; \theta) = \frac{1}{2\theta} I_{[-\theta, \theta]}(x) \qquad \text{for} \quad \theta > 0.$$

(*a*) Find a maximum-likelihood estimator of θ.
(*b*) Suppose $n = 1$, so that you have only one observation, say $X = X_1$. Clearly X is a sufficient statistic. Is X a minimal sufficient statistic? Is X complete?

34 Let $X_1, \ldots, X_n$ be a random sample from the negative exponential density

$$f(x; \theta) = \theta e^{-\theta x} I_{[0, \infty)}(x).$$

(*a*) Find the uniformly minimum-variance unbiased estimator of var $[X_1]$ if such exists.
(*b*) Find an unbiased estimator of $1/\theta$ based only on $Y_1^{(n)} = \min [X_1, \ldots, X_n]$. Is your sequence of estimators mean-squared-error consistent?

35 Let $X_1, \ldots, X_n$ be a random sample from the density

$$f(x; \theta) = \frac{\log \theta}{\theta - 1} \theta^x I_{(0, 1)}(x), \qquad \theta > 1.$$

(*a*) Find a complete sufficient statistic if there is one.
(*b*) Find a function of θ for which there exists an unbiased estimator whose variance coincides with the Cramér-Rao lower bound if such exists.

36 Show that

$$\mathscr{E}_\theta\left[\left\{\frac{\partial}{\partial\theta} \log f(X; \theta)\right\}^2\right] = -\mathscr{E}_\theta\left[\frac{\partial^2}{\partial\theta^2} \log f(X; \theta)\right].$$

37 Let $X_1, \ldots, X_n$ be a random sample from the density

$$f(x; \theta) = e^{-(x-\theta)} \exp(-e^{-(x-\theta)}),$$

where $-\infty < \theta < \infty$.

(*a*) Find a method-of-moments estimator of θ.
(*b*) Find a maximum-likelihood estimator of θ.
(*c*) Find a complete sufficient statistic.
(*d*) Find the Cramér-Rao lower bound for unbiased estimators of θ.
(*e*) Is there a function of θ for which there exists an unbiased estimator, the variance of which coincides with the Cramér-Rao lower bound? If so, find it.
*(*f*) Show that $\Gamma'(n)/\Gamma(n) - \log(\sum e^{-X_i})$ is the UMVUE of θ.

38 Let $X_1, \ldots, X_n$ denote a random sample from

$$f(x;\theta) = f_\theta(x) = \theta f_1(x) + (1-\theta) f_0(x),$$

where $0 \leq \theta \leq 1$ and $f_1(\cdot)$ and $f_0(\cdot)$ are known densities.

(*a*) Estimate θ by the method of moments.

(*b*) For $n = 2$, find a maximum-likelihood estimator of θ.

(*c*) Find the Cramér-Rao lower bound for the variance of unbiased estimators of θ.

39 Suppose that $a(\cdot)$ and $b(\cdot)$ are two nonnegative functions such that $f(x;\theta) = a(\theta)b(x)I_{(0,\theta)}(x)$ is a probability density function for each $\theta > 0$.

(*a*) What is a maximum-likelihood estimator of θ?

(*b*) Is there a complete sufficient statistic? If so, find it.

(*c*) Is there an UMVUE of θ? If so, find it.

40 Let $X_1, \ldots, X_n$ be a random sample from $N(\theta, \theta)$, $\theta > 0$.

(*a*) Find a complete sufficient statistic if such exists.

(*b*) Argue that $\bar{X}$ is not an UMVUE of θ.

(*c*) Is θ either a location or scale parameter?

41 Let $X_1, \ldots, X_n$ be a random sample from $N(\theta, \theta^2)$, $-\infty < \theta < \infty$.

(*a*) Is there a unidimensional sufficient statistic?

(*b*) Find a two-dimensional sufficient statistic.

(*c*) Is $\bar{X}$ an UMVUE of θ? {HINT: Find an unbiased estimator of θ based on S^2; call it T^*. Find a constant a to minimize var $[a\bar{X} + (1-a)T^*]$.}

(*d*) Is θ either a location or scale parameter?

42 Let $X_1, \ldots, X_n$ be a random sample of size n from the density

$$f(x;\theta) = \frac{1}{\theta} I_{[\theta, 2\theta]}(x), \qquad \theta > 0.$$

(*a*) Find a maximum-likelihood estimator of θ.

(*b*) We know that Y_1 and Y_n are jointly sufficient. Are they jointly complete?

(*c*) Find the Pitman estimator for the scale parameter θ.

(*d*) For a and b constant (they may depend on n), find an unbiased estimator of θ of the form $aY_1 + bY_n$ satisfying $P[Y_n/2 \leq aY_1 + bY_n \leq Y_1] = 1$ if such exists. Why is $P[Y_n/2 \leq aY_1 + bY_n \leq Y_1] = 1$ desirable?

43 Let $Z_1, \ldots, Z_n$ be a random sample from $N(0, \theta^2)$, $\theta > 0$. Define $X_i = |Z_i|$, and consider estimation of θ and θ^2 on the basis of the random sample $X_1, \ldots, X_n$.

(*a*) Find the UMVUE of θ^2 if such exists.

(*b*) Find an estimator of θ^2 that has uniformly smaller mean-squared error than the estimator that you found in part (*a*).

(*c*) Find the UMVUE of θ if such exists.

(*d*) Find the Pitman estimator for the scale parameter θ.

(*e*) Does the estimator that you found in part (*d*) have uniformly smaller mean-squared error than the estimator that you found in part (*c*)?

44 Let $X_1, \ldots, X_n$ be a random sample from

$$f(x; \theta) = e^{-(x-\theta)} I_{[\theta, \infty)}(x) \qquad \text{for } -\infty < \theta < \infty.$$

(*a*) Find a sufficient statistic.
(*b*) Find a maximum-likelihood estimator of θ.
(*c*) Find a method-of-moments estimator of θ.
(*d*) Is there a complete sufficient statistic? If so, find it.
(*e*) Find the UMVUE of θ if one exists.
(*f*) Find the Pitman estimator for the location parameter θ.
(*g*) Using the prior density $g(\theta) = e^{-\theta} I_{(0, \infty)}(\theta)$, find the posterior Bayes estimator of θ.

45 Let $X_1, \ldots, X_n$ be a random sample from $f(x \mid \theta) = \theta x^{\theta-1} I_{(\theta, 1)}(x)$, where $\theta > 0$. Assume that the prior distribution of Θ is given by

$$g_\Theta(\theta) = [1/\Gamma(r)] \lambda^r \theta^{r-1} e^{-\theta\lambda} I_{(0, \infty)}(\theta),$$

where r and λ are known.
(*a*) What is the posterior distribution of Θ?
(*b*) Find the Bayes estimator of θ with respect to the given gamma prior distribution using a squared-error loss function.

46 Let X be a single observation from the density $f(x \mid \theta) = (2x/\theta^2) I_{(0, \theta)}(x)$, where $\theta > 0$. Assume that Θ has a uniform prior distribution over the interval (0, 1). For the loss function $\ell(t; \theta) = \theta^2(t - \theta)^2$, find the Bayes estimator of θ.

47 Let $X_1, X_2, \ldots, X_n$ be a random sample of size n from the following discrete density:

$$f(x; \theta) = \binom{2}{x} \theta^x (1 - \theta)^{2-x} I_{\{0, 1, 2\}}(x),$$

where $\theta > 0$.
(*a*) Is there a unidimensional sufficient statistic? If so, is it complete?
(*b*) Find a maximum-likelihood estimator of $\theta^2 = P[X_i = 2]$. Is it unbiased?
(*c*) Find an unbiased estimator of θ whose variance coincides with the corresponding Cramér-Rao lower bound if such exists. If such an estimate does not exist, prove that it does not.
(*d*) Find a uniformly minimum-variance unbiased estimator of θ^2 if such exists.
(*e*) Using the squared-error loss function find a Bayes estimator of θ with respect to the beta prior distribution

$$g(\theta) = \frac{1}{B(a, b)} \theta^{a-1} (1 - \theta)^{b-1} I_{(0, 1)}(\theta).$$

(*f*) Using the squared-error loss function, find a minimax estimator of θ.
(*g*) Find a mean-squared error consistent estimator of θ^2.

48 Let $X_1, \ldots, X_n$ be a random sample from a Poisson density

$$\frac{e^{-\theta}\theta^x}{x!} I_{\{0, 1, \ldots\}}(x),$$

where $\theta > 0$. For a squared-error loss function find the Bayes estimator of θ for a gamma prior distribution. Find the posterior distribution of Θ. Find the posterior Bayes estimator of $\tau(\theta) = P[X_i = 0]$.

49 Let $X_1, \ldots, X_n$ be a random sample from $f(x \mid \theta) = (1/\theta)I_{(0, \theta)}(x)$, where $\theta > 0$. For the loss function $(t - \theta)^2/\theta^2$ and a prior distribution proportional to $\theta^{-\alpha}I_{(1, \infty)}(\theta)$ find the Bayes estimator of θ.

50 Let $X_1, \ldots, X_n$ be a random sample from the Bernoulli distribution. Using the squared-error loss function, find that estimator of θ which has minimum area under its risk function.

51 Let $X_1, \ldots, X_n$ be a random sample from the geometric density

$$f(x; \theta) = \theta(1 - \theta)^x I_{\{0, 1, \ldots\}}(x)$$

where $0 < \theta < 1$.

(*a*) Find a method-of-moments estimator of θ.

(*b*) Find a maximum-likelihood estimator of θ.

(*c*) Find a maximum-likelihood estimator of the mean.

(*d*) Find the Cramér-Rao lower bound for the variance of unbiased estimators of $1 - \theta$.

(*e*) Is there a function of θ for which there exists an unbiased estimator the variance of which coincides with the Cramér-Rao lower bound? If so, find it.

(*f*) Find the UMVUE of $(1 - \theta)/\theta$ if such exists.

(*g*) Find the UMVUE of θ if such exists.

(*h*) Assume a uniform prior distribution and find the posterior distribution of Θ. For a squared-error loss function, find the Bayes estimator of θ with respect to a uniform prior distribution.

52 Let θ be the true I.Q. of a certain student. To measure his I.Q., the student takes a test, and it is known that his test scores are normally distributed with mean μ and standard deviation 5.

(*a*) The student takes the I.Q. test and gets a score of 130. What is the maximum-likelihood estimate of θ?

(*b*) Suppose that it is known that I.Q.'s of students of a certain age are distributed normally with mean 100 and variance 225; that is, $\Theta \sim N(100, 225)$. Let X denote a student's test score [X is distributed $N(\theta, 25)$]. Find the posterior distribution of Θ given $X = x$. What is the posterior Bayes estimate of the student's I.Q. if $X = 130$.

**53* Let $X_1, \ldots, X_n$ be a random sample from the density

$$f(x; \alpha, \theta) = f(x; \alpha, \beta) = \beta\, e^{-(1/\beta)(x-\alpha)} I_{[\alpha, \infty)}(x),$$

where $-\infty < \alpha < \infty$ and $\beta > 0$. Show that Y_1 and $\sum X_i$ are jointly sufficient. It can be shown that Y_1 and $\sum (X_i - Y_1)$ are jointly complete and independent of each other. Using such results, find the estimator of (α, β) that has an ellipsoid of concentration that is contained in the ellipsoid of concentration of any other unbiased estimator of (α, β). ($Y_1 = \min [X_1, \ldots, X_n]$.)

54 Let $X_1, \ldots, X_n$ be a random sample from the density

$$f(x; \alpha, \theta) = (1 - \theta)\theta^{x-\alpha} I_{\{\alpha, \alpha+1, \ldots\}}(x),$$

where $-\infty < \alpha < \infty$ and $0 < \theta < 1$.

(*a*) Find a two-dimensional set of sufficient statistics.

*(*b*) Find the maximum-likelihood estimator of (α, θ).

55 Let $X_1, \ldots, X_n$ be a random sample from the density

$$f(x; \theta) = \frac{2x}{\theta} I_{(0,\theta]}(x) + \frac{2(1-x)}{1-\theta} I_{(\theta,1]}(x),$$

where $0 \le \theta \le 1$.

(*a*) Estimate θ by the method of moments.

(*b*) Find the maximum-likelihood estimator of θ for $n = 1$ and $n = 2$.

(*c*) For $n = 1$ find a complete sufficient statistic if such exists. Find a UMVUE of θ for $n = 1$ if such exists.

*(*d*) Find the maximum-likelihood estimator of θ.

VIII

PARAMETRIC INTERVAL ESTIMATION

1 INTRODUCTION AND SUMMARY

Chapter VII dealt with the *point estimation* of a parameter, or more precisely, point estimation of a value of a function of a parameter. Such point estimates are quite useful, yet they leave something to be desired. In all those cases when the point estimator under consideration had a probability density function, the probability that the estimator actually equaled the value of the parameter being estimated was 0. (The probability that a continuous random variable equals any one value is 0.) Hence, it seems desirable that a point estimate should be accompanied by some measure of the possible error of the estimate. For instance, a point estimate might be accompanied by some interval about the point estimate together with some measure of assurance that the true value of the parameter lies within the interval. Instead of making the inference of estimating the true value of the parameter to be a point, we might make the inference of estimating that the true value of the parameter is contained in some interval. We then speak of *interval estimation*, which is to be the subject of this chapter.

Like point estimation, the problem of interval estimation is twofold. First, there is the problem of finding interval estimators, and, second, there is the problem of determining good, or optimum, interval estimators. The considerations of these two problems that will appear in this chapter will be incomplete. Further considerations will be presented at the end of the next chapter on testing hypotheses. The mathematics of *interval estimation* and *hypotheses testing* are closely related. Either concept could be used to introduce the other. In this book, we have decided to introduce interval estimation first, right after our presentation of point estimation, then introduce hypotheses testing, and finally point out the close mathematical relationship between the two.

The introduction to interval estimation that appears in this chapter will not be as thorough as was our discussion of point estimation in the last chapter. One should not infer from this that interval estimation is less important since in practice the opposite is usually true. It is just easier to present the basic theory of point estimation. No concerted effort will be given to the problem of finding optimum interval estimators. The chapter will be divided into six main sections, the first being this introductory section. Section 2 will be devoted to confidence intervals, where the notion is introduced and defined. One method of finding confidence intervals will also be given as well as some idea as to what an optimum confidence interval might be. Section 3 will consider several examples of confidence intervals that are associated with sampling from the normal distribution. Such discussion will hinge on the results of Sec. 4 of Chap. VI. Several general methods of finding confidence intervals are given in Sec. 4; another method, which utilizes the theory of hypotheses testing, will be given at the end of Chap. IX. A brief discussion of large-sample confidence intervals appears in Sec. 5, and Sec. 6 presents another type of interval estimation, namely, Bayesian interval estimation.

2 CONFIDENCE INTERVALS

2.1 An Introduction to Confidence Intervals

In practice, estimates are often given in the form of the estimate plus or minus a certain amount. For instance, an electric charge may be estimated to be $(4.770 \pm .005)10^{-10}$ electrostatic unit with the idea that the first factor is very unlikely to be outside the range 4.765 to 4.775. A cost accountant for a publishing company in trying to allow for all factors which enter into the cost of producing a certain book (actual production costs, proportion of plant overhead, proportion of executive salaries, etc.) may estimate the cost to be 83 ± 4.5 cents per volume with the implication that the correct cost very probably lies between

78.5 and 87.5 cents per volume. The Bureau of Labor Statistics may estimate the number of unemployed in a certain area to be $2.4 \pm .3$ million at a given time, feeling rather sure that the actual number is between 2.1 and 2.7 million. What we are saying is that in practice one is quite accustomed to seeing estimates in the form of intervals.

In order to give precision to these ideas, we shall consider a particular example. Suppose that a random sample (1.2, 3.4, .6, 5.6) of four observations is drawn from a normal population with an unknown mean μ and a known standard deviation 3. The maximum-likelihood estimate of μ is the mean of the sample observations:

$$\bar{x} = 2.7.$$

We wish to determine upper and lower limits which are rather certain to contain the true unknown parameter value between them.

In general, for samples of size 4 from the given distribution the quantity

$$Z = \frac{\bar{X} - \mu}{\frac{3}{2}}$$

will be normally distributed with mean 0 and unit variance. $\bar{X}$ is the sample mean, and $\frac{3}{2}$ is $\sigma/\sqrt{n}$. Thus the quantity Z has a density

$$f_Z(z) = \phi(z) = \frac{1}{\sqrt{2\pi}} e^{-\frac{1}{2}z^2},$$

which is independent of the true value of the unknown parameter; so we can compute the probability that Z will be between any two arbitrarily chosen numbers. Thus, for example,

$$P[-1.96 < Z < 1.96] = \int_{-1.96}^{1.96} \phi(z)\, dz = .95. \tag{1}$$

In this relation the inequality $-1.96 < Z$, or

$$-1.96 < \frac{\bar{X} - \mu}{\frac{3}{2}},$$

is equivalent to the inequality

$$\mu < \bar{X} + \tfrac{3}{2}(1.96) = \bar{X} + 2.94,$$

and the inequality

$$Z < 1.96$$

is equivalent to

$$\mu > \bar{X} - 2.94.$$

We may therefore rewrite Eq. (1) in the form

$$P[\bar{X} - 2.94 < \mu < \bar{X} + 2.94] = .95, \tag{2}$$

and substituting 2.7 for $\bar{X}$ we obtain the interval

$$(-.24, 5.64).$$

It is at this point that a certain abuse of language takes place since the random interval $(\bar{X} - 2.94, \bar{X} + 2.94)$ and the interval $(-.24, 5.64)$ are each called a *confidence interval*, or more precisely a 95 percent confidence interval. [The interval $(-.24, 5.64)$ is the value of the random interval $(\bar{X} - 2.94, \bar{X} + 2.94)$ when $\bar{X} = 2.7$.] The meaning of Eq. (2) is the following: The probability that the *random interval* $(\bar{X} - 2.94, \bar{X} + 2.94)$ covers the unknown true mean μ is .95. That is, if samples of size 4 were repeatedly drawn from the normal population and if the random interval $(\bar{X} - 2.94, \bar{X} + 2.94)$ were computed for each sample, then the relative frequency of those intervals that contain the true unknown mean μ would approach 95 percent. We therefore have considerable *confidence* that the observed interval, here $(-.24, 5.64)$, covers the true mean. The measure of our *confidence* is .95 because *before* the sample was drawn .95 was the probability that the interval that we were going to construct would cover the true mean. .95 is called the *confidence coefficient.*

Similarly, intervals with any desired degree of confidence between 0 and 1 can be obtained. Thus, since

$$P[-2.58 < Z < 2.58] = .99,$$

a 99 percent confidence interval for the true mean is obtained by converting the inequalities as before to get

$$P[\bar{X} - 3.87 < \mu < \bar{X} + 3.87] = .99$$

and then substituting 2.7 for $\bar{X}$ to get the interval $(-1.17, 6.57)$.

It is to be observed that there are, in fact, many possible intervals with the same probability (with the same confidence coefficient). Thus, for example, since

$$P[-1.68 < Z < 2.70] = .95,$$

another 95 percent confidence interval for μ is given by the interval $(-1.35, 5.22)$. This interval is inferior to the one obtained before because its length 6.57 is greater than the length 5.88 of the interval $(-.24, 5.64)$; it gives less precise information about the location of μ. Any two numbers a and b such that

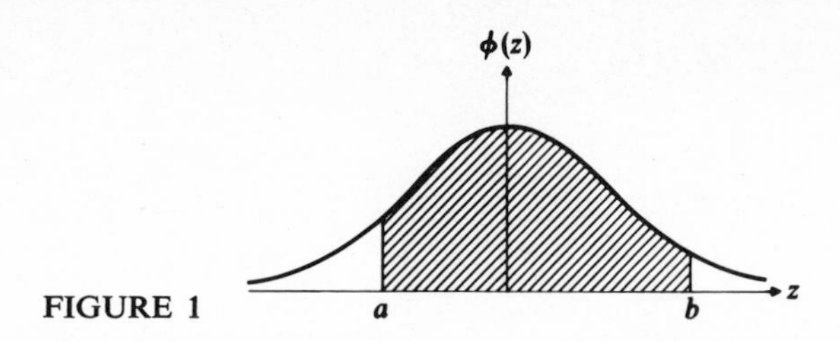

FIGURE 1

95 percent of the area under $\phi(z)$ lies between a and b will determine a 95 percent confidence interval. Ordinarily one would want the confidence interval to be as short as possible, and it is made so by making a and b as close together as possible because the relation $P[a < Z < b] = .95$ gives rise to a confidence interval of length $(\sigma/\sqrt{n})(b - a)$. The distance $b - a$ will be minimized for a fixed area when $\phi(a) = \phi(b)$, as is evident on referring to Fig. 1. If the point b is moved a short distance to the left, the point a will need to be moved a lesser distance to the left in order to keep the area the same; this operation decreases the length of the interval and will continue to do so as long as $\phi(b) < \phi(a)$. Since $\phi(z)$ is symmetrical about $z = 0$ in the present example, the minimum value of $b - a$ for a fixed area occurs when $b = -a$. Thus for $\bar{x} = 2.7$, $(-.24, 5.64)$ gives the shortest 95 percent confidence interval, and $(-1.17, 6.57)$ gives the shortest 99 percent confidence interval for μ.

In most problems it is not possible to construct confidence intervals which are shortest for a given confidence coefficient. In these cases one may wish to find a confidence interval which has the shortest expected length or is such that the probability that the confidence interval covers a value μ^* is minimized, where $\mu^* \neq \mu$.

The method of finding a confidence interval that has been illustrated in the example above is a general method. The method entails finding, if possible, a function (the quantity Z above) of the sample and the parameter to be estimated which has a distribution independent of the parameter and any other parameters. Then any probability statement of the form $P[a < Z < b] = \gamma$ for known a and b, where Z is the function, will give rise to a probability statement about the parameter that we hope can be rewritten to give a confidence interval. This method, or technique, is fully described in Subsec. 2.3 below. This technique is applicable in many important problems, but in others it is not because in these others it is either impossible to find functions of the desired form or it is impossible to rewrite the derived probability statements. These latter problems can be dealt with by a more general technique to be described in Sec. 4.

The idea of interval estimation can be extended to include simultaneous

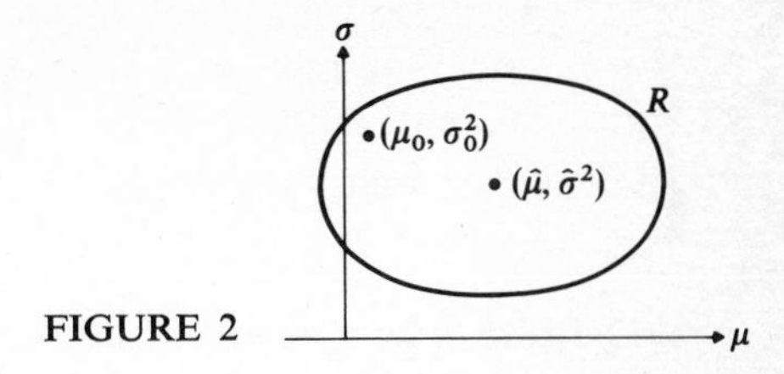

FIGURE 2

estimation of several parameters. Thus the two parameters of the normal distribution may be estimated by some plane region R in the so-called parameter space, that is, the space of all possible combinations of values of μ and σ^2. A 95 percent *confidence region* is a region constructible from the sample such that if samples were repeatedly drawn and a region constructed for each sample, 95 percent of those regions in a long-term relative-frequency sense would include the true parameter point (μ_0, σ_0^2)(see Fig. 2).

Confidence intervals and regions provide good illustrations of uncertain inferences. In Eq. (2) the inference is made that the interval $-.24$ to 5.64 covers the true parameter value, but that statement is not made categorically. A measure, .05, of the uncertainty of the inference is an essential part of the statement.

2.2 Definition of Confidence Interval

In the previous subsection we attempted to give some feel for the concept of a confidence interval by discussing a simple example. In this subsection we define, in general, what a confidence interval is and in the next subsection describe one method of finding confidence intervals.

We assume that we have a random sample $X_1, \ldots, X_n$ from a density $f(\cdot\,; \theta)$ parameterized by θ. Previously, in Chap. VII, we considered point estimates of say $\tau(\theta)$, some real function of θ. Now we look for an interval estimate of $\tau(\theta)$.

Definition 1 Confidence interval Let $X_1, \ldots, X_n$ be a random sample from the density $f(\cdot\,; \theta)$. Let $T_1 = \mathscr{t}_1(X_1, \ldots, X_n)$ and $T_2 = \mathscr{t}_2(X_1, \ldots, X_n)$ be two statistics satisfying $T_1 \leq T_2$ for which $P_\theta[T_1 < \tau(\theta) < T_2] \equiv \gamma$, where γ does not depend on θ; then the random interval (T_1, T_2) is called a 100γ *percent confidence interval for* $\tau(\theta)$; γ is called the *confidence coefficient*; and T_1 and T_2 are called the *lower* and *upper confidence limits*, respectively, *for* $\tau(\theta)$. A value (t_1, t_2) of the random interval (T_1, T_2) is also called a 100γ *percent confidence interval for* $\tau(\theta)$. ////

We note that one or the other, but not both, of the two statistics $t_1(X_1, \ldots, X_n)$ and $t_2(X_1, \ldots, X_n)$ may be constant; that is, one of the two end points of the random interval (T_1, T_2) may be constant.

Definition 2 One-sided confidence interval Let $X_1, \ldots, X_n$ be a random sample from the density $f(\cdot\,; \theta)$. Let $T_1 = t_1(X_1, \ldots, X_n)$ be a statistic for which $P_\theta[T_1 < \tau(\theta)] \equiv \gamma$; then T_1 is called a *one-sided lower confidence limit for* $\tau(\theta)$. Similarly, let $T_2 = t_2(X_1, \ldots, X_n)$ be a statistic for which $P_\theta[\tau(\theta) < T_2] \equiv \gamma$; then T_2 is called a *one-sided upper confidence limit for* $\tau(\theta)$. (γ does not depend on θ.) ////

EXAMPLE 1 Let $X_1, \ldots, X_n$ be a random sample from $f(x; \theta) = \phi_{\theta, 9}(x)$. Set $T_1 = t_1(X_1, \ldots, X_n) = \bar{X} - 6/\sqrt{n}$ and $T_2 = t_2(X_1, \ldots, X_n) = \bar{X} + 6/\sqrt{n}$; then (T_1, T_2) constitutes a random interval and is a confidence interval for $\tau(\theta) = \theta$ with confidence coefficient $\gamma = P_\theta[\bar{X} - 6/\sqrt{n} < \theta < \bar{X} + 6/\sqrt{n}] = P_\theta[-2 < (\bar{X} - \theta)/(3/\sqrt{n}) < 2] = \Phi(2) - \Phi(-2) = .9772 - .0228 = .9544$. Also, if a random sample of 25 observations has a sample mean of, say, 17.5, then the interval $(17.5 - 6/\sqrt{25},\ 17.5 + 6/\sqrt{25})$ is also called a 95.44 percent confidence interval for θ. ////

Remark If a confidence interval for θ has been determined, then, in essence, a whole family of confidence intervals has been determined. More specifically, for a given 100γ percent confidence interval estimator of θ a 100γ percent confidence-interval estimator of $\tau(\theta)$ can be obtained, where $\tau(\cdot)$ is *any* strictly monotone function. For example if $\tau(\cdot)$ is a monotone, increasing function and $(T_1 = t_1(X_1, \ldots, X_n), T_2 = t_2(X_1, \ldots, X_n))$ is a 100γ percent confidence interval for θ, then $(\tau(T_1), \tau(T_2))$ is a 100γ percent confidence interval for $\tau(\theta)$ since

$$P_\theta[\tau(T_1) < \tau(\theta) < \tau(T_2)] = P_\theta[T_1 < \theta < T_2] = \gamma. \qquad ////$$

As was the case in point estimation, our problem is twofold: First, we need methods of finding a confidence interval, and, second, we need criteria for comparing competing confidence intervals or for assessing the goodness of a confidence interval. In the next subsection, we will describe one method of finding confidence intervals and call it the *pivotal-quantity method*.

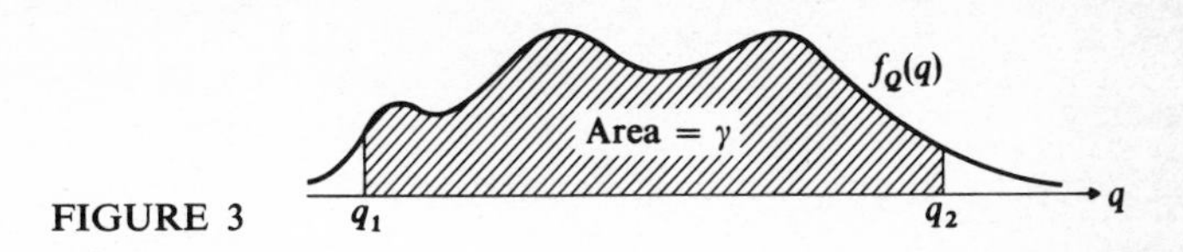

FIGURE 3

2.3 Pivotal Quantity

As before, we assume a random sample $X_1, \ldots, X_n$ from some density $f(\cdot; \theta)$ parameterized by θ. Our object is to find a confidence-interval estimate of $\tau(\theta)$, a real-valued function of θ. θ itself may be vector-valued.

Definition 3 Pivotal quantity Let $X_1, \ldots, X_n$ be a random sample from the density $f(\cdot; \theta)$. Let $Q = q(X_1, \ldots, X_n; \theta)$; that is, let Q be a function of $X_1, \ldots, X_n$ and θ. If Q has a distribution that does not depend on θ, then Q is defined to be a *pivotal quantity*. ////

EXAMPLE 2 Let $X_1, \ldots, X_n$ be a random sample from $f(x; \theta) = \phi_{\theta,9}(x)$. $\bar{X} - \theta$ is a pivotal quantity since $\bar{X} - \theta$ is normally distributed with mean 0 and variance $9/n$. Also $(\bar{X} - \theta)/(3/\sqrt{n})$ has a standard normal distribution and, hence, is a pivotal quantity. On the other hand, $\bar{X}/\theta$ is not a pivotal quantity since $\bar{X}/\theta$ is normally distributed with mean unity and variance $9/\theta^2 n$, which depends on θ. ////

Our hope is to utilize a pivotal quantity to obtain a confidence interval.

Pivotal-quantity method If $Q = q(X_1, \ldots, X_n; \theta)$ is a pivotal quantity and has a probability density function, then for any fixed $0 < \gamma < 1$ there will exist q_1 and q_2 depending on γ such that $P[q_1 < Q < q_2] = \gamma$. Now, if for each possible sample value $(x_1, \ldots, x_n)$, $q_1 < q(x_1, \ldots, x_n; \theta) < q_2$ if and only if $t_1(x_1, \ldots, x_n) < \tau(\theta) < t_2(x_1, \ldots, x_n)$ for functions t_1 and t_2 (not depending on θ), then (T_1, T_2) is a 100γ percent confidence interval for $\tau(\theta)$, where $T_i = t_i(X_1, \ldots, X_n)$, $i = 1, 2$.

Before illustrating the pivotal-quantity method with a simple example we make several comments. First, q_1 and q_2 are independent of θ since the distribution of Q is. Second, for any fixed γ there are many possible pairs of numbers q_1 and q_2 that can be selected so that $P[q_1 < Q < q_2] = \gamma$. See Fig. 3. Different pairs of q_1 and q_2 will produce different t_1 and t_2. We should want to select that pair of q_1 and q_2 that will make t_1 and t_2 close together in some sense. For instance, if $t_2(X_1, \ldots, X_n) - t_1(X_1, \ldots, X_n)$, which is the length of the confidence

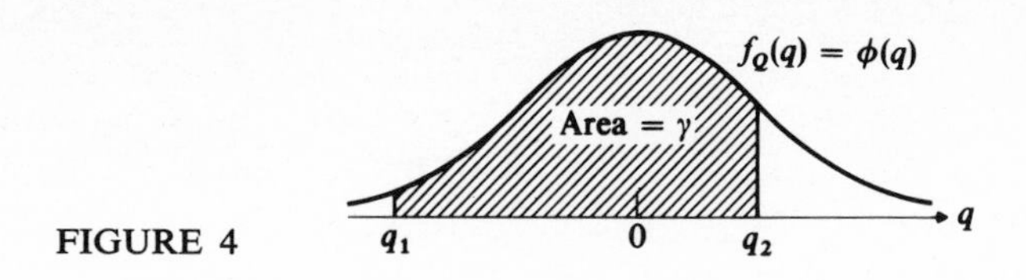

FIGURE 4

interval, is not random, then we might select that pair of q_1 and q_2 that makes the length of the interval smallest; or if the length of the confidence interval is random, then we might select that pair of q_1 and q_2 that makes the *average* length of the interval smallest.

As a third and final comment, note that the essential feature of the pivotal-quantity method is that the inequality $\{q_1 < \mathscr{q}(x_1, \ldots, x_n; \theta) < q_2\}$ can be rewritten, or inverted or "pivoted," as $\{t_1(x_1, \ldots, x_n) < \tau(\theta) < t_2(x_1, \ldots, x_n)\}$ for any possible sample value $x_1, \ldots, x_n$. [This last comment indicates that "pivotal quantity" may be a misnomer since according to our definition $Q = \mathscr{q}(X_1, \ldots, X_n; \theta)$ may be a pivotal quantity, yet it may be impossible to "pivot" it.]

EXAMPLE 3 Let $X_1, \ldots, X_n$ be a random sample from $\phi_{\theta,1}(x)$. Consider estimating $\tau(\theta) = \theta$. $Q = \mathscr{q}(X_1, \ldots, X_n; \theta) = (\bar{X} - \theta)/(\sqrt{1/n})$ has a standard normal distribution and, hence, is a pivotal quantity. $f_Q(q) = \phi(q)$. For given γ there exist q_1 and q_2 such that $P[q_1 < Q < q_2] = \gamma$ (in fact, there exist many such q_1 and q_2). See Fig. 4.

Now $\{q_1 < (\bar{x} - \theta)/\sqrt{1/n} < q_2\}$ if and only if $\{\bar{x} - q_2\sqrt{1/n} < \theta < \bar{x} - q_1\sqrt{1/n}\}$; so $(\bar{X} - q_2\sqrt{1/n}, \bar{X} - q_1\sqrt{1/n})$ is a 100γ percent confidence interval for θ. The length of the confidence interval is given by $(\bar{X} - q_1\sqrt{1/n}) - (\bar{X} - q_2\sqrt{1/n}) = (q_2 - q_1)\sqrt{1/n}$; so the length will be made smallest by selecting q_1 and q_2 so that $q_2 - q_1$ is a minimum under the restriction that $\gamma = P[q_1 < Q < q_2] = \Phi(q_2) - \Phi(q_1)$, and $q_2 - q_1$ will be a minimum if $q_1 = -q_2$, as can be seen from Fig. 4. ////

The steps in the pivotal-quantity method of finding a confidence interval are two: First, find a pivotal quantity, and, second, invert it. We will comment further on techniques for finding pivotal quantities in Sec. 4. The method is thoroughly exploited in the next section.

3 SAMPLING FROM THE NORMAL DISTRIBUTION

Let $X_1, \ldots, X_n$ be a random sample from the normal distribution with mean μ and variance σ^2. The first three subsections of this section are generated by the cases (i) confidence interval for μ only, (ii) confidence interval for σ^2 only, and (iii) simultaneous confidence interval for μ and σ. The fourth subsection considers a confidence interval for the difference between two means.

3.1 Confidence Interval for the Mean

There are really two cases to consider depending on whether or not σ^2 is known. We leave the case σ^2 known as an exercise. (The technique is given in Example 3.) We want a confidence interval for μ when σ^2 is unknown. In our general discussion in Sec. 2 above our parameter was denoted by θ. Here $\theta = (\mu, \sigma)$, and $\tau(\theta) = \mu$. We need a pivotal quantity. $(\bar{X} - \mu)/(\sigma/\sqrt{n})$ has a standard normal distribution; so it is a pivotal quantity, but $\{q_1 < (\bar{x} - \mu)/(\sigma/\sqrt{n}) < q_2\}$ cannot be inverted to give $\{t_1(x_1, \ldots, x_n) < \mu < t_2(x_1, \ldots, x_n)\}$ for any statistics t_1 and t_2. The problem with $(\bar{X} - \mu)/(\sigma/\sqrt{n})$ seems to be the presence of σ. We look for a pivotal quantity involving only μ. We know that

$$\frac{(\bar{X} - \mu)/(\sigma/\sqrt{n})}{\sqrt{\sum (X_i - \bar{X})^2/(n-1)\sigma^2}} = \frac{\bar{X} - \mu}{S/\sqrt{n}}$$

has a t distribution with $n - 1$ degrees of freedom. [Recall that $S^2 = \sum (X_i - \bar{X})^2/(n-1)$.] So $(\bar{X} - \mu)/(S/\sqrt{n})$ has a density that is independent of μ and σ^2; hence it is a pivotal quantity. Now one has $\{q_1 < (\bar{x} - \mu)/(s/\sqrt{n}) < q_2\}$ if and only if $\{\bar{x} - q_2(s/\sqrt{n}) < \mu < \bar{x} - q_1(s/\sqrt{n})\}$, where q_1 and q_2 are such that $P[q_1 < (\bar{X} - \mu)/(S/\sqrt{n}) < q_2] = \gamma$; therefore, $(\bar{X} - q_2(S/\sqrt{n}), \bar{X} - q_1(S/\sqrt{n}))$ is a 100γ percent confidence interval for μ. The length of this confidence interval is $(q_2 - q_1)(S/\sqrt{n})$, which is random. For any given sample the length will be minimized if q_1 and q_2 are selected so that $q_2 - q_1$ is a minimum. A little reflection will convince one that q_1 and q_2 should be symmetrically selected about 0, or the following argument can be advanced. We seek to minimize

$$L = \frac{S}{\sqrt{n}}(q_2 - q_1)$$

subject to

$$\int_{q_1}^{q_2} f_T(t)\, dt = \gamma, \tag{3}$$

where $f_T(t)$ is the density of the t distribution with $n-1$ degrees of freedom. Equation (3) gives q_2 as a function of q_1, and differentiating Eq. (3) with respect to q_1 yields

$$f_T(q_2)\frac{dq_2}{dq_1} - f_T(q_1) = 0.$$

To minimize L, we set $dL/dq_1 = 0$; that is,

$$\frac{dL}{dq_1} = \frac{S}{\sqrt{n}}\left(\frac{dq_2}{dq_1} - 1\right) = 0,$$

but

$$\frac{S}{\sqrt{n}}\left(\frac{dq_2}{dq_1} - 1\right) = \frac{S}{\sqrt{n}}\left(\frac{f_T(q_1)}{f_T(q_2)} - 1\right) = 0$$

if and only if $f_T(q_1) = f_T(q_2)$, which implies that $q_1 = q_2$ [in which case $\int_{q_1}^{q_2} f_T(t)\,dt \neq \gamma$] or $q_1 = -q_2$. $q_1 = -q_2$ is the desired solution, and such q_1 and q_2 can be readily obtained from a table of the t distribution.

3.2 Confidence Interval for the Variance

Again there are two cases depending on whether or not μ is assumed known, and again we leave the case μ known as an exercise. We want a confidence interval for σ^2 when μ is unknown. We need a pivotal quantity that can be inverted. We know that

$$Q = \frac{\sum (X_i - \overline{X})^2}{\sigma^2} = \frac{(n-1)S^2}{\sigma^2}$$

has a chi-square distribution with $n-1$ degrees of freedom; hence Q is a pivotal quantity. Also, one has

$$\left\{q_1 < \frac{(n-1)s^2}{\sigma^2} < q_2\right\}$$

if and only if

$$\left\{\frac{(n-1)s^2}{q_2} < \sigma^2 < \frac{(n-1)s^2}{q_1}\right\}$$

so

$$\left(\frac{(n-1)S^2}{q_2}, \frac{(n-1)S^2}{q_1}\right)$$

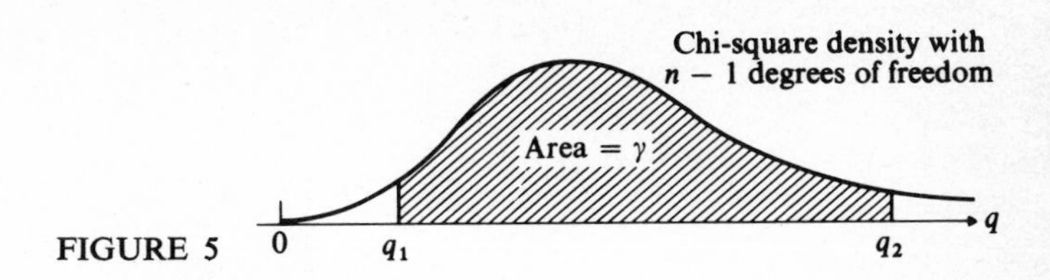

FIGURE 5

is a 100γ percent confidence interval for σ^2, where q_1 and q_2 are given by $P[q_1 < Q < q_2] = \gamma$. See Fig. 5.

q_1 and q_2 are often selected so that $P[Q < q_1] = P[Q > q_2] = (1 - \gamma)/2$. Such a confidence interval is sometimes referred to as the *equal-tails* confidence interval for σ^2. q_1 and q_2 can be obtained from a table of the chi-square distribution. Again, we might be interested in selecting q_1 and q_2 so as to minimize the length, say L, of the confidence interval.

$$L = (n - 1)S^2 \left(\frac{1}{q_1} - \frac{1}{q_2}\right).$$

Let $f_Q(q)$ be a chi-square density with $n - 1$ degrees of freedom; then differentiating

$$\int_{q_1}^{q_2} f_Q(q)\, dq = \gamma$$

with respect to q_1 yields

$$\frac{dq_2}{dq_1} f_Q(q_2) - f_Q(q_1) = 0,$$

and so

$$\frac{dL}{dq_1} = (n - 1)S^2\left(-\frac{1}{q_1^2} + \frac{1}{q_2^2}\frac{dq_2}{dq_1}\right) = (n - 1)S^2\left(-\frac{1}{q_1^2} + \frac{1}{q_2^2}\frac{f_Q(q_1)}{f_Q(q_2)}\right) = 0,$$

which implies that $q_1^2 f_Q(q_1) = q_2^2 f_Q(q_2)$. The length of the confidence interval will be minimized if q_1 and q_2 are selected so that

$$q_1^2 f_Q(q_1) = q_2^2 f_Q(q_2)$$

subject to

$$\int_{q_1}^{q_2} f_Q(q)\, dq = \gamma.$$

A solution for q_1 and q_2 can be obtained by trial and error or numerical integration.

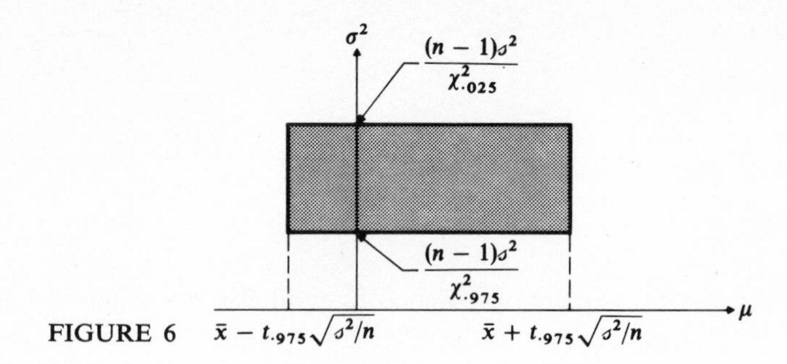

FIGURE 6

We might note that for any q_1 and q_2 satisfying

$$\int_{q_1}^{q_2} f_Q(q)\,dq = \gamma,$$

$$\left(\sqrt{\frac{(n-1)S^2}{q_2}}, \sqrt{\frac{(n-1)S^2}{q_1}}\right)$$

is a 100γ percent confidence interval for σ.

3.3 Simultaneous Confidence Region for the Mean and Variance

In constructing a region for the joint estimation of the mean μ and variance σ^2 of a normal distribution, one might at first be inclined to use Subsec. 3.1 and 3.2 above. That is, for example, one might construct a confidence region as in Fig. 6 by using the two relations

$$P[\bar{X} - t_{.975}\sqrt{S^2/n} < \mu < \bar{X} + t_{.975}\sqrt{S^2/n}] = .95 \tag{4}$$

and

$$P\left[\frac{(n-1)S^2}{\chi^2_{.975}} < \sigma^2 < \frac{(n-1)S^2}{\chi^2_{.025}}\right] = .95, \tag{5}$$

where $t_{.975}$ is the .975th quantile point of the t distribution with $n-1$ degrees of freedom and $\chi^2_{.025}$ and $\chi^2_{.975}$ are the .025th quantile point and .975th quantile point, respectively, of the chi-square distribution with $n-1$ degrees of freedom. The region displayed in Fig. 6 does indeed give a confidence region for (μ, σ^2), but we do not know what its corresponding confidence coefficient is. [It is not $.95^2$ since the two events given in Eqs. (4) and (5) are not independent.]

A confidence region, whose corresponding confidence coefficient can be

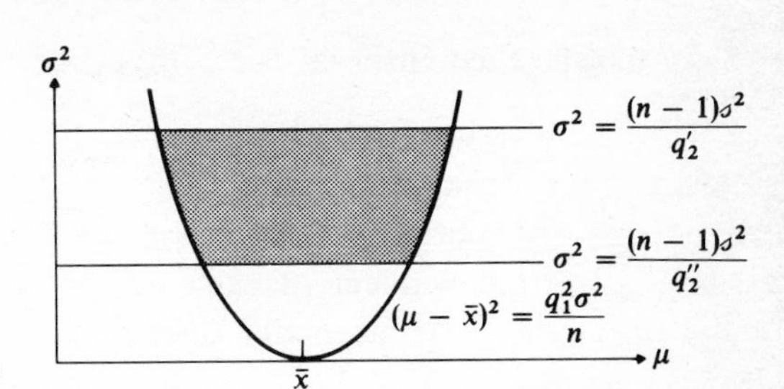

FIGURE 7

readily evaluated, may be set up, however, using the independence of $\bar{X}$ and S^2. Since

$$Q_1 = \frac{\bar{X} - \mu}{\sigma/\sqrt{n}} \quad \text{and} \quad Q_2 = \frac{(n-1)S^2}{\sigma^2}$$

are each pivotal quantities, we may find numbers q_1, q_2', and q_2'' such that

$$P\left[-q_1 < \frac{\bar{X} - \mu}{\sigma/\sqrt{n}} < q_1\right] = \gamma_1 \tag{6}$$

and

$$P\left[q_2' < \frac{(n-1)S^2}{\sigma^2} < q_2''\right] = \gamma_2. \tag{7}$$

Also, since Q_1 and Q_2 are independent, we have the joint probability

$$P\left[-q_1 < \frac{\bar{X} - \mu}{\sigma/\sqrt{n}} < q_1;\, q_2' < \frac{(n-1)S^2}{\sigma^2} < q_2''\right] = \gamma_1\gamma_2. \tag{8}$$

The four inequalities in Eq. (8) determine a region in the parameter space, which is easily found by plotting its boundaries. One merely replaces the inequality signs by equality signs and plots each of the four resulting equations as functions of μ and σ^2. A region such as the shaded area in Fig. 7 will result.

We might note that a confidence region for (μ, σ) could be obtained in exactly the same way; the equations would be plotted as functions of σ instead of σ^2, and the parabola in Fig. 7 would become a pair of straight lines given by $\mu = \bar{x} \pm q_1\sigma/\sqrt{n}$ intersecting at $\bar{x}$ on the μ axis.

The region that we have constructed does not have minimum area for fixed γ_1 and γ_2. Its advantage is that it is easily constructible from existing tables and it will differ but little from the region of minimum area unless the sample size is small. The region of minimum area is roughly elliptic in shape and difficult to construct.

3.4 Confidence Interval for Difference in Means

Let $X_1, \ldots, X_m$ be a random sample of size m from a normal distribution with mean μ_1 and variance σ^2, and let $Y_1, \ldots, Y_n$ be a random sample of size n from a normal distribution with mean μ_2 and variance σ^2. Assume that the two samples are independent of each other. We want a confidence interval for $\mu_2 - \mu_1$. $\bar{Y} - \bar{X}$ is normally distributed with mean $\mu_2 - \mu_1$ and variance $\sigma^2/n + \sigma^2/m$. $\sum (X_i - \bar{X})/\sigma^2$ is chi-square-distributed with $m - 1$ degrees of freedom, and $\sum (Y_i - \bar{Y})^2/\sigma^2$ is chi-square-distributed with $n - 1$ degrees of freedom; hence

$$\frac{\sum (X_i - \bar{X})^2}{\sigma^2} + \frac{\sum (Y_i - \bar{Y})^2}{\sigma^2}$$

is chi-square-distributed with $m + n - 2$ degrees of freedom. Finally,

$$\begin{aligned} Q &= \frac{[(\bar{Y} - \bar{X}) - (\mu_2 - \mu_1)]/\sqrt{\sigma^2/m + \sigma^2/n}}{\sqrt{[\sum (X_i - \bar{X})^2 + \sum (Y_i - \bar{Y})^2]/\sigma^2(m + n - 2)}} \\ &= \frac{(\bar{Y} - \bar{X}) - (\mu_2 - \mu_1)}{\sqrt{(1/m + 1/n)\,[\sum (X_i - \bar{X})^2 + \sum (Y_i - \bar{Y})^2]/(m + n - 2)}} \\ &= \frac{(\bar{Y} - \bar{X}) - (\mu_2 - \mu_1)}{\sqrt{(1/m + 1/n)(S_p^2)}} \end{aligned}$$

has a t distribution with $m + n - 2$ degrees of freedom. Thus it follows that $\gamma = P[-t_{(1+\gamma)/2} < Q < t_{(1+\gamma)/2}]$, where $t_{(1+\gamma)/2}$ is the $[(1 + \gamma)/2]$th quantile point of the t distribution with $m + n - 2$ degrees of freedom. S_p^2 is an unbiased estimator of the common variance σ^2. (The subscript p can be thought of as an abbreviation for "pooled"; S_p^2 is a *pooled* estimator of σ^2, the two samples being pooled together.) Now

$$-t_{(1+\gamma)/2} < \frac{(\bar{y} - \bar{x}) - (\mu_2 - \mu_1)}{\sqrt{(1/m + 1/n)s_p^2}} < t_{(1+\gamma)/2}$$

if and only if

$$(\bar{y} - \bar{x}) - t_{(1+\gamma)/2}\sqrt{\left(\frac{1}{m} + \frac{1}{n}\right)s_p^2} < \mu_2 - \mu_1 < (\bar{y} - \bar{x}) + t_{(1+\gamma)/2}\sqrt{\left(\frac{1}{m} + \frac{1}{n}\right)s_p^2};$$

hence

$$\left((\bar{Y} - \bar{X}) - t_{(1+\gamma)/2}\sqrt{\left(\frac{1}{m} + \frac{1}{n}\right)S_p^2},\ (\bar{Y} - \bar{X}) + t_{(1+\gamma)/2}\sqrt{\left(\frac{1}{m} + \frac{1}{n}\right)S_p^2}\right) \tag{9}$$

is a 100γ percent confidence interval for $\mu_2 - \mu_1$.

We assumed above that we had two independent random samples. Now assume that $(X_1, Y_1), \ldots, (X_n, Y_n)$ is a random sample from the bivariate normal distribution with parameters given by $\mu_1 = \mathscr{E}[X]$, $\mu_2 = \mathscr{E}[Y]$, $\sigma_1^2 = \text{var}[X]$, $\sigma_2^2 = \text{var}[Y]$, and $\rho = \text{cov}\,[X, Y]/\sigma_1\sigma_2$. The object is to find a confidence-interval estimate of $\mu_2 - \mu_1$. Let $D_i = Y_i - X_i$, $i = 1, \ldots, n$; then $D_1, \ldots, D_n$ are independent and identically distributed random variables with common normal distribution having mean $\mu_D = \mu_2 - \mu_1$ and variance $\sigma_D^2 = \sigma_1^2 + \sigma_2^2 - 2\rho\sigma_1\sigma_2$. Pretending that $D_1, \ldots, D_n$ is our random sample and proceeding as in Subsec. 3.1, we obtain the following 100γ percent confidence interval for $\mu_2 - \mu_1$:

$$\left(\bar{D} - t_{(1+\gamma)/2}\sqrt{\frac{\sum (D_i - \bar{D})^2}{n(n-1)}},\ \bar{D} + t_{(1+\gamma)/2}\sqrt{\frac{\sum (D_i - \bar{D})^2}{n(n-1)}}\right) \tag{10}$$

where $t_{(1+\gamma)/2}$ is the $[(1 + \gamma)/2]$th quantile point of the t distribution with $n - 1$ degrees of freedom. The above obtained confidence interval for $\mu_2 - \mu_1$ is often referred to as the confidence interval for the difference in means for *paired* observations. The ith X observation is paired with the ith Y observation.

4 METHODS OF FINDING CONFIDENCE INTERVALS

In this section we will discuss two methods of obtaining a confidence interval. A third method will be presented in Chap. IX.

4.1 Pivotal-quantity Method

We described the pivotal-quantity method of finding confidence intervals in Subsec. 2.3, but we left unanswered the question of whether or not a pivotal quantity would actually exist for a given problem. The following remark gives a partial answer to this question.

Remark If $X_1, \ldots, X_n$ is a random sample from $f(\cdot\,; \theta)$, for which the corresponding cumulative distribution function $F(x; \theta)$ is continuous in x, then, by the probability integral transform, $F(X_i; \theta)$ has a uniform distribution over the interval (0, 1). Hence $-\log F(X_i; \theta)$ has the density $e^{-u}I_{(0,\infty)}(u)$ since $P[-\log F(X_i; \theta) \geq u] = P[\log F(X_i; \theta) \leq -u] = P[F(X_i; \theta) \leq e^{-u}] = e^{-u}$ for $u > 0$. Finally $-\sum \log F(X_i; \theta)$ has a gamma distribution with parameters n and 1; that is,

$$P\left[-\log q_2 < -\sum_{i=1}^{n}\log F(X_i;\theta) < -\log q_1\right]$$
$$= \int_{-\log q_2}^{-\log q_1}\frac{1}{\Gamma(n)}z^{n-1}e^{-z}\,dz$$
$$= P\left[q_1 < \prod_{i=1}^{n}F(X_i;\theta) < q_2\right] \quad \text{for} \quad 0 < q_1 < q_2 < 1. \tag{11}$$

So

$$\prod_{i=1}^{n}F(X_i;\theta), \qquad \text{or} \quad -\sum_{i=1}^{n}\log F(X_i;\theta),$$

is a pivotal quantity. ////

The remark shows that any time that we sample from a population having a continuous cumulative distribution function, a pivotal quantity exists. Note that as of yet we have no assurance that the pivotal quantity exhibited by the remark can be utilized to find a confidence interval. If, however, $F(x;\theta)$ is monotone in θ for each x, then $\prod_{i=1}^{n} F(x_i;\theta)$ is also monotone in θ for each $x_1, \ldots, x_n$, and such monotonicity allows one to find a confidence interval for θ. We see from Fig. 8 that $q_1 < \prod_{i=1}^{n} F(x_i;\theta) < q_2$ if and only if $\ell_1(x_1, \ldots, x_n) < \theta < \ell_2(x_1, \ldots, x_n)$, where ℓ_1 and ℓ_2 are defined as indicated.

EXAMPLE 4 Let $X_1, \ldots, X_n$ be a random sample from the density $f(x;\theta) = \theta x^{\theta-1}I_{(0,1)}(x)$; then $F(x;\theta)x^{\theta} = I_{(0,1)}(x) + I_{[1,\infty)}(x)$. If q_1 and q_2 are selected [see Eq. (11)] so that

$$\gamma = P\left[q_1 < \prod_{i=1}^{n}F(X_i;\theta) < q_2\right]$$
$$= P\left[q_1 < \prod_{i=1}^{n}X_i^{\theta} < q_2\right]$$
$$= P\left[\log q_1 < \theta\log\prod_{i=1}^{n}X_i < \log q_2\right]$$
$$= P\left[-\log q_2 < -\theta\log\prod_{i=1}^{n}X_i < -\log q_1\right]$$
$$= P\left[\frac{\log q_2}{\log\prod_{i=1}^{n}X_i} < \theta < \frac{\log q_1}{\log\prod_{i=1}^{n}X_i}\right],$$

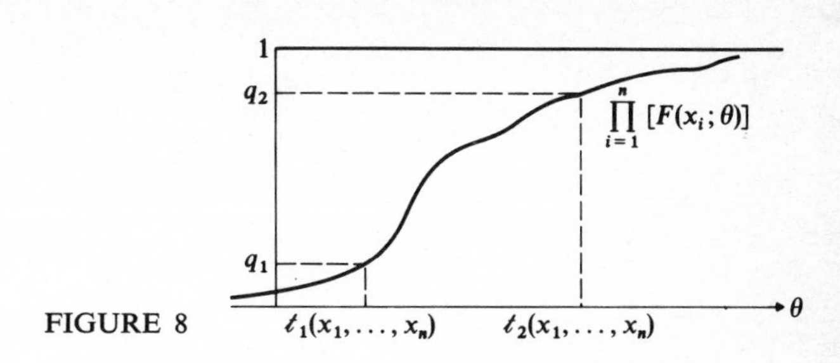

FIGURE 8

then

$$\left(\frac{\log q_2}{\log \prod_{i=1}^{n} X_i}, \frac{\log q_1}{\log \prod_{i=1}^{n} X_i} \right)$$

is a 100γ percent confidence interval for θ. ////

We conclude this subsection with two further comments regarding the existence of pivotal quantities. First, if θ is a location parameter, then $X_i - \theta$ has a distribution independent of θ by definition and, hence, is a pivotal quantity as are a variety of other random quantities, including $\sum X_i - n\theta$, $Y_j - \theta$, $Y_1 + Y_n - 2\theta$, etc. Second, if θ is a scale parameter, then by definition X_i/θ is distributed independently of θ and, hence, is a pivotal quantity as are $\sum X_i/\theta$, Y_j/θ, etc.

4.2 Statistical Method

As usual, we assume that we have a random sample $X_1, \ldots, X_n$ from the density $f(\cdot\,; \theta_0)$. We further assume that the parameter θ_0 is real and that the parameter space $\bar{\Theta}$ is some interval. (In this subsection, we will let θ_0 denote the true parameter value.) We seek an interval estimate of θ_0 itself. Let $T = \ell(X_1, \ldots, X_n)$ be some statistic. The statistic T can be selected in several ways. For instance, if a sufficient statistic (unidimensional) exists, then T could be taken to be a sufficient statistic; or if no sufficient statistic exists, T could be taken to be a point estimator, possibly the maximum-likelihood estimator, of θ_0. The actual choice of T might depend on the ease with which the operations that need to be performed to obtain the confidence interval can be performed. One of those operations will be the determination of the density of T.

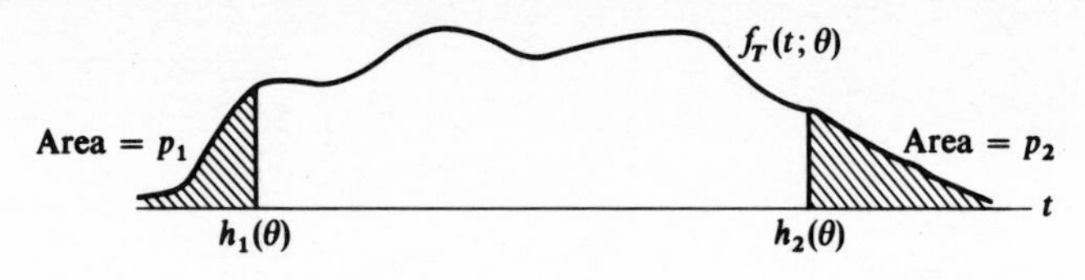

FIGURE 9

Let $f_T(t;\ \theta)$ denote the density of T. We will proceed as though T is a continuous random variable; although the technique will also work for T as a discrete random variable. We can define two functions, say $h_1(\theta)$ and $h_2(\theta)$, as follows:

$$\int_{-\infty}^{h_1(\theta)} f_T(t;\ \theta)\, dt = p_1 \qquad \text{and} \qquad \int_{h_2(\theta)}^{\infty} f_T(t;\ \theta)\, dt = p_2, \tag{12}$$

where p_1 and p_2 are two fixed numbers satisfying $0 < p_1$, $0 < p_2$, and $p_1 + p_2 < 1$. See Fig. 9.

$h_1(\theta)$ and $h_2(\theta)$ can be plotted as functions of θ. We will assume that both $h_1(\cdot)$ and $h_2(\cdot)$ are strictly monotone, and for our sketch we will assume that they are monotone, increasing functions. We know that $h_1(\theta) < h_2(\theta)$. See Fig. 10.

Let t_0 denote an observed value of T; that is, $t_0 = t(x_1, \ldots, x_n)$ for an observed random sample $x_1, \ldots, x_n$. Plot the value of t_0 on the vertical axis in Fig. 10, and then find v_1 and v_2 as indicated. For any possible value of t_0, a corresponding v_1 and v_2 can be obtained, so v_1 and v_2 are functions of t_0; denote these by $v_1 = v_1(t_0)$ and $v_2 = v_2(t_0)$. The interval (V_1, V_2) will turn out to be a $100(1 - p_1 - p_2)$ percent confidence interval for θ_0. To argue that this is so, let us repeat Fig. 10 as Fig. 11 and add to it. (Figure 10 indicates the method of finding the confidence interval.)

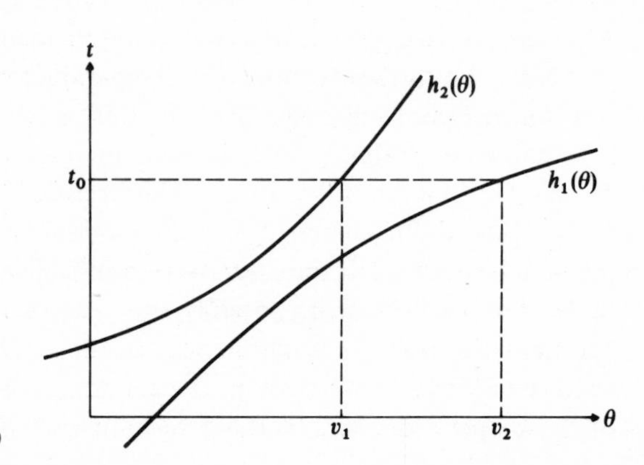

FIGURE 10

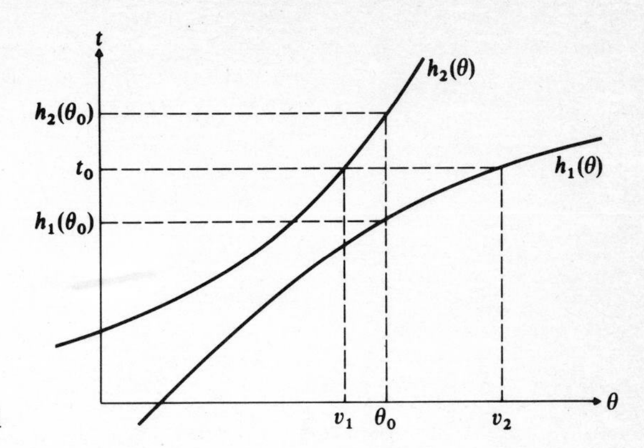

FIGURE 11

We see from Fig. 11 that $h_1(\theta_0) < t_0 = \ell(x_1, \ldots, x_n) < h_2(\theta_0)$ if and only if $v_1 = v_1(x_1, \ldots, x_n) < \theta_0 < v_2 = v_2(x_1, \ldots, x_n)$ for any possible observed sample $(x_1, \ldots, x_n)$. But by definition of $h_1(\cdot)$ and $h_2(\cdot)$,

$$P_{\theta_0}[h_1(\theta_0) < \ell(X_1, \ldots, X_n) < h_2(\theta_0)] = 1 - p_1 - p_2;$$

so

$$P_{\theta_0}[v_1(X_1, \ldots, X_n) < \theta_0 < v_2(X_1, \ldots, X_n)] = 1 - p_1 - p_2;$$

that is, as stated, (V_1, V_2) is a $100(1 - p_1 - p_2)$ percent confidence interval for θ_0, where $V_i = v_i(X_1, \ldots, X_n)$ for $i = 1, 2$.

We might note that the above procedure would work even if $h_1(\cdot)$ and $h_2(\cdot)$ were not monotone functions, only then we would obtain a confidence region (often in the form of a set of intervals) instead of a confidence interval.

EXAMPLE 5 Let $X_1, \ldots, X_n$ be a random sample from the density $f(x; \theta_0) = (1/\theta_0)I_{(0, \theta_0)}(x)$. We want a confidence interval for θ_0. $Y_n = \max [X_1, \ldots, X_n]$ is known to be a sufficient statistic; it is also the maximum-likelihood estimator of θ_0. We will use Y_n as our statistic T that appears in the above discussion; then

$$f_T(t; \theta) = n\left(\frac{t}{\theta}\right)^{n-1} \frac{1}{\theta} I_{(0, \theta)}(t).$$

For given p_1 and p_2, find $h_1(\theta)$ and $h_2(\theta)$. $p_1 = \int_0^{h_1(\theta)} nt^{n-1}\theta^{-n}\, dt$ implies that $\int_\theta^{h_1(\theta)} t^{n-1}\, dt = \theta^n p_1/n$, which in turn implies $[h_1(\theta)]^n/n = \theta^n p_1/n$, or finally $h_1(\theta) = \theta p_1^{1/n}$. Similarly, $p_2 = \int_{h_2(\theta)}^{\theta} nt^{n-1}\theta^{-n}\, dt$ implies that $\theta^n - [h_2(\theta)]^n = \theta^n p_2$ or $h_2(\theta) = \theta(1 - p_2)^{1/n}$. See Fig. 12, which is Fig. 10 for the example at hand.

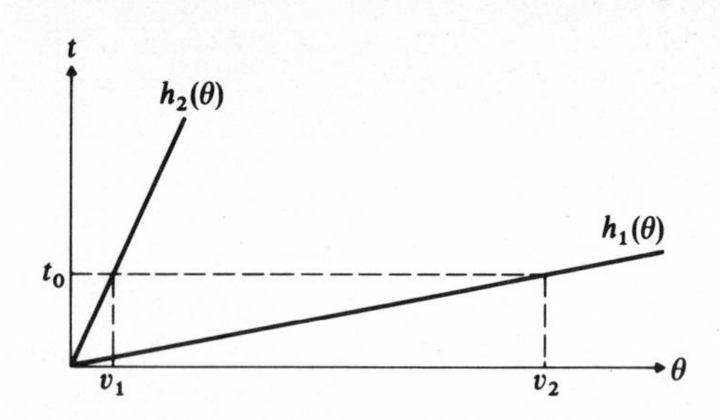

FIGURE 12

For observed $t_0 = \max[x_1, \ldots, x_n]$, v_1 is such that $h_2(v_1) = t_0$; that is, $h_2(v_1) = v_1(1 - p_2)^{1/n} = t_0$ or $v_1 = t_0(1 - p_2)^{-1/n}$. Similarly, $v_2 = t_0 p^{-1/n}$. So a $100(1 - p_1 - p_2)$ percent confidence interval for θ_0 is given by $(Y_n(1 - p_2)^{-1/n}, Y_n p_1^{-1/n})$. We could worry about selecting p_1 and p_2 so that the confidence interval is shortest subject to the restriction that $1 - p_1 - p_2 = \gamma$. The length of the confidence interval is

$$L = Y_n[p_1^{-1/n} - (1 - p_2)^{-1/n}]$$

and so the length will be shortest if p_1 and p_2 are picked so as to minimize $p_1^{-1/n} - (1 - p_2)^{-1/n}$ subject to $1 - p_1 - p_2 = \gamma$ and $0 < p_1 + p_2 < 1$, which is accomplished by picking $p_2 = 0$ and $p_1 = 1 - \gamma$.

We might note that Y_n/θ is a pivotal quantity and a confidence interval for θ can be obtained more easily using the pivotal-quantity method. ////

We observe in the example above, and in general for that matter, that $h_1(\theta)$ *and* $h_2(\theta)$ *are really not needed.* For a given observed value $t_0 = t(x_1, \ldots, x_n)$ of the statistic T, we need to find $v_1 = v_1(x_1, \ldots, x_n)$ and $v_2 = v_2(x_1, \ldots, x_n)$. v_2 can be found by solving for θ in the equation

$$p_1 = \int_{-\infty}^{h_1(\theta)=t_0} f_T(t; \theta)\, dt; \qquad (13)$$

v_2 is the solution; v_1 can be found by solving for θ in the equation

$$p_2 = \int_{h_2(\theta)=t_0}^{\infty} f_T(t; \theta)\, dt; \qquad (14)$$

v_1 is the solution.

We mentioned at the outset that the method would work for discrete random variables as well as for continuous random variables. Then the integrals in Eqs. (12) to (14) would need to be replaced by summations. Two popular discrete density functions are the Bernoulli and Poisson. One could be interested in confidence-interval estimates of the parameters in each. In Example 6 to follow we will consider the Bernoulli density function; the Poisson case is left as an exercise.

EXAMPLE 6 Let $X_1, \ldots, X_n$ be a random sample from the Bernoulli density; that is, $P[X = 1] = \theta_0 = 1 - P[X = 0]$. We know that $T = \sum X_i$ is a sufficient statistic; furthermore, T has a binomial distribution; that is, $P[T = t] = \binom{n}{t}\theta_0^t(1 - \theta_0)^{n-t}$ for $t = 0, 1, \ldots, n$. We want a confidence-interval estimate for θ_0. Suppose we observe $T = t_0$ (necessarily an integer). According to Eqs. (13) and (14) we need to solve for θ in each of the equations

$$p_1 = \sum_{t=0}^{t_0} \binom{n}{t} \theta^t(1 - \theta)^{n-t}$$

and

$$p_2 = \sum_{t=t_0}^{n} \binom{n}{t} \theta^t(1 - \theta)^{n-t}.$$

To actually solve these equations, a table of the binomial distribution is useful. If $p_1 = .0509$, $p_2 = .0159$, $n = 20$, and if $T = 4$ is observed then the 93.33 percent confidence interval (.05, .40) for θ_0 is obtained. ////

5 LARGE-SAMPLE CONFIDENCE INTERVALS

We have seen in our studies of point estimation that it is sometimes possible to find a sequence of estimators, say $T_n = t_n(X_1, \ldots, X_n)$, of θ in a density $f(\cdot\,; \theta)$ that are asymptotically normally distributed about θ; that is, T_n is approximately normally distributed with mean θ and variance, say, $\sigma_n^2(\theta)$, where $\sigma_n^2(\theta)$ indicates that the variance is a function of θ (since it will ordinarily depend on θ) and the sample size n. In particular, we have seen in Sec. 9 of Chap. VII that for large samples the maximum-likelihood estimator, say $\hat{\Theta}_n = \hat{\vartheta}_n(X_1, \ldots, X_n)$, for a parameter θ in a density $f(\cdot\,; \theta)$ is approximately normally distributed about θ under rather general conditions. The large-sample variance of the maximum-likelihood estimator was seen to be, say,

$$\sigma_n^2(\theta) = \frac{1}{n\mathscr{E}_\theta[\{(\partial/\partial\theta) \log f(X; \theta)\}^2]} = \frac{-1}{n\mathscr{E}_\theta[(\partial^2/\partial\theta^2) \log f(X; \theta)]}. \tag{15}$$

When such a sequence of asymptotically normally distributed estimators $\{T_n\}$ of θ exists, it is sometimes possible to obtain approximate confidence intervals quite easily. $(T_n - \theta)/\sigma_n(\theta)$ can be treated as an approximate pivotal quantity, and, therefore, for large sample size a confidence interval with an approximate confidence coefficient γ may be determined by converting the inequalities

$$-z < \frac{T_n - \theta}{\sigma_n(\theta)} < z, \tag{16}$$

where $z = z_{(1+\gamma)/2}$ is defined by $\Phi(z_{(1+\gamma)/2}) = (1 + \gamma)/2$ or $\Phi(z) - \Phi(-z) = \gamma$. The above described method will always work to find a large-sample confidence interval provided the inequality $-z < (T_n - \theta)/\sigma_n(\theta) < z$ can be inverted.

EXAMPLE 7 Let $X_1, \ldots, X_n$ be a random sample from the density $f(x; \theta) = \theta e^{-\theta x} I_{(0,\infty)}(x)$. We know (see Example 51 in Chap. VII) that the maximum-likelihood estimator of θ, which is $1/\overline{X}_n$, has an asymptotic normal distribution with mean θ and variance equal to

$$\sigma_n^2(\theta) = \frac{1}{n\mathscr{E}_\theta[\{(\partial/\partial\theta)\log f(X; \theta)\}^2]} = \frac{\theta^2}{n}.$$

Therefore,

$$\begin{aligned}\gamma &\approx P\left[-z < \frac{1/\overline{X}_n - \theta}{\sqrt{\theta^2/n}} < z\right]\\ &= P\left[\frac{-z\theta}{\sqrt{n}} < \frac{1}{\overline{X}_n} - \theta < \frac{z\theta}{\sqrt{n}}\right]\\ &= P\left[\frac{1/\overline{X}_n}{1 + z/\sqrt{n}} < \theta < \frac{1/\overline{X}_n}{1 - z/\sqrt{n}}\right],\end{aligned}$$

and hence

$$\left(\frac{1/\overline{X}_n}{1 + z/\sqrt{n}}, \frac{1/\overline{X}_n}{1 - z/\sqrt{n}}\right)$$

is a large-sample confidence interval for θ with an approximate confidence coefficient γ, where z is given by $\Phi(z) - \Phi(-z) = \gamma$. ////

EXAMPLE 8 Consider sampling from the Bernoulli distribution with parameter $\theta = P[X = 1] = 1 - P[X = 0]$. The maximum-likelihood estimator of θ is $\hat{\Theta} = \bar{X}_n$, and it has variance $\sigma_n^2(\theta) = \theta(1 - \theta)/n$. An approximate 100γ percent confidence interval for θ is obtained by converting the inequalities in

$$P\left[-z < \frac{\hat{\Theta} - \theta}{\sqrt{\theta(1-\theta)/n}} < z\right] \approx \gamma$$

to get

$$P\left[\frac{2n\hat{\Theta} + z^2 - z\sqrt{4n\hat{\Theta} + z^2 - 4n\hat{\Theta}^2}}{2(n + z^2)} < \theta < \frac{2n\hat{\Theta} + z^2 + z\sqrt{4n\hat{\Theta} + z^2 - 4n\hat{\Theta}^2}}{2(n + z^2)}\right] \approx \gamma. \tag{17}$$

These expressions for the limits may be simplified since in deriving the large-sample distribution certain terms containing the factor $1/\sqrt{n}$ are neglected; that is, the asymptotic normal distribution is correct only to within error terms of size a constant times $1/\sqrt{n}$. We may therefore neglect terms of this order in the limits in Eq. (17) without appreciably affecting the accuracy of the approximation. This means simply that we may omit all the z^2 terms in Eq. (17) because they always occur added to a term with factor n and will be negligible relative to n when n is large to within the degree of approximation that we are assuming. Thus Eq. (17) may be rewritten as

$$P\left[\hat{\Theta} - z\sqrt{\frac{\hat{\Theta}(1-\hat{\Theta})}{n}} < \theta < \hat{\Theta} + z\sqrt{\frac{\hat{\Theta}(1-\hat{\Theta})}{n}}\right] \approx \gamma. \tag{18}$$

In particular,

$$P\left[\hat{\Theta} - 1.96\sqrt{\frac{\hat{\Theta}(1-\hat{\Theta})}{n}} < \theta < \hat{\Theta} + 1.96\sqrt{\frac{\hat{\Theta}(1-\hat{\Theta})}{n}}\right] \approx .95$$

gives an approximate 95 percent confidence interval for θ for large samples. ////

We may observe that Eq. (18) is just the expression that would have been obtained had $\hat{\Theta}$ been substituted for θ in $\sigma_n^2(\theta)$. The substitution would imply that

$$\frac{\hat{\Theta} - \theta}{\sqrt{\hat{\Theta}(1-\hat{\Theta})/n}}$$

is approximately normally distributed with mean 0 and unit variance. It is, in fact, true in general that in the asymptotic normal distribution of a maximum-likelihood estimator $\hat{\Theta}$ the variance $\sigma_n^2(\theta)$ may be replaced by its estimator $\sigma_n^2(\hat{\Theta})$ without appreciably affecting the accuracy of the approximation. We shall not prove this fact but shall use it because it greatly simplifies the conversion of the inequalities that is required to get the large-sample confidence intervals. For instance,

$$P\left[-z < \frac{\hat{\Theta} - \theta}{\sigma_n(\hat{\Theta})} < z\right] \approx \gamma$$

readily converts to give

$$P[\hat{\Theta} - z\sigma_n(\hat{\Theta}) < \theta < \hat{\Theta} + z\sigma_n(\hat{\Theta})] \approx \gamma, \tag{19}$$

where $\hat{\Theta}$ is the asymptotically normally distributed maximum-likelihood estimator of θ, $\sigma_n(\hat{\Theta})$ in this expression is the maximum-likelihood estimator of $\sigma_n(\theta)$ (which is the large-sample standard deviation of $\hat{\Theta}$), and z is given by $\Phi(z) - \Phi(-z) = \gamma$.

We noted in Sec. 9 of Chap. VII that under regularity conditions the joint distribution of the maximum-likelihood estimators of the components of a k-dimensional parameter is asymptotically normally distributed. Although we will not so argue, such a result could be used to obtain a large-sample confidence region.

The large-sample confidence intervals presented in this section have an optimum property which we shall point out but not prove. Recall that in the earlier sections, particularly Sec. 3, of this chapter we were concerned with finding the shortest interval for a given probability. Loosely speaking, an analogous optimum property of large-sample confidence intervals based on maximum-likelihood estimators is this: *Large-sample confidence intervals based on the maximum-likelihood estimator will be shorter, on the average, than intervals determined by any other estimator.*

6 BAYESIAN INTERVAL ESTIMATES

In Sec. 7 of Chap. VII we examined what is called Bayes estimation. There we assumed that a random sample, say $X_1, \ldots, X_n$, from some density $f(\cdot\,; \theta) = f(\cdot \mid \theta)$ was available, where the form of the function $f(\cdot \mid \cdot)$ was known and the fixed value of θ was unknown. We further assumed that the unknown fixed value of θ was the value of a random variable Θ with known density, denoted by $g_\Theta(\cdot)$ and called the prior density of Θ. We then used this additional knowledge of a known prior density to define the posterior distribution of Θ, and from this posterior distribution we defined the posterior Bayes *point* estimator of θ. In this section we use this same posterior distribution of Θ to arrive at an *interval* estimator of θ.

If $f(\cdot \mid \theta)$ is the density sampled from and $g_\Theta(\cdot)$ is the prior density of Θ, then the posterior density of Θ given $(X_1, \ldots, X_n) = (x_1, \ldots, x_n)$ is [recall Eq. (19) of Chap. VII]

$$f_{\Theta \mid X_1 = x_1, \ldots, X_n = x_n}(\theta \mid x_1, \ldots, x_n) = \frac{\left[\prod_1^n f(x_i \mid \theta)\right] g_\Theta(\theta)}{\int \left[\prod_1^n f(x_i \mid \theta)\right] g_\Theta(\theta)\, d\theta}. \tag{20}$$

For fixed γ, any interval, say (t_1, t_2), satisfying

$$\int_{t_1}^{t_2} f_{\Theta \mid X_1 = x_1, \ldots, X_n = x_n}(\theta \mid x_1, \ldots, x_n)\, d\theta = \gamma \tag{21}$$

is defined to be a 100γ percent *Bayesian interval estimate* of θ. In practice, one would naturally pick those t_1 and t_2 satisfying Eq. (21) for which $t_2 - t_1$ is smallest. Note that $t_i = t_i(x_1, \ldots, x_n)$; that is, t_i is some function of the observations $x_1, \ldots, x_n$.

EXAMPLE 9 Let $X_1, \ldots, X_n$ be a random sample from the normal density with mean θ and variance 1. Assume that Θ has a normal density with mean x_0 and variance 1. Consider estimating θ. We saw in Example 44 of Chap. VII that the posterior distribution of Θ is normal with mean $\sum_0^n x_i/(n+1)$ and variance $1/(n+1)$. We seek t_1 and t_2 satisfying

$$\begin{aligned} \gamma &= \int_{t_1}^{t_2} f_{\Theta \mid X_1 = x_1, \ldots, X_n = x_n}(\theta \mid x_1, \ldots, x_n)\, d\theta \\ &= \Phi\left(\frac{t_2 - \sum_0^n x_i/(n+1)}{\sqrt{1/(n+1)}}\right) - \Phi\left(\frac{t_1 - \sum_0^n x_i/(n+1)}{\sqrt{1/(n+1)}}\right). \end{aligned} \tag{22}$$

If z is such that $\Phi(z) - \Phi(-z) = \gamma$, then

$$t_2 = \frac{\sum_0^n x_i}{n+1} + z\sqrt{\frac{1}{n+1}} \qquad \text{and} \qquad t_1 = \frac{\sum_0^n x_i}{n+1} - z\sqrt{\frac{1}{n+1}}$$

gives the shortest 100γ percent Bayesian interval estimate of θ. Note that the corresponding 100γ percent confidence-interval estimate of θ is given by $\left(\sum_1^n x_i/n - z\sqrt{1/n}, \sum_1^n x_i/n + z\sqrt{1/n}\right)$. The only difference in the results of the two methods for this example is that the sample size seems to increase by 1 and the apparent "additional observation" is the mean of the assumed prior normal distribution. ////

PROBLEMS

1 Let X be a single observation from the density

$$f(x; \theta) = \theta x^{\theta - 1} I_{(0, 1)}(x),$$

where $\theta > 0$.

(*a*) Find a pivotal quantity, and use it to find a confidence-interval estimator of θ.

(*b*) Show that $(Y/2, Y)$ is a confidence interval for θ. Find its confidence coefficient. Also, find a better confidence interval for θ. Define $Y = -1/\log X$.

2 Let $X_1, \ldots, X_n$ be a random sample from $N(\theta, \theta)$, $\theta > 0$. Give an example of a pivotal quantity, and use it to obtain a confidence-interval estimator of θ.

3 Suppose that T_1 is a 100γ percent lower confidence limit for $\tau(\theta)$ and T_2 is a 100γ percent upper confidence limit for $\tau(\theta)$. Further assume that $P_\theta[T_1 < T_2] = 1$. Find a $100(2\gamma - 1)$ percent confidence interval for $\tau(\theta)$. (Assume $\gamma > \frac{1}{2}$.)

4 Let $X_1, \ldots, X_n$ denote a random sample from $f(x; \theta) = I_{(\theta - \frac{1}{2}, \theta + \frac{1}{2})}(x)$. Let $Y_1 < \cdots < Y_n$ be the corresponding ordered sample. Show that (Y_1, Y_n) is a confidence interval for θ. Find its confidence coefficient.

5 Let $X_1, \ldots, X_n$ be a random sample from $f(x; \theta) = \theta e^{-\theta x} I_{(0, \infty)}(x)$.

(*a*) Find a 100γ percent confidence interval for the mean of the population.

(*b*) Do the same for the variance of the population.

(*c*) What is the probability that these intervals cover the true mean and true variance, simultaneously?

(*d*) Find a confidence-interval estimator of $e^{-\theta} = P[X > 1]$.

(*e*) Find a pivotal quantity based only on Y_1, and use it to find a confidence-interval estimator of θ. ($Y_1 = \min [X_1, \ldots, X_n]$.)

6 X is a single observation from $\theta e^{-\theta x} I_{(0, \infty)}(x)$, where $\theta > 0$.

(*a*) $(X, 2X)$ is a confidence interval for $1/\theta$. What is its confidence coefficient?

(*b*) Find another confidence interval for $1/\theta$ that has the same coefficient but smaller expected length.

7 Let X_1, X_2 denote a random sample of size 2 from $N(\theta, 1)$. Let $Y_1 < Y_2$ be the corresponding ordered sample.

(*a*) Determine γ in $P[Y_1 < \theta < Y_2] = \gamma$. Find the expected length of the interval (Y_1, Y_2).

(*b*) Find that confidence-interval estimator for θ using $\bar{X} - \theta$ as a pivotal quantity that has a confidence coefficient γ, and compare the length with the expected length in part (*a*).

8 Consider random sampling from a normal distribution with mean μ and variance σ^2.

(*a*) Derive a confidence interval estimator of μ when σ^2 is known.

(*b*) Derive a confidence interval estimator of σ^2 when μ is known.

9 Find a 90 percent confidence interval for the mean of a normal distribution with $\sigma = 3$ given the sample (3.3, −.3, −.6, −.9). What would be the confidence interval if σ were unknown?

10 The breaking strengths in pounds of five specimens of manila rope of diameter $\frac{3}{16}$ inch were found to be 660, 460, 540, 580, and 550.

(*a*) Estimate the mean breaking strength by a 95 percent confidence interval assuming normality.

(*b*) Estimate the point at which only 5 percent of such specimens would be expected to break.

(*c*) Estimate σ^2 by a 90 percent confidence interval; also σ.

(*d*) Plot an 81 percent confidence region for the joint estimation of μ and σ^2; for μ and σ.

11 A sample was drawn from each of five populations assumed to be normal with the same variance. The values of $(n-1)S^2 = \sum (X_i - \bar{X})^2$ and n, the sample size, were

S^2:	40	30	20	42	50
n:	6	4	3	7	8

Find 98 percent confidence limits for the common variance.

12 Develop a method for estimating the ratio of variances of two normal populations by a confidence interval.

13 What is the probability that the length of a t confidence interval for μ when sampling from a normal distribution will be less than σ for samples of size 20?

14 In sampling from a normal population compare the average length of the two confidence intervals for the mean μ when (*a*) σ is known and (*b*) σ is unknown.

15 Show that the length and the variance of the length of the t confidence interval for μ when sampling from a normal population approach 0 with increasing sample size.

16 In sampling from a normal population with both μ and σ unknown, how large a sample must be drawn to make the probability .95 that a 90 percent confidence interval for μ will have length less than $\sigma/5$?

17 Show that the length of the confidence interval for σ of a normal population approaches 0 with increasing sample size.

18 To test two promising new lines of hybrid corn under normal farming conditions, a seed company selected eight farms at random in Iowa and planted both lines in experimental plots on each farm. The yields (converted to bushels per acre) for the eight locations were

Line *A*:	86	87	56	93	84	93	75	79
Line *B*:	80	79	58	91	77	82	74	66

Assuming that the two yields are jointly normally distributed, estimate the difference between the mean yields by a 95 percent confidence interval.

19 $X_1, \ldots, X_n$ is a random sample from $(1/\theta)x^{(1-\theta)/\theta}I_{[0,1]}(x)$, where $\theta > 0$. Find the 100γ percent confidence interval for θ. Find its expected length. Find the limiting expected length of your confidence interval. Find n such that $P[\text{length} \leq \delta\theta] \geq \rho$ for fixed δ and ρ. (You may use the central-limit theorem.)

20 Develop a method for estimating the parameter of the Poisson distribution by a confidence interval.

21 Find a good 100γ percent confidence interval for θ when sampling from $f(x; \theta) = I_{(\theta - \frac{1}{2}, \theta + \frac{1}{2})}(x)$.

22 Find a good 100γ percent confidence interval for θ when sampling from $f(x; \theta) = (2x/\theta^2)I_{(0, \theta)}(x)$, where $\theta > 0$.

23 One head and two tails resulted when a coin was tossed three times. Find a 90 percent confidence interval for the probability of a head.

24 Suppose that 175 heads and 225 tails resulted from 400 tosses of a coin. Find a 90 percent confidence interval for the probability of a head. Find a 99 percent confidence interval. Does this appear to be a true coin?

25 Let $X_1, \ldots, X_n$ be a random sample from $f(x; \theta) = f(x; \mu, \sigma) = \phi_{\mu, \sigma^2}(x)$. Define $\tau(\theta)$ by $\int_{\tau(\theta)}^{\infty} \phi_{\mu, \sigma^2}(x)\, dx = \alpha$($\alpha$ is fixed). Recall what the UMVUE of $\tau(\theta)$ is. Find a 100γ percent confidence interval for $\tau(\theta)$. (If you cannot find an exact 100γ percent confidence interval, find an approximate one).

26 Let $X_1, \ldots, X_n$ be a random sample from $f(x; \theta) = \phi_{\theta, 1}(x)$. Assume that the prior distribution of Θ is $N(\mu_0, \sigma_0^2)$, μ_0 and σ_0^2 known. Find a 100γ percent Bayesian interval estimator of θ, and compare it with the corresponding confidence interval.

27 Let $X_1, \ldots, X_n$ be a random sample from $f(x \mid \theta) = \theta x^{\theta - 1}I_{(0, 1)}(x)$, where $\theta > 0$. Assume that the prior distribution of Θ is given by

$$g_\Theta(\theta) = \frac{1}{\Gamma(r)} \lambda^r \theta^{r-1} e^{-\lambda\theta} I_{(0, \infty)}(\theta),$$

where r and λ are known. Find a 95 percent Bayesian interval estimator of θ.

**28* Let X denote the life in hours of a radioactive particle. Suppose X has a density

$$f(x; \theta) = \theta e^{-\theta x} I_{(0, \infty)}(x).$$

A random sample of n particles is put under observation, but the experiment is to stop when the kth particle has expired; i.e., it is intended not to wait until all the particles have ceased activity but only until k of them (k fixed in advance) have done so. The data consist of the k measurements $Y_1, \ldots, Y_k$ and $n - k$ measurements known only to exceed Y_k, where Y_i is the lifetime of the ith particle to expire. Find the maximum-likelihood estimator of the mean lifetime $1/\theta$. Also find a confidence-interval estimator of $1/\theta$.

IX

TESTS OF HYPOTHESES

1 INTRODUCTION AND SUMMARY

There are two major areas of statistical inference: the estimation of parameters and the testing of hypotheses. We shall study the second of these two areas in this chapter. Our aim will be to develop general methods for testing hypotheses and to apply those methods to some common problems. The methods developed will be of further use in later chapters.

In experimental research, the object is sometimes merely to estimate parameters. Thus one may wish to estimate the yield of a new hybrid line of corn. But more often the ultimate purpose will involve some use of the estimate. One may wish, for example, to compare the yield of the new line with that of a standard line and perhaps recommend that the new line replace the standard line if it appears superior. This is a common situation in research. One may wish to determine whether a new method of sealing light bulbs will increase the life of the bulbs, whether a new germicide is more effective in treating a certain infection than a standard germicide, whether one method of preserving foods is better than another insofar as retention of vitamins is concerned, and so on.

Using the light-bulb example as an illustration, let us suppose that the average life of bulbs made under a standard manufacturing procedure is 1400 hours. It is desired to test a new procedure for manufacturing the bulbs. The statistical model here is this: We are dealing with two populations of light bulbs: those made by the standard process and those made by the proposed process. We know (from numerous past investigations) that the mean of the first population is about 1400. The question is whether the mean of the second population is greater than or less than 1400. Traditionally, to answer this type of question, we set up the hypothesis that one mean is greater than the other mean. Then, on the basis of a sample from the population of the proposed process we shall either accept or reject the hypothesis.

For our example, we formulate the hypothesis that the proposed process is no better than the standard process. Generally we hope that the hypothesis will be rejected. To test the hypothesis, a number of bulbs are made by the new process and their lives measured. Suppose that the mean of this sample of observations is 1550 hours. The indication is that the new process is better, but suppose that the estimate of the standard deviation of the mean $\hat{\sigma}/\sqrt{n}$ is 125 (n being the sample size). Then a 95 percent confidence interval for the mean of the second population (assuming normality) is roughly 1300 to 1800 hours. The sample mean 1550 could very easily have come from a population with mean 1400. We have no strong grounds for rejecting the hypothesis. If, on the other hand, $\hat{\sigma}/\sqrt{n}$ were 25, then we could very confidently reject the hypothesis and pronounce the proposed manufacturing process to be superior.

The testing of hypotheses is seen to be closely related to the problem of estimation. It will be instructive, however, to develop the theory of testing independently of the theory of estimation, at least in the beginning.

In order to conveniently talk about testing of hypotheses, we need to introduce some language and notation and give some definitions. As was the case when we studied estimation, we will assume that we can obtain a random sample $X_1, \ldots, X_n$ from some density $f(\cdot\,; \theta)$. A statistical hypothesis will be a hypothesis about the distribution of the population.

Definition 1 Statistical hypothesis A *statistical hypothesis* is an assertion or conjecture about the distribution of one or more random variables. If the statistical hypothesis completely specifies the distribution, then it is called *simple*; otherwise, it is called *composite*. ////

Notation To denote a statistical hypothesis, we will use a script capital $\mathscr{H}$ followed by a colon that in turn is followed by the assertion that specifies the hypothesis. ////

EXAMPLE 1 Let $X_1, \ldots, X_n$ be a random sample from $f(x; \theta) = \phi_{\theta, 25}(x)$. The statistical hypothesis that the mean of this normal population is less than or equal to 17 is denoted as follows: $\mathscr{H}: \theta \leq 17$. Such a hypothesis is composite; it does not completely specify the distribution. On the other hand, the hypothesis $\mathscr{H}: \theta = 17$ is simple since it completely specifies the distribution. ////

Definition 2 Test of a statistical hypothesis A *test* of a statistical hypothesis $\mathscr{H}$ is a rule or procedure for deciding whether to reject $\mathscr{H}$. ////

Notation Let us use a capital upsilon Υ to denote a test. ////

EXAMPLE 2 Let $X_1, \ldots, X_n$ be a random sample from $f(x; \theta) = \phi_{\theta, 25}(x)$. Consider $\mathscr{H}: \theta \leq 17$. One possible test Υ is as follows: Reject $\mathscr{H}$ if and only if $\bar{X} > 17 + 5/\sqrt{n}$. ////

A test can be either *randomized* or *nonrandomized*. The test Υ given in Example 2 above is an example of a nonrandomized test. Another possible test, say Υ', of $\mathscr{H}$ in Example 2 is the following: Toss a coin, and reject $\mathscr{H}$ if and only if a head appears. Such is an example of a randomized test. Although we will make little use of randomized tests in this book, we do include their definition. Definitions of both nonrandomized and randomized tests follow.

Notation As in previous chapters, we let $\mathfrak{X}$ denote the sample space of observations, or the potential data set; that is, $\mathfrak{X} = \{(x_1, \ldots, x_n): (x_1, \ldots, x_n)$ is a possible value of $(X_1, \ldots, X_n)\}$. ////

Definition 3 Nonrandomized test and critical region Let a test Υ of a statistical hypothesis $\mathscr{H}$ be defined as follows: Reject $\mathscr{H}$ if and only if $(x_1, \ldots, x_n) \in C_\Upsilon$, where C_Υ is a subset of $\mathfrak{X}$; then Υ is called a *nonrandomized test*, and C_Υ is called the *critical region* of the test Υ. ////

EXAMPLE 3 Let $X_1, \ldots, X_n$ be a random sample from $f(x; \theta) = \phi_{\theta, 25}(x)$. $\mathfrak{X}$ is euclidean n space. Consider $\mathscr{H}: \theta \leq 17$ and the test Υ: Reject $\mathscr{H}$ if and only if $\bar{x} > 17 + 5/\sqrt{n}$. Then Υ is nonrandomized, and $C_\Upsilon = \{(x_1, \ldots, x_n): \bar{x} > 17 + 5/\sqrt{n}\}$. ////

Remark A nonrandomized test Υ of $\mathscr{H}$ is a decomposition of $\mathfrak{X}$ into C_Υ and $\bar{C}_\Upsilon$ such that if $(x_1, \ldots, x_n) \in C_\Upsilon$, $\mathscr{H}$ is rejected. So a nonrandomized test is specified by its corresponding critical region. ////

Definition 4 Randomized test A test Υ of a hypothesis $\mathscr{H}$ is defined to be a *randomized test* if Υ is defined by the function $\psi_\Upsilon(x_1, \ldots, x_n) = P[\mathscr{H}$ is rejected$\,|\,(x_1, \ldots, x_n)$ is observed$]$. The function $\psi_\Upsilon(\cdot, \ldots, \cdot)$ is called the *critical function* of the test Υ. ////

The actual performing of a nonrandomized test Υ of $\mathscr{H}$ is straightforward; one observes a random sample, say $x_1, \ldots, x_n$, checks to see whether the observed sample falls in the critical region, and rejects $\mathscr{H}$ when it does. On the other hand, to perform a randomized test Υ of $\mathscr{H}$, one first observes the random sample, say $x_1, \ldots, x_n$, then evaluates $\psi_\Upsilon(x_1, \ldots, x_n)$, and finally observes the result of some auxiliary Bernoulli trial that has $\psi_\Upsilon(x_1, \ldots, x_n)$ as its probability of success, and if the Bernoulli trial results in a success, then $\mathscr{H}$ is rejected. Since the performance of the auxiliary Bernoulli trial is extraneous to the actual testing problem, that is, it does not depend on the data $x_1, \ldots, x_n$ of the experiment, one might reasonably wonder why its result should be the deciding factor in accepting or rejecting the hypothesis. It is for this reason that randomized tests are not often employed in practice; when they are, usually the sample space $\mathfrak{X}$ is decomposed into three sets, one where the hypothesis is accepted, another where the hypothesis is rejected, and the third where "randomization" takes place. This third region is often the boundary between the acceptance and rejection region and/or a region where it is not easy to decide whether to accept or reject. The following example may help in understanding randomized tests.

EXAMPLE 4 Let $X_1, \ldots, X_{10}$ be a random sample of size 10 from $f(x; \theta) = \theta^x(1-\theta)^{1-x}$ for $x = 0$ or 1. Suppose we want to test the hypothesis $\mathscr{H}: \theta < \frac{1}{2}$. A possible test Υ is the following: Reject $\mathscr{H}$ if $\sum_1^{10} x_i > 5$, accept $\mathscr{H}$ if $\sum_1^{10} x_i < 5$, and decide between rejecting and accepting by tossing a fair coin if $\sum_1^{10} x_i = 5$. (The tossing of the coin is the auxiliary Bernoulli trial.) Such a test Υ partitions $\mathfrak{X}$ into three regions, say A, B, and C, where

$$A = \left\{(x_1, \ldots, x_{10}): \sum_1^{10} x_i < 5\right\},$$

$$B = \left\{(x_1, \ldots, x_{10}): \sum_1^{10} x_i = 5\right\},$$

and

$$C = \left\{(x_1, \ldots, x_{10}) : \sum_1^{10} x_i > 5\right\}.$$

The critical function of test Υ is given by

$$\psi_\Upsilon(x_1, \ldots, x_{10}) = \begin{cases} 1 & \text{if } (x_1, \ldots, x_{10}) \in C \\ 1/2 & \text{if } (x_1, \ldots, x_{10}) \in B \\ 0 & \text{if } (x_1, \ldots, x_{10}) \in A. \end{cases}$$ ////

Remark We saw that a nonrandomized test was specified by its critical region. Likewise, a randomized test is specified by its critical function. In fact any function $\psi(\cdot, \ldots, \cdot)$ with domain $\mathfrak{X}$ and counterdomain the interval [0, 1] is a possible critical function and defines a randomized test. ////

The following remark shows that a nonrandomized test is a particular case of a randomized test.

Remark If test Υ has a critical function defined by

$$\psi_\Upsilon(x_1, \ldots, x_n) = \left.\begin{cases} 1 & \text{if } (x_1, \ldots, x_n), \in C_\Upsilon \\ 0 & \text{otherwise} \end{cases}\right\} = I_{C_\Upsilon}(x_1, \ldots, x_n),$$

then Υ is a nonrandomized test with a critical region C_Υ. ////

As we mentioned earlier, we will not make extensive use of randomized tests. Theorem 1 below requires their use; other than that, their only use will be in obtaining tests of exact size (see Definition 7), and then only for sampling from discrete distributions.

In many hypotheses-testing problems two hypotheses are discussed: The first, the hypothesis being tested, is called the *null* hypothesis, denoted by $\mathscr{H}_0$, and the second is called the *alternative* hypothesis, denoted by $\mathscr{H}_1$. The thinking is that if the null hypothesis is false, then the alternative hypothesis is true, and vice versa. We often say that $\mathscr{H}_0$ is tested against, or versus, $\mathscr{H}_1$. If the null hypothesis $\mathscr{H}_0$ is not rejected, we say that $\mathscr{H}_0$ is *accepted.* With this kind of thinking, two types of errors can be made.

Definition 5 Types of error and size of error Rejection of $\mathscr{H}_0$ when it is true is called a *Type I error*, and acceptance of $\mathscr{H}_0$ when it is false is called a *Type II error*. The *size of a Type I error* is defined to be the probability that a Type I error is made, and similarly the *size of a Type II error* is the probability that a Type II error is made. ////

If the distribution from which the sample was obtained is parameterized by θ, where $\theta \in \overline{\Theta}$, then associated with any test is a power function, defined as in Definition 6.

Definition 6 Power function Let Υ be a test of the null hypothesis $\mathscr{H}_0$. The *power function* of the test Υ, denoted by $\pi_\Upsilon(\theta)$, is defined to be the probability that $\mathscr{H}_0$ is rejected when the distribution from which the sample was obtained was parameterized by θ. ////

The power function will play the same role in hypothesis testing that mean-squared error played in estimation. It will usually be our standard in assessing the goodness of a test or in comparing two competing tests. An *ideal* power function, of course, is a function that is 0 for those θ corresponding to the null hypothesis and is unity for those θ corresponding to the alternative hypothesis. The idea is that you do not want to reject $\mathscr{H}_0$ if $\mathscr{H}_0$ is true and you do want to reject $\mathscr{H}_0$ when $\mathscr{H}_0$ is false.

Remark $\pi_\Upsilon(\theta) = P_\theta[\text{reject } \mathscr{H}_0]$, where θ is the true value of the parameter. If Υ is a nonrandomized test, then $\pi_\Upsilon(\theta) = P_\theta[(X_1, \ldots, X_n) \in C_\Upsilon]$, where C_Υ is the critical region associated with test Υ. If Υ is a randomized test with critical function $\psi_\Upsilon(\cdot, \ldots, \cdot)$, then

$$\begin{aligned}
\pi_\Upsilon(\theta) &= P_\theta[\text{reject } \mathscr{H}_0] \\
&= \int \cdots \int P[\text{reject } \mathscr{H}_0 \mid x_1, \ldots, x_n] f_{X_1, \ldots, X_n}(x_1, \ldots, x_n; \theta) \prod_{i=1}^{n} dx_i \\
&= \int \cdots \int \psi_\Upsilon(x_1, \ldots, x_n) f_{X_1, \ldots, X_n}(x_1, \ldots, x_n; \theta) \prod_{i=1}^{n} dx_i \\
&= \mathscr{E}_\theta[\psi_\Upsilon(X_1, \ldots, X_n)].
\end{aligned}$$

The argument is similar for discrete random variables. ////

EXAMPLE 5 Let $X_1, \ldots, X_n$ be a random sample from $f(x; \theta) = \phi_{\theta, 25}(x)$.

Consider $\mathscr{H}_0: \theta \le 17$ and the test Υ: Reject if and only if $\overline{X} > 17 + 5/\sqrt{n}$.

$$\pi_\Upsilon(\theta) = P_\theta\left[\overline{X} > 17 + \frac{5}{\sqrt{n}}\right] = 1 - \Phi\left(\frac{17 + 5/\sqrt{n} - \theta}{5/\sqrt{n}}\right).$$

For $n = 25$, $\pi_\Upsilon(\theta)$ is sketched in Fig. 1.

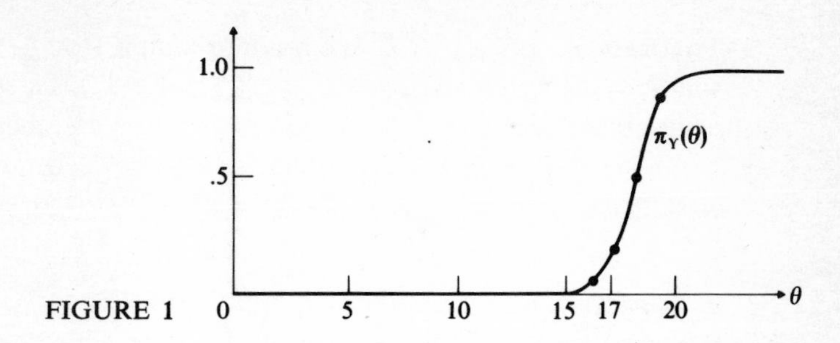

FIGURE 1

The power function is useful in telling how good a particular test is. In this example, if θ is greater than about 20, the test Υ is almost certain to reject $\mathscr{H}_0$, as it should. And if θ is less than about 16, the test Υ is almost certain not to reject $\mathscr{H}_0$, as it should. On the other hand, if $17 < \theta < 18$ (so $\mathscr{H}_0$ is false), the test Υ has less than half a chance of rejecting $\mathscr{H}_0$. ////

Definition 7 Size of test Let Υ be a test of the hypothesis $\mathscr{H}_0: \theta \in \bar{\Theta}_0$, where $\bar{\Theta}_0 \subset \bar{\Theta}$; that is, $\bar{\Theta}_0$ is a subset of the parameter space $\bar{\Theta}$. The *size of the test* Υ of $\mathscr{H}_0$ is defined to be $\sup_{\theta \in \bar{\Theta}_0} [\pi_\Upsilon(\theta)]$. The size of the test for a nonrandomized test is also referred to as the *size of the critical region.* ////

Remark Many writers use the terms "significance level" and "size of test" interchangeably. We, however, will avoid use of the term "significance level," intending to reserve its use for *tests of significance*, a type of statistical inference that is closely related to hypothesis testing. Tests of significance will not be considered in this book; the interested reader is referred to Ref. [37]. ////

EXAMPLE 6 Let $X_1, \ldots, X_n$ be a random sample from $f(x; \theta) = \phi_{\theta, 25}(x)$. Consider the $\mathscr{H}_0: \theta \leq 17$ and the test Υ: Reject $\mathscr{H}_0$ if $\bar{X} > 17 + 5/\sqrt{n}$.

$\bar{\Theta}_0 = \{\theta: \theta \leq 17\}$ and the size of the test Υ is $\sup_{\theta \in \bar{\Theta}_0} [\pi_\Upsilon(\theta)]$

$$= \sup_{\theta \leq 17} \left[1 - \Phi\left(\frac{17 + 5/\sqrt{n} - \theta}{5/\sqrt{n}}\right)\right] = 1 - \Phi(1) \approx .159. \qquad ////$$

In our study of point estimation, we found that for certain considerations we could restrict attention to estimators that were functions of sufficient statistics only. The same is true for testing hypotheses when the power function is used as a basis of comparing tests, as the following theorem shows.

Theorem 1 If $X_1, \ldots, X_n$ is a random sample from $f(x; \theta)$, where $\theta \in \underline{\overline{\Theta}}$, and $S_1 = \mathscr{s}_1(X_1, \ldots, X_n), \ldots, S_r = \mathscr{s}_r(X_1, \ldots, X_n)$ is a set of sufficient statistics, then for any test Υ with critical function ψ_Υ, there exists a test, say Υ', and corresponding critical function, say $\psi_{\Upsilon'}$, depending only on the set of sufficient statistics which satisfies $\pi_\Upsilon(\theta) = \pi_{\Upsilon'}(\theta)$ for all $\theta \in \underline{\overline{\Theta}}$.

PROOF Define $\psi_{\Upsilon'}(s_1, \ldots, s_r) = \mathscr{E}[\psi_\Upsilon(X_1, \ldots, X_n) | S_1 = s_1, \ldots, S_r = s_r]$; then $\psi_{\Upsilon'}$ is a critical function. Furthermore, $\pi_{\Upsilon'}(\theta) = \mathscr{E}_\theta[\psi_{\Upsilon'}(S_1, \ldots, S_r)] = \mathscr{E}_\theta[\mathscr{E}[\psi_\Upsilon(X_1, \ldots, X_n) | S_1, \ldots, S_r]] = \mathscr{E}_\theta[\psi_\Upsilon(X_1, \ldots, X_n)] = \pi_\Upsilon(\theta)$. ////

The theorem shows that given any test, another test which depends only on a set of sufficient statistics can be found, and this new test has a power function identical to the power function of the original test. So, in our search for good tests we need only look among tests that depend on sufficient statistics.

We have introduced some of the language of testing in the above. The problem of testing is like estimation in the sense that it is twofold: First, a method of finding a test is needed, and, second, some criteria for comparing competing tests are desirable. Although we will be interested in both aspects of the problem, we will not discuss them in that order. First we will consider, in Sec. 2, the problem of testing a simple null hypothesis against a simple alternative. Two approaches will be assumed. The first will use the power function as a basis for setting goodness criteria for tests, and the second will use a loss function. The Neyman-Pearson lemma is stated and proved. It will turn out that all those tests, which are best in some sense, will be of the form of a simple likelihood-ratio, which is defined.

Tests of composite hypotheses will be discussed in Sec. 3. The section will commence, in Subsec. 3.1, with a discussion of the generalized likelihood-ratio principle and the generalized likelihood-ratio test. This principle plays a central role in testing, just as maximum likelihood played a central role in estimation. It is a technique for arriving at a test that in general will be a good test, just as maximum likelihood led to an estimator that in general was quite a good estimator. For a book of the level of this book, it is probably the most important concept in testing. The notion of uniformly most powerful tests will be introduced in Subsec. 3.2, and several methods that are sometimes useful in finding such tests will be presented. Unbiasedness and invariance in estimation are two methods of restricting the class of estimators with the hope of finding a best estimator within the restricted class. These two concepts play essentially the same role in testing; they are methods of restricting the totality of possible tests with the hope of finding a best test within the restricted class. We will discuss only unbiasedness, and it only briefly in Subsec. 3.3. Subsection 3.4 will summarize several methods of finding tests of composite hypotheses.

Section 4 will be devoted to consideration of various hypotheses and tests that arise in sampling from a normal distribution. Section 5 will consider tests that fall within a category of tests generally labeled chi-square tests. Included will be the asymptotic distribution of the generalized likelihood-ratio, goodness-of-fit tests, tests of the equality of two or more distributions, and tests of independence in contingency tables. Section 6 will give the promised discussion of the connection between tests of hypotheses and interval estimation. The chapter will end with an introduction to sequential tests of hypotheses in Sec. 7.

The reader will note that our discussion of tests of hypotheses is not as thorough as that of estimation. Both testing and estimation will be used in later chapters, especially in Chap. X. Also, a number of the nonparametric techniques that will be presented in Chap. XI will be tests of hypotheses.

We stated at the beginning of this section that testing of hypotheses is one major area of statistical inference. A type of statistical inference that is closely related (in fact so closely related that many writers do not make a distinction) to hypothesis testing is that of *significance testing*. The concept of significance testing has important use in applied problems; however, we will not consider it in this book. The interested reader is referred to Ref. [37].

2 SIMPLE HYPOTHESIS VERSUS SIMPLE ALTERNATIVE

2.1 Introduction

In this section we consider testing a simple null hypothesis against a simple alternative hypothesis. This case is actually not very useful in applied statistics, but it will serve the purpose of introducing us to the theory of testing hypotheses.

We assume that we have a sample that came from one of two completely specified distributions. Our object is to determine which one. More precisely, assume that a random sample $X_1, \ldots, X_n$ came from the density $f_0(x)$ or $f_1(x)$ and we want to test $\mathscr{H}_0 : X_i$ distributed as $f_0(\cdot)$, abbreviated $X_i \sim f_0(\cdot)$, versus $\mathscr{H}_1 : X_i \sim f_1(\cdot)$. If we had only one observation x_1 and $f_0(\cdot)$ and $f_1(\cdot)$ were as in Fig. 2, one might quite rationally decide that the observation came from $f_0(\cdot)$ if $f_0(x_1) > f_1(x_1)$ and, conversely, decide that the observation came from $f_1(\cdot)$ if $f_1(x_1) > f_0(x_1)$. This simple intuitive method of obtaining a test can be expanded into a family of tests that, as we shall see, will contain some good tests.

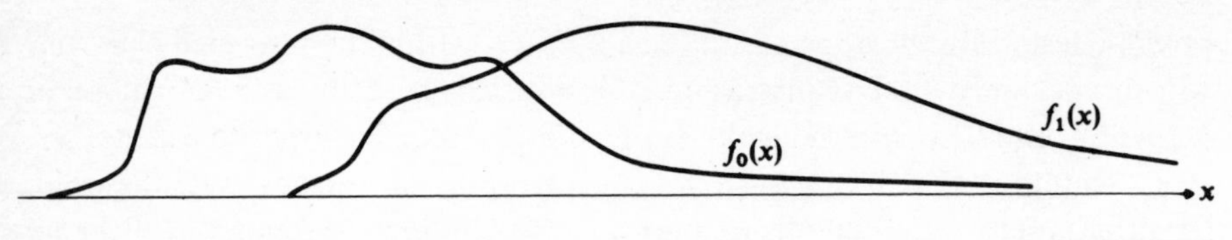

FIGURE 2

Definition 8 Simple likelihood-ratio test Let $X_1, \ldots, X_n$ be a random sample from either $f_0(\cdot)$ or $f_1(\cdot)$. A test Υ of $\mathscr{H}_0$: $X_i \sim f_0(\cdot)$ versus $\mathscr{H}_1$: $X_i \sim f_1(\cdot)$ is defined to be a *simple likelihood-ratio test* if Υ is defined by

$$\begin{aligned} &\text{Reject } \mathscr{H}_0 \quad \text{if } \lambda < k, \\ &\text{Accept } \mathscr{H}_0 \quad \text{if } \lambda > k, \\ &\text{Either accept } \mathscr{H}_0, \text{ reject } \mathscr{H}_0, \text{ or randomize} \quad \text{if } \lambda = k, \end{aligned} \tag{1}$$

where

$$\lambda = \lambda(x_1, \ldots, x_n) = \frac{\prod_{i=1}^{n} f_0(x_i)}{\prod_{i=1}^{n} f_1(x_i)} = \frac{L_0(x_1, \ldots, x_n)}{L_1(x_1, \ldots, x_n)} = \frac{L_0}{L_1}$$

and k is a nonnegative constant. [$L_j = L_j(x_1, \ldots, x_n)$ is the likelihood function for sampling from the density $f_j(\cdot)$.] ////

For each different k we have a different test. For a fixed k the test says to reject $\mathscr{H}_0$ if the ratio of likelihoods is small; that is, reject $\mathscr{H}_0$ if it is more likely (L_1 is large compared to L_0) that the sample came from $f_1(\cdot)$ than from $f_0(\cdot)$. Such a test certainly has intuitive appeal. In fact, one might suspect that an optimum test will have to be the form of a simple likelihood-ratio test.

Optimality of a test of a simple hypothesis versus a simple alternative can be approached in two ways. One way, using the power of the test to set goodness criteria, is discussed in Subsec. 2.2, and the other way, using a loss function and a decision-theoretical approach, is considered in Subsec. 2.3.

2.2 Most Powerful Test

Let $X_1, \ldots, X_n$ be a random sample from the density $f_0(\cdot)$ or the density $f_1(\cdot)$. Let us write $f_0(x) = f(x; \theta_0)$ and $f_1(x) = f(x; \theta_1)$; then $X_1, \ldots, X_n$ is a random sample from one or the other member of the parametric family $\{f(x; \theta): \theta = \theta_0$ or $\theta = \theta_1\}$. $\underline{\Theta} = \{\theta_0, \theta_1\}$ is a parameter space with only two points in it. θ_0 and

θ_1 are known. We want to test $\mathscr{H}_0 : \theta = \theta_0$ versus $\mathscr{H}_1 : \theta = \theta_1$. Corresponding to any test Υ of $\mathscr{H}_0$ versus $\mathscr{H}_1$ is its power function $\pi_\Upsilon(\theta)$. A good test is a test for which $\pi_\Upsilon(\theta_0) = P[\text{reject } \mathscr{H}_0 \mid \mathscr{H}_0 \text{ is true}]$ is small (ideally 0) and $\pi_\Upsilon(\theta_1) = P[\text{reject } \mathscr{H}_0 \mid \mathscr{H}_0 \text{ is false}]$ is large (ideally unity). One might reasonably use the two values $\pi_\Upsilon(\theta_0)$ and $\pi_\Upsilon(\theta_1)$ to set up criteria for defining a best test. $\pi_\Upsilon(\theta_0) =$ size of Type I error, and $1 - \pi_\Upsilon(\theta_1) = P[\text{accept } \mathscr{H}_0 \mid \mathscr{H}_0 \text{ is false}] =$ size of Type II error; so our goodness criterion might concern making the two error sizes small. For example, one might define as best that test which has the smallest sum of the error sizes. Another method of defining a best test, made precise in the following definition, is to fix the size of the Type I error and to minimize the size of the Type II error.

Definition 9 Most powerful test A test Υ^* of $\mathscr{H}_0$: $\theta = \theta_0$ versus $\mathscr{H}_1 : \theta = \theta_1$ is defined to be a *most powerful test of size* α $(0 < \alpha < 1)$ if and only if:

(i) $\pi_{\Upsilon^*}(\theta_0) = \alpha$. (2)

(ii) $\pi_{\Upsilon^*}(\theta_1) \geq \pi_\Upsilon(\theta_1)$ for any other test Υ for which $\pi_\Upsilon(\theta_0) \leq \alpha$. (3)

////

A test Υ^* is most powerful of size α if it has size α and if among all other tests of size α or less it has the largest power. Or a test Υ^* is most powerful of size α if it has the size of its Type I error equal to α and has smallest size Type II error among all other tests with size of Type I error α or less.

The justification for fixing the size of the Type I error to be α (usually small and often taken as .05 or .01) seems to arise from those testing situations where the two hypotheses are formulated in such a way that one type of error is more serious than the other. The hypotheses are stated so that the Type I error is the more serious, and hence one wants to be certain that it is small.

The following theorem is useful in finding a most powerful test of size α. The statement of the theorem as given here, as well as the proof, considers only nonrandomized tests. We might note that the statement and proof of the theorem can be altered to include all randomized tests.

Theorem 2 Neyman-Pearson lemma Let $X_1, \ldots, X_n$ be a random sample from $f(x; \theta)$, where θ is one of the two known values θ_0 or θ_1, and let $0 < \alpha < 1$ be fixed.

Let k^* be a positive constant and C^* be a subset of $\mathfrak{X}$ which satisfy:

(i) $P_{\theta_0}[(X_1, \ldots, X_n) \in C^*] = \alpha$. (4)

$$\text{(ii)} \quad \lambda = \frac{L(\theta_0; x_1, \ldots, x_n)}{L(\theta_1; x_1, \ldots, x_n)} = \frac{L_0}{L_1} \le k^* \qquad \text{if } (x_1, \ldots, x_n) \in C^* \tag{5}$$

and $\lambda \ge k^*$ if $(x_1, \ldots, x_n) \in \bar{C}^*$.

Then, the test Υ^* corresponding to the critical region C^* is a most powerful test of size α of $\mathscr{H}_0$: $\theta = \theta_0$ versus $\mathscr{H}_1$: $\theta = \theta_1$. [Recall that $L_j = L(\theta_j; x_1, \ldots, x_n) = \prod_{i=1}^{n} f(x_i; \theta_j)$ for $j = 0$ or 1 and $\bar{C}^*$ is the complement of C^*; that is, $\bar{C}^* = \mathfrak{X} - C^*$.]

PROOF Suppose that k^* and C^* satisfying conditions (i) and (ii) exist. If there is no other test of size α or less, then Υ^* is automatically most powerful. Let Υ be another test of size α or less, and let C be its corresponding critical region. We have $P_{\theta_0}[(X_1, \ldots, X_n) \in C] \le \alpha$. We must show that $\pi_{\Upsilon^*}(\theta_1) \ge \pi_{\Upsilon}(\theta_1)$ to complete the proof. {For any subset R of $\mathfrak{X}$, let us abbreviate $\int \cdots \int_R \left[\prod_{i=1}^{n} f(x_i; \theta_j)\, dx_i \right]$ as $\int_R L_j$ for $j = 0, 1$. Our notation indicates that $f_0(\cdot)$ and $f_1(\cdot)$ are probability density functions. The same proof holds for discrete density functions.} Showing that $\pi_{\Upsilon^*}(\theta_1) \ge \pi_{\Upsilon}(\theta_1)$ is equivalent to showing that $\int_{C^*} L_1 \ge \int_C L_1$. See Fig. 3.

Now $\int_{C^*} L_1 - \int_C L_1 = \int_{C^*\bar{C}} L_1 - \int_{C\bar{C}^*} L_1 \ge (1/k^*) \int_{C^*\bar{C}} L_0 - (1/k^*) \int_{C\bar{C}^*} L_0$ since $L_1 \ge L_0/k^*$ on C^* (hence also on $C^*\bar{C}$) and $L_1 \le L_0/k^*$, or $-L_1 \ge -L_0/k^*$, on $\bar{C}^*$ (hence also on $C\bar{C}^*$). But $(1/k^*)(\int_{C^*\bar{C}} L_0 - \int_{C\bar{C}^*} L_0) = (1/k^*)(\int_{C^*\bar{C}} L_0 + \int_{C^*C} L_0 - \int_{C^*C} L_0 - \int_{C\bar{C}^*} L_0) = (1/k^*)(\int_{C^*} L_0 - \int_C L_0) = (1/k^*)(\alpha - \text{size of test } \Upsilon) \ge 0$; so $\int_{C^*} L_1 - \int_C L_1 \ge 0$, as was to be shown. ////

We comment that k^* and C^* satisfying conditions (i) and (ii) do not always exist and then the theorem, as stated, would not give a most powerful size-α test. However, whenever $f_0(\cdot)$ and $f_1(\cdot)$ are probability density functions, a k^* and C^* will exist. Although the theorem does not explicitly say how to find k^* and C^*, implicitly it does since the form of the test, that is, the critical region, is given by Eq. (5). In practice, even though k^* and C^* do exist, often it is not necessary to find them. Instead the inequality $\lambda \le k^*$ for $(x_1, \ldots, x_n) \in C^*$ is manipulated into an equivalent inequality that is easier to work with, and the actual test is then expressed in terms of the new inequality. The following examples should help clarify the above.

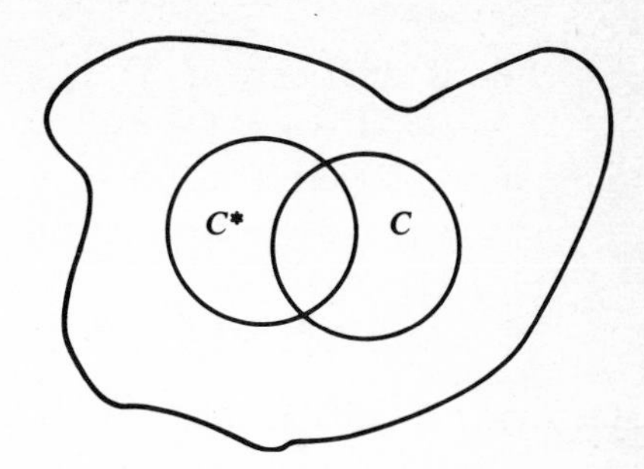

FIGURE 3

EXAMPLE 7 Let $X_1, \ldots, X_n$ be a random sample from $f(x; \theta) = \theta e^{-\theta x} I_{(0, \infty)}(x)$, where $\theta = \theta_0$ or $\theta = \theta_1$. θ_0 and θ_1 are known fixed numbers, and for concreteness we assume that $\theta_1 > \theta_0$. We want to test $\mathscr{H}_0: \theta = \theta_0$ versus $\mathscr{H}_1: \theta = \theta_1$. Now $L_0 = \theta_0^n \exp(-\theta_0 \sum x_i)$, $L_1 = \theta_1^n \exp(-\theta_1 \sum x_i)$, and according to the Neyman-Pearson lemma the most powerful test will have the form: Reject $\mathscr{H}_0$ if $\lambda \le k^*$ or if $(\theta_0/\theta_1)^n \exp[-(\theta_0 - \theta_1) \sum x_i] \le k^*$, which is equivalent to

$$\sum x_i \le \frac{1}{\theta_1 - \theta_0} \log_e \left[\left(\frac{\theta_1}{\theta_0} \right)^n k^* \right] = k' \text{ (say)},$$

where k' is just a constant. The inequality $\lambda \le k^*$ has been simplified and expressed as the equivalent inequality $\sum x_i \le k'$. Condition (i) is $\alpha = P_{\theta_0}[\text{reject } \mathscr{H}_0] = P_{\theta_0}[\sum X_i \le k']$. We know that $\sum X_i$ has a gamma distribution with parameters n and θ; hence

$$P_{\theta_0}[\sum X_i \le k'] = \int_0^{k'} \frac{1}{\Gamma(n)} \theta_0^n x^{n-1} e^{-x\theta_0} \, dx = \alpha,$$

an equation in k', from which k' can be determined; and the most powerful test of size α of $\mathscr{H}_0: \theta = \theta_0$ versus $\mathscr{H}_1: \theta = \theta_1$, $\theta_1 > \theta_0$ is this: Reject $\mathscr{H}_0$ if $\sum x_i \le k'$, where k' is the αth quantile point of the gamma distribution with parameters n and θ_0. ////

EXAMPLE 8 Let $X_1, \ldots, X_n$ be a random sample from $f(x; \theta) = \theta^x (1 - \theta)^{1-x} I_{\{0, 1\}}(x)$, where $\theta = \theta_0$ or $\theta = \theta_1$. We want to test $\mathscr{H}_0: \theta = \theta_0$ versus $\mathscr{H}_1: \theta = \theta_1$, where, say, $\theta_0 < \theta_1$. $L_0 = \theta_0^{\Sigma x_i}(1 - \theta_0)^{n - \Sigma x_i}$, $L_1 = \theta_1^{\Sigma x_i}(1 - \theta_1)^{n - \Sigma x_i}$, and so $\lambda \le k^*$ if and only if

$$\theta_0^{\Sigma x_i}(1 - \theta_0)^{n - \Sigma x}/\theta_1^{\Sigma x_i}(1 - \theta_1)^{n - \Sigma x_i} \le k^*,$$

if and only if

$$\left[\frac{\theta_0(1 - \theta_1)}{(1 - \theta_0)\theta_1} \right]^{\Sigma x_i} \left(\frac{1 - \theta_0}{1 - \theta_1} \right)^n \le k^*,$$

or if and only if $\sum x_i \geq k'$, where k' is a constant. {Note that $\log_e [\theta_0(1-\theta_1)/(1-\theta_0)\theta_1] < 0$.} So a most powerful test would be of the form: Reject $\mathscr{H}_0$ if $\sum x_i$ is large. For definiteness, let us take $\theta_0 = \frac{1}{4}$, $\theta_1 = \frac{3}{4}$, and $n = 10$. We must find k' so that

$$\alpha = P_{\theta_0=1/4}[\text{reject } \mathscr{H}_0] = P_{\theta_0=1/4}\left[\sum X_i \geq k'\right] = \sum_{i=k'}^{10} \binom{10}{i} \left(\frac{1}{4}\right)^i \left(\frac{3}{4}\right)^{10-i}.$$

If $\alpha = .0197$, then $k' = 6$, and if $\alpha = .0781$, then $k' = 5$. For $\alpha = .05$, there is no critical region C^* and constant k^* of the form given in the Neyman-Pearson lemma. In this example our random variables are discrete, and for discrete random variables it is not possible to find a k^* and C^* satisfying conditions (i) and (ii) for an arbitrary fixed $0 < \alpha < 1$. In practice, one is usually content to change the size of the test to some α for which a test given by the Neyman-Pearson lemma can be found. We might note, however, that a most powerful test of size α does exist. The test would be a randomized test. For the example at hand, if we take $\alpha = .05$, the randomized test with critical function

$$\psi(x_1, \ldots, x_{10}) = \begin{cases} 1 & \text{if } \sum x_i \geq 6 \\ \dfrac{.05 - .0197}{.0584} & \text{if } \sum x_i = 5 \\ 0 & \text{if } \sum x_i \leq 4 \end{cases}$$

is the most powerful test of size $\alpha = .05$. ////

In closing this subsection, we note that a most powerful test of size α, given by the Neyman-Pearson lemma, is necessarily a simple likelihood-ratio test.

2.3 Loss Function

As in the last subsection, we assume that we have a random sample $X_1, \ldots, X_n$ from one or the other of the two completely known densities $f_0(\cdot) = f(\cdot\,; \theta_0)$ and $f_1(\cdot) = f(\cdot\,; \theta_1)$. On the basis of the observed sample we have to decide from which of the two densities the sample came; that is, we test $\mathscr{H}_0: \theta = \theta_0$ versus $\mathscr{H}_1: \theta = \theta_1$. We can make one of two decisions, say d_0 or d_1, where d_j is the decision that $f_j(\cdot)$ is the density from which the sample came, $j = 0, 1$. We assume that a *loss function* is available.

Definition 10 Loss function In testing $\mathscr{H}_0: \theta = \theta_0$ versus $\mathscr{H}_1: \theta = \theta_1$, define $\ell(d_i; \theta_j)$ = loss incurred when decision d_i is made and θ_j is the true parameter value for $i = 0, 1$ and $j = 0, 1$, where d_i is the decision of deciding that hypothesis $\mathscr{H}_i$ is correct. We will adopt the convention that $\ell(d_i; \theta_i) = 0$ for $i = 0, 1$ and $\ell(d_i; \theta_j) > 0$ for $i \neq j$. ////

The values of the loss function are the amounts that are lost if decision d_i is made when θ_j was correct. With our convention, nothing is lost if the right decision is made, and a positive amount is lost if a wrong decision is made. If we think in terms of a (nonrandomized) test Υ having a critical region C_Υ, then decision d_1 is made if the observed sample $(x_1, \ldots, x_n)$ belongs to C_Υ, and decision d_0 is made if $(x_1, \ldots, x_n)$ belongs to $\bar{C}_\Upsilon$. A test can be thought of as a *decision function* since for a given observed sample the test tells us which decision to make. We do not consider the problem of selecting an appropriate loss function, and hence we will always assume that an appropriate loss function has been prescribed. Note that if decision d_1 is made when θ_0 is correct, then a Type I error is made, and if decision d_0 is made when θ_1 is correct, then a Type II error is made.

In comparing two tests we naturally prefer that test which has smaller loss, and among all tests we would prefer that test which has smallest loss. However, seldom will there exist one test that has smallest loss for both possible decisions and for both θ_0 and θ_1. This motivates the defining of *average loss*, and by continuing to borrow language from decision theory we define the *risk function* of a test.

Definition 11 Risk function For a random sample $X_1, \ldots, X_n$ from $f(\cdot; \theta_0)$ or $f(\cdot; \theta_1)$, let Υ be a test of $\mathscr{H}_0: \theta = \theta_0$ versus $\mathscr{H}_1: \theta = \theta_1$ having a critical region C_Υ. For a given loss function $\ell(\cdot; \cdot)$, the *risk function* of Υ, denoted by $\mathscr{R}_\Upsilon(\theta)$, is defined to be the expected loss; that is

$$\mathscr{R}_\Upsilon(\theta) = \int_{C_\Upsilon}\cdots\int \ell(d_1; \theta)\left[\prod_{i=1}^{n} f(x_i; \theta)\, dx_i\right] + \int_{\bar{C}_\Upsilon}\cdots\int \ell(d_0; \theta)\left[\prod_{i=1}^{n} f(x_i; \theta)\, dx_i\right]. \quad ////$$

Remark

$$\begin{aligned}\mathscr{R}_\Upsilon(\theta) &= \ell(d_1; \theta)P_\theta[(X_1, \ldots, X_n) \in C_\Upsilon] + \ell(d_0; \theta)P_\theta[(X_1, \ldots, X_n) \in \bar{C}_\Upsilon] \\ &= \ell(d_1; \theta)\pi_\Upsilon(\theta) + \ell(d_0; \theta)[1 - \pi_\Upsilon(\theta)];\end{aligned} \tag{6}$$

that is, the risk function is a linear function of the power function; the coefficients in the linear function are determined by the values of the loss function. Since θ assumes only two values, $\mathscr{R}_\Upsilon(\theta)$ can take on only two values, which are

$$\mathscr{R}_\Upsilon(\theta_0) = \ell(d_1; \theta_0)\pi_\Upsilon(\theta_0) \qquad \text{and} \qquad \mathscr{R}_\Upsilon(\theta_1) = \ell(d_0; \theta_1)[1 - \pi_\Upsilon(\theta_1)]. \tag{7}$$

////

Our object is to select that test which has smallest risk, but, unfortunately, such a test will seldom exist. The difficulty is that the risk takes on two values, and a test that minimizes both of these values simultaneously over all possible tests does not exist except in rare situations. (See Prob. 3.) Not being able to find a test with smallest risk, we resort to another less desirable criterion, that of minimizing the largest value of the risk function.

Definition 12 Minimax test A test Υ_m of $\mathscr{H}_0: \theta = \theta_0$ versus $\mathscr{H}_1: \theta = \theta_1$ is defined to be *minimax* if and only if

$$\max [\mathscr{R}_{\Upsilon_m}(\theta_0), \mathscr{R}_{\Upsilon_m}(\theta_1)] \le \max [\mathscr{R}_{\Upsilon}(\theta_0), \mathscr{R}_{\Upsilon}(\theta_1)]$$

for any other test Υ. ////

The following theorem is sometimes useful in finding a minimax test. (As with Theorem 2, we state the theorem and proof in terms of nonrandomized tests.)

Theorem 3 For a random sample $X_1, \ldots, X_n$ from $f(\cdot\,; \theta_0)$ or $f(\cdot\,; \theta_1)$, consider testing $\mathscr{H}_0: \theta = \theta_0$ versus $\mathscr{H}_1: \theta = \theta_1$. If a test Υ_m has a critical region given by $C_m = \{(x_1, \ldots, x_n): \lambda \le k_m\}$, where k_m is a positive constant such that $\mathscr{R}_{\Upsilon_m}(\theta_0) = \mathscr{R}_{\Upsilon_m}(\theta_1)$, then Υ_m is minimax. Recall that $\lambda = L_0/L_1 = [\prod_{i=1}^{n} f(x_i; \theta_0)]/[\prod_{i=1}^{n} f(x_i; \theta_1)]$.

PROOF We will assume that $f(\cdot\,; \theta_0)$ and $f(\cdot\,; \theta_1)$ are probability density functions. The proof for discrete density functions is similar.

Let Υ be any other test with a critical region C_Υ which satisfies $\mathscr{R}_\Upsilon(\theta_0) \le \mathscr{R}_{\Upsilon_m}(\theta_0)$. [Note that if $\mathscr{R}_\Upsilon(\theta_0) > \mathscr{R}_{\Upsilon_m}(\theta_0)$, then Υ would not even be a candidate for minimax.] We have $\mathscr{R}_\Upsilon(\theta_0) \le \mathscr{R}_{\Upsilon_m}(\theta_0)$, $\ell(d_1; \theta_0)\pi_\Upsilon(\theta_0) \le \ell(d_1; \theta_0)\pi_{\Upsilon_m}(\theta_0)$, or $\pi_\Upsilon(\theta_0) \le \pi_{\Upsilon_m}(\theta_0)$; that is, Υ has size less than or equal to that of Υ_m. But, by the Neyman-Pearson lemma, Υ_m is the most powerful test of size $\pi_{\Upsilon_m}(\theta_0)$; hence $\pi_\Upsilon(\theta_1) \le \pi_{\Upsilon_m}(\theta_1)$, $1 - \pi_\Upsilon(\theta_1) \ge 1 - \pi_{\Upsilon_m}(\theta_1)$, $\ell(d_0; \theta_1)[1 - \pi_\Upsilon(\theta_1)] \ge \ell(d_0; \theta_1)[1 - \pi_{\Upsilon_m}(\theta_1)]$, or $\mathscr{R}_\Upsilon(\theta_1) \ge \mathscr{R}_{\Upsilon_m}(\theta_1)$; so, we have $\max [\mathscr{R}_{\Upsilon_m}(\theta_0), \mathscr{R}_{\Upsilon_m}(\theta_1)] = \mathscr{R}_{\Upsilon_m}(\theta_1) \le \mathscr{R}_\Upsilon(\theta_1) \le \max [\mathscr{R}_\Upsilon(\theta_0), \mathscr{R}_\Upsilon(\theta_1)]$; that is, Υ_m is minimax. ////

EXAMPLE 9 Let $X_1, \ldots, X_n$ be a random sample from $f(x; \theta) = \theta e^{-\theta x} I_{(0, \infty)}(x)$. For $\theta_1 > \theta_0$, test $\mathscr{H}_0: \theta = \theta_0$ versus $\mathscr{H}_1: \theta = \theta_1$. In Example 7, we found the most powerful size-α test. We seek now to find the minimax test for a loss function given by $\ell(d_1; \theta_0) = a$ and $\ell(d_0; \theta_1) = b$. According to Theorem 3, the minimax test Υ_m is given by $C_m = \{(x_1, \ldots, x_n): \lambda \le k_m\}$

where k_m is such that $\mathscr{R}_{\Upsilon_m}(\theta_0) = \mathscr{R}_{\Upsilon_m}(\theta_1)$. $\{\lambda \leq k_m\}$ can be rewritten as $\{\sum x_i \leq k\}$ for some constant k, and $\mathscr{R}_{\Upsilon_m}(\theta_0) = \mathscr{R}_{\Upsilon_m}(\theta_1)$ if and only if $a\pi_{\Upsilon_m}(\theta_0) = b[1 - \pi_{\Upsilon_m}(\theta_1)]$; so we seek k such that $aP_{\theta_0}[\sum X_i \leq k] = bP_{\theta_1}[\sum X_i > k]$. k is given as the solution to

$$a \int_0^k \frac{1}{\Gamma(n)} \theta_0^n x^{n-1} e^{-\theta_0 x}\, dx = b \int_k^\infty \frac{1}{\Gamma(n)} \theta_1^n x^{n-1} e^{-\theta_1 x}\, dx. \qquad ////$$

Before leaving minimax tests we make two comments: First, if $f_0(\cdot)$ and $f_1(\cdot)$ are discrete density functions, then there may not exist a k_m such that $\mathscr{R}_{\Upsilon_m}(\theta_0) = \mathscr{R}_{\Upsilon_m}(\theta_1)$ unless randomized tests are allowed; and, second, a minimax test as given in Theorem 3 was a simple likelihood-ratio test.

In the above we assumed that $X_1, \ldots, X_n$ was a random sample from $f(\cdot\,; \theta)$, where $\theta = \theta_0$ or θ_1, and for each θ, $f(\cdot\,; \theta)$ is completely known. We also assumed that we had an appropriate loss function. Now, we further assume that θ_0 and θ_1 are the possible values of a random variable Θ and that we know the distribution of Θ, which is called the *prior* distribution, just as in our considerations of Bayes estimation. Θ is discrete, taking on only two values θ_0 and θ_1; so the prior distribution of Θ is completely given by, say, g, where $g = P[\Theta = \theta_1] = 1 - P[\Theta = \theta_0]$. We mentioned above that, in general, a test with smallest risk function for both arguments does not exist. Now that we have a prior distribution for the two arguments of the risk function, we can define an average risk and seek that test with smallest average risk.

Definition 13 Bayes test A test Υ_g of $\mathscr{H}_0: \theta = \theta_0$ versus $\mathscr{H}_1: \theta = \theta_1$ is defined to be a *Bayes test* with respect to the prior distribution given by $g = P[\Theta = \theta_1]$ if and only if

$$(1 - g)\mathscr{R}_{\Upsilon_g}(\theta_0) + g\mathscr{R}_{\Upsilon_g}(\theta_1) \leq (1 - g)\mathscr{R}_{\Upsilon}(\theta_0) + g\mathscr{R}_{\Upsilon}(\theta_1) \qquad (8)$$

for any other test Υ. ////

To find a Bayes test we seek a critical region C_g that minimizes

$$\begin{aligned}(1 - g)\mathscr{R}_{\Upsilon}(\theta_0) + g\mathscr{R}_{\Upsilon}(\theta_1) &= (1 - g)\ell(d_1; \theta_0)\pi_{\Upsilon}(\theta_0) + g\ell(d_0; \theta_1)[1 - \pi_{\Upsilon}(\theta_1)] \\ &= (1 - g)\ell(d_1; \theta_0) \textstyle\int_C L_0 + g\ell(d_0; \theta_1) \int_{\bar{C}} L_1\end{aligned}$$

as a function of the region C. Now

$$\begin{aligned}&(1 - g)\ell(d_1; \theta_0) \textstyle\int_C L_0 + g\ell(d_0; \theta_1) \int_{\bar{C}} L_1 \\ &\qquad = g\ell(d_0; \theta_1) + \textstyle\int_C [(1 - g)\ell(d_1; \theta_0)L_0 - g\ell(d_0; \theta_1)L_1], \qquad (9)\end{aligned}$$

which is minimized if C is defined to be all $(x_1, \ldots, x_n)$ for which the last integrand of Eq. (9) is negative; that is

$$C_g = \{(x_1, \ldots, x_n): (1-g)\ell(d_1; \theta_0)L_0 - g\ell(d_0; \theta_1)L_1 < 0\}. \tag{10}$$

We have proved the following theorem.

Theorem 4 The Bayes test Υ_g of $\mathscr{H}_0$: $\theta = \theta_0$ versus $\mathscr{H}_1$: $\theta = \theta_1$ with respect to a prior distribution given by $g = P[\Theta = \theta_1]$ has a critical region defined by

$$C_g = \left\{(x_1, \ldots, x_n): \lambda < \frac{g\ell(d_0; \theta_1)}{(1-g)\ell(d_1; \theta_0)}\right\}. \tag{11}$$

////

We note that once again a good test, in this case a Bayes test, turns out to be a simple likelihood-ratio test. The exact form of the Bayes test is given by Eq. (11).

EXAMPLE 10 Let $X_1, \ldots, X_n$ be a random sample from $f(x; \theta) = \theta e^{-\theta x} I_{(0, \infty)}(x)$. Test $\mathscr{H}_0$: $\theta = \theta_0$ versus $\mathscr{H}_1$: $\theta = \theta_1$. The critical region of a Bayes test is given by

$$C_g = \left\{(x_1, \ldots, x_n): \frac{\theta_0^n \exp(-\theta_0 \sum x_i)}{\theta_1^n \exp(-\theta_1 \sum x_i)} < \frac{g\ell(d_0; \theta_1)}{(1-g)\ell(d_1; \theta_0)}\right\}$$

$$= \left\{(x_1, \ldots, x_n): \sum x_i < \frac{1}{\theta_1 - \theta_0} \log_e \frac{g\theta_1^n \ell(d_0; \theta_1)}{(1-g)\theta_0^n \ell(d_1; \theta_0)}\right\}$$

$$\text{for } \theta_1 > \theta_0.$$

////

3 COMPOSITE HYPOTHESES

In Sec. 2 above we considered testing a simple hypothesis against a simple alternative. We return now to the more general hypotheses-testing problem, that of testing composite hypotheses. We assume that we have a random sample from $f(x; \theta)$, $\theta \in \underline{\overline{\Theta}}$, and we want to test $\mathscr{H}_0$: $\theta \in \underline{\overline{\Theta}}_0$ versus $\mathscr{H}_1 : \theta \in \underline{\overline{\Theta}}_1$, where $\underline{\overline{\Theta}}_0 \subset \underline{\overline{\Theta}}$, $\underline{\overline{\Theta}}_1 \subset \underline{\overline{\Theta}}$, and $\underline{\overline{\Theta}}_0$ are $\underline{\overline{\Theta}}_1$ are disjoint. Usually $\underline{\overline{\Theta}}_1 = \underline{\overline{\Theta}} - \underline{\overline{\Theta}}_0$. We begin by discussing a general method of constructing a test.

3.1 Generalized Likelihood-ratio Test

For a random sample $X_1, \ldots, X_n$ from a density $f(x; \theta)$, $\theta \in \underline{\overline{\Theta}}$, we seek a test of $\mathscr{H}_0: \theta \in \underline{\overline{\Theta}}_0$ versus $\mathscr{H}_1: \theta \in \underline{\overline{\Theta}}_1 = \underline{\overline{\Theta}} - \underline{\overline{\Theta}}_0$.

Definition 14 Generalized likelihood-ratio Let $L(\theta; x_1, \ldots, x_n)$ be the likelihood function for a sample $X_1, \ldots, X_n$ having joint density $f_{X_1, \ldots, X_n}(x_1, \ldots, x_n; \theta)$ where $\theta \in \underline{\overline{\Theta}}$. The *generalized likelihood-ratio*, denoted by λ or λ_n, is defined to be

$$\lambda = \lambda_n = \lambda(x_1, \ldots, x_n) = \frac{\sup_{\theta \in \underline{\overline{\Theta}}_0} L(\theta; x_1, \ldots, x_n)}{\sup_{\theta \in \underline{\overline{\Theta}}} L(\theta; x_1, \ldots, x_n)}. \tag{12}$$

////

Note that λ is a function of $x_1, \ldots, x_n$, namely $\lambda(x_1, \ldots, x_n)$. When the observations are replaced by their corresponding random variables $X_1, \ldots, X_n$, then we write Λ for λ; that is, $\Lambda = \lambda(X_1, \ldots, X_n)$. Λ is a function of the random variables $X_1, \ldots, X_n$ and is itself a random variable. In fact, Λ is a statistic since it does not depend on unknown parameters.

Several further notes follow: (i) Although we used the same symbol λ to denote the simple likelihood-ratio, the generalized likelihood-ratio does not reduce to the simple likelihood-ratio for $\underline{\overline{\Theta}} = \{\theta_0, \theta_1\}$. (ii) λ given by Eq. (12) necessarily satisfies $0 \leq \lambda \leq 1$; $\lambda \geq 0$ since we have a ratio of nonnegative quantities, and $\lambda \leq 1$ since the supremum taken in the denominator is over a larger set of parameter values than that in the numerator; hence the denominator cannot be smaller than the numerator. (iii) The parameter θ can be vector-valued. (iv) The denominator of Λ is the likelihood function evaluated at the maximum-likelihood estimator. (v) In our considerations of the generalized likelihood-ratio, often the sample $X_1, \ldots, X_n$ will be a *random* sample from a density $f(x; \theta)$ where $\theta \in \underline{\overline{\Theta}}$.

The values λ of the statistic Λ are used to formulate a test of $\mathscr{H}_0: \theta \in \underline{\overline{\Theta}}_0$ versus $\mathscr{H}_1: \theta \in \underline{\overline{\Theta}} - \underline{\overline{\Theta}}_0$ by employing the *generalized likelihood-ratio test principle*, which states that $\mathscr{H}_0$ is to be rejected if and only if $\lambda \leq \lambda_0$, where λ_0 is some fixed constant satisfying $0 \leq \lambda_0 \leq 1$. (The constant λ_0 is often specified by fixing the size of the test.) Λ is the test statistic. The generalized likelihood-ratio test makes good intuitive sense since λ will tend to be small when $\mathscr{H}_0$ is not true, since then the denominator of λ tends to be larger than the numerator. In general, a generalized likelihood-ratio test will be a good test; although there are examples where the generalized likelihood-ratio test makes a poor showing compared to other tests. One possible drawback of the test is that it is sometimes difficult to find $\sup L(\theta; x_1, \ldots, x_n)$; another is that it can be difficult to find the distribution of Λ which is required to evaluate the power of the test.

EXAMPLE 11 Let $X_1, \ldots, X_n$ be a random sample from $f(x; \theta) = \theta e^{-\theta x} I_{(0, \infty)}(x)$, where $\underline{\Theta} = \{\theta; \theta > 0\}$. Test $\mathscr{H}_0: \theta \le \theta_0$ versus $\mathscr{H}_1: \theta > \theta_0$.

$$\sup_{\theta \in \underline{\Theta}} L[(\theta; x_1, \ldots, x_n)] = \sup_{\theta > 0} [\theta^n \exp(-\theta \sum x_i)] = \left(\frac{n}{\sum x_i}\right)^n e^{-n},$$

and

$$\sup_{\theta \in \underline{\Theta}_0} [L(\theta; x_1, \ldots, x_n)] = \sup_{0 < \theta \le \theta_0} [\theta^n \exp(-\theta \sum x_i)]$$

$$= \begin{cases} \left(\dfrac{n}{\sum x_i}\right)^n e^{-n} & \text{if } \dfrac{n}{\sum x_i} \le \theta_0 \\ \theta_0^n \exp(-\theta_0 \sum x_i) & \text{if } \dfrac{n}{\sum x_i} > \theta_0. \end{cases}$$

Hence

$$\lambda = \begin{cases} 1 & \text{if } \dfrac{n}{\sum x_i} \le \theta_0 \\ \dfrac{\theta_0^n \exp(-\theta_0 \sum x_i)}{(n/\sum x_i)^n e^{-n}} & \text{if } \dfrac{n}{\sum x_i} > \theta_0. \end{cases} \tag{13}$$

If $0 < \lambda_0 < 1$, then a generalized likelihood-ratio test is given by the following: Reject $\mathscr{H}_0$ if $\lambda \le \lambda_0$, or

$$\text{reject } \mathscr{H}_0 \text{ if } \frac{n}{\sum x_i} > \theta_0 \quad \text{and} \quad \left(\frac{\theta_0 \sum x_i}{n}\right)^n \exp(-\theta_0 \sum x_i + n) \le \lambda_0, \tag{14}$$

or reject $\mathscr{H}_0$ if $\theta_0 \bar{x} < 1$ and $(\theta_0 \bar{x})^n e^{-n(\theta_0 \bar{x} - 1)} \le \lambda_0$. Write $y = \theta_0 \bar{x}$, and note that $y^n e^{-n(y-1)}$ has a maximum for $y = 1$. Hence $y < 1$, and $y^n e^{-n(y-1)} \le \lambda_0$ if and only if $y \le k$, where k is a constant satisfying $0 < k < 1$. See Fig. 4.

We see that a generalized likelihood-ratio test reduces to the following:

$$\text{Reject } \mathscr{H}_0 \text{ if and only if } \theta_0 \bar{x} < k, \text{ where } 0 < k < 1; \tag{15}$$

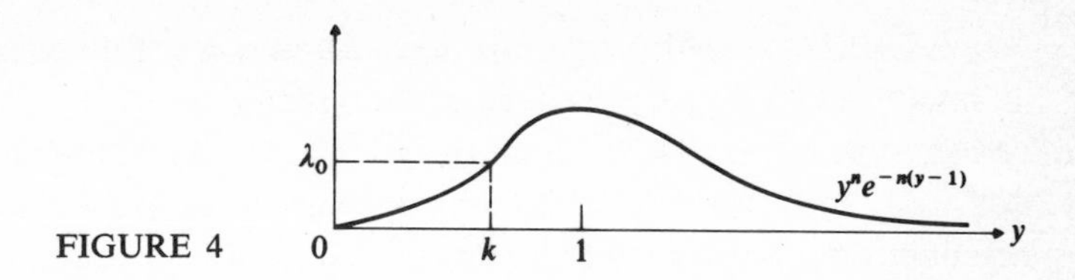

FIGURE 4

that is, reject $\mathscr{H}_0$ if $\bar{x}$ is less than some fraction of $1/\theta_0$. If that generalized likelihood-ratio test having size α is desired, k is obtained as the solution to the equation

$$\alpha = P_{\theta_0}[\theta_0 \bar{X} < k] = P_{\theta_0}[\theta_0 \sum X_i < nk] = \int_0^{nk} \frac{1}{\Gamma(n)} u^{n-1} e^{-u} \, du.$$

(Note that $P_\theta[\theta_0 \bar{X} < k] \le P_{\theta_0}[\theta_0 \bar{X} < k]$ for $\theta \le \theta_0$.) ////

We note that in the above example the first form of the test, as given in Eq. (14), is rather messy, yet after some manipulation the test reduces to a very simple form as given in Eq. (15). Such a pattern often appears in dealing with generalized likelihood-ratio tests—their first form is often foreboding, yet the tests often simplify into some nice form. We will observe this again in Sec. 4 below when we consider tests concerning sampling from the normal distribution.

We might note, by considering the factorization criterion, that a generalized likelihood-ratio test must necessarily depend only on minimal sufficient statistics.

In Sec. 5 below, a large-sample distribution of the generalized likelihood-ratio is given. This will provide us with a method of obtaining tests with approximate size α.

3.2 Uniformly Most Powerful Tests

In Subsec. 3.1 above we exhibited a method of obtaining a family of tests of $\mathscr{H}_0$: $\theta \in \underline{\overline{\Theta}}_0$ versus $\mathscr{H}_1$: $\theta \in \underline{\overline{\Theta}} - \underline{\overline{\Theta}}_0$. We now define one optimum property that such a test may possess. It is defined in terms of the power function $\pi_\Upsilon(\theta)$ and the size of the test.

Definition 15 Uniformly most powerful test A test Υ^* of $\mathscr{H}_0$: $\theta \in \underline{\overline{\Theta}}_0$ versus $\mathscr{H}_1$: $\theta \in \underline{\overline{\Theta}} - \underline{\overline{\Theta}}_0$ is defined to be a *uniformly most powerful* size-α test if and only if:

(i) $\sup_{\theta \in \underline{\overline{\Theta}}_0} \pi_{\Upsilon^*}(\theta) = \alpha$.

(ii) $\pi_{\Upsilon^*}(\theta) \ge \pi_\Upsilon(\theta)$ for all $\theta \in \underline{\overline{\Theta}} - \underline{\overline{\Theta}}_0$ and for any test Υ with size less than or equal to α. ////

A test Υ^* is uniformly most powerful of size α if it has size α and if among all tests of size less than or equal to α it has the largest power for *all* alternative values of θ. The adverb "uniformly" refers to "all" alternative θ values. A uniformly most powerful test does not exist for all testing problems, but when one does exist, we can see that it is quite a nice test since among all tests of size α or less it has the greatest chance of rejecting $\mathscr{H}_0$ whenever it should.

EXAMPLE 12 Let $X_1, \ldots, X_n$ be a random sample from $f(x; \theta) = \theta e^{-\theta x} I_{(0, \infty)}(x)$, where $\bar{\Theta} = \{\theta: \theta \geq \theta_0\}$. Find a uniformly most powerful test of $\mathscr{H}_0: \theta = \theta_0$ versus $\mathscr{H}_1: \theta > \theta_0$. For fixed $\theta_1 > \theta_0$, we determined in Example 7 that the most powerful test of $\mathscr{H}_0: \theta = \theta_0$ versus $\mathscr{H}_1: \theta = \theta_1$ was given by the following: Reject $\mathscr{H}_0$ if $\sum x_i \leq k'$, where k' was given as a solution to the equation

$$\alpha = \int_0^{k'} \frac{1}{\Gamma(n)} \theta_0^n x^{n-1} e^{-\theta_0 x}\, dx.$$

Such a test was given by the Neyman-Pearson lemma. Note that the test in no way depends on θ_1 except that $\theta_1 > \theta_0$; hence, we would get the same most powerful test for *any* $\theta_1 > \theta_0$, and thus the test is actually uniformly most powerful! ////

The above example provides us with an example of a situation where a uniformly most powerful test can be obtained using the Neyman-Pearson lemma. That same technique can be used to find uniformly most powerful tests in more general situations, such as those given in Theorems 5 and 6 below, which are given without proof.

Theorem 5 Let $X_1, \ldots, X_n$ be a random sample from the density $f(x; \theta)$, $\theta \in \bar{\Theta}$, where $\bar{\Theta}$ is some interval. Assume that $f(x; \theta) = a(\theta)b(x) \exp[c(\theta)d(x)]$, and set $t(x_1, \ldots, x_n) = \sum_1^n d(x_i)$.

(i) If $c(\theta)$ is a monotone, increasing function in θ and if there exists k^* such that $P_{\theta_0}[t(X_1, \ldots, X_n) > k^*] = \alpha$, then the test Υ^* with a critical region $C^* = \{(x_1, \ldots, x_n): t(x_1, \ldots, x_n) > k^*\}$ is a uniformly most powerful size-α test of $\mathscr{H}_0: \theta \leq \theta_0$ versus $\mathscr{H}_1: \theta > \theta_0$ or of $\mathscr{H}_0: \theta = \theta_0$ versus $\mathscr{H}_1: \theta > \theta_0$.

(ii) If $c(\theta)$ is a monotone, decreasing function in θ and if there exists k^* such that $P_{\theta_0}[t(X_1, \ldots, X_n) < k^*] = \alpha$, then the test Υ^* with a critical region $C^* = \{(x_1, \ldots, x_n): t(x_1, \ldots, x_n) < k^*\}$ is a uniformly most powerful size-α test of $\mathscr{H}_0: \theta \le \theta_0$ versus $\mathscr{H}_1: \theta > \theta_0$ or of $\mathscr{H}_0: \theta = \theta_0$ versus $\mathscr{H}_1: \theta > \theta_0$. ////

EXAMPLE 13 Let $X_1, \ldots, X_n$ be a random sample from $f(x; \theta) = \theta e^{-\theta x} I_{(0, \infty)}(x)$, where $\underline{\Theta} = \{\theta: \theta > 0\}$. Test $\mathscr{H}_0: \theta \le \theta_0$ versus $\mathscr{H}_1: \theta > \theta_0$. $f(x; \theta) = \theta I_{(0, \infty)}(x) \exp(-\theta x) = a(\theta) b(x) \exp[c(\theta) d(x)]$; so $t(x_1, \ldots, x_n) = \sum_1^n x_i$, and $c(\theta) = -\theta$. $c(\theta)$ is a monotone, decreasing function; so by (ii) of Theorem 5 a uniformly most powerful test is given by the following: Reject $\mathscr{H}_0$ if and only if $\sum x_i < k^*$, where k^* is given as a solution to

$$\alpha = P_{\theta_0}[\sum X_i < k^*] = \int_0^{k^*} \frac{1}{\Gamma(n)} \theta_0^n u^{n-1} e^{-\theta_0 u} \, du. \qquad ////$$

Definition 16 Monotone likelihood-ratio A family of densities $\{f(x; \theta): \theta \in \underline{\Theta}, \underline{\Theta}$ an interval$\}$ is said to have a *monotone likelihood-ratio* if there exists a statistic, say $T = t(X_1, \ldots, X_n)$, such that the ratio $L(\theta'; x_1, \ldots, x_n)/L(\theta''; x_1, \ldots, x_n)$ is either a nonincreasing function of $t(x_1, \ldots, x_n)$ for every $\theta' < \theta''$ or a nondecreasing function of $t(x_1, \ldots, x_n)$ for every $\theta' < \theta''$. ////

Note that in the term "monotone likelihood-ratio" the likelihood-ratio is not a generalized likelihood-ratio; it is a ratio of two likelihood functions.

EXAMPLE 14 If $\{f(x; \theta): \theta \in \underline{\Theta}\} = \{\theta e^{-\theta x} I_{(0, \infty)}(x): \theta > 0\}$, then

$$\frac{L(\theta'; x_1, \ldots, x_n)}{L(\theta''; x_1, \ldots, x_n)} = \frac{(\theta')^n \exp(-\theta' \sum x_i)}{(\theta'')^n \exp(-\theta'' \sum x_i)} = \left(\frac{\theta'}{\theta''}\right)^n \exp[-(\theta' - \theta'') \sum x_i]$$

which is a monotone, increasing function in $\sum x_i$. ////

EXAMPLE 15 If $\{f(x; \theta): \theta \in \underline{\Theta}\} = \{(1/\theta) I_{(0, \theta)}(x): \theta > 0\}$, then

$$\frac{L(\theta'; x_1, \ldots, x_n)}{L(\theta''; x_1, \ldots, x_n)} = \frac{(1/\theta')^n \prod_{i=1}^n I_{(0, \theta')}(x_i)}{(1/\theta'')^n \prod_{i=1}^n I_{(0, \theta'')}(x_i)} = \frac{(1/\theta')^n I_{(0, \theta')}(y_n)}{(1/\theta'')^n I_{(0, \theta'')}(y_n)}$$

$$= \begin{cases} \left(\dfrac{\theta''}{\theta'}\right)^n & \text{for } 0 < y_n < \theta' \\ 0 & \text{for } \theta' \le y_n < \theta'', \end{cases}$$

which is a monotone, nonincreasing function in $y_n = \max [x_1, \ldots, x_n]$. [Note that y_n cannot fall outside of the interval $(0, \theta'')$ when θ is either θ' or θ''.] ////

Theorem 6 Let $X_1, \ldots, X_n$ be a random sample from $f(x; \theta)$, where $\overline{\Theta}$ is some interval. Assume that the family of densities $\{f(x; \theta): \theta \in \overline{\Theta}\}$ has a monotone likelihood-ratio in the statistic $t(X_1, \ldots, X_n)$:

(i) If the monotone likelihood-ratio is nondecreasing in $t(x_1, \ldots, x_n)$ and if k^* is such that $P_{\theta_0}[t(X_1, \ldots, X_n) < k^*] = \alpha$, then the test corresponding to the critical region $C^* = \{(x_1, \ldots, x_n): t(x_1, \ldots, x_n) < k^*\}$ is a uniformly most powerful test of size α of $\mathscr{H}_0: \theta \le \theta_0$ versus $\mathscr{H}_1: \theta > \theta_0$.

(ii) If the monotone likelihood-ratio is nonincreasing in $t(x_1, \ldots, x_n)$ and if k^* is such that $P_{\theta_0}[t(X_1, \ldots, X_n) > k^*] = \alpha$, then the test corresponding to the critical region $C^* = \{(x_1, \ldots, x_n): t(x_1, \ldots, x_n) > k^*\}$ is a uniformly most powerful test of size α of $\mathscr{H}_0: \theta \le \theta_0$ versus $\mathscr{H}_1: \theta > \theta_0$. ////

EXAMPLE 16 Let $X_1, \ldots, X_n$ be a random sample from $f(x; \theta) = (1/\theta)I_{(0, \theta)}(x)$, where $\theta > 0$. Test $\mathscr{H}_0: \theta \le \theta_0$ versus $\mathscr{H}_1: \theta > \theta_0$. We saw in Example 15 that the family of densities has a monotone, nonincreasing likelihood-ratio in $t(x_1, \ldots, x_n) = y_n = \max [x_1, \ldots, x_n]$. According to (ii) of Theorem 6, a uniformly most powerful size-α test is given by the following: Reject $\mathscr{H}_0$ if $y_n > k^*$, where k^* is given as the solution to

$$\alpha = P_{\theta_0}[Y_n > k^*] = \int_{k^*}^{\theta_0} n\left(\frac{y}{\theta_0}\right)^{n-1} \frac{dy}{\theta_0} = \frac{1}{\theta_0^n}[\theta_0^n - (k^*)^n] = 1 - \left(\frac{k^*}{\theta_0}\right)^n,$$

which implies that $k^* = \theta_0 \sqrt[n]{1 - \alpha}$. ////

Several comments are in order. First, the null hypothesis was stated as $\theta \le \theta_0$ in both Theorems 5 and 6; if it had been stated as $\theta \ge \theta_0$, the two theorems would remain valid provided the inequalities that define the critical regions were reversed. Second, Theorem 5 is a consequence of Theorem 6. Third, the theorems consider only *one-sided* hypotheses.

This completes our brief study of uniformly most powerful tests. We have seen that a uniformly most powerful test exists for one-sided hypotheses if the density sampled from has a monotone likelihood-ratio in some statistic. There are many hypothesis-testing problems for which no uniformly most powerful test exists. One method of restricting the class of tests, with the hope and intention of finding an optimum test within the restricted class, is to consider *unbiasedness* of tests, to be defined in the next subsection.

3.3 Unbiased Tests

There are many hypotheses-testing problems for which a uniformly most powerful test does not exist. In these cases it may be possible to restrict the class of tests and find a uniformly most powerful test in the restricted class. One such class that has some merit is the class of *unbiased* tests.

Definition 17 Unbiased tests A test Υ of the null hypothesis $\mathscr{H}_0: \theta \in \overline{\underline{\Theta}}_0$ against the alternative hypothesis $\mathscr{H}_1: \theta \in \overline{\underline{\Theta}}_1$ is an *unbiased test* if and only if

$$\sup_{\theta \in \overline{\underline{\Theta}}_0} \pi_T(\theta) \leq \inf_{\theta \in \overline{\underline{\Theta}}_1} \pi_T(\theta). \qquad ////$$

Consequently in an unbiased test the probability of rejecting $\mathscr{H}_0$ when it is false is at least as large as the probability of rejecting $\mathscr{H}_0$ when it is true. In many respects this seems to be a reasonable restriction to place on a test. If within this restricted class a test exists that is uniformly most powerful, then we have a *uniformly most powerful unbiased* test. An elaborate theory has been developed for finding uniformly most powerful unbiased tests, but we will not study it. See [16].

3.4 Methods of Finding Tests

We presented in Subsec. 3.1 the generalized likelihood-ratio principle; it provides us with one method of obtaining tests of hypotheses. In Subsec. 3.2 we gave a method of finding a uniformly most powerful test for certain testing problems. There are still other methods of finding tests. One method is sketched on pages 456 to 459 of Subsec. 5.4 of this chapter. Another method, which might be called the *confidence-interval method*, is to use a confidence interval to obtain a test. For instance, if it is desired to test $\mathscr{H}_0: \theta = \theta_0$ versus $\mathscr{H}_1: \theta \neq \theta_0$, then we might compute a confidence-interval estimate of θ from the data, and if the interval contains θ_0, accept $\mathscr{H}_0$, and otherwise, reject it. If the confidence interval had a confidence coefficient γ, then the resulting test would have size $1 - \gamma$. We will say more about how confidence intervals might be used to obtain tests in Sec. 6 below.

A useful and intuitive technique for obtaining tests is the following: Discover some statistic which behaves differently under the two hypotheses, and utilize the different behavior to design a test. As an illustration, consider testing $\mathscr{H}_0\colon \theta \le \theta_0$ versus $\mathscr{H}_1\colon \theta > \theta_0$, where the sample $X_1, \ldots, X_n$ is selected from the density $f(x; \theta) = \phi_{\theta,1}(x)$. The statistic $\bar{X}$ has a normal distribution with mean θ and variance $1/n$; hence the statistic $\bar{X}$ will tend to be smaller when $\mathscr{H}_0$ is true than when $\mathscr{H}_0$ is false. The statistic $\bar{X}$ behaves differently under the two hypotheses. A reasonable test, then, would be to reject $\mathscr{H}_0$ for $\bar{X}$ large; that is, reject $\mathscr{H}_0$ if $\bar{X} > k$, where k is determined by, say, fixing the size of the test. (We know from Subsec. 3.2 above that such a test is uniformly most powerful.) To employ this technique, a statistic has to be discovered which will behave differently under the two hypotheses. There are various ways of approaching the task of discovering such a statistic. For instance, if a sufficient statistic exists, then it is a natural candidate to try; or a good estimator, such as a maximum-likelihood estimator, of the parameter or parameters that are used to specify the hypotheses is another possibility for the needed statistic. In the above simple illustration $\bar{X}$ was all these since $\bar{X}$ is the maximum-likelihood estimator of θ as well as being a sufficient statistic. We make frequent use of this intuitive technique for obtaining tests in the remaining sections.

EXAMPLE 17 Let $X_1, \ldots, X_n$ be a random sample from a Poisson distribution with mean θ. Suppose that it is desired to test that the mean is a fixed value, say θ_0; that is, test $\mathscr{H}_0\colon \theta = \theta_0$ versus $\mathscr{H}_1\colon \theta \ne \theta_0$. We know that $\bar{X}$ is the maximum-likelihood estimator of θ and that $\bar{X}$ will tend to be distributed about θ_0 if $\mathscr{H}_0$ is true. Consequently, the following test seems reasonable: Accept $\mathscr{H}_0$ if $c_1 < \bar{x} < c_2$, and otherwise reject it, where c_1 and c_2 are selected so that the test will have a desired size. To be specific, let $n = 10$ and $\theta_0 = 1$. The test given by "Accept $\mathscr{H}_0$ if and only if $.4 < \bar{x} < 1.6$" has size given by

$$1 - P_{\theta=1}[.4 < \bar{X} < 1.6] = 1 - P_{\theta=1}[4 < \sum X_i < 16]$$
$$= 1 - \sum_{5}^{15} \frac{e^{-10}10^j}{j!} \approx .078. \qquad ////$$

A test that has been quite extensively applied in various fields of science is $\mathscr{H}_0\colon \theta = \theta_0$ against $\mathscr{H}_1\colon \theta \ne \theta_0$. For example, let θ be the mean difference of yields between two varieties of wheat. It is often suggested that it is desirable to test the hypothesis $\mathscr{H}_0\colon \theta = 0$ against $\mathscr{H}_1\colon \theta \ne 0$, that is, to test if the two varieties are different in their mean yields. However, in this situation, and many others where θ can vary continuously in some interval, it is inconceivable that θ is exactly equal to 0 (that the varieties are identical in their mean yields). Yet this is what the test is stating: Are the two mean yields identical (to one ten-billionth of a bushel, etc.)? In many cases it seems more realistic for an experimenter to select an interval about θ_0, say $\theta_1 \leq \theta_0 \leq \theta_2$, and test $\mathscr{H}_0: \theta_1 \leq \theta \leq \theta_2$ against the alternative $\mathscr{H}_1: \theta < \theta_1$ or $\theta > \theta_2$. For example, it may be feasible to set $\theta_1 = -\frac{1}{2}$ and $\theta_2 = \frac{1}{2}$ in the above illustration and test if the difference of the mean yields of the two varieties is between $-\frac{1}{2}$ bushel and $+\frac{1}{2}$ bushel against the alternative that it is not in this interval. A test that is uniformly most powerful for the above hypothesis may be difficult or impossible to devise, but if $f(x; \theta)$ is a density with a single parameter, then the maximum-likelihood estimator $\hat{\Theta}$ may sometimes be used to construct a test and the power of this test compared with the ideal power function for a test of size α. A test of the following form may be used for some densities: Reject $\mathscr{H}_0$ if $\hat{\theta}$ is not in some interval, say (c_1, c_2), and accept $\mathscr{H}_0$ if $\hat{\theta}$ is in the interval, where c_1 and c_2 are chosen so that the test has size α. Often c_1 and c_2 can be chosen so that

$$\int_{c_1}^{c_2} f_{\hat{\Theta}}(\hat{\theta}; \theta_1)\, d\hat{\theta} = \int_{c_1}^{c_2} f_{\hat{\Theta}}(\hat{\theta}; \theta_2)\, d\hat{\theta} = 1 - \alpha,$$

where $f_{\hat{\Theta}}(\hat{\theta}; \theta)$ is the density of $\hat{\Theta}$ when θ is the parameter. The power function of this test is

$$\pi(\theta) = 1 - \int_{c_1}^{c_2} f_{\hat{\Theta}}(\hat{\theta}; \theta)\, d\hat{\theta} \qquad \text{for } \theta \text{ in } \overline{\Theta}.$$

This power function can be compared with the ideal power function, and if it does not deviate further from the ideal than the experimenter can tolerate, the test may be useful even though it may not be a uniformly most powerful test.

Let us illustrate the above with a simple example.

EXAMPLE 18 Let $X_1, \ldots, X_n$ be a random sample from $\phi_{\theta,1}(x)$. Test $\mathscr{H}_0: 1 \leq \theta \leq 2$ versus $\mathscr{H}_1: \theta < 1$ or $\theta > 2$. $\overline{X}$ is the maximum-likelihood estimator of θ; it has a normal distribution with mean θ and variance $1/n$. According to the above we would like to select c_1 and c_2 so that

$$1 - \alpha = \int_{c_1}^{c_2} \phi_{1,1/n}(x)\, dx = \int_{c_1}^{c_2} \phi_{2,1/n}(x)\, dx.$$

We have

$$\Phi\left(\frac{c_2 - 1}{\sqrt{1/n}}\right) - \Phi\left(\frac{c_1 - 1}{\sqrt{1/n}}\right) = \Phi\left(\frac{c_2 - 2}{\sqrt{1/n}}\right) - \Phi\left(\frac{c_1 - 2}{\sqrt{1/n}}\right) = 1 - \alpha,$$

and we can see from Fig. 5 that $c_1 = \frac{3}{2} - d$ and $c_2 = \frac{3}{2} + d$, where d is given by, say,

$$\Phi\left(\frac{d + \frac{1}{2}}{\sqrt{1/n}}\right) - \Phi\left(\frac{\frac{1}{2} - d}{\sqrt{1/n}}\right) = 1 - \alpha.$$

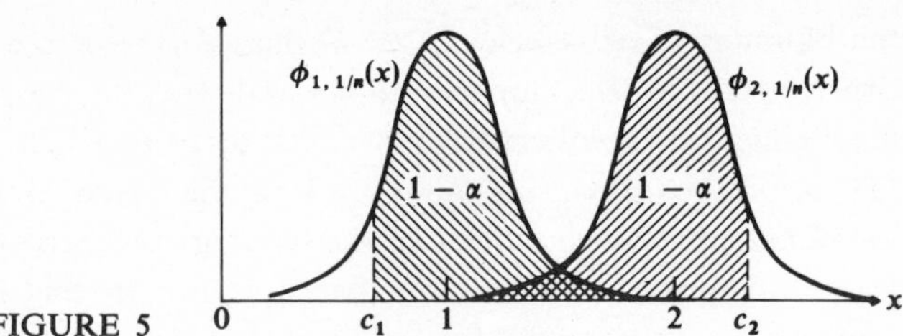

FIGURE 5

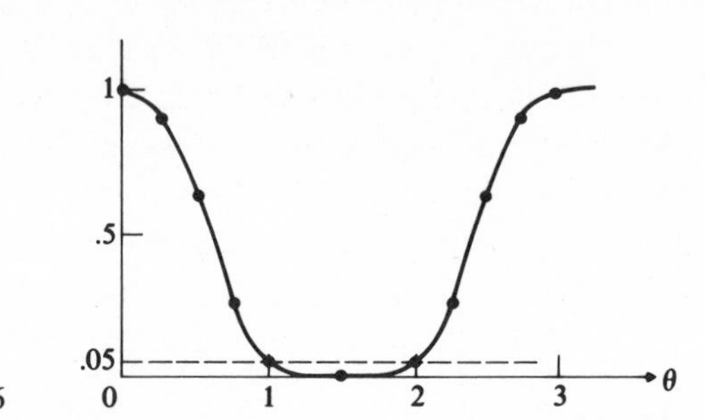

FIGURE 6

For example, if $\alpha = .05$ and $n = 16$, then $d \approx .911$; so $c_1 \approx .589$, and $c_2 \approx 2.411$. The power function is given by

$$\pi(\theta) = 1 - P_\theta[c_1 < \bar{X} < c_2] = 1 - P_\theta[.589 < \bar{X} < 2.411]$$

and is sketched in Fig. 6. ////

4 TESTS OF HYPOTHESES—SAMPLING FROM THE NORMAL DISTRIBUTION

A number of the foregoing ideas are well illustrated by common practical testing problems—those problems of testing hypotheses concerning the parameters of normal distributions. The section is subdivided into four subsections, the first two dealing with just one normal population and the last two dealing with several normal populations.

4.1 Tests on the Mean

We shall assume that we have a random sample of n observations $X_1, \ldots, X_n$ from a normal population with mean μ and variance σ^2, and we will be interested in testing hypotheses about μ. There is quite a variety of hypotheses about the mean μ that can be formulated; we begin by considering one-sided hypotheses.

$\mathscr{H}_0: \mu \le \mu_0$ **versus** $\mathscr{H}_1: \mu > \mu_0$ In testing $\mathscr{H}_0: \mu \le \mu_0$ versus $\mathscr{H}_1: \mu > \mu_0$ there are two cases to consider depending on whether or not σ^2 is assumed known. If σ^2 is assumed known, our parameter space is the real line, and we are testing a one-sided hypothesis; so we have hope of finding a uniformly most powerful test. Since σ^2 is assumed known, it is a known constant; hence

$$f(x; \theta) = \phi_{\mu, \sigma^2}(x) = \frac{1}{\sqrt{2\pi}\sigma} e^{-\frac{1}{2}[(x-\mu)/\sigma]^2}$$

$$= \frac{1}{\sqrt{2\pi}\sigma} e^{-\frac{1}{2}(\mu/\sigma)^2} e^{-\frac{1}{2}(x/\sigma)^2} e^{(\mu/\sigma)x},$$

which is a member of the exponential family with

$$a(\mu) = \frac{1}{\sqrt{2\pi}\sigma} e^{-\frac{1}{2}(\mu/\sigma)^2}, \qquad b(x) = e^{-\frac{1}{2}(x/\sigma)^2}, \qquad c(\mu) = \frac{\mu}{\sigma}, \qquad \text{and} \qquad d(x) = x.$$

The conditions for Theorem 5 are satisfied; so the uniformly most powerful size-α test is given by the following: Reject $\mathscr{H}_0$ if $t(x_1, \ldots, x_n) = \sum_1^n x_i > k^*$, where k^* is given as a solution to $P_{\mu_0}[\sum X_i > k^*] = \alpha$. Now $\alpha = P_{\mu_0}[\sum X_i > k^*] = 1 - \Phi((k^* - n\mu_0)/\sqrt{n}\sigma)$; so $(k^* - n\mu_0)/\sqrt{n}\sigma = z_{1-\alpha}$, where $z_{1-\alpha}$ is the $(1-\alpha)$th quantile of the standard normal distribution. The test becomes the following: Reject $\mathscr{H}_0$ if $\sum x_i > n\mu_0 + \sqrt{n}\sigma z_{1-\alpha}$, or reject $\mathscr{H}_0$ if $\bar{x} > \mu_0 + (\sigma/\sqrt{n})z_{1-\alpha}$.

If σ^2 is assumed unknown, then testing $\mathscr{H}_0: \mu \le \mu_0$ versus $\mathscr{H}_1: \mu > \mu_0$ is equivalent to testing $\mathscr{H}_0: \theta \in \overline{\Theta}_0$ versus $\mathscr{H}_1: \theta \notin \overline{\Theta}_0$, where $\theta = (\mu, \sigma^2)$, $\overline{\Theta} = \{(\mu, \sigma^2): -\infty < \mu < \infty; \sigma^2 > 0\}$, and $\overline{\Theta}_0 = \{(\mu, \sigma^2): \mu \le \mu_0; \sigma^2 > 0\}$. To obtain a test, we could use the generalized likelihood-ratio principle, or we could find some statistic that behaves differently under the two hypotheses and base our test on it. Such a statistic is $T = (\bar{X} - \mu_0)/(S/\sqrt{n})$, where $\bar{X}$ is the sample mean and S^2 is the sample variance. Since T would tend to be larger for $\mu > \mu_0$ than for $\mu \le \mu_0$, a test based on T is given by the following: Reject $\mathscr{H}_0$ if T is large; that is, reject $\mathscr{H}_0$ if $T > k$. If $\mu = \mu_0$, then T has a t distribution with $n-1$ degrees of freedom; so k can be determined by setting $\alpha = P_{\mu=\mu_0}[T > k]$, which implies that $k = t_{1-\alpha}(n-1)$, the $(1-\alpha)$th quantile of a t distribution with $n-1$ degrees of freedom. It can be shown that the test derived here is a generalized likelihood-ratio test having size α.

$\mathscr{H}_0: \mu = \mu_0$ **versus** $\mathscr{H}_1: \mu \ne \mu_0$ Again, we have two cases to consider depending on whether or not σ^2 is assumed known. For σ^2 known, we know that $(\bar{X} - z_{(1+\gamma)/2}(\sigma/\sqrt{n}), \bar{X} + z_{(1+\gamma)/2}(\sigma/\sqrt{n}))$ is a 100γ percent confidence interval for μ, where $z_{(1+\gamma)/2}$ is the $[(1+\gamma)/2]$th quantile of the standard normal distribution. A possible test is given by the following: Reject $\mathscr{H}_0$ if the confidence interval does not contain μ_0. Such a test has size $1-\gamma$ since

$$P_{\mu=\mu_0}\left[\bar{X} - z_{(1+\gamma)/2}\frac{\sigma}{\sqrt{n}} < \mu_0 < \bar{X} + z_{(1+\gamma)/2}\frac{\sigma}{\sqrt{n}}\right] = \gamma.$$

If σ^2 is assumed unknown, we could obtain a test, similar to the one above, using the 100γ percent confidence interval

$$\left(\bar{X} - t_{(1+\gamma)/2}(n-1)\frac{S}{\sqrt{n}},\ \bar{X} + t_{(1+\gamma)/2}(n-1)\frac{S}{\sqrt{n}}\right).$$

Instead, let us find a generalized likelihood-ratio test.

$$L(\mu, \sigma^2; x_1, \ldots, x_n) = \left(\frac{1}{\sqrt{2\pi}\sigma}\right)^n \exp\left[-\frac{1}{2}\sum\left(\frac{x_i-\mu}{\sigma}\right)^2\right],$$

$\bar{\Theta}_0 = \{(\mu, \sigma^2): \mu = \mu_0; \sigma^2 > 0\}$, and $\bar{\Theta} = \{(\mu, \sigma^2): -\infty < \mu < \infty; \sigma^2 > 0\}$. We have already seen that the values of μ and σ^2 which maximize $L(\mu, \sigma^2; x_1, \ldots, x_n)$ in $\bar{\Theta}$ are $\hat{\mu} = \bar{x}$ and $\hat{\sigma}^2 = (1/n)\sum(x_i - \bar{x})^2$; so

$$\sup_{\bar{\Theta}} L(\mu, \sigma^2; x_1, \ldots, x_n) = \left[\frac{n}{2\pi\sum(x_i-\bar{x})^2}\right]^{n/2} e^{-n/2}.$$

To maximize L over $\bar{\Theta}_0$, we put $\mu = \mu_0$, and the only remaining parameter is σ^2; the value of σ^2 which then maximizes L is readily found to be $\sigma^2 = (1/n)\sum(x_i-\mu_0)^2$, which gives

$$\sup_{\bar{\Theta}_0} L(\mu, \sigma^2; x_1, \ldots, x_n) = \left[\frac{n}{2\pi\sum(x_i-\mu_0)^2}\right]^{n/2} e^{-n/2}.$$

The generalized likelihood-ratio is then

$$\begin{aligned}\lambda &= \left[\frac{\sum(x_i-\bar{x})^2}{\sum(x_i-\mu_0)^2}\right]^{n/2} = \left[\frac{\sum(x_i-\bar{x})^2}{\sum(x_i-\bar{x}+\bar{x}-\mu_0)^2}\right]^{n/2} \\ &= \left[\frac{\sum(x_i-\bar{x})^2}{\sum(x_i-\bar{x})^2 + n(\bar{x}-\mu_0)^2}\right]^{n/2} = \left[\frac{1}{1+n(\bar{x}-\mu_0)^2/\sum(x_i-\bar{x})^2}\right]^{n/2}. \qquad (16)\end{aligned}$$

We note now that λ is a monotonic function of

$$t^2 = t^2(x_1, \ldots, x_n) = \frac{n(n-1)(\bar{x}-\mu_0)^2}{\sum(x_i-\bar{x})^2} = \left[\frac{\bar{x}-\mu_0}{\sqrt{\sum(x_i-\bar{x})^2/(n-1)n}}\right]^2,$$

and so a critical region of the form $\lambda \leq \lambda_0$ is equivalent to a critical region of the form $t^2(x_1, \ldots, x_n) \geq k^2$. A generalized likelihood-ratio test is then given by the following: Reject $\mathscr{H}_0$ if and only if

$$T^2 = \left[\frac{\bar{X} - \mu_0}{S/\sqrt{n}}\right]^2 \geq k^2,$$

or accept $\mathscr{H}_0$ if and only if $-k < T < k$. Since T has a t distribution with $n - 1$ degrees of freedom when $\mu = \mu_0$, if k is selected so that $\int_{-k}^{k} f_T(t; n - 1)\, dt = 1 - \alpha$, then our test will have size α. k is given by $t_{1-\alpha/2}(n - 1)$, the $(1 - \alpha/2)$th quantile of a t distribution with $n - 1$ degrees of freedom. We might note that this size-α test obtained by using the generalized likelihood-ratio principle is the same size-α test that we obtained above using the confidence-interval method of obtaining tests with the confidence interval

$$\left(\bar{X} - t_{(1+\gamma)/2}(n - 1)\frac{S}{\sqrt{n}}, \bar{X} + t_{(1+\gamma)/2}(n - 1)\frac{S}{\sqrt{n}}\right),$$

where $\gamma = 1 - \alpha$. Although we will not prove it, the test that we have obtained is uniformly most powerful unbiased.

We have found tests on the mean of a normal distribution for both one-sided and two-sided hypotheses. One might note that the one-sided null hypothesis $\mu \leq \mu_0$ could be reversed and comparable results obtained. There are other hypotheses about the mean that could be formulated, such as $\mathscr{H}_0$: $\mu_1 \leq \mu \leq \mu_2$ versus $\mathscr{H}_1$: $\mu < \mu_1$ or $\mu > \mu_2$.

4.2 Tests on the Variance

As in the last subsection, we shall assume that we have a random sample of size n from a normal population with mean μ and variance σ^2. We will be interested in testing hypotheses about σ^2.

$\mathscr{H}_0: \sigma^2 \leq \sigma_0^2$ versus $\mathscr{H}_1: \sigma^2 > \sigma_0^2$ There are two cases to consider depending on whether or not μ is assumed known. If μ is known, then our parameter space is an interval, and our hypothesis is one-sided; so we have a chance of finding a uniformly most powerful size-α test.

$$f(x; \theta) = f(x; \sigma^2) = \frac{1}{\sqrt{2\pi}\sigma} e^{-(1/2\sigma^2)(x-\mu)^2},$$

which is a member of the exponential family with $a(\sigma^2) = (2\pi\sigma^2)^{-\frac{1}{2}}$, $b(x) = 1$, $c(\sigma^2) = -1/2\sigma^2$, and $d(x) = (x - \mu)^2$. [μ is known; so $d(x)$ is a function of x only.] $c(\sigma^2)$ is a monotone, increasing function in σ^2; so, by Theorem 5, the test with critical region $= \{(x_1, \ldots, x_n): \sum (x_i - \mu)^2 > k^*\}$ is uniformly most powerful of size α, where k^* is given by $P_{\sigma^2=\sigma_0^2}[\sum (X_i - \mu)^2 > k^*] = \alpha$, which implies that $k^* = \sigma_0^2 \chi_{1-\alpha}^2(n)$, where $\chi_{1-\alpha}^2(n)$ is the $(1-\alpha)$th quantile point of the chi-square distribution with n degrees of freedom.

If μ is unknown, a test can be found using the statistic $V = \sum (X_i - \bar{X})^2/\sigma_0^2$. V will tend to be larger for $\sigma^2 > \sigma_0^2$ than for $\sigma^2 \leq \sigma_0^2$; so a reasonable test would be to reject $\mathscr{H}_0$ for V large. If $\sigma^2 = \sigma_0^2$, then V has a chi-square distribution with $n-1$ degrees of freedom, and $P_{\sigma^2=\sigma_0^2}[V > \chi_{1-\alpha}^2(n-1)] = \alpha$, where $\chi_{1-\alpha}^2(n-1)$ is the $(1-\alpha)$th quantile of a chi-square distribution with $n-1$ degrees of freedom. It can be shown that the test given by the following: Reject $\mathscr{H}_0$ if and only if $\sum (X_i - \bar{X})^2/\sigma_0^2 > \chi_{1-\alpha}^2(n-1)$ is a generalized likelihood-ratio test of size α.

$\mathscr{H}_0$: $\sigma^2 = \sigma_0^2$ versus $\mathscr{H}_1$: $\sigma^2 \neq \sigma_0^2$ We leave the case μ assumed known as an exercise. For μ unknown, so that $\overline{\Theta}_0 = \{(\mu, \sigma): -\infty < \mu < \infty;\ \sigma^2 = \sigma_0^2\}$, we can find a size-$\alpha$ test using the confidence-interval method. In Subsec. 3.2 of Chap. VIII, we found the following 100γ percent confidence interval for σ^2:

$$\left(\frac{(n-1)S^2}{q_2}, \frac{(n-1)S^2}{q_1}\right),$$

where q_1 and q_2 are quantile points of a chi-square distribution with $n-1$ degrees of freedom, say $f_Q(q; n-1)$, satisfying

$$\int_{q_1}^{q_2} f_Q(q; n-1)\, dq = \gamma.$$

A size-$(\alpha = 1 - \gamma)$ test is given by the following: Accept $\mathscr{H}_0$ if and only if σ_0^2 is contained in the above confidence interval. It is left as an exercise to show that for a particular pair of q_1 and q_2 the test of size α derived by the confidence-interval method is in fact the generalized likelihood-ratio test of size α.

4.3 Tests on Several Means

In this subsection we will consider testing hypotheses regarding the means of two or more normal populations. We begin with a test of the equality of two means.

Equality of two means In many situations it is necessary to compare two means when neither is known. If, for example, one wished to compare two proposed new processes for manufacturing light bulbs, one would have to base the comparison on estimates of both process means. In comparing the yield of a new line of hybrid corn with that of a standard line, one would also have to use estimates of both mean yields because it is impossible to state the mean yield of the standard line for the given weather conditions under which the new line would be grown. It is necessary to compare the two lines by planting them in the same season and on the same soil type and thereby obtain estimates of the mean yields for both lines under similar conditions. Of course the comparison is thus specialized; a complete comparison of the two lines would require tests over a period of years on a variety of soil types.

The general problem is this: We have two normal populations—one with a random variable X_1, which has a mean μ_1 and variance σ_1^2, and the other with a random variable X_2, which has a mean μ_2 and variance σ_2^2. On the basis of two samples, one from each population, we wish to test the null hypothesis

$$\mathscr{H}_0: \mu_1 = \mu_2, \sigma_1^2 > 0, \sigma_2^2 > 0 \qquad \text{versus} \qquad \mathscr{H}_1: \mu_1 \neq \mu_2, \sigma_1^2 > 0, \sigma_2^2 > 0.$$

The parameter space $\underline{\Theta}$ here is four-dimensional; a joint distribution of X_1 and X_2 is specified when values are assigned to the four quantities $(\mu_1, \mu_2, \sigma_1^2, \sigma_2^2)$. The subspace $\underline{\Theta}_0$ is three-dimensional because values for only three quantities $(\mu, \sigma_1^2, \sigma_2^2)$ need be specified in order to specify completely the joint distribution under the hypothesis that $\mu_1 = \mu_2 = \mu$, say.

We shall suppose that there are n_1 observations $(X_{11}, X_{12}, \ldots, X_{1n_1})$ in the sample from the first population and n_2 observations $(X_{21}, X_{22}, \ldots, X_{2n_2})$ from the second. The likelihood function is

$$L(\mu_1, \mu_2, \sigma_1^2, \sigma_2^2; x_{11}, \ldots, x_{1n_1}, x_{21}, \ldots, x_{2n_2}) = L$$

$$= \left(\frac{1}{2\pi\sigma_1^2}\right)^{n_1/2} \exp\left[-\frac{1}{2}\sum_1^{n_1}\left(\frac{x_{1i} - \mu_1}{\sigma_1}\right)^2\right]\left(\frac{1}{2\pi\sigma_2^2}\right)^{n_2/2} \exp\left[-\frac{1}{2}\sum_1^{n_2}\left(\frac{x_{2j} - \mu_2}{\sigma_2}\right)^2\right],$$

and its maximum in $\underline{\Theta}$ is readily seen to be

$$\sup_{\underline{\Theta}} L = \left[\frac{n_1}{2\pi\sum_1^{n_1}(x_{1i} - \bar{x}_1)^2}\right]^{n_1/2}\left[\frac{n_2}{2\pi\sum_1^{n_2}(x_{2j} - \bar{x}_2)^2}\right]^{n_2/2} e^{-n_1/2}e^{-n_2/2}.$$

If we put μ_1 and μ_2 equal to μ, say, and try to maximize L with respect to μ, σ_1^2, and σ_2^2, it will be found that the estimate of μ is given as the root of a cubic equation and will be a very complex function of the observations. The resulting generalized likelihood-ratio λ will therefore be a complicated function, and to find its distribution is a tedious task indeed and involves the ratio of the two variances. This makes it impossible to determine a critical region $0 < \lambda < k$

for a given probability of a Type I error because the ratio of the population variances is assumed unknown. A number of special devices can be employed in an attempt to circumvent this difficulty, but we shall not pursue the problem further here. For large samples the following criterion may be used: The root of the cubic equation can be computed in any instance by numerical methods, and λ can then be calculated; furthermore, as we shall see in Sec. 5 below, the quantity $-2\log\Lambda$ has approximately the chi-square distribution with one degree of freedom, and hence a test that would reject for $-2\log\lambda$ large could be devised.

When it can be assumed that the two populations have the same variance, the problem becomes relatively simple. The parameter space $\overline{\Theta}$ is then three-dimensional with coordinates (μ_1, μ_2, σ^2), while $\overline{\Theta}_0$ for the null hypothesis $\mu_1 = \mu_2 = \mu$ is two-dimensional with coordinates (μ, σ^2). In $\overline{\Theta}$ we find that the maximum-likelihood estimates of μ_1, μ_2, and σ^2 are, respectively, $\bar{x}_1$, $\bar{x}_2$, and

$$\frac{1}{n_1+n_2}\left[\sum_1^{n_1}(x_{1i}-\bar{x}_1)^2+\sum_1^{n_2}(x_{2j}-\bar{x}_2)^2\right];$$

so

$$\sup_{\overline{\Theta}} L=\left\{\frac{n_1+n_2}{2\pi[\sum(x_{1i}-\bar{x}_1)^2+\sum(x_{2j}-\bar{x}_2)^2]}\right\}^{(n_1+n_2)/2}e^{-(n_1+n_2)/2}.$$

In $\overline{\Theta}_0$, the maximum-likelihood estimates of μ and σ^2 are

$$\hat{\mu}=\frac{1}{n_1+n_2}\left(\sum_1^{n_1}x_{1i}+\sum_1^{n_2}x_{2j}\right)=\frac{n_1\bar{x}_1+n_2\bar{x}_2}{n_1+n_2}\qquad\text{for }\mu$$

and

$$\frac{1}{n_1+n_2}\left[\sum(x_{1i}-\hat{\mu})^2+\sum(x_{2j}-\hat{\mu})^2\right]$$

$$=\frac{1}{n_1+n_2}\left[\sum(x_{1i}-\bar{x}_1)^2+\sum(x_{2j}-\bar{x}_2)^2+\frac{n_1n_2}{n_1+n_2}(\bar{x}_1-\bar{x}_2)^2\right]\qquad\text{for }\sigma^2,$$

which gives

$$\sup_{\overline{\Theta}_0} L$$

$$=\left[\frac{n_1+n_2}{2\pi\left[\sum(x_{1i}-\bar{x}_1)^2+\sum(x_{2j}-\bar{x}_2)^2+\frac{n_1n_2}{n_1+n_2}(\bar{x}_1-\bar{x}_2)^2\right]}\right]^{(n_1+n_2)/2}$$

$$\times\, e^{-(n_1+n_2)/2}.$$

Finally,

$$\lambda = \left(1 + \frac{[n_1 n_2/(n_1 + n_2)](\bar{x}_1 - \bar{x}_2)^2}{\sum (x_{1i} - \bar{x}_1)^2 + \sum (x_{2j} - \bar{x}_2)^2}\right)^{-(n_1+n_2)/2} \tag{17}$$

This last expression is very similar to the corresponding one obtained in Subsec. 4.1, and it turns out that this test can also be performed in terms of a quantity which has the t distribution. We know that $\bar{X}_1$ and $\bar{X}_2$ are independently normally distributed with means μ_1 and μ_2 and with variances σ^2/n_1 and σ^2/n_2. Also it is readily seen that $\bar{X}_1 - \bar{X}_2$ is normally distributed with mean $\mu_1 - \mu_2$ and variance $\sigma^2(1/n_1 + 1/n_2)$. Under the null hypothesis the mean of $\bar{X}_1 - \bar{X}_2$ will be 0. The quantities $\sum (X_{1i} - \bar{X}_1)^2/\sigma^2$ and $\sum (X_{2j} - \bar{X}_2)^2/\sigma^2$ are independently distributed as chi-square distributions with $n_1 - 1$ and $n_2 - 1$ degrees of freedom, respectively; hence their sum has the chi-square distribution with $n_1 + n_2 - 2$ degrees of freedom. Since under the null hypothesis

$$Z = \frac{\bar{X}_1 - \bar{X}_2}{\sigma\sqrt{1/n_1 + 1/n_2}}$$

is normally distributed with mean 0 and unit variance, the quantity

$$T = \frac{\sqrt{n_1 n_2/(n_1 + n_2)}(\bar{X}_1 - \bar{X}_2)}{\sqrt{[\sum (X_{1i} - \bar{X}_1)^2 + \sum (X_{2j} - \bar{X}_2)^2]/(n_1 + n_2 - 2)}} \tag{18}$$

has the t distribution with $n_1 + n_2 - 2$ degrees of freedom. [Note that we do have independence of the numerator and denominator in Eq. (18).] The generalized likelihood-ratio is

$$\lambda = \left[\frac{1}{1 + [t^2/(n_1 + n_2 - 2)]}\right]^{(n_1+n_2)/2}, \tag{19}$$

and its distribution is determined by the t distribution. The test would, of course, be done in terms of T rather than λ. A 5 percent critical region for T is $T^2 > [t_{.975}(n_1 + n_2 - 2)]^2$, where $t_{.975}(n_1 + n_2 - 2)$ is the .975th quantile of the t distribution with $n_1 + n_2 - 2$ degrees of freedom.

If we want to test $\mathscr{H}_0\colon \mu_1 = \mu_2$ versus $\mathscr{H}_1\colon \mu_1 > \mu_2$ or $\mathscr{H}_0\colon \mu_1 \leq \mu_2$ versus $\mathscr{H}_1\colon \mu_1 > \mu_2$, a size-$\alpha$ test is given by the following: Reject $\mathscr{H}_0$ if and only if $T > t_{1-\alpha}(n_1 + n_2 - 2)$, where T is defined in Eq. (18) and $t_{1-\alpha}(n_1 + n_2 - 2)$ is the $(1 - \alpha)$th quantile of the t distribution with $n_1 + n_2 - 2$ degrees of freedom.

Equality of several means The test presented above can be extended from just two normal populations to k normal populations. We assume that we have available k random samples, one from each of k normal populations; that is, let $X_{j1}, \ldots, X_{jn_j}$ be a random sample of size n_j from the jth normal population, $j = 1, \ldots, k$. Assume that the jth population has mean μ_j and variance σ^2. Further assume that the k random samples are independent. Our object is to test the null hypothesis that all the population means are the same versus the alternative that not all the means are equal. We seek a generalized likelihood-ratio test. The likelihood function is given by

$$L(\mu_1, \ldots, \mu_k, \sigma^2; x_{11}, \ldots, x_{1n_1}, \ldots, x_{k1}, \ldots, x_{kn_k})$$

$$= \prod_{j=1}^{k} \prod_{i=1}^{n_j} \frac{1}{\sqrt{2\pi\sigma}} e^{-\frac{1}{2}[(x_{ji}-\mu_j)/\sigma]^2}$$

$$= (2\pi\sigma^2)^{-n/2} \exp\left[-\frac{1}{2\sigma^2} \sum_{j=1}^{k} \sum_{i=1}^{n_j} (x_{ji} - \mu_j)^2\right],$$

where $n = \sum_{j=1}^{k} n_j$.

The parameter space $\underline{\Theta}$ is $(k+1)$-dimensional with coordinates $(\mu_1, \ldots, \mu_k, \sigma^2)$, and $\underline{\Theta}_0$, the collection of points in the parameter space corresponding to the null hypothesis, is two-dimensional with coordinates (μ, σ^2), where $\mu = \mu_1 = \cdots = \mu_k$. In $\underline{\Theta}$, the maximum-likelihood estimates of $\mu_1, \ldots, \mu_k, \sigma^2$ are given by

$$\hat{\mu}_j = \bar{x}_{j.} = \frac{1}{n_j} \sum_{i=1}^{n_j} x_{ji}, \qquad j = 1, \ldots, k,$$

and

$$\hat{\sigma}^2_{\underline{\Theta}} = \frac{1}{n} \sum_{j=1}^{k} \sum_{i=1}^{n_j} (x_{ji} - \bar{x}_{j.})^2; \tag{20}$$

hence,

$$\sup_{\underline{\Theta}} L = \left[\frac{2\pi \sum_j \sum_i (x_{ji} - \bar{x}_{j.})^2}{n}\right]^{-n/2} e^{-n/2}.$$

In $\underline{\Theta}_0$, the maximum-likelihood estimates of μ and σ^2 are

$$\hat{\mu} = \bar{\bar{x}} = \frac{1}{n} \sum_{j=1}^{k} \sum_{i=1}^{n_j} x_{ji} \qquad \text{and} \qquad \hat{\sigma}^2_{\underline{\Theta}_0} = \frac{1}{n} \sum_{j=1}^{k} \sum_{i=1}^{n_j} (x_{ji} - \bar{\bar{x}})^2,$$

and so

$$\sup_{\underline{\Theta}_0} L = \left[\frac{2\pi \sum_j \sum_i (x_{ji} - \bar{\bar{x}})^2}{n}\right]^{-n/2} e^{-n/2}.$$

The generalized likelihood-ratio is then

$$\lambda = \frac{\sup\limits_{\underline{\Theta}_0} L}{\sup\limits_{\underline{\Theta}} L} = \left[\frac{\sum\limits_j \sum\limits_i (x_{ji} - \bar{\bar{x}})^2}{\sum\limits_j \sum\limits_i (x_{ji} - \bar{x}_{j.})^2}\right]^{-n/2}$$

$$= \left[\frac{\sum\limits_j \sum\limits_i (x_{ji} - \bar{x}_{j.} + \bar{x}_{j.} - \bar{\bar{x}})^2}{\sum\limits_j \sum\limits_i (x_{ji} - \bar{x}_{j.})^2}\right]^{-n/2}$$

$$= \left[\frac{\sum\limits_j \sum\limits_i (x_{ji} - \bar{x}_{j.})^2 + \sum\limits_j n_j(\bar{x}_{j.} - \bar{\bar{x}})^2}{\sum\limits_j \sum\limits_i (x_{ji} - \bar{x}_{j.})^2}\right]^{-n/2}$$

$$= \left[1 + \frac{k-1}{n-k} \frac{\sum\limits_j n_j(\bar{x}_{j.} - \bar{\bar{x}})^2/(k-1)}{\sum\limits_j \sum\limits_i (x_{ji} - \bar{x}_{j.})^2/(n-k)}\right]^{-n/2}$$

A generalized likelihood-ratio test is given by the following: Reject $\mathscr{H}_0$ if and only if $\lambda \le \lambda_0$. But $\lambda \le \lambda_0$ if and only if

$$r = \frac{\sum\limits_{j=1}^{k} n_j(\bar{x}_{j.} - \bar{\bar{x}})^2/(k-1)}{\sum\limits_j \sum\limits_i (x_{ji} - \bar{x}_{j.})^2/(n-k)} \ge \text{some constant, say } c. \tag{21}$$

The ratio r is sometimes called the *variance ratio*, or F ratio. The constant c is determined so that the test will have size α; that is, c is selected so that $P[R \ge c \mid \mathscr{H}_0] = \alpha$. Note that $\bar{X}_{j.}$ is independent of $\sum\limits_i (X_{ji} - \bar{X}_{j.})^2$ and, hence, the numerator of Eq. (21) is independent of the denominator. Also, under $\mathscr{H}_0$, note that the numerator divided by σ^2 has a chi-square distribution with $k-1$ degrees of freedom, and the denominator divided by σ^2 has a chi-square distribution with $n-k$ degrees of freedom. Consequently, if $\mathscr{H}_0$ is true, R has an F distribution with $k-1$ and $n-k$ degrees of freedom; so the constant c is the $(1-\alpha)$th quantile of the F distribution with $k-1$ and $n-k$ degrees of freedom.

The testing problem considered above is often referred to as a *one-way analysis of variance*. In some experimental situations, an experimenter is interested in determining whether or not various possible treatments affect the yield. For example, one might be interested in finding out whether various types of fertilizer applications affect the yield of a certain crop. The different treatments correspond to the different populations, and when we test that there is no population difference, we are testing that there is no "treatment" effect. The term "analysis of variance" is explained if we note that the denominator of the ratio in Eq. (21) is an estimate of the variation within populations and the numerator is an estimate of the variation between populations when means are equal. We are analyzing variance to test equality of means.

4.4 Tests on Several Variances

Two variances Given random samples from each of two normal populations with means and variances (μ_1, σ_1^2) and (μ_2, σ_2^2), we may test hypotheses about the two variances. We will consider testing:

(i) $\mathscr{H}_0: \sigma_1^2 \le \sigma_2^2$ versus $\mathscr{H}_1: \sigma_1^2 > \sigma_2^2$
(ii) $\mathscr{H}_0: \sigma_1^2 \ge \sigma_2^2$ versus $\mathscr{H}_1: \sigma_1^2 < \sigma_2^2$
(iii) $\mathscr{H}_0: \sigma_1^2 = \sigma_2^2$ versus $\mathscr{H}_1: \sigma_1^2 \ne \sigma_2^2$

If $X_{11}, \ldots, X_{1n_1}$ is a random sample from a normal density with mean μ_1 and variance σ_1^2, if $X_{21}, \ldots, X_{2n_2}$ is a random sample from a normal density with mean μ_2 and variance σ_2^2, and if the two samples are independent, then we know that

$$\frac{\sum (X_{1i} - \bar{X}_1)^2/(n_1 - 1)\sigma_1^2}{\sum (X_{2i} - \bar{X}_2)^2/(n_2 - 1)\sigma_2^2}$$

has the F distribution with $n_1 - 1$ and $n_2 - 1$ degrees of freedom, and, in particular, the statistic

$$R = \frac{(n_2 - 1) \sum (X_{1i} - \bar{X}_1)^2}{(n_1 - 1) \sum (X_{2i} - \bar{X}_2)^2} \tag{22}$$

has the F distribution with $n_1 - 1$ and $n_2 - 1$ degrees of freedom when $\sigma_1^2 = \sigma_2^2$. Note that the statistic R tends to be large when $\sigma_1^2 > \sigma_2^2$ and small when $\sigma_1^2 < \sigma_2^2$, and so we can capitalize on this different behavior to formulate tests for the hypotheses (i) to (iii). For instance, in testing $\mathscr{H}_0: \sigma_1^2 \le \sigma_2$ versus $\mathscr{H}_1: \sigma_1^2 > \sigma_2^2$, we would reject $\mathscr{H}_0$ for large R, or a size-α test is given by the following: Reject $\mathscr{H}_0$ if and only if R exceeds $F_{1-\alpha}(n_1 - 1, n_2 - 1)$, the $(1 - \alpha)$th quantile of the F distribution with $n_1 - 1$ and $n_2 - 1$ degrees of freedom. Similarly, a test of $\mathscr{H}_0: \sigma_1^2 \ge \sigma_2^2$ versus $\mathscr{H}_1: \sigma_1^2 < \sigma_2^2$ is given by the following: Reject $\mathscr{H}_0$ if and only if R is less than $F_\alpha(n_1 - 1, n_2 - 1)$, the αth quantile of the F distribution with $n_1 - 1$ and $n_2 - 1$ degrees of freedom. A test of $\mathscr{H}_0: \sigma_1^2 = \sigma_2^2$ versus $\mathscr{H}_1: \sigma_1^2 \ne \sigma_2^2$ should be *two-tailed*; that is, $\mathscr{H}_0$ should be rejected for small or large R. In other words, a test is given by the following: Accept $\mathscr{H}_0$ if and only if $k_1 < R < k_2$, where k_1 and k_2 are selected so that the test will have size α. It is customary to make the two tails have equal areas of $\alpha/2$ (although this is not quite the best test); then $k_1 = F_{\alpha/2}(n_1 - 1, n_2 - 1)$, and $k_2 = F_{1-\alpha/2}(n_1 - 1, n_2 - 1)$.

We might mention that the above defined tests can all be derived using the generalized likelihood-ratio principle.

Equality of several variances Let $X_{j1}, \ldots, X_{jn_j}$ be a random sample of size n_j from a normal population with mean μ_j and variance $\sigma_j^2, j = 1, \ldots, k$. Assume that the k samples are independent. Our object is to test the null hypothesis $\mathscr{H}_0: \sigma_1^2 = \sigma_2^2 = \cdots = \sigma_k^2$ against the alternative that not all variances are equal. The likelihood function

$$L(\mu_1, \ldots, \mu_k, \sigma_1^2, \ldots, \sigma_k^2; x_{11}, \ldots, x_{1n_1}, \ldots, x_{k1}, \ldots, x_{kn_k})$$

$$= \prod_{j=1}^{k} \prod_{i=1}^{n_j} \frac{1}{\sqrt{2\pi}\,\sigma_j} e^{-\frac{1}{2}[(x_{ji}-\mu_j)/\sigma_j]^2}$$

and the maximum-likelihood estimates of $\mu_j, \sigma_j^2, j = 1, \ldots, k$ are given by

$$\hat{\mu}_j = \frac{1}{n_j} \sum_{i=1}^{n_j} x_{ji} = \bar{x}_{j.}$$

and

$$\hat{\sigma}_j^2 = \frac{1}{n_j} \sum_{i=1}^{n_j} (x_{ji} - \bar{x}_{j.})^2.$$

The null hypothesis states that all σ_j^2 are equal. Let σ^2 denote their common value; then $\overline{\Theta}_0 = \{(\mu_1, \ldots, \mu_k, \sigma^2): -\infty < \mu_j < \infty; \sigma^2 > 0\}$, and the maximum-likelihood estimates of $\mu_1, \ldots, \mu_k, \sigma^2$ over $\overline{\Theta}_0$ are given by

$$\hat{\mu}_j = \bar{x}_{j.}, \qquad j = 1, \ldots, k,$$

and

$$\hat{\sigma}^2 = \frac{1}{\sum n_j} \sum_{j=1}^{k} \sum_{i=1}^{n_j} (x_{ji} - \bar{x}_{j.})^2 = \frac{\sum n_j \hat{\sigma}_j^2}{\sum n_j}.$$

Therefore,

$$\lambda = \frac{\sup\limits_{\overline{\Theta}_0} L}{\sup\limits_{\overline{\Theta}} L} = \frac{\left(\frac{1}{\hat{\sigma}^2}\right)^{\Sigma n_j/2} \exp\left(-\sum n_j/2\right)}{\prod\limits_{j=1}^{k} \left(\frac{1}{\hat{\sigma}_j^2}\right)^{n_j/2} \exp\left(-\sum n_j/2\right)}$$

$$= \frac{\prod\limits_{j=1}^{k} (\hat{\sigma}_j^2)^{n_j/2}}{\left(\sum n_j \hat{\sigma}_j^2 / \sum n_j\right)^{\Sigma n_j/2}}.$$

A generalized likelihood-ratio test is given by the following: Reject $\mathscr{H}_0$ if and only if $\lambda \leq \lambda_0$. We would like to determine the size of the test for any constant

λ_0 or find λ_0 so that the test has size α, but, unfortunately, the distribution of the generalized likelihood-ratio is intractable. An approximate size-α test can be obtained for large n_j since it can be proved that $-2 \log \Lambda$ is approximately distributed as a chi-square distribution with $k - 1$ degrees of freedom. According to the generalized likelihood-ratio principle $\mathscr{H}_0$ is to be rejected for small λ; hence $\mathscr{H}_0$ should be rejected here for large $-2 \log \lambda$; that is, the critical region of the approximate test should be the right tail. So the approximate size-α test is the following: Reject $\mathscr{H}_0$ if and only if $-2 \log \lambda > \chi_{1-\alpha}(k - 1)$, the $(1 - \alpha)$th quantile of the chi-square distribution with $k - 1$ degrees of freedom. (Several other approximations to the distribution of the likelihood-ratio statistic have been given, and some exact tests are also available.)

5 CHI-SQUARE TESTS

In this section we present a number of tests of hypotheses that one way or another involve the chi-square distribution. Included will be the asymptotic distribution of the generalized likelihood-ratio, goodness-of-fit tests, and tests concerning contingency tables. The material in this section will be presented with an aim of merely finding tests of certain hypotheses, and it will not be presented in such a way that concern is given to the optimality of the test. Thus, the power functions of the derived tests will not be discussed.

5.1 Asymptotic Distribution of Generalized Likelihood-ratio

On two occasions in Sec. 4 we found that the distribution of the generalized likelihood-ratio was intractable, and both times we indicated that an approximate test could be obtained by using an asymptotic distribution of the generalized likelihood-ratio. The following theorem, which we shall not be able to prove because of the advanced character of its proof, gives the asymptotic distribution of the generalized likelihood-ratio.

Theorem 7 Let $X_1, \ldots, X_n$ be a sample with joint density $f_{X_1, \cdots, X_n}(\cdot, \ldots, \cdot; \theta)$, where $\theta = (\theta_1, \ldots, \theta_k)$, that is assumed to satisfy quite general regularity conditions. Suppose that the parameter space $\overline{\Theta}$ is k-dimensional. In testing the hypothesis

$$\mathscr{H}_0: \theta_1 = \theta_1^0, \ldots, \theta_r = \theta_r^0, \theta_{r+1}, \ldots, \theta_k,$$

where $\theta_1^0, \ldots, \theta_r^0$ are known and $\theta_{r+1}, \ldots, \theta_k$ are left unspecified, $-2 \log \Lambda_n$ is approximately distributed as a chi-square distribution with r degrees of freedom when $\mathscr{H}_0$ is true and the sample size n is large. ////

We have assumed that $1 \leq r \leq k$ in the above theorem. If $r = k$, then all parameters are specified and none is left unspecified. The parameter space $\overline{\underline{\Theta}}$ is k-dimensional, and since $\mathscr{H}_0$ specifies the value of r of the components of $(\theta_1, \ldots, \theta_k)$, the dimension of $\overline{\underline{\Theta}}_0$ is $k - r$. Thus, the degrees of freedom of the asymptotic chi-square distribution in Theorem 7 can be thought of in two ways: first, as the number of parameters specified by $\mathscr{H}_0$ and, second, as the difference in the dimensions of $\overline{\underline{\Theta}}$ and $\overline{\underline{\Theta}}_0$.

Recall that Λ_n is the random variable which has values

$$\lambda_n = \sup_{\overline{\underline{\Theta}}_0} L(\theta_1, \ldots, \theta_k; x_1, \ldots, x_n) / \sup_{\overline{\underline{\Theta}}} L(\theta_1, \ldots, \theta_k; x_1, \ldots, x_n),$$

which in turn is the generalized likelihood-ratio for a sample of size n. $\overline{\underline{\Theta}}_0$ is that subset of $\overline{\underline{\Theta}}$ that is specified by $\mathscr{H}_0$. The generalized likelihood-ratio principle dictates that $\mathscr{H}_0$ is to be rejected for λ_n small, but since $-2 \log \lambda_n$ increases as λ_n decreases, a test that is equivalent to a generalized likelihood-ratio test is one that rejects for $-2 \log \lambda_n$ large. Now, since the theorem gives an approximate distribution for the values $-2 \log \lambda_n$ when $\mathscr{H}_0$ is true, a test with approximate size α is given by the following:

$$\text{Reject } \mathscr{H}_0 \text{ if and only if } -2 \log \lambda_n > \chi^2_{1-\alpha}(r),$$

where $\chi^2_{1-\alpha}(r)$ is the $(1 - \alpha)$th quantile of the chi-square distribution with r degrees of freedom. Note that the degrees of freedom r is the number of components of the parameter space that are specified by the null hypothesis.

Because of the specific form of the null hypothesis in the theorem, it may appear that the result is not too widely applicable. The null hypothesis of the theorem specifies the values of a subset of the k components of the k-dimensional parameter space, and not many null hypotheses are of that form. However, often the density can be *reparameterized* so that the null hypothesis is of the form given in the theorem. We illustrate with two examples.

EXAMPLE 19 Recall that in Subsec. 4.3 we discussed testing $\mathscr{H}_0$: $\mu_1 = \mu_2$, $\sigma_1^2 > 0$, $\sigma_2^2 > 0$ versus $\mathscr{H}_1$: $\mu_1 \neq \mu_2$, $\sigma_1^2 > 0$, $\sigma_2^2 > 0$, where μ_1 and σ_1^2 are the mean and variance of one normal population and μ_2 and σ_2^2 are the mean and variance of another. Here the parameter space is four-dimensional, and although $\mathscr{H}_0$ does not appear to be of the form given in Theorem 7, we can reparameterize to make it of that form. Let $\theta_1 = \mu_1 - \mu_2$, $\theta_2 = \mu_2$, $\theta_3 = \sigma_1^2$, and $\theta_4 = \sigma_2^2$. In terms of the reparameterization, $\mathscr{H}_0$ becomes $\mathscr{H}_0$: $\theta_1 = \theta_1^0 = 0$, θ_2, θ_3, θ_4; that is, the component θ_1 is specified to be 0, and the remaining three components are unspecified. The theorem is now applicable for the reparameterization; that is, the asymptotic distribution of $-2 \log \Lambda'$ is known (and is the chi-square distribution with one degree of freedom) for $\mathscr{H}_0$ true, where Λ' is the generalized likelihood-ratio obtained under the reparameterization. However, because of the invariance property of maximum-likelihood estimators, the generalized likelihood-ratio Λ' obtained under the reparameterization is the same as the generalized likelihood-ratio Λ obtained before reparameterization. ////

EXAMPLE 20 In Subsec. 4.4 we tested $\mathscr{H}_0$: $\sigma_1^2 = \cdots = \sigma_m^2$, $\mu_1, \ldots, \mu_m$, where μ_j and σ_j^2 were, respectively, the mean and variance of the jth normal population, $j = 1, \ldots, m$. (In Subsec. 4.4, k was used instead of m.) If we make the following reparameterization, $\mathscr{H}_0$ will have the desired form of Theorem 7:

$$\theta_1 = \frac{\sigma_1^2}{\sigma_m^2}, \ldots, \theta_{m-1} = \frac{\sigma_{m-1}^2}{\sigma_m^2}, \theta_m = \sigma_m^2, \theta_{m+1} = \mu_1, \ldots, \theta_{m+m} = \mu_m.$$

Now $\mathscr{H}_0$ becomes $\mathscr{H}_0$: $\theta_1 = 1, \ldots, \theta_{m-1} = 1, \theta_m, \theta_{m+1}, \ldots, \theta_{2m}$; that is, the first $m - 1$ components are specified to be 1 and the remaining are unspecified. Theorem 7 is now applicable, and, again, because of the invariance property of maximum-likelihood estimates, the generalized likelihood-ratio obtained before and after reparameterization are the same; hence the asymptotic distribution of $-2 \log \Lambda$, as claimed in Subsec. 4.4, is the chi-square distribution with $m - 1$ degrees of freedom when $\mathscr{H}_0$ is true. ////

5.2 Chi-square Goodness-of-fit Test

We commence this section with an example of a testing problem that involves the specification of the parameters of a multinomial distribution. It is hoped that this example will help motivate the presentation of the goodness-of-fit test.

If a population has a multinomial density

$$f(x_1, \ldots, x_k; p_1, \ldots, p_k) = \prod_{j=1}^{k+1} p_j^{x_j}, \qquad (23)$$

where $x_j = 0$ or $1, j = 1, \ldots, k + 1; 0 \le p_j \le 1, j = 1, \ldots, k + 1; \sum_{j=1}^{k+1} x_j = 1;$ and $\sum_{j=1}^{k+1} p_j = 1$ (as would be the case in sampling with replacement from a population of individuals who could be classified into $k + 1$ classes or categories), a common problem is that of testing whether the probabilities p_j have specified numerical values. Thus, for instance, the result of casting a die may be classified into one of six classes, and on the basis of a sample of observations we may wish to test whether the die is true, that is, whether $p_j = \frac{1}{6}$ for $j = 1, \ldots, 6$. One can also think in terms of independent, repeated trials, where each trial can result in any one of $k + 1$ outcomes, called *classes* or *categories*. The density in Eq. (23) then gives the density for the outcome of one trial. The result of one trial can be represented by the multivariate random variable $(X_1, \ldots, X_k)$, where X_j is unity if the trial results in category j and is 0 otherwise. p_j is the probability that a trial results in category j. Now if we independently repeat the trial n times, we have n observations of the multivariate random variable $(X_1, \ldots, X_k)$; we can display them as

$$(X_{11}, \ldots, X_{1k}), (X_{21}, \ldots, X_{2k}), \ldots, (X_{n1}, \ldots, X_{nk}).$$

If we let $N_j = \sum_{i=1}^{n} X_{ij}$, then the random variable N_j is the number of the n trials resulting in category j. We know that $(N_1, \ldots, N_k)$ has a multinomial distribution. (See Example 5 in Subsec. 2.2 of Chap. IV.)

To test the null hypothesis $\mathscr{H}_0$: $p_j = p_j^0, j = 1, \ldots, k + 1$, where p_j^0 are given probabilities summing to unity, we hope to employ the generalized likelihood-ratio principle. The likelihood function is given by

$$L = L(p_1, \ldots, p_k; x_{11}, \ldots, x_{1k}, \ldots, x_{n1}, \ldots, x_{nk}) = \prod_{i=1}^{n} \prod_{j=1}^{k+1} p_j^{x_{ij}}. \tag{24}$$

The parameter space $\overline{\Theta}$ has k dimensions (given k of the $k + 1$ p_j's, the remaining one is determined by $\sum p_j = 1$), while $\overline{\Theta}_0$ is a point. It is readily found that L is maximized in $\overline{\Theta}$ when

$$p_j = \sum_{i=1}^{n} \frac{x_{ij}}{n} = \frac{n_j}{n},$$

where n_j is a value of the random variable N_j. Hence,

$$\sup_{\overline{\Theta}} L = \frac{1}{n^n} \prod_{j=1}^{k+1} n_j^{n_j}.$$

The maximum of L over $\overline{\Theta}_0$ is its only value $\prod_{j=1}^{k+1} (p_j^0)^{n_j}$, and so the generalized likelihood-ratio is

$$\lambda = n^n \prod_{j=1}^{k+1} \left(\frac{p_j^0}{n_j}\right)^{n_j}.$$

A generalized likelihood-ratio test is given by the following: Reject $\mathscr{H}_0$ if and only if $\lambda < \lambda_0$, where the constant λ_0 is chosen to give the desired probability of a Type I error. For small n, the distribution of the generalized likelihood-ratio may be tabulated directly in order to determine λ_0; for large values of n, we may use Theorem 7, which states that $-2 \log \Lambda$ has approximately the chi-square distribution with k degrees of freedom. The chi-square approximation is surprisingly good even if n is small provided that $k > 2$.

Another test which is still commonly used for testing $\mathscr{H}_0$ was proposed (by Karl Pearson) before the general theory of testing hypotheses was developed. This test uses the statistic

$$Q_k^0 = \sum_{j=1}^{k+1} \frac{(N_j - np_j^0)^2}{np_j^0}, \tag{25}$$

which tends to be small when $\mathscr{H}_0$ is true and large when $\mathscr{H}_0$ is false. Note that N_j is the observed number of trial outcomes resulting in category j and np_j^0 is the expected number when $\mathscr{H}_0$ is true. It can be easily shown (see Prob. 39) that

$$\mathscr{E}[Q_k^0] = \sum_{j=1}^{k+1} \frac{1}{np_j^0}[np_j(1 - p_j) + n^2(p_j - p_j^0)^2], \tag{26}$$

where the p_j are the true parameters. If $\mathscr{H}_0$ is true, then $\mathscr{E}[Q_k^0] = \sum (1 - p_j^0) = k + 1 - 1 = k$. The following theorem gives a limiting distribution for Q_k^0 when the null hypothesis $\mathscr{H}_0$ is true.

Theorem 8 Let the possible outcomes of a certain random experiment be decomposed into $k + 1$ mutually exclusive sets, say $A_1, \ldots, A_{k+1}$. Define $p_j = P[A_j]$, $j = 1, \ldots, k + 1$. In n independent repetitions of the random experiment, let N_j denote the number of outcomes belonging to set A_j, $j = 1, \ldots, k + 1$, so that $\sum_{j=1}^{k+1} N_j = n$. Then

$$Q_k = \sum_{j=1}^{k+1} \frac{(N_j - np_j)^2}{np_j} \tag{27}$$

has as a limiting distribution, as n approaches infinity, the chi-square distribution with k degrees of freedom. ////

We will not prove the above theorem, but we will indicate its proof for $k = 1$. What needs to be demonstrated is that for each argument x, $F_{Q_k}(x)$ converges to $F_{\chi^2(k)}(x)$ as $n \to \infty$, where $F_{Q_k}(\cdot)$ is the cumulative distribution function of the random quantity Q_k and $F_{\chi^2(k)}(\cdot)$ is the cumulative distribution function of a chi-square random variable having k degrees of freedom. (Note that $k + 1$, the number of groups, is held fixed, and n, the sample size, is increasing.) If $k = 1$, then

$$Q_k = Q_1 = \frac{(N_1 - np_1)^2}{np_1} + \frac{(N_2 - np_2)^2}{np_2} = \frac{(N_1 - np_1)^2}{np_1}$$

$$+ \frac{(n - N_1 - n + np_1)^2}{n(1 - p_1)} = \frac{(N_1 - np_1)^2}{np_1(1 - p_1)}.$$

We know that N_1 has a binomial distribution with parameters n and p_1 and that $Y_n = (N_1 - np_1)/\sqrt{np_1(1 - p_1)}$ has a limiting standard normal distribution; hence, since the square of a standard normal random variable has a chi-square distribution with one degree of freedom, we suspect that $Y_n^2 = Q_1$ has a limiting chi-square distribution with one degree of freedom, and such can be easily shown to be the case, which would give a proof of Theorem 8 for $k = 1$.

Theorem 8 gives the limiting distribution for the *statistic*

$$Q_k^0 = \sum_{j=1}^{k+1} \frac{(N_j - np_j^0)^2}{np_j^0}$$

when the null hypothesis $\mathscr{H}_0$: $p_j = p_j^0$, $j = 1, \ldots, k + 1$, is true. Thus a test of $\mathscr{H}_0$: $p_j = p_j^0$, $j = 1, \ldots, k + 1$, which has approximate size α, is given by the following:

$$\text{Reject } \mathscr{H}_0 \text{ if and only if } Q_k^0 > \chi^2_{1-\alpha}(k),$$

the $(1 - \alpha)$th quantile of the chi-square distribution with k degrees of freedom. We now have two large-sample tests of the null hypothesis $\mathscr{H}_0$: $p_j = p_j^0$, $i = 1, \ldots, k + 1$, the one just defined, which uses Theorem 8, and the other given in terms of the generalized likelihood-ratio, which uses Theorem 7. It can, in fact, be shown that the two tests are equivalent for large samples.

EXAMPLE 21 Mendelian theory indicates that the shape and color of a certain variety of pea ought to be grouped into four groups, "round and yellow," "round and green," "angular and yellow," and "angular and green," according to the ratios 9/3/3/1. For $n = 556$ peas, the following were observed (the last column gives the expected number):

Round and yellow	315	312.75
Round and green	108	104.25
Angular and yellow	101	104.25
Angular and green	32	34.75

A size-.05 test of the null hypothesis $\mathscr{H}_0$: $p_1 = \frac{9}{16}$, $p_2 = \frac{3}{16}$, $p_3 = \frac{3}{16}$, and $p_4 = \frac{1}{16}$ is given by the following:

Reject $\mathscr{H}_0$ if and only if $Q_3^0 = \sum_1^4 \frac{(N_j - np_j^0)^2}{np_j^0}$ exceeds $\chi^2_{1-\alpha}(k) = \chi^2_{.95}(3) =$ 7.81.

The observed Q_3^0 is

$$\frac{(315 - 312.75)^2}{312.75} + \frac{(108 - 104.25)^2}{104.25} + \frac{(101 - 104.25)^2}{104.25} + \frac{(32 - 34.75)^2}{34.75} \approx .470,$$

and so there is good agreement with the null hypothesis; that is, there is a good fit of the data to the model. ////

Theorem 8 can be generalized to the case where the probabilities p_j may depend on unknown parameters. The generalization is given in the next theorem.

Theorem 9 Let the possible outcomes of a certain random experiment be decomposed into $k + 1$ mutually exclusive sets, say $A_1, \ldots, A_{k+1}$. Define $p_j = P[A_j]$, $j = 1, \ldots, k + 1$, and assume that p_j depends on r unknown parameters $\theta_1, \ldots, \theta_r$, so that $p_j = \not{p}_j(\theta_1, \ldots, \theta_r)$, $j = 1, \ldots, k + 1$. In n independent repetitions of the random experiment, let N_j denote the number of outcomes belonging to set A_j, $j = 1, \ldots, k + 1$, so that $\sum_{j=1}^{k+1} N_j = n$. Let $\hat{\Theta}_1, \ldots, \hat{\Theta}_r$ be BAN estimators (e.g., maximum-likelihood estimators) of $\theta_1, \ldots, \theta_r$ based on $N_1, \ldots, N_k$. Then, under certain general regularity conditions on the $\not{p}_j$'s,

$$Q'_k = \sum_{j=1}^{k+1} \frac{(N_j - n\hat{P}_j)^2}{n\hat{P}_j} \tag{28}$$

has a limiting distribution that is the chi-square distribution with $k - r$ degrees of freedom, where $\hat{P}_j = \not{p}_j(\hat{\Theta}_1, \ldots, \hat{\Theta}_r)$, $j = 1, \ldots, k + 1$. ////

The proof of Theorem 9 is beyond the scope of this book. The limiting distribution given in Theorem 9 differs from the limiting distribution given in Theorem 8 only in the number of degrees of freedom. In Theorem 8 there are k degrees of freedom, and in Theorem 9 there are $k - r$ degrees of freedom; the number of degrees of freedom has been reduced by one for each parameter that is estimated from the data.

No mention of hypothesis testing is made in the statement of Theorem 9. However, we will show now how the results of the theorem can be used to obtain a goodness-of-fit test. Suppose that it is desired to test that a random sample $X_1, \ldots, X_n$ came from a density $f(x; \theta_1, \ldots, \theta_r)$, where $\theta_1, \ldots, \theta_r$ are unknown parameters but the function f is known. The null hypothesis is the composite hypothesis $\mathscr{H}_0$: X_i has density $f(x; \theta_1, \ldots, \theta_r)$ for some $\theta_1, \ldots, \theta_r$. The null hypothesis states that the random sample came from the parametric family of densities that is specified by $f(\cdot\,; \theta_1, \ldots, \theta_r)$. If the range of the random variable X_i is decomposed into $k + 1$ subsets, say $A_1, \ldots, A_{k+1}$, if $p_j = P[X_i \in A_j]$, and if N_j = number of X_i's falling in A_j, then, according to Theorem 9,

$$Q'_k = \sum_1^{k+1} \frac{(N_j - n\hat{P}_j)^2}{n\hat{P}_j}$$

is approximately distributed as the chi-square distribution with $k - r$ degrees of freedom if n is large and $\mathscr{H}_0$ is true, where $\hat{P}_j = p_j(\hat{\Theta}_1, \ldots, \hat{\Theta}_r)$ and $\hat{\Theta}_i$ is a maximum-likelihood estimator of θ_i, $i = 1, \ldots, r$, obtained from the statistics $N_1, \ldots, N_k$. {Note that $p_j(\theta_1, \ldots, \theta_r) = P[X_i \in A_j]$, which for a continuous random variable X_i equals $\int_{A_j} f(x; \theta_1, \ldots, \theta_r)\, dx$.} Hence, a test of $\mathscr{H}_0$ can be obtained by rejecting $\mathscr{H}_0$ if and only if the statistic Q'_k is large; that is, reject $\mathscr{H}_0$ if and only if Q'_k exceeds $\chi^2_{1-\alpha}(k - r)$, where $\chi^2_{1-\alpha}(k - r)$ is the $(1 - \alpha)$th quantile of the chi-square distribution with $k - r$ degrees of freedom. Such a test is called a goodness-of-fit test since it tests whether or not the observations $x_1, \ldots, x_n$ fit, or are consistent with, the assumption that they are observations from the density $f(x; \theta_1, \ldots, \theta_r)$.

In the above, the θ_i, for $i = 1, \ldots, r$, were estimated by using the statistics $N_1, \ldots, N_k$ rather than $X_1, \ldots, X_n$. The statistics $N_1, \ldots, N_k$ give the number of x observations falling in each of the A_j subsets or groups. In practice, often the values of the X_i's are not recorded, and then the group totals $N_1, \ldots, N_k$ constitute the available information. If, however, the observations $X_1, \ldots, X_n$ were available, then one could estimate θ_i, $i = 1, \ldots, r$, more efficiently by using, say, maximum-likelihood estimators based on $X_1 \ldots, X_n$. When such estimators are used, the limiting distribution of Q'_k is no longer a chi-square distribution with $k - r$ degrees of freedom; instead, the limiting distribution of Q'_k is bounded between a chi-square distribution with $k - r$ degrees of freedom and a chi-square distribution with k degrees of freedom. In a sense, some of the "lost" r degrees of freedom are recouped by efficiently estimating $\theta_1, \ldots, \theta_r$. For a proof of Theorem 9 and further discussion of the above, the reader is referred to Kendall and Stuart [14].

EXAMPLE 22 Suppose it is desired to test the hypothesis that an observed random sample $x_1, \ldots, x_n$ has been drawn from some normal population. Let the n sample values $x_1, \ldots, x_n$ be grouped into $k + 1$ classes. For example, the jth class could be taken as all those observations falling in the interval $(z_{j-1}, z_j]$, $j = 1, \ldots, k + 1$, for some $z_0 < z_1 < z_2 < \cdots < z_k < z_{k+1}$, where $z_0 = -\infty$ and $z_{k+1} = +\infty$. Then

$$p_j = p_j(\mu, \sigma^2) = \int_{z_{j-1}}^{z_j} \phi_{\mu,\sigma^2}(x)\,dx = \Phi\left(\frac{z_j - \mu}{\sigma}\right) - \Phi\left(\frac{z_{j-1} - \mu}{\sigma}\right).$$

Let $\hat{\mu}$ and $\hat{\sigma}$ be the maximum-likelihood estimates of μ and σ based on $n_1, \ldots, n_k$, where n_j is the number of observations falling in the jth interval. Then,

$$\hat{p}_j = \Phi\left(\frac{z_j - \hat{\mu}}{\hat{\sigma}}\right) - \Phi\left(\frac{z_{j-1} - \hat{\mu}}{\hat{\sigma}}\right)$$

can be determined from the sample, and so the value

$$q'_k = \sum_{j=1}^{k+1} \frac{(n_j - n\hat{p}_j)^2}{n\hat{p}_j}$$

of Q'_k can also be obtained from the sample. The hypothesis that the sample came from a normal population would be rejected at the α level if $q'_k > \chi^2_{1-\alpha}(k-2)$. If, on the other hand, μ and σ were obtained from maximum-likelihood estimators based on $X_1, \ldots, X_n$, then the asymptotic distribution of Q'_k would fall between a chi-square distribution with $k - 2$ degrees of freedom and a chi-square distribution with k degrees of freedom. The hypothesis would be rejected if $q'_k > C$, where C falls between $\chi^2_{1-\alpha}(k-2)$ and $\chi^2_{1-\alpha}(k)$. Note that for k large there is little difference between $\chi^2_{1-\alpha}(k-2)$ and $\chi^2_{1-\alpha}(k)$. ////

5.3 Test of the Equality of Two Multinomial Distributions and Generalizations

A problem that is of great practical importance is that of testing whether several random samples can be considered as *drawn from the same population*. For instance, in Subsec. 4.3 we tested whether several assumed normal populations could be considered the same normal population. In this subsection we first indicate a test of the hypothesis that two multinomial populations can be considered the same and then indicate some generalizations. Suppose that there are $k + 1$ groups associated with each of the two multinomial populations. Let the first population have associated probabilities $p_{11}, p_{12}, \ldots, p_{1k}, p_{1,k+1}$ and the second $p_{21}, p_{22}, \ldots, p_{2k}, p_{2,k+1}$. It is desired to test $\mathscr{H}_0: p_{1j} = p_{2j}\ (= p_j$, say), $j = 1, \ldots, k + 1$. For a sample of size n_1 from the first population, let N_{1j} denote the number of outcomes in group j, $j = 1, \ldots, k + 1$. Similarly, let N_{2j} denote the number of outcomes in group j of a sample of size n_2 from the second population. (Here we are assuming that the sample sizes n_1 and n_2 are known.) We know that

$$\sum_{j=1}^{k+1} \frac{(N_{ij} - n_i p_{ij})^2}{n_i p_{ij}}$$

has a limiting chi-square distribution with k degrees of freedom for $i = 1$ and 2; hence

$$\sum_{i=1}^{2} \sum_{j=1}^{k+1} \frac{(N_{ij} - n_i p_{ij})^2}{n_i p_{ij}}$$

has a limiting chi-square distribution with $2k$ degrees of freedom if the two random samples are independent. If $\mathscr{H}_0$ is true, then

$$Q_{2k} = \sum_{i=1}^{2} \sum_{j=1}^{k+1} \frac{(N_{ij} - n_i p_j)^2}{n_i p_j} \tag{29}$$

has a limiting chi-square distribution with $2k$ degrees of freedom. If $\mathscr{H}_0$ specifies the values p_j, then Q_{2k} is a statistic and can be used as a test statistic. On the other hand, if the p_j defined by $\mathscr{H}_0$ are unknown, then they have to be estimated. If $\mathscr{H}_0$ is true, the two samples can be considered as one random sample of size $n_1 + n_2$ from a multinomial population with probabilities $p_1, \ldots, p_{k+1}$. Maximum-likelihood estimators of the p_j are then $(N_{1j} + N_{2j})/(n_1 + n_2)$, $j = 1, \ldots, k$, and if the p_j in Eq. (29) are replaced by their maximum-likelihood estimators, we then obtain

$$Q'_{2k} = \sum_{i=1}^{2} \sum_{j=1}^{k+1} \frac{[N_{ij} - n_i(N_{1j} + N_{2j})/(n_1 + n_2)]^2}{n_i(N_{1j} + N_{2j})/(n_1 + n_2)}. \tag{30}$$

It can be shown that Q'_{2k} has a limiting chi-square distribution with $2k - k = k$ degrees of freedom. (This result is not a direct corollary of Theorem 9; it would, however, be a corollary of a generalization of Theorem 9 from one to two populations.) Again the degrees of freedom of the limiting distribution of Q'_{2k} have been reduced by unity for each parameter estimated.

Another test of the homogeneity of two multinomial populations can be derived by finding the generalized likelihood-ratio Λ and employing Theorem 7 to obtain the limiting distribution of $-2 \log \Lambda$. (Reparameterization is required before Theorem 7 can be employed directly.) The details of finding such a test are left as an exercise.

EXAMPLE 23 In an opinion survey regarding a certain political issue there was some question as to whether or not the eligible voters under 25 years of age might view the issue differently from those over 25. Fifteen hundred individuals of those over 25 were interviewed, and 1000 of those under 25 were interviewed with the following results (the data are obviously artificial to facilitate calculations):

	Opposed	Undecided	Favor	Total
Under 25	400	100	500	1000
Over 25	600	400	500	1500
Total	1000	500	1000	2500

Test the null hypothesis that there is no evidence of difference of opinion due to the different age grouping; that is, test $\mathscr{H}_0$: $p_{1j} = p_{2j} = p_j$, $j = 1, 2, 3$. p_1 and p_2 need to be estimated. We can calculate the value of the statistic given in Eq. (30) as follows:

$$\frac{(400 - 1000 \cdot 1000/2500)^2}{1000 \cdot 1000/2500} + \frac{(100 - 1000 \cdot 500/2500)^2}{1000 \cdot 500/2500}$$
$$+ \frac{(500 - 1000 \cdot 1000/2500)^2}{1000 \cdot 1000/2500} + \frac{(600 - 1500 \cdot 1000/2500)^2}{1500 \cdot 1000/2500}$$
$$+ \frac{(400 - 1500 \cdot 500/2500)^2}{1500 \cdot 500/2500} + \frac{(500 - 1500 \cdot 1000/2500)^2}{1500 \cdot 1000/2500} = 125.$$

The 99 percent quantile point for the chi-square distribution with two degrees of freedom is only 9.21; so there is strong evidence that the two age groups have different opinions on the political issue. ////

The technique presented in this subsection can be generalized in two directions. First, a test of the *homogeneity* of several, rather than just two, multinomial populations can be obtained, and, second, a test of the hypothesis that several given samples are drawn from the same population of a specified type (such as the Poisson, the gamma, etc.) can be obtained using a procedure similar to that above. We illustrate with an example.

EXAMPLE 24 One hundred observations were drawn from each of two Poisson populations with the following results:

	0	1	2	3	4	5	6	7	8	9 or more	Total
Population 1	11	25	28	20	9	3	3	0	1	0	100
Population 2	13	27	28	17	11	1	2	1	0	0	100
Total	24	52	56	37	20	11					200

Is there strong evidence in the data to support the contention that the two Poisson populations are different? That is, test the hypothesis that the two populations are the same. This hypothesis can be tested in a variety of ways. We first use the chi-square technique mentioned above. We group the data into six groups, the last including all digits greater than 4, as indicated in the above table. If the two populations are the same, we have to estimate one parameter, namely, the mean of the common Poisson distribution. The maximum-likelihood estimate is the sample mean, which is

$$\frac{0(24) + 1(52) + 2(56) + 3(37) + 4(20) + 5(4) + 6(5) + 7(1) + 8(1)}{200}$$

$$= \frac{420}{200} = 2.1.$$

The expected number in each group of each population is given by

0	1	2	3	4	5 or more
12.25	25.72	27.00	18.90	9.92	6.21

The value of the statistic in Eq. (29), where $n_i p_j$ is replaced by the estimates given in the above table, can be calculated. It is approximately 1.68. The degrees of freedom should be $2k - 1$ (one parameter is estimated), which is 9. The test indicates that there is no reason to suspect that the two assumed Poisson populations are different Poisson populations. ////

We mentioned earlier that there are several methods of testing the null hypothesis considered here. For example the generalized likelihood-ratio principle and employment of Theorem 7 yield a test that the student may find instructive to find for himself.

5.4 Tests of Independence in Contingency Tables

A *contingency table* is a multiple classification; for example, in a public opinion survey the individuals interviewed may be classified according to their attitude on a political proposal and according to sex to obtain a table of the form

	Favor	Oppose	Undecided
Men	1154	475	243
Women	1083	442	362

This is a 2×3 contingency table. The individuals are classified by two criteria, one having two categories and the other three categories. The six distinct classifications are called *cells*. A three-way contingency table would have been obtained had the individuals been further classified according to a third criterion, say, according to an annual-income group. If there were five income groups set up (such as under \$2000, \$2000 to \$4000, . . .), the contingency table would be called a $2 \times 3 \times 5$ table and would have 30 cells into which a person might be put. It is often quite convenient to think of the cells as cubes in a block two units wide, three units long, and five units deep. If the individuals were still further classified into eight geographic locations, one would have a four-way ($2 \times 3 \times 5 \times 8$) contingency table with 240 cells in a four-dimensional block with edges two, three, five, and eight units long. A contingency table provides a convenient display of the data for ultimately investigating suspected relationships. Thus one may suspect that men and women will react differently to a certain political proposal, in which case one would construct such a table as the one above and test the null hypothesis that their attitudes were independent of their sex. To consider another example, a geneticist may suspect that susceptibility to a certain disease is heritable. He would classify a sample of individuals according to (i) whether or not they ever had the disease, (ii) whether or not their fathers had the disease, and (iii) whether or not their mothers had the disease. In the resulting $2 \times 2 \times 2$ contingency table he would test the null hypothesis that classification (i) was independent of (ii) and (iii). Again a medical research worker might suspect a certain environmental condition favored a given disease and classify individuals according to (i) whether or not they ever had the disease,

(ii) whether or not they were subject to the condition. An industrial engineer could use a contingency table to discover whether or not two kinds of defects in a manufactured product were due to the same underlying cause or to different causes. It is apparent that the technique can be a very useful tool in any field of research.

Two-way contingency tables We shall suppose that n individuals or items are classified according to two criteria A and B, that there are r classifications $A_1, A_2, \ldots, A_r$ in A and s classifications $B_1, B_2, \ldots, B_s$ in B, and that the number of individuals belonging to A_i *and* B_j is N_{ij}. We have then an $r \times s$ contingency table with cell frequencies N_{ij} and $\sum N_{ij} = n$:

	B_1	B_2	B_3	$\cdots$	B_s
A_1	N_{11}	N_{12}	N_{13}	$\cdots$	N_{1s}
A_2	N_{21}	N_{22}	N_{23}	$\cdots$	N_{2s}
A_3	N_{31}	N_{32}	N_{33}	$\cdots$	N_{3s}
$\vdots$					
A_r	N_{r1}	N_{r2}	N_{r3}	$\cdots$	N_{rs}

(31)

As a further notation we shall denote the row totals by $N_{i.}$ and the column totals by $N_{.j}$; that is,

$$N_{i.} = \sum_j N_{ij} \qquad \text{and} \qquad N_{.j} = \sum_i N_{ij}.$$

Of course,

$$\sum_i N_{i.} = \sum_j N_{.j} = n.$$

We shall now set up a probability model for the problem with which we wish to deal. The n individuals will be regarded as a sample of size n from a multinomial population with probabilities p_{ij} ($i = 1, 2, \ldots, r$; $j = 1, 2, \ldots, s$). The probability density function for a single observation is

$$f(x_{11}, x_{12}, \ldots, x_{rs}; p_{11}, \ldots, p_{rs}) = \prod_{i,j} p_{ij}^{x_{ij}}, \tag{32}$$

where

$$x_{ij} = 0 \quad \text{or} \quad 1 \quad \text{and} \quad \sum_{i,j} x_{ij} = 1.$$

We wish to test the null hypothesis that the A and B classifications are independent, i.e., that the probability that an individual falls in B_j is not affected by the A class to which the individual happens to belong. Using the symbolism of Chap. I, we would write

$$P[B_j | A_i] = P[B_j] \qquad \text{and} \qquad P[A_i | B_j] = P[A_i]$$

or

$$P[A_i \cap B_j] = P[A_i]P[B_j].$$

If we denote the marginal probabilities $P[A_i]$ by $p_{i.}$ $(i = 1, 2, \ldots, r)$ and the marginal probabilities $P[B_j]$ by $p_{.j}$ $(j = 1, 2, \ldots, s)$, the null hypothesis is simply

$$\mathscr{H}_0\colon p_{ij} = p_{i.}p_{.j}; \qquad \sum p_{i.} = 1, \sum p_{.j} = 1. \tag{33}$$

When the null hypothesis is not true, there is said to be *interaction* between the two criteria of classification.

The complete parameter space $\bar{\Theta}$ for the distribution of $N_{11}, \ldots, N_{rs}$ has $rs - 1$ dimensions (having specified all but one of the p_{ij}, the remaining one is fixed by $\sum_{i,j} p_{ij} = 1$), while under $\mathscr{H}_0$ we have a parameter space $\bar{\Theta}_0$ with $r - 1 + s - 1$ dimensions. (The null hypothesis is specified by $p_{i.}$, $i = 1, \ldots, r$, and $p_{.j}$, $j = 1, \ldots, s$, but there are only $r - 1 + s - 1$ dimensions because $\sum p_{i.} = 1$ and $\sum p_{.j} = 1$.) The likelihood for a sample of size n is

$$L = \prod_{i,j} p_{ij}^{n_{ij}}$$

and its maximum in $\bar{\Theta}$ occurs when

$$\hat{p}_{ij} = \frac{n_{ij}}{n}.$$

In $\bar{\Theta}_0$,

$$L = \prod_{i,j} (p_{i.}p_{.j})^{n_{ij}} = \left(\prod_i p_{i.}^{n_{i.}}\right)\left(\prod_j p_{.j}^{n_{.j}}\right), \tag{34}$$

and its maximum occurs at

$$\hat{p}_{i.} = \frac{n_{i.}}{n} \qquad \text{and} \qquad \hat{p}_{.j} = \frac{n_{.j}}{n}. \tag{35}$$

The generalized likelihood-ratio is therefore

$$\lambda = \frac{\left(\prod_i n_{i.}^{n_{i.}}\right)\left(\prod_j n_{.j}^{n_{.j}}\right)}{n^n \prod_{i,j} n_{ij}^{n_{ij}}}. \tag{36}$$

The distribution of Λ under the null hypothesis is not unique because the hypothesis is composite and the exact distribution of Λ does involve the unknown parameters $p_{i.}$ and $p_{.j}$; hence, it is very difficult to solve for λ_0 in $\sup_{\overline{\Theta}_0} P_\theta[\Lambda \leq \lambda_0] = \alpha$. For large samples we do have a test, however, because $-2 \log \Lambda$ is in that case approximately distributed as a chi-square random variable with

$$rs - 1 - (r + s - 2) = (r - 1)(s - 1)$$

degrees of freedom and on the basis of this distribution a unique critical region for λ may be determined. The degrees of freedom $rs - 1 - (r + s - 2)$ is obtained by subtracting $r + s - 2$, which is the dimension of $\overline{\Theta}_0$, from $rs - 1$, which is the dimension of $\overline{\Theta}$. Also, $(r - 1)(s - 1)$ is the number of parameters specified by $\mathscr{H}_0$. (See Theorem 7 and the comment following it.) Actually, the null hypothesis $\mathscr{H}_0$: $p_{ij} = p_{i.}p_{.j}$ is not of the form required by Theorem 7; so it might be instructive to consider the necessary reparameterization. For convenience, let us take $r = s = 2$. Now $\overline{\Theta} = \{(\theta_1, \theta_2, \theta_3) = (p_{11}, p_{12}, p_{21})$: $p_{11} \geq 0$; $p_{12} \geq 0$; $p_{21} \geq 0$; and $p_{11} + p_{12} + p_{21} \leq 1\}$. Let $\overline{\Theta}'$ with points $(\theta_1', \theta_2', \theta_3')$ denote the reparameterized space, where $\theta_1' = p_{11} - p_{1.}p_{.1}$, $\theta_2' = p_{1.}$, and $\theta_3' = p_{.1}$. It can be easily demonstrated that $\overline{\Theta}'$ is a one-to-one transformation of $\overline{\Theta}$. Also, the null hypothesis $\mathscr{H}_0$: $p_{11} = p_{1.}p_{.1}$, $p_{12} = p_{1.}(1 - p_{.1})$, and $p_{21} = (1 - p_{1.})p_{.1}$ in the original parameter space $\overline{\Theta}$ becomes $\mathscr{H}_0'$: $\theta_1' = 0$ and θ_2' and θ_3' unspecified in the reparameterized space $\overline{\Theta}'$. [Note that $p_{12} = p_{1.}(1 - p_{.1})$ is equivalent to $p_{1.} - p_{11} = p_{1.}(1 - p_{.1})$, which is equivalent to $p_{11} - p_{1.}p_{.1} = 0$. Similarly for $p_{21} = (1 - p_{1.})p_{.1}$.] $\mathscr{H}_0'$ is of the form required by Theorem 7. In general, a point in the $(rs - 1)$-dimensional parameter space $\overline{\Theta}$ can be conveniently displayed as

$$\begin{vmatrix} p_{11} & p_{12} & \cdots & p_{1,s-1} & p_{1s} \\ p_{21} & p_{22} & \cdots & p_{2,s-1} & p_{2s} \\ \vdots & \vdots & & \vdots & \vdots \\ p_{r-1,1} & p_{r-1,2} & \cdots & p_{r-1,s-1} & p_{r-1,s} \\ p_{r1} & p_{r2} & \cdots & p_{r,s-1} & \end{vmatrix},$$

and a point in the reparameterized space $\overline{\Theta}'$ can be displayed as

$$\begin{vmatrix} p_{11} - p_{1.}p_{.1} & p_{12} - p_{1.}p_{.2} & \cdots & p_{1,s-1} - p_{1.}p_{.,s-1} & p_{1.} \\ p_{21} - p_{2.}p_{.1} & p_{22} - p_{2.}p_{.2} & \cdots & p_{2,s-1} - p_{2.}p_{.,s-1} & p_{2.} \\ \vdots & \vdots & & \vdots & \vdots \\ p_{r-1,1} - p_{r-1,.}p_{.1} & p_{r-1,2} - p_{r-1,.}p_{.2} & \cdots & p_{r-1,s-1} - p_{r-1,.}p_{.,s-1} & p_{r-1,.} \\ p_{.1} & p_{.2} & \cdots & p_{.,s-1} & \end{vmatrix}.$$

In casting about for a test which may be used when the sample is not large, we may inquire how it is that a test criterion comes to have a unique distribution for large samples when the distribution actually depends on unknown parameters which may have any values in certain ranges. The answer is that the parameters are not really unknown; they can be estimated, and their estimates approach their true values as the sample size increases. In the limit as n becomes infinite, the parameters are known exactly, and it is at that point that the distribution of Λ actually becomes unique. It is unique because a particular point in $\overline{\Theta}_0$ is selected as the true parameter point, so that the N_{ij} are given a unique distribution, and the distribution of Λ is then determined by this distribution.

It would appear reasonable to employ a similar procedure to set up a test for small samples, i.e., to define a distribution for Λ by using the estimates for the unknown parameters. In the present problem, since the estimates of the $p_{i.}$ and $p_{.j}$ are given by Eq. (35), we might just substitute those values in the distribution function of the N_{ij} and use the distribution to obtain a distribution for Λ. However, we should still be in trouble; the critical region would depend on the marginal totals $N_{i.}$ and $N_{.j}$; hence the probability of a Type I error would vary from sample to sample for any fixed critical region $0 < \lambda < \lambda_0$.

There is a way out of this difficulty, which is well worth investigation because of its own interest and because the problem is important in applied statistics. Let us denote the joint density of all the N_{ij} briefly by $f(n_{ij})$, the marginal density of all the $N_{i.}$ and $N_{.j}$ by $g(n_{i.}, n_{.j})$, and the conditional density of the N_{ij}, given the marginal totals, by

$$f(n_{ij} | n_{i.}, n_{.j}) = \frac{f(n_{ij})}{g(n_{i.}, n_{.j})}.$$

Under the null hypothesis, this conditional distribution happens to be independent of the unknown parameters (as we shall show presently); the estimators $N_{i.}/n$ and $N_{.j}/n$ form a sufficient set of statistics for the $p_{i.}$ and $p_{.j}$. This fact will enable us to construct a test.

The joint density of the N_{ij} is simply the multinomial distribution

$$f(n_{ij}) = f(n_{11}, n_{12}, \ldots, n_{rs}) = \frac{n!}{\prod_{i,j} n_{ij}!} \prod_{i,j} p_{ij}^{n_{ij}} \tag{37}$$

n $\overline{\Theta}$, and in $\overline{\Theta}_0$ (we are interested in the distribution of Λ under $\mathscr{H}_0$) this becomes

$$f(n_{11}, n_{12}, \ldots, n_{rs}) = \frac{n!}{\prod_{i,j} n_{ij}!} \left(\prod_i p_{i.}^{n_{i.}}\right)\left(\prod_j p_{.j}^{n_{.j}}\right). \tag{38}$$

To obtain the desired conditional distribution, we must first find the distribution of the $N_{i.}$ and $N_{.j}$, and this is accomplished by summing Eq. (38) over all sets of n_{ij} such that

$$\sum_i n_{ij} = n_{.j} \qquad \text{and} \qquad \sum_j n_{ij} = n_{i.}. \tag{39}$$

For fixed marginal totals, only the factor $1/\prod n_{ij}!$ in Eq. (38) is involved in the sum; so we have, in effect, to sum that factor over all n_{ij} subject to Eq. (39). The desired sum is given by comparing the coefficients of $\prod_i x_i^{n_{i.}}$ in the expression

$$(x_1 + \cdots + x_r)^{n_{.1}}(x_1 + \cdots + x_r)^{n_{.2}} \cdots (x_1 + \cdots + x_r)^{n_{.s}} = (x_1 + \cdots + x_r)^n. \tag{40}$$

On the right-hand side the coefficient of $\prod x_i^{n_{i.}}$ is simply

$$\frac{n!}{\prod_i n_{i.}!}. \tag{41}$$

On the left-hand side there are terms with coefficients of the form

$$\frac{n_{.1}!}{\prod_i n_{i1}!}\,\frac{n_{.2}!}{\prod_i n_{i2}!} \cdots \frac{n_{.s}!}{\prod_i n_{is}!} = \frac{\prod_j n_{.j}!}{\prod_{i,j} n_{ij}!}, \tag{42}$$

where n_{ij} is the exponent of x_i in the jth multinomial. In this expression the n_{ij} satisfy conditions of Eq. (39); the first condition is satisfied in view of the multinomial theorem, while the second is satisfied because we require the exponent of x_i in these terms to be $n_{i.}$. The sum of all such coefficients, Eq. (42), must equal Eq. (41); hence, we may write

$$\sum \frac{1}{\prod n_{ij}!} = \frac{n!}{\left(\prod_i n_{i.}! \prod_j n_{.j}!\right)}. \tag{43}$$

This is precisely the sum that we require because there is obviously one and only one coefficient of the form of Eq. (42) on the left of Eq. (40) for every possible contingency table, Eq. (31), with given marginal totals. The distribution of the $N_{i.}$ and $N_{.j}$ is, therefore,

$$g(n_{i.}, n_{.j}) = \frac{(n!)^2}{(\prod n_{i.}!)(\prod n_{.j}!)} (\prod p_{i.}^{n_{i.}})(\prod p_{.j}^{n_{.j}}), \tag{44}$$

which shows, incidentally, that the $N_{i.}$ are distributed independently of the $N_{.j}$ under $\mathscr{H}_0$; this is unexpected because $N_{1.}$ and $N_{.1}$, for example, have the random variable N_{11} in common!

The conditional distribution of the N_{ij}, given the marginal totals, is obtained by dividing Eq. (38) by Eq. (44) to obtain

$$f(n_{11}, n_{12}, \ldots, n_{rs} \mid n_{1.}, n_{2.}, \ldots, n_{.s}) = \frac{(\prod n_{i.}!)(\prod n_{.j}!)}{n!\prod n_{ij}!}, \tag{45}$$

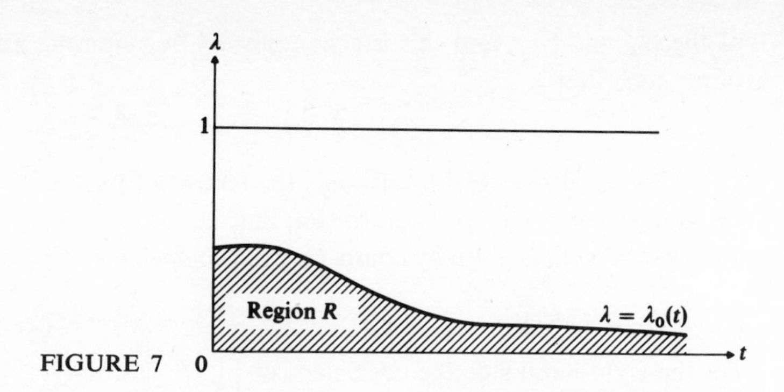

FIGURE 7

which, happily, does not involve the unknown parameters and shows that the estimators are sufficient.

To see how a test may be constructed, let us consider the general situation in which a test statistic Λ for some test has a distribution $f_\Lambda(\lambda; \theta)$ which involves an unknown parameter θ. If θ has a sufficient statistic, say T, then the joint density of Λ and T may be written

$$f_{\Lambda, T}(\lambda, t; \theta) = f_{\Lambda|T}(\lambda|t) f_T(t; \theta),$$

and the conditional density of Λ, given T, will not involve θ. Using the conditional distribution, we may find a number, say $\lambda_0(t)$, for every t such that

$$\int_0^{\lambda_0(t)} f_{\Lambda|T}(\lambda|t)\, d\lambda = .05, \tag{46}$$

for example. In the λt-plane the curve $\lambda = \lambda_0(t)$ together with the line $\lambda = 0$ will determine a region R. See Fig. 7. The probability that a sample will give rise to a pair of values (λ, t) which correspond to a point in R is exactly .05 because

$$\begin{aligned} P[(\Lambda, T) \in R] &= \int_{-\infty}^{\infty} \int_0^{\lambda_0(t)} f_{\Lambda, T}(\lambda, t; \theta)\, d\lambda\, dt \\ &= \int_{-\infty}^{\infty} \left[\int_0^{\lambda_0(t)} f_{\Lambda|T}(\lambda|t)\, d\lambda\right] f_T(t; \theta)\, dt \\ &= \int_{-\infty}^{\infty} .05 f_T(t; \theta)\, dt \\ &= .05. \end{aligned}$$

Hence we may test the hypothesis by using T in conjunction with Λ. The critical region is a plane region instead of an interval $0 < \lambda < \lambda_0$; it is such a region that, whatever the unknown value of θ may be, the Type I error has a specified probability. The test in any given situation actually amounts to a conditional test; we observe T and then perform the test by using the interval $0 < \lambda < \lambda_0(t)$ using the conditional distribution of Λ, given T. It is to be observed that this device cannot be employed unless there is a sufficient statistic for θ.

The above technique is obviously applicable when θ is a set of parameters rather than a single parameter and has a set of sufficient statistics. In particular, the technique may be employed to test the null hypothesis of a two-way contingency table using Eq. (36) to define λ. One merely uses the conditional distribution of Eq. (45) and determines an interval $0 < \lambda < \lambda_0(n_{i.}; n_{.j})$ which has the desired probability of a Type I error for the observed marginal totals.

In applications of this test one is confronted with a very tedious computation in determining the distribution of Λ unless r, s, and the marginal totals are quite small. It can be shown, however, that the large-sample approximation may be used without appreciable error except when both r and s equal 2. In the latter instance, other simplifying approximations have been developed (see, for example, Fisher and Yates, "Tables for Statisticians and Biometricians," Oliver & Boyd Ltd., Edinburgh or London, 1938), but we shall not explore the problem that far.

Another test of the $\mathscr{H}_0$ given in Eq. (33) is obtained if the distribution in Eq. (45) is replaced by its multivariate normal approximation since then it can be shown that the statistic

$$Q = \sum_{i,j} \frac{[N_{ij} - n(N_{i.}/n)(N_{.j}/n)]^2}{n(N_{i.}/n)(N_{.j}/n)} \tag{47}$$

has approximately the chi-square distribution with $rs - 1 - (r - 1 + s - 1) = (r - 1)(s - 1)$ degrees of freedom. The test criterion is to reject $\mathscr{H}_0$ for large Q. This is the criterion first proposed (by Karl Pearson) for testing the hypothesis, and it differs from $-2 \log \lambda$ by terms of order $1/\sqrt{n}$. The two criteria are therefore essentially equivalent unless n is small. The argument that Q is a reasonable test statistic is entirely analogous to that used in Subsec. 5.2 above to justify Eq. (25). The statistic Q of Eq. (47) has intuitive appeal. N_{ij} is the observed number in the ijth cell, and $n(N_{i.}/n)(N_{.j}/n)$ is an estimator of the expected number in the ijth cell when $\mathscr{H}_0$ is true. Thus, Q will tend to be small for $\mathscr{H}_0$ true and large for $\mathscr{H}_0$ false.

Three-way contingency tables If the elements of a population can be classified according to three criteria A, B, and C with classifications A_i $(i = 1, 2, \ldots, s_1)$, B_j $(j = 1, 2, \ldots, s_2)$, and C_k $(k = 1, 2, \ldots, s_3)$, a sample of n individuals may be classified in a three-way $s_1 \times s_2 \times s_3$ contingency table. We shall let p_{ijk}

represent the probabilities associated with the individual cells and N_{ijk} be the numbers of sample elements in the individual cells, and, as before, marginal totals will be indicated by replacing the summed index by a dot; thus

$$N_{i.k} = \sum_{j=1}^{s_2} N_{ijk} \qquad \text{and} \qquad N_{..k} = \sum_{i=1}^{s_1} \sum_{j=1}^{s_2} N_{ijk}. \tag{48}$$

There are four hypotheses that may be tested in connection with this table. We may test whether all three criteria are mutually independent, in which case the null hypothesis is

$$p_{ijk} = p_{i..}\, p_{.j.}\, p_{..k}, \tag{49}$$

where $p_{i..} = \sum_j \sum_k p_{ijk}$, $p_{.j.} = \sum_i \sum_k p_{ijk}$, and $p_{..k} = \sum_i \sum_j p_{ijk}$; or we may test whether any one of the three criteria is independent of the other two. Thus to test whether the B classification is independent of A and C, we set up the null hypothesis

$$p_{ijk} = p_{i.k}\, p_{.j.}, \tag{50}$$

where $p_{i.k} = \sum_j p_{ijk}$.

The procedure for testing these hypotheses is entirely analogous to that for the two-way tables. The likelihood of the sample is

$$L = \prod_{i,j,k} p_{ijk}^{n_{ijk}}, \tag{51}$$

where

$$\sum_{i,j,k} p_{ijk} = 1 \qquad \text{and} \qquad \sum_{i,j,k} n_{ijk} = n.$$

In $\underline{\Theta}$ the maximum of L occurs when

$$\hat{p}_{ijk} = \frac{n_{ijk}}{n},$$

so that

$$\sup_{\underline{\Theta}} L = \frac{1}{n^n} \prod_{i,j,k} n_{ijk}^{n_{ijk}}. \tag{52}$$

To test the null hypothesis in Eq. (50), for example, we make the substitution of Eq. (50) into Eq. (51) and maximize L with respect to the $p_{i.k}$ and $p_{.j.}$ to find

$$\hat{p}_{i.k} = \frac{n_{i.k}}{n} \qquad \text{and} \qquad \hat{p}_{.j.} = \frac{n_{.j.}}{n},$$

and

$$\sup_{\underline{\Theta}_0} L = \frac{1}{n^{2n}} \left(\prod_{i,k} n_{i.k}^{n_{i.k}} \right) \left(\prod_j n_{.j.}^{n_{.j.}} \right). \tag{53}$$

The generalized likelihood-ratio λ is given by the quotient of Eqs. (52) and (53), and in large samples $-2 \log \Lambda$ has the chi-square distribution with

$$s_1 s_2 s_3 - 1 - [(s_1 s_3 - 1) + s_2 - 1] = (s_1 s_3 - 1)(s_2 - 1)$$

degrees of freedom. Again the large-sample distribution is quite adequate for many purposes. $(s_1 s_3 - 1) + (s_2 - 1)$ is the dimension of $\overline{\underline{\Theta}}_0$, and $s_1 s_2 s_3 - 1$ is the dimension of $\overline{\underline{\Theta}}$.

A test statistic analogous to that given in Eq. (47) for testing independence in a 2×2 contingency table can also be derived. For testing $\mathscr{H}_0$: A and C classifications are independent of the B classification, such a test statistic is

$$Q = \sum_i \sum_j \sum_k \frac{[N_{ijk} - n(N_{i.k}/n)(N_{.j.}/n)]^2}{n(N_{i.k}/n)(N_{.j.}/n)}. \tag{54}$$

Under $\mathscr{H}_0$, Q has an asymptotic chi-square distribution with $s_1 s_2 s_3 - 1 - (s_1 s_3 - 1) - (s_2 - 1) = (s_1 s_3 - 1)(s_2 - 1)$ degrees of freedom. Again, the statistic Q of Eq. (54) has intuitive appeal since N_{ijk} is the observed number in cell ijk and $n(N_{i.k}/n)(N_{.j.}/n)$ is an estimator of the expected number when $\mathscr{H}_0$ is true.

6 TESTS OF HYPOTHESES AND CONFIDENCE INTERVALS

In Subsec. 3.4 above we noted that a confidence interval for a unidimensional parameter θ could be used to obtain a test of $\mathscr{H}_0$: $\theta = \theta_0$ versus $\mathscr{H}_1$: $\theta \neq \theta_0$. In this section we will further explore that concept and show that one can reverse the operation; that is, one can use a family of tests of $\mathscr{H}_0$: $\theta = \theta_0$ versus $\mathscr{H}_1$: $\theta \neq \theta_0$ (the family is generated by varying θ_0) to obtain a confidence interval for θ. Our considerations in this section will not be very thorough; our intent is merely to present an introduction to the usefulness of the close relationship between hypothesis testing and confidence intervals.

Our discussion can be made somewhat more general if we speak in terms of *confidence sets* rather than *confidence intervals*. As usual, let $\mathfrak{X}$ denote the sample space, $\overline{\underline{\Theta}}$ the parameter space, and $(x_1, \ldots, x_n)$ the observed sample.

Definition 18 Confidence set A family of subsets of the parameter space $\overline{\underline{\Theta}}$ indexed by $(x_1, \ldots, x_n) \in \mathfrak{X}$, denoted by $\vartheta = \{\overline{\underline{\Theta}}(x_1, \ldots, x_n)$: $\overline{\underline{\Theta}}(x_1, \ldots, x_n) \subseteq \overline{\underline{\Theta}}$; $(x_1, \ldots, x_n) \in \mathfrak{X}\}$, is defined to be a *family of confidence sets with confidence coefficient* γ if and only if

$$P_\theta[\overline{\underline{\Theta}}(X_1, \ldots, X_n) \text{ contains } \theta] = \gamma \qquad \text{for all } \theta \in \overline{\underline{\Theta}}. \tag{55}$$

////

It should be emphasized that any member, say $\overline{\Theta}(x_1, \ldots, x_n)$, of the family of confidence sets is a subset of $\overline{\Theta}$, the parameter space. $\overline{\Theta}(X_1, \ldots, X_n)$ is a *random* subset; for any possible value, say $(x_1, \ldots, x_n)$, of $(X_1, \ldots, X_n)$, $\overline{\Theta}(X_1, \ldots, X_n)$ takes on the value $\overline{\Theta}(x_1, \ldots, x_n)$, a member of the family ϑ. To aid in the interpretation of the probability statement in Eq. (55), note that for a fixed (yet arbitrary) θ "$\overline{\Theta}(X_1, \ldots, X_n)$ contains θ" is an event [it is the event that the random interval $\overline{\Theta}(X_1, \ldots, X_n)$ contains the fixed θ] and the θ that appears as a subscript in P_θ is the θ that indexes the distribution of the X_i's appearing in $\overline{\Theta}(X_1, \ldots, X_n)$.

For instance, suppose $X_1, \ldots, X_n$ is a random sample from $N(\theta, 1)$. $\overline{\Theta} = \{\theta\colon -\infty < \theta < \infty\}$. Let the subset $\overline{\Theta}(x_1, \ldots, x_n)$ be the interval $(\bar{x} - z/\sqrt{n}, \bar{x} + z/\sqrt{n})$, where z is given by $\Phi(z) - \Phi(-z) = \gamma$; then the family of subsets $\vartheta = \{\overline{\Theta}(x_1, \ldots, x_n)\colon \overline{\Theta}(x_1, \ldots, x_n) = (\bar{x} - z/\sqrt{n}, \bar{x} + z/\sqrt{n})\}$ is a family of confidence sets with a confidence coefficient γ since

$$\begin{aligned} P_\theta[\overline{\Theta}(X_1, \ldots, X_n) \text{ contains } \theta] &= P_\theta\left[\overline{X} - \frac{z}{\sqrt{n}} < \theta < \overline{X} + \frac{z}{\sqrt{n}}\right] \\ &= P_\theta\left[-z < \frac{\overline{X} - \theta}{1/\sqrt{n}} < z\right] = \gamma \qquad \text{for all } \theta \in \overline{\Theta}. \end{aligned}$$

The family ϑ is a family of confidence intervals for θ having a confidence coefficient γ. In general, then, a confidence interval is an example of a confidence set.

Confidence sets can be constructed from tests of hypotheses, as we now show. Let Υ_{θ_0} be a size-α test (nonrandomized) of the null hypothesis $\mathscr{H}_0$: $\theta = \theta_0$, and let $\mathfrak{X}(\theta_0)$ be the acceptance region of the test Υ_{θ_0}. [The acceptance region is the set complement of the critical region; that is, if the critical region is given by $C(\theta_0)$, then $\mathfrak{X}(\theta_0) = \mathfrak{X} - C(\theta_0)$.] Note that $\mathfrak{X}(\theta_0)$ is a subset of $\mathfrak{X}$ indexed by θ_0. Since the test Υ_{θ_0} has size α,

$$P_{\theta_0}[(X_1, \ldots, X_n) \in \mathfrak{X}(\theta_0)] = 1 - \alpha.$$

If we now vary θ_0 over $\overline{\Theta}$ and for each θ_0 we have a test Υ_{θ_0}, then we get a family of acceptance regions, namely, $\{\mathfrak{X}(\theta_0)\colon \theta_0 \in \overline{\Theta}\}$. $\mathfrak{X}(\theta_0)$ is the acceptance region of test Υ_{θ_0}. One can now define

$$\overline{\Theta}(x_1, \ldots, x_n) = \{\theta_0\colon (x_1, \ldots, x_n) \in \mathfrak{X}(\theta_0)\}. \tag{56}$$

Clearly $\overline{\Theta}(x_1, \ldots, x_n)$ is a subset of $\overline{\Theta}$. Furthermore, the family $\{\overline{\Theta}(x_1, \ldots, x_n)\}$ is a family of confidence sets with a confidence coefficient $\gamma = 1 - \alpha$ since one has $\{\overline{\Theta}(X_1, \ldots, X_n)$ contains $\theta_0\}$ if and only if one has $\{(X_1, \ldots, X_n) \in \mathfrak{X}(\theta_0)\}$, and so

$$P_{\theta_0}[\overline{\Theta}(X_1, \ldots, X_n) \text{ contains } \theta_0] = P_{\theta_0}[(X_1, \ldots, X_n) \in \mathfrak{X}(\theta_0)] = 1 - \alpha.$$

EXAMPLE 25 Let $X_1, \ldots, X_n$ be a random sample from $N(\theta, 1)$, and consider testing $\mathscr{H}_0: \theta = \theta_0$. A test with size α is given by the following: Reject $\mathscr{H}_0$ if and only if $|\bar{x} - \theta_0| \geq z/\sqrt{n}$, where z is defined by $\Phi(z) - \Phi(-z) = 1 - \alpha$. The acceptance region of this test is given by

$$\mathfrak{X}(\theta_0) = \left\{(x_1, \ldots, x_n): \theta_0 - \frac{z}{\sqrt{n}} < \bar{x} < \theta_0 + \frac{z}{\sqrt{n}}\right\}.$$

We can now define, as in Eq. (56),

$$\begin{aligned}\underline{\overline{\Theta}}(x_1, \ldots, x_n) &= \{\theta_0: (x_1, \ldots, x_n) \in \mathfrak{X}(\theta_0)\} \\ &= \left\{\theta_0: \theta_0 - \frac{z}{\sqrt{n}} < \bar{x} < \theta_0 + \frac{z}{\sqrt{n}}\right\} \\ &= \left\{\theta_0: \bar{x} - \frac{z}{\sqrt{n}} < \theta_0 < \bar{x} + \frac{z}{\sqrt{n}}\right\}.\end{aligned}$$

$\underline{\overline{\Theta}}(x_1, \ldots, x_n)$ is a confidence set (in fact a confidence interval) with a confidence coefficient $\gamma = 1 - \alpha$. ////

The general procedure exhibited above shows how tests of hypotheses can be used to generate or construct confidence sets. The procedure is reversible; that is, a given family of confidence sets can be "reverted" to give a test of hypothesis. Specifically, for a given family $\{\underline{\overline{\Theta}}(x_1, \ldots, x_n)\}$ of confidence sets with a confidence coefficient γ, if we defined

$$\mathfrak{X}(\theta_0) = \{(x_1, \ldots, x_n): \theta_0 \in \underline{\overline{\Theta}}(x_1, \ldots, x_n)\}, \tag{57}$$

then the nonrandomized test with acceptance region $\mathfrak{X}(\theta_0)$ is a test of $\mathscr{H}_0: \theta = \theta_0$ with size $\alpha = 1 - \gamma$.

The usefulness of the strong relationship between tests of hypotheses and confidence sets is exemplified not only in the fact that one can be used to construct the other but also in the result that often an optimal property of one carries over to the other. That is, if one can find a test that is optimal in some sense, then the corresponding constructed confidence set is also optimal in some sense, and conversely. We will not study the very interesting theoretical result alluded to in the previous sentence, but we will give the following in order to give some idea of the types of optimality that can be expected. (See the more advanced books of Ref. 16 and Ref. 19 for a detailed discussion.) An optimum property of confidence sets is given in the following definition.

Definition 19 Uniformly most accurate A family $\{\underline{\overline{\Theta}}^*(x_1, \ldots, x_n)\}$ of confidence sets with a confidence coefficient γ is defined to be a *uniformly most accurate* family of confidence sets at a confidence coefficient γ if for any other family $\{\underline{\overline{\Theta}}(x_1, \ldots, x_n)\}$ of confidence sets with a coefficient γ

$$P_\theta[\underline{\overline{\Theta}}^*(X_1, \ldots, X_n) \text{ contains } \theta'] \le P_\theta[\underline{\overline{\Theta}}(X_1, \ldots, X_n) \text{ contains } \theta']$$

for all θ and θ'. ////

Definition 19 is saying that $\underline{\overline{\Theta}}^*(X_1, \ldots, X_n)$ is less likely to contain an incorrect θ' than is $\underline{\overline{\Theta}}(X_1, \ldots, X_n)$, whereas both $\underline{\overline{\Theta}}^*(X_1, \ldots, X_n)$ and $\underline{\overline{\Theta}}(X_1, \ldots, X_n)$ have the same probability of containing the correct θ. As you may have guessed, uniformly most accurate confidence sets rarely exist. However, uniformly most accurate confidence sets within restricted classes of confidence sets could also be defined, and then one could be hopeful of the existence of such optimal confidence sets. A general type of result that derives from the close relationship between tests of hypotheses and confidence sets is the following: If Υ^* is a uniformly most powerful size-α test of $\mathscr{H}_0: \theta = \theta_0$ within some restricted class of tests, then the confidence set corresponding to Υ^* is uniformly most accurate with coefficient $\gamma = 1 - \alpha$ within some restricted class of confidence sets. With such a result one can see how an optimality of a test can be transferred to an optimality of a corresponding confidence set, and therein lies the real utility of the close relationship between hypotheses testing and confidence sets.

7 SEQUENTIAL TESTS OF HYPOTHESES

7.1 Introduction

Sequential analysis refers to techniques for testing hypotheses or estimating parameters when the sample size is not fixed in advance but is determined during the course of the experiment by criteria which depend on the observations as they occur. In this section we propose to consider, and then only briefly, one form of sequential analysis, namely, the *sequential probability ratio test*.

In Sec. 2 above we considered testing the simple null hypothesis $\mathscr{H}_0: \theta = \theta_0$ versus the simple alternative hypothesis $\mathscr{H}_1: \theta = \theta_1$. It was shown (Neyman-Pearson lemma) that for samples of fixed size n, the test which minimized the size, say β, of the Type II error for fixed size, say α, of the Type I error was a simple likelihood-ratio test. That is, for fixed n and α, β was minimized. Suppose now that it is desired to fix both α and β in advance and then find that simple likelihood-ratio test having minimum sample size n and having size of Type I error equal to α and size of Type II error β. The solution of such a problem is illustrated in the following example.

EXAMPLE 26 A manufacturer of a certain component, say, an oil seal, knows the history of his current manufacturing process. He knows, for instance, that the distribution of lifetimes of the seals now being manufactured is, say, $N(100, 100)$. A new manufacturing process is suggested; the manufacturer wants to continue with his present manufacturing process if the new process is not better (longer mean lifetime), yet he also wants to be quite certain to switch to the new process if the new process increases the mean lifetime by, say, 5 percent. He proposes to take a sample of observations of lifetimes of seals made by the new process and then from the sample decide whether or not the process has longer mean life. He models the experiment by assuming that the random variable X, representing the lifetime of a seal manufactured using the new process, is distributed as $N(\theta, 100)$, and he wants to test $\mathscr{H}_0$: $\theta \leq 100$ versus $\mathscr{H}_1$: $\theta > 100$. He fixes his error sizes and wants to determine the sample size n so that, say,

$$.01 = \alpha = P_{\theta=100}[\text{reject } \mathscr{H}_0] \qquad \text{and} \qquad .05 = \beta = P_{\theta=105}[\text{accept } \mathscr{H}_0].$$

That is, he seeks to determine n so that there is only a 1 percent chance of rejecting that the new process is no better than the old when it is not, yet there is a 95 percent chance of rejecting that the mean lifetime of the new process is less than 100 when in fact it is 5 percent larger. It can be shown that the simple likelihood-ratio test is equivalent to the test of rejecting $\mathscr{H}_0$ for large $\bar{X}_n$. Thus he seeks to determine n and k so that

$$.01 = P_{\theta=100}[\bar{X}_n > k] \qquad \text{and} \qquad .05 = P_{\theta=105}[\bar{X}_n \leq k]$$

or

$$.01 = 1 - \Phi\left(\frac{k-100}{10/\sqrt{n}}\right) \qquad \text{and} \qquad .05 = \Phi\left(\frac{k-105}{10/\sqrt{n}}\right).$$

$$\Phi\left(\frac{k-100}{10/\sqrt{n}}\right) = .99$$

implies

$$\frac{k-100}{10/\sqrt{n}} \approx 2.326,$$

and

$$\Phi\left(\frac{k-105}{10/\sqrt{n}}\right) = .05$$

implies

$$\frac{k - 105}{10/\sqrt{n}} \approx -1.645;$$

which together imply that $100 + 10(2.326)/\sqrt{n} \approx 105 - 10(1.645)/\sqrt{n})$ or $n \approx 63.08$; so a sample of size 64 is needed. ////

Referring to the above example, the following considerations make sequential analysis interesting both from the theoretical and practical viewpoint. In drawing the 64 observations to test $\mathscr{H}_0$, it is possible that among the first few observations, say 20, 30, or 40, the evidence is quite sufficient relative to α and β for accepting or rejecting $\mathscr{H}_0$, and then observing additional observations would be a waste of time and effort. In other words, the possibility is raised that, by constructing the test in a fashion which permits termination of the sampling at any observation, one can test $\mathscr{H}_0$ with fixed error sizes α and β and yet do so with fewer than 64 observations on an average. This is in fact the case; although it may at first appear surprising in view of the fact that the best test for fixed sample size requires 64 observations. The saving in observations is often quite large, sometimes as much as 50 percent! We will study such a sequential procedure in the remaining subsections.

7.2 Definition of Sequential Probability Ratio Test

Consider testing a simple null hypothesis against a simple alternative hypothesis. In other words, suppose a sample can be drawn from one of two distributions (it is not known which one) and it is desired to test that the sample came from one distribution against the possibility that it came from the other. If X_1, X_2, ... denotes the random variables, we want to test $\mathscr{H}_0$: $X_i \sim f_0(\cdot)$ versus $\mathscr{H}_1$: $X_i \sim f_1(\cdot)$. The simple likelihood-ratio test was of the following form:

$$\text{Reject } \mathscr{H}_0 \text{ if } \lambda = \frac{L_0}{L_1} \leq k \text{ for some constant } k > 0.$$

The sequential test that we propose to consider employs the likelihood-ratios sequentially. Define

$$\lambda_m = \lambda_m(x_1, \ldots, x_m) = \frac{L_0(x_1, \ldots, x_m)}{L_1(x_1, \ldots, x_m)} = \frac{L_0(m)}{L_1(m)} = \frac{\prod_{i=1}^{m} f_0(x_i)}{\prod_{i=1}^{m} f_1(x_i)}$$

for $m = 1, 2, \ldots,$ and compute sequentially $\lambda_1, \lambda_2, \ldots,$. For fixed k_0 and k_1 satisfying $0 < k_0 < k_1$, adopt the following procedure: Take observation x_1 and compute λ_1; if $\lambda_1 \leq k_0$, reject $\mathscr{H}_0$; if $\lambda_1 \geq k_1$, accept $\mathscr{H}_0$; and if $k_0 < \lambda_1 < k_1$, take observation x_2, and compute λ_2. If $\lambda_2 \leq k_0$, reject $\mathscr{H}_0$; if $\lambda_2 \geq k_1$, accept $\mathscr{H}_0$; and if $k_0 < \lambda_2 < k_1$, observe x_3, etc. The idea is to continue sampling as long as $k_0 < \lambda_j < k_1$ and stop as soon as $\lambda_m \leq k_0$ or $\lambda_m \geq k_1$, rejecting $\mathscr{H}_0$ if $\lambda_m \leq k_0$ and accepting $\mathscr{H}_0$ if $\lambda_m \geq k_1$. The critical region of the described sequential test can be defined as $C = \bigcup_{n=1}^{\infty} C_n$, where

$$C_n = \{(x_1, \ldots, x_n): k_0 < \lambda_j(x_1, \ldots, x_j) < k_1, j = 1, \ldots, n-1,$$
$$\lambda_n(x_1, \ldots, x_n) \leq k_0\}. \tag{58}$$

A point in C_n indicates that $\mathscr{H}_0$ is to be rejected for a sample of size n. Similarly, the acceptance region can be defined as $A = \bigcup_{n=1}^{\infty} A_n$, where

$$A_n = \{(x_1, \ldots, x_n): k_0 < \lambda_j(x_1, \ldots, x_j) < k_1, j = 1, \ldots, n-1,$$
$$\lambda_n(x_1, \ldots, x_n) \geq k_1\}. \tag{59}$$

Definition 20 Sequential probability ratio test For fixed $0 < k_0 < k_1$, a test as described above is defined to be a *sequential probability ratio test*. ////

When we considered the simple likelihood-ratio test for fixed sample size n, we determined k so that the test would have preassigned size α. We now want to determine k_0 and k_1 so that the sequential probability ratio test will have preassigned α and β for its respective sizes of the Type I and Type II errors. Note that

$$\alpha = P[\text{reject } \mathscr{H}_0 \mid \mathscr{H}_0 \text{ is true}] = \sum_{n=1}^{\infty} \int_{C_n} L_0(n) \tag{60}$$

and

$$\beta = P[\text{accept } \mathscr{H}_0 \mid \mathscr{H}_0 \text{ is false}] = \sum_{n=1}^{\infty} \int_{A_n} L_1(n), \tag{61}$$

where, as before, $\int_{C_n} L_0(n)$ is a shortened notation for $\underset{C_n}{\int \cdots \int} \left[\prod_{i=1}^{n} f_0(x_i)\, dx_i\right]$. For fixed α and β, Eqs. (60) and (61) are two equations in the two unknowns k_0 and k_1. (Both A_n and C_n are defined in terms of k_0 and k_1.) A solution of these two equations would give the sequential probability ratio test having the desired preassigned error sizes α and β. As might be anticipated, the actual determination of k_0 and k_1 from Eqs. (60) and (61) can be a major computational project. In practice, they are seldom determined that way because a very simple and accurate approximation is available and is given in the next subsection.

We note that the sample size of a sequential probability ratio test is a random variable. The procedure says to continue sampling until $\lambda_n = \lambda_n(x_1, \ldots, x_n)$ first falls outside the interval (k_0, k_1). The actual sample size then depends on which x_i's are observed; it is a function of the random variables $X_1, X_2, \ldots$ and consequently is itself a random variable. Denote it by N. Ideally we would like to know the distribution of N or at least the expectation of N. (The procedure, as defined, seemingly allows for the sampling to continue indefinitely, meaning that N could be infinite. Although we will not so prove, it can be shown that N is finite with probability 1.) One way of assessing the performance of the sequential probability ratio test would be to evaluate the expected sample size that is required under each hypothesis. The following theorem, given without proof (see Lehmann [16]), states that the sequential probability ratio test is an optimal test if performance is measured using expected sample size.

Theorem 10 The sequential probability ratio test with error sizes α and β minimizes both $\mathscr{E}[N \mid \mathscr{H}_0$ is true$]$ and $\mathscr{E}[N \mid \mathscr{H}_1$ is true$]$ among all tests (sequential or not) which satisfy the following: $P[\mathscr{H}_0$ is rejected $\mid \mathscr{H}_0$ is true$] \leq \alpha$, $P[\mathscr{H}_0$ is accepted $\mid \mathscr{H}_0$ is false$] \leq \beta$, and the expected sample size is finite. ////

Note that in particular the sequential probability ratio test requires fewer observations on the average than does the fixed-sample-size test that has the same error sizes. In Subsec. 7.4 we will evaluate the expected sample size for the example given in the introduction in which 64 observations were required for a fixed-sample-size test with preassigned α and β.

7.3 Approximate Sequential Probability Ratio Test

We noted above that the determination of k_0 and k_1 that defines that particular sequential probability ratio test which has error sizes α and β is in general computationally quite difficult. The following remark gives an approximation to k_0 and k_1.

Remark Let k_0 and k_1 be defined so that the sequential probability ratio test corresponding to k_0 and k_1 has error sizes α and β; then k_0 and k_1 can be approximated by, say, k_0' and k_1', where

$$k_0' = \frac{\alpha}{1-\beta} \qquad \text{and} \qquad k_1' = \frac{1-\alpha}{\beta}. \tag{62}$$

PROOF $\left(\text{Assume } \sum_{n=1}^{\infty} P[N = n \mid \mathscr{H}_i] = 1 \text{ for } i = 0, 1.\right)$

$$\alpha = P[\text{reject } \mathscr{H}_0 \mid \mathscr{H}_0 \text{ is true}] = \sum_{n=1}^{\infty} \int_{C_n} L_0(n) \le \sum_{n=1}^{\infty} \int_{C_n} k_0 L_1(n)$$

$$= k_0 \sum_{n=1}^{\infty} \int_{C_n} L_1(n) = k_0 P[\text{reject } \mathscr{H}_0 \mid \mathscr{H}_1 \text{ is true}]$$

$$= k_0(1 - \beta),$$

and hence $k_0 \ge \alpha/(1 - \beta)$. Also

$$1 - \alpha = P[\text{accept } \mathscr{H}_0 \mid \mathscr{H}_0 \text{ is true}]$$

$$= \sum_{n=1}^{\infty} \int_{A_n} L_0(n) \ge \sum_{n=1}^{\infty} \int_{A_n} k_1 L_1(n)$$

$$= k_1 P[\text{accept } \mathscr{H}_0 \mid \mathscr{H}_1 \text{ is true}] = k_1 \beta,$$

and hence $k_1 \le (1 - \alpha)/\beta$. Note that the approximations $k_0' = \alpha/(1 - \beta)$ and $k_1' = (1 - \alpha)/\beta$ satisfy

$$k_0' = \frac{\alpha}{1 - \beta} \le k_0 < k_1 \le \frac{1 - \alpha}{\beta} = k_1'. \tag{63}$$

////

Remark Let α' and β' be the error sizes of the sequential probability ratio test defined by k_0' and k_1' given in Eq. (62). Then $\alpha' + \beta' \le \alpha + \beta$.

PROOF Let A' and C' (with corresponding A_n' and C_n') denote the acceptance and critical regions of the sequential probability ratio test defined by k_0' and k_1'. Then

$$\alpha' = \sum_{n=1}^{\infty} \int_{C_n'} L_0(n) \le \frac{\alpha}{1 - \beta} \sum_{n=1}^{\infty} \int_{C_n'} L_1(n) = \frac{\alpha}{1 - \beta}(1 - \beta'),$$

and

$$1 - \alpha' = \sum_{n=1}^{\infty} \int_{A_n'} L_0(n) \ge \frac{1 - \alpha}{\beta} \sum_{n=1}^{\infty} \int_{A_n'} L_1(n) = \frac{1 - \alpha}{\beta} \beta';$$

hence $\alpha'(1 - \beta) \le \alpha(1 - \beta')$, and $(1 - \alpha)\beta' \le (1 - \alpha')\beta$, which together imply that $\alpha'(1 - \beta) + (1 - \alpha)\beta' \le \alpha(1 - \beta') + (1 - \alpha')\beta$ or $\alpha' + \beta' \le \alpha + \beta$.

////

Naturally, one would prefer to use that sequential probability ratio test having the desired preassigned error sizes α and β; however, since it is difficult to find the k_0 and k_1 corresponding to such a sequential probability ratio test, instead one can use that sequential probability ratio test defined by k_0' and k_1' of Eq. (62) and be assured that the sum of the error sizes α' and β' is less than or equal to the sum of the desired error sizes α and β.

7.4 Approximate Expected Sample Size of Sequential Probability Ratio Test

The procedure used in performing a sequential probability ratio test is to continue sampling as long as $k_0 < \lambda_m < k_1$ and stop sampling as soon as $\lambda_m \leq k_0$ or $\lambda_m \geq k_1$. If $z_i = \log_e [f_0(x_i)/f_1(x_i)]$, an equivalent test is given by the following: Continue sampling as long as $\log_e k_0 < \sum_1^m z_i < \log_e k_1$, and stop sampling as soon as $\sum_1^m z_i \leq \log_e k_0$ (and then reject $\mathscr{H}_0$) or $\sum_1^m z_i \geq \log_e k_1$ (and then accept $\mathscr{H}_0$). As before, let N be the random variable denoting the sample size of the sequential probability ratio test, and let $Z_i = \log_e [f_0(X_i)/f_1(X_i)]$. Equation (64), given in the following theorem, is useful in finding an approximate expected sample size of the sequential probability ratio test.

Theorem 11 Wald's equation Let $Z_1, Z_2, \ldots, Z_n, \ldots$ be independent identically distributed random variables satisfying $\mathscr{E}[|Z_i|] < \infty$. Let N be an integer-valued random variable whose value n depends only on the values of the first n z_i's. Suppose $\mathscr{E}[N] < \infty$. Then

$$\mathscr{E}[Z_1 + \cdots + Z_N] = \mathscr{E}[N] \cdot \mathscr{E}[Z_i]. \tag{64}$$

PROOF

$$\begin{aligned}
\mathscr{E}[Z_1 + \cdots + Z_N] &= \mathscr{E}[\mathscr{E}[Z_1 + \cdots + Z_N | N]] \\
&= \sum_{n=1}^{\infty} \mathscr{E}[Z_1 + \cdots + Z_n | N = n] P[N = n] \\
&= \sum_{n=1}^{\infty} \sum_{i=1}^{n} \mathscr{E}[Z_i | N = n] P[N = n] \\
&= \sum_{i=1}^{\infty} \sum_{n=i}^{\infty} \mathscr{E}[Z_i | N = n] P[N = n] \\
&= \sum_{i=1}^{\infty} \mathscr{E}[Z_i | N \geq i] P[N \geq i] \\
&= \sum_{i=1}^{\infty} \mathscr{E}[Z_i] P[N \geq i] \\
&= \mathscr{E}[Z_i] \sum_{i=1}^{\infty} P[N \geq i] \\
&= \mathscr{E}[Z_i] \mathscr{E}[N].
\end{aligned}$$

($\mathscr{E}[Z_i] = \mathscr{E}[Z_i | N \geq i]$ since the event $\{N \geq i\}$ depends only on $Z_1, \ldots, Z_{i-1}$ and hence is independent of Z_i. Also $\mathscr{E}[N] = \sum_{i=1}^{\infty} P[N \geq i]$ follows from Eq. (6) of Chap. II.) ////

If the sequential probability ratio test leads to rejection of $\mathscr{H}_0$, then the random variable $Z_1 + \cdots + Z_N \leq \log_e k_0$, but $Z_1 + \cdots + Z_N$ is close to $\log_e k_0$ since $Z_1 + \cdots + Z_N$ first became less than or equal to $\log_e k_0$ at the Nth observation; hence $\mathscr{E}[Z_1 + \cdots + Z_N] \approx \log_e k_0$. Similarly, if the test leads to acceptance, $\mathscr{E}[Z_1 + \cdots + Z_N] \approx \log_e k_1$; hence $\mathscr{E}[Z_1 + \cdots + Z_N] \approx \rho \log_e k_0 + (1 - \rho) \log_e k_1$, where $\rho = P[\mathscr{H}_0 \text{ is rejected}]$. Using

$$\mathscr{E}[N] = \frac{\mathscr{E}[Z_1 + \cdots + Z_N]}{\mathscr{E}[Z_i]} \approx \frac{\rho \log_e k_0 + (1 - \rho) \log_e k_1}{\mathscr{E}[Z_i]}$$

we obtain

$$\begin{aligned}\mathscr{E}[N \mid \mathscr{H}_0 \text{ is true}] &\approx \frac{\alpha \log_e k_0 + (1 - \alpha) \log_e k_1}{\mathscr{E}[Z_i \mid \mathscr{H}_0 \text{ is true}]} \\ &\approx \frac{\alpha \log_e [\alpha/(1 - \beta)] + (1 - \alpha) \log_e [(1 - \alpha)/\beta]}{\mathscr{E}[Z_i \mid \mathscr{H}_0 \text{ is true}]}, \qquad (65)\end{aligned}$$

and

$$\begin{aligned}\mathscr{E}[N \mid \mathscr{H}_0 \text{ is false}] &\approx \frac{(1 - \beta) \log_e k_0 + \beta \log_e k_1}{\mathscr{E}[Z_i \mid \mathscr{H}_0 \text{ is false}]} \\ &\approx \frac{(1 - \beta) \log_e [\alpha/(1 - \beta)] + \beta \log_e [(1 - \alpha)/\beta]}{\mathscr{E}[Z_i \mid \mathscr{H}_0 \text{ is false}]}. \qquad (66)\end{aligned}$$

EXAMPLE 27 Consider sampling from $N(\theta, \sigma^2)$, where σ^2 is assumed known. Test $\mathscr{H}_0\colon \theta = \theta_0$ versus $\mathscr{H}_1\colon \theta = \theta_1$. Now

$$\begin{aligned} z_i &= \log_e \frac{f_0(x_i)}{f_1(x_i)} \\ &= \log_e \left[\frac{(1/\sqrt{2\pi}\sigma)e^{-\frac{1}{2}[(x_i - \theta_0)/\sigma]^2}}{(1/\sqrt{2\pi}\sigma)e^{-\frac{1}{2}[(x_i - \theta_1)/\sigma]^2}} \right] \\ &= -\frac{1}{2\sigma^2}[(x_i - \theta_0)^2 - (x_i - \theta_1)^2] \\ &= -\frac{1}{2\sigma^2}[(\theta_0^2 - \theta_1^2) - 2x_i(\theta_0 - \theta_1)]; \end{aligned}$$

hence

$$\mathscr{E}[Z_i | \mathscr{H}_0 \text{ is true}] = -\frac{1}{2\sigma^2}[(\theta_0^2 - \theta_1^2) - 2\theta_0(\theta_0 - \theta_1)]$$

$$= \frac{1}{2\sigma^2}(\theta_1 - \theta_0)^2,$$

and

$$\mathscr{E}[Z_i | \mathscr{H}_0 \text{ is false}] = -\frac{1}{2\sigma^2}(\theta_1 - \theta_0)^2.$$

For $\alpha = .01$, $\beta = .05$, $\sigma^2 = 100$, $\theta_0 = 100$, and $\theta_1 = 105$ (as in Example 26), Eq. (65) reduces to

$$\mathscr{E}[N | \mathscr{H}_0 \text{ is true}] \approx \frac{.01 \log_e(.01/.95) + .99 \log_e(.99/.05)}{25/200} \approx 24$$

and

$$\mathscr{E}[N | \mathscr{H}_0 \text{ is false}] \approx 34.$$

The average sample sizes of 24 and 34 for the sequential probability ratio test compare to a sample size of 64 for the fixed-sample-size test. ////

PROBLEMS

1 Let X have a Bernoulli distribution, where $P[X = 1] = \theta = 1 - P[X = 0]$.

(*a*) For a random sample of size $n = 10$, test $\mathscr{H}_0: \theta \leq \frac{1}{2}$ versus $\mathscr{H}_1: \theta > \frac{1}{2}$. Use the critical region $\{\sum x_i \geq 6\}$.

(i) Find the power function, and sketch it.

(ii) What is the size of this test?

(*b*) For a random sample of size $n = 10$:

(i) Find the most powerful size-α ($\alpha = .0547$) test of $\mathscr{H}_0: \theta = \frac{1}{2}$ versus $\mathscr{H}_1: \theta = \frac{1}{4}$.

(ii) Find the power of the most powerful test at $\theta = \frac{1}{4}$.

(*c*) For a random sample of size 10, test $\mathscr{H}_0: \theta = \frac{1}{2}$ versus $\mathscr{H}_1: \theta = \frac{1}{4}$.

(i) Find the minimax test for the loss function $0 = \ell(d_0; \theta_0) = \ell(d_1; \theta_1)$, $\ell(d_0; \theta_1) = 1719$, $\ell(d_1; \theta_0) = 2241$.

(ii) Compare the maximum risk of the minimax test with the maximum risk of the most powerful test given in part (*b*).

(*d*) Again, for a sample of size 10, test $\mathscr{H}_0: \theta = \frac{1}{2}$ versus $\mathscr{H}_1: \theta = \frac{1}{4}$. Use the above loss function to find the Bayes test corresponding to prior probabilities given by

$$g = \frac{3^4}{(1719/2241)(\frac{2}{3})^{10} + 3^4}.$$

2 Let X have the density $f(x;\theta) = \theta x^{\theta-1} I_{(0,1)}(x)$.

(*a*) To test $\mathscr{H}_0: \theta \leq 1$ versus $\mathscr{H}_1: \theta > 1$, a sample of size 2 was selected, and the critical region $C = \{(x_1, x_2): 3/4x_1 \leq x_2\}$ was used. Find the power function and size of this test.

(*b*) For a random sample of size 2, find the most powerful size-$[\alpha = \frac{1}{2}(1 - \ln 2)]$ test of $\mathscr{H}_0: \theta = 1$ versus $\mathscr{H}_1: \theta = 2$.

(*c*) Are the tests that you obtained in parts (*a*) and (*b*) unbiased?

(*d*) For a random sample of size 2, find the minimax test of $\mathscr{H}_0: \theta_0 = 1$ versus $\mathscr{H}_1: \theta_1 = 2$ using the loss function $\ell(d_0; \theta_0) = \ell(d_1; \theta_1) = 0$, $\ell(d_0; \theta_1) = 1 - \log_e 2$, $\ell(d_1; \theta_0) = \frac{1}{2} + \log_e 2$.

(*e*) For a random sample of size n, find the Bayes test corresponding to prior probabilities given by $g = \frac{2}{3}$ of $\mathscr{H}_0: \theta = 1$ versus $\mathscr{H}_1: \theta = 2$ using the loss function $\ell(d_0; \theta_0) = \ell(d_1; \theta_1) = 0$, $\ell(d_0; \theta_1) = 1$, $\ell(d_1; \theta_0) = 2$.

(*f*) Test $\mathscr{H}_0: \theta = 1$ versus $\mathscr{H}_1: \theta = 2$ using a sample of size 2. Let α = size of Type I error and β = size of Type II error. Find the test that minimizes the largest of α and β.

3 Let $\overline{\Theta} = \{1, 2\}$, and suppose you have one observation from the density $I_{(\theta - \frac{1}{2}, \theta + \frac{1}{2})}(x)$. Show that a test that has uniformly smallest risk among all tests exists, and find it.

4 Let X be a *single* observation from the density

$$f(x;\theta) = \theta x^{\theta-1} I_{(0,1)}(x),$$

where $\theta > 0$.

(*a*) In testing $\mathscr{H}_0: \theta \leq 1$ versus $\mathscr{H}_1: \theta > 1$, find the power function and size of the test given by the following: Reject $\mathscr{H}_0$ if and only if $X \geq \frac{1}{2}$.

(*b*) Find a most powerful size-α test of $\mathscr{H}_0: \theta = 2$ versus $\mathscr{H}_1: \theta = 1$.

(*c*) For the loss function given by $\ell(d_0; 2) = \ell(d_1; 1) = 0$, $\ell(d_0; 1) = \ell(d_1; 2) = 1$, find the minimax test of $\mathscr{H}_0: \theta = 2$ versus $\mathscr{H}_1: \theta = 1$.

(*d*) Is there a uniformly most powerful size-α test of $\mathscr{H}_0: \theta \geq 2$ versus $\mathscr{H}_1: \theta < 2$? If so, what is it?

(*e*) Among all possible simple likelihood-ratio tests of $\mathscr{H}_0: \theta = 2$ versus $\mathscr{H}_1: \theta = 1$, find that test that minimizes $\alpha + \beta$, where α and β are the respective sizes of the Type I and Type II errors.

(*f*) Find the generalized likelihood-ratio test of size α of $\mathscr{H}_0: \theta = 1$ versus $\mathscr{H}_1: \theta \neq 1$.

5 Let X be a *single* observation from the density $f(x;\theta) = (2\theta x + 1 - \theta) I_{[0,1]}(x)$, where $-1 \leq \theta \leq 1$.

(*a*) Find the most powerful size-α test of $\mathscr{H}_0: \theta = 0$ versus $\mathscr{H}_1: \theta = 1$. (Your test should be expressed in terms of α.)

(*b*) To test $\mathscr{H}_0: \theta \leq 0$ versus $\mathscr{H}_1: \theta > 0$, the following procedure was used: Reject $\mathscr{H}_0$ if X exceeds $\frac{1}{2}$. Find the power and size of this test.

(*c*) Is there a uniformly most powerful size-α test of $\mathscr{H}_0: \theta \leq 0$ versus $\mathscr{H}_1: \theta > 0$? If so, what is it?

(*d*) What is the generalized likelihood-ratio test of $\mathscr{H}_0: \theta = 0$ versus $\mathscr{H}_1: \theta \neq 0$?

(*e*) Among all possible simple likelihood-ratio tests of $\mathscr{H}_0: \theta = 0$ versus $\mathscr{H}_1: \theta = 1$ find that test which minimizes $\alpha + \beta$, where α and β are the respective sizes of the Type I and Type II errors.

(*f*) Given a set of observations, all of which fall between 0 and 1, indicate how you would test the hypothesis that the observations came from the density $f(x; \theta)$.

6 Let $X_1, \ldots, X_n$ denote a random sample from $f(x; \theta) = (1/\theta)I_{(0, \theta)}(x)$, and let $Y_1, \ldots, Y_n$ be the corresponding ordered sample. To test $\mathscr{H}_0: \theta = \theta_0$ versus $\mathscr{H}_1: \theta \neq \theta_0$, the following test was used: Accept $\mathscr{H}_0$ if $\theta_0(\sqrt[n]{\alpha}) \leq Y_n \leq \theta_0$; otherwise reject.

(*a*) Find the power function for this test, and sketch it.

(*b*) Find another (nonrandomized) test that has the same size as the given test, and show that the given test is more powerful (for all alternative θ) than the test you found.

7 Let $X_1, \ldots, X_n$ denote a random sample from

$$f(x; \theta) = (1/\theta)x^{(1-\theta)/\theta}I_{(0, 1)}(x).$$

Test $\mathscr{H}_0: \theta \leq \theta_0$ versus $\mathscr{H}_1: \theta > \theta_0$.

(*a*) For a sample of size *n*, find a uniformly most powerful (UMP) size-α test if such exists.

(*b*) Take $n = 2$, $\theta_0 = 1$, and $\alpha = .05$, and sketch the power function of the UMP test.

8 Let $X_1, \ldots, X_n$ be a random sample from the Poisson distribution

$$f(x; \theta) = \frac{e^{-\theta}\theta^x}{x!} I_{\{0, 1, 2, \ldots\}}(x).$$

(*a*) Find the UMP test of $\mathscr{H}_0: \theta = \theta_0$ versus $\mathscr{H}_1: \theta > \theta_0$, and sketch the power function for $\theta_0 = 1$ and $n = 25$. (Use the central-limit theorem. Pick $\alpha = .05$.)

(*b*) Test $\mathscr{H}_0: \theta = \theta_0$ versus $\mathscr{H}_1: \theta \neq \theta_0$. Find the general form of the critical region corresponding to the test arrived at using the generalized likelihood-ratio principle. (The critical region should be defined in terms of $\sum X_i$.)

(*c*) A reasonable test of $\mathscr{H}_0: \theta = \theta_0$ versus $\mathscr{H}_1: \theta \neq \theta_0$ would be the following: Reject if $|\bar{X} - \theta_0| \geq K$. For $\alpha = .05$, find *K* so that $P[\text{reject } \mathscr{H}_0 | \mathscr{H}_0] = .05$. (Assume that *n* is large enough so that the central-limit theorem can be used to find an approximation to *K*.)

9 Let $\overline{\Theta} = \{\theta_0, \theta_1\}$. Show that any test arrived at using the generalized likelihood-ratio principle is equivalent to a simple likelihood-ratio test.

10 To test $\mathscr{H}_0: \theta \leq 1$ versus $\mathscr{H}_1: \theta > 1$ on the basis of two observations, say X_1 and X_2, from the uniform distribution on $(0, \theta)$, the following test was used:

$$\text{Reject } \mathscr{H}_0 \quad \text{if} \quad X_1 + X_2 \geq 1.$$

(*a*) Find the power function of the above test, and note its size. [Recall that $X_1 + X_2$ has a triangular distribution on $(0, 2\theta)$.]

(*b*) Find another test that has the same size as the given test but has greater power for some $\theta > 1$ if such exists. If such does not exist, explain why.

11 Let $X_1, \ldots, X_n$ be a random sample of size n from $f(x; \theta) = \theta^2 x e^{-\theta x} I_{(0, \infty)}(x)$.

(*a*) In testing $\mathscr{H}_0\colon \theta \leq 1$ versus $\mathscr{H}_1\colon \theta > 1$ for $n = 1$ (a sample of size 1) the following test was used: Reject $\mathscr{H}_0$ if and only if $X_1 \leq 1$. Find the power function and size of this test.

(*b*) Find a most powerful size-α test of $\mathscr{H}_0\colon \theta = 1$ versus $\mathscr{H}_1\colon \theta = 2$.

(*c*) Does there exist a uniformly most powerful size-α test of $\mathscr{H}_0\colon \theta \leq 1$ versus $\mathscr{H}_1\colon \theta > 1$? If so, what is it?

(*d*) In testing $\mathscr{H}_0\colon \theta = 1$ versus $\mathscr{H}_1\colon \theta = 2$, among all simple likelihood-ratio tests find that test which minimizes the sum of the sizes of the Type I and Type II errors. You may take $n = 1$.

12 Let $X_1, \ldots, X_n$ be a random sample from the uniform distribution over the interval $(\theta, \theta + 1)$. To test $\mathscr{H}_0\colon \theta = 0$ versus $\mathscr{H}_1\colon \theta > 0$, the following test was used: Reject $\mathscr{H}_0$ if and only if $Y_n \geq 1$ or $Y_1 \geq k$, where k is a constant.

(*a*) Determine k so that the test will have size α.

(*b*) Find the power function of the test you obtained in part (*a*).

(*c*) Prove or disprove: If k is selected so that the test has size α, then the given test is uniformly most powerful of size α.

13 Let $X_1, \ldots, X_m$ be a random sample from the density $\theta_1 x^{\theta_1 - 1} I_{(0, 1)}(x)$, and let $Y_1, \ldots, Y_n$ be a random sample from the density $\theta_2 y^{\theta_2 - 1} I_{(0, 1)}(y)$. Assume that the samples are independent. Set $U_i = -\log_e X_i$, $i = 1, \ldots, m$, and $V_j = -\log_e Y_j$, $j = 1, \ldots, n$.

(*a*) Find the generalized likelihood-ratio for testing $\mathscr{H}_0\colon \theta_1 = \theta_2$ versus $\mathscr{H}_1\colon \theta_1 \neq \theta_2$.

(*b*) Show that the generalized likelihood-ratio test can be expressed in terms of the statistic

$$T = \frac{\sum U_i}{\sum U_i + \sum V_j}.$$

(*c*) If $\mathscr{H}_0$ is true, what is the distribution of T? (You do not have to derive it if you know the answer.) Does the distribution of T depend on $\theta = \theta_1 = \theta_2$ given that $\mathscr{H}_0$ is true?

14 Find a generalized likelihood-ratio test of size α for testing $\mathscr{H}_0\colon \theta \leq 1$ versus $\mathscr{H}_1\colon \theta > 1$ on the basis of a random sample $X_1, \ldots, X_n$ from $f(x; \theta) = \theta e^{-\theta x} I_{(0, \infty)}(x)$.

15 Let X be a *single* observation from the density $f(x; \theta) = (1 + \theta) x^{\theta} I_{(0, 1)}(x)$, where $\theta > -1$.

(*a*) Find the most powerful size-α test of $\mathscr{H}_0\colon \theta = 0$ versus $\mathscr{H}_1\colon \theta = 1$.

(*b*) Is there a uniformly most powerful size-α test of $\mathscr{H}_0\colon \theta \leq 0$ versus $\mathscr{H}_1\colon \theta > 0$? If so, what is it?

(*c*) Among all possible simple likelihood-ratio tests of $\mathscr{H}_0$: $\theta = 0$ versus $\mathscr{H}_1$: $\theta = 1$ find a test which minimizes $2\alpha + \beta$, where α and β are the respective sizes of the Type I and Type II errors.

(*d*) Find a generalized likelihood-ratio test of $\mathscr{H}_0$: $\theta = 0$ versus $\mathscr{H}_1$: $\theta \neq 0$.

16 Let $X_1, \ldots, X_m$ be a random sample from $\theta_1 e^{-\theta_1 x} I_{(0, \infty)}(x)$, and let $Y_1, \ldots, Y_n$ be a random sample from $\theta_2 e^{-\theta_2 y} I_{(0, \infty)}(y)$. Assume that the samples are independent.

(*a*) Find the generalized likelihood-ratio for testing $\mathscr{H}_0$: $\theta_1 = \theta_2$ versus $\mathscr{H}_1$: $\theta_1 \neq \theta_2$.

(*b*) Show that the generalized likelihood-ratio test can be expressed in terms of the statistic $T = \sum X_i / (\sum X_i + \sum Y_j)$. Argue (or show) that the distribution of T does not depend on $\theta = \theta_1 = \theta_2$ when $\mathscr{H}_0$ is true.

17 Use the confidence-interval technique to derive a test of $\mathscr{H}_0$: $\mu_1 = \mu_2$ versus $\mathscr{H}_1$: $\mu_1 \neq \mu_2$ in sampling from the bivariate normal distribution. Such a test is often called a *paired t test*. (See the last paragraph in Subsec. 3.4 in Chap. VIII.)

18 Given the sample $(-.2, -.9, -.6, .1)$ from a normal population with unit variance, test whether the population mean is less than 0 at the .05 level (i.e., with probability .05 of a Type I error). That is, test $\mathscr{H}_0$: $\mu \leq 0$ at the .05 level versus $\mathscr{H}_1$: $\mu > 0$.

19 Given the sample $(-4.4, 4.0, 2.0, -4.8)$ from a normal population with variance 4 and the sample $(6.0, 1.0, 3.2, -.4)$ from a normal population with variance 5, test at the .05 level that the means differ by no more than one unit. Plot the power function for this test. Plot the ideal power function.

20 A metallurgist made four determinations of the melting point of manganese: 1269, 1271, 1263, and 1265 degrees centigrade. Test the hypothesis that the mean μ of this population is within 5 degrees centigrade of the published value of 1260. Use $\alpha = .05$. (Assume normality and $\sigma^2 = 5$.)

21 Plot the power function for a test of the null hypothesis $\mathscr{H}_0$: $-1 < \mu < 1$ for a normal distribution with known variance using sample sizes 1, 4, 16, and 64. (Use the standard deviation σ as the unit of measurement on the μ axis and .05 probability of Type I error.) Plot the ideal power function.

22 Let $X_1, \ldots, X_n$ be a random sample of size n from a normal density with known variance. What is the best critical region for testing the null hypothesis that the mean is 6 against the alternative that the mean is 4?

23 Derive a test of $\mathscr{H}_0$: $\sigma^2 < 10$ against $\mathscr{H}_1$: $\sigma^2 \geq 10$ for a sample of size n from a normal population with a mean of 0.

24 In testing between two values μ_0 and μ_1 for the mean of a normal population, show that the probabilities for both types of error can be made arbitrarily small by taking a sufficiently large sample.

25 A cigarette manufacturer sent each of two laboratories presumably identical samples of tobacco. Each made five determinations of the nicotine content in milligrams as follows: (i) 24, 27, 26, 21, and 24 and (ii) 27, 28, 23, 31, and 26. Were the two laboratories measuring the same thing? (Assume normality and a common variance.)

26 The metallurgist of Prob. 20, after assessing the magnitude of the various errors that might accrue in his experimental technique, decided that his measurements should have a standard deviation of 2 degrees centigrade or less. Are the data consistent with this supposition at the .05 level? (That is, test $\mathcal{H}_0$: $\sigma \leq 2$.)

27 Test the hypothesis that the two samples of Prob. 19 came from populations with the same variance. Use $\alpha = .05$.

28 The power function for a test that the means of two normal populations are equal depends on the values of the two means μ_1 and μ_2 and is therefore a surface. But the value of the function depends only on the difference $\theta = \mu_1 - \mu_2$, so that it can be adequately represented by a curve, say $\beta(\theta)$. Plot $\beta(\theta)$ when samples of 4 are drawn from one population with variance 2 and samples of 2 are drawn from another population with variance 3 for tests at the .01 level.

29 Given the samples (1.8, 2.9, 1.4, 1.1) and (5.0, 8.6, 9.2) from normal populations, test whether the variances are equal at the .05 level.

30 Given a sample of size 100 with $\bar{X} = 2.7$ and $\sum (X_i - \bar{X})^2 = 225$, test the null hypothesis $\mathcal{H}_0$: $\mu = 3$ and $\sigma^2 = 2.5$ at the .01 level, assuming that the population is normal.

31 Using the sample of Prob. 30, test the hypothesis that $\mu = \sigma^2$ at the .01 level.

32 Using the sample of Prob. 30, test at the 0.1 level whether the .95 quantile point, say $\xi = \xi_{.95}$, of the population distribution is 3 relative to alternatives $\xi < 3$. Recall that ξ is such that $\int_{-\infty}^{\xi} f(x)\, dx = .95$, where $f(x)$ is the population density; it is, of course, $\mu + 1.645\sigma$ in the present instance where the distribution is assumed to be normal.

33 A sample of size n is drawn from each of k normal populations with the same variance. Derive the generalized likelihood-ratio test for testing the hypothesis that the means are all 0. Show that the test is a function of a ratio which has the F distribution.

34 Derive the generalized likelihood-ratio test for testing whether the correlation of a bivariate normal distribution is 0.

35 If $X_1, X_2, \ldots, X_n$ are observations from normal populations with known variances $\sigma_1^2, \sigma_2^2, \ldots, \sigma_n^2$, how would one test whether their means were all equal?

36 A newspaper in a certain city observed that driving conditions were much improved in the city because the number of fatal automobile accidents in the past year was 9 whereas the average number per year over the past several years was 15. Is it possible that conditions were more hazardous than before? Assume that the number of accidents in a given year has a Poisson distribution.

37 Six 1-foot specimens of insulated wire were tested at high voltage for weak spots in the insulation. The numbers of such weak spots were found to be 2, 0, 1, 1, 3, and 2. The manufacturer's quality standard states that there are less than 120 such defects per 100 feet. Is the batch from which these specimens were taken worse than the standard at the .05 level? (Use the Poisson distribution.)

38 Consider sampling from the normal distribution with unknown mean and variance:

(*a*) Find a generalized likelihood-ratio test of $\mathscr{H}_0: \sigma^2 \leq \sigma_0^2$ versus $\mathscr{H}_1: \sigma^2 > \sigma_0^2$.

(*b*) Find a generalized likelihood-ratio test of $\mathscr{H}_0: \sigma^2 = \sigma_0^2$ versus $\mathscr{H}_1: \sigma^2 \neq \sigma_0^2$.

39 (*a*) Suppose $(N_1, \ldots, N_k)$ is multinomially distributed with parameters n, $p_1, \ldots, p_k$, where $N_{k+1} = n - N_1 - \cdots - N_k$ and $p_{k+1} = 1 - \sum_1^k p_j$. Theorem 8 states that

$$Q = Q_k = \sum_{j=1}^{k+1} \frac{(N_j - np_j)^2}{np_j}$$

has a limiting chi-square distribution. Find the exact mean and variance of Q.

(*b*) Let $(N_1, \ldots, N_k)$ be distributed as in part (*a*). Define

$$Q_k^0 = \sum_{j=1}^{k+1} \frac{(N_j - np_j^0)^2}{np_j^0}.$$

[See Eq. (25).] Find $\mathscr{E}[Q_k^0]$. [See Eq. (26).] Is $\mathscr{E}[Q_k^0]$ for $p_1 = p_1^0, \ldots, p_{k+1} = p_{k+1}^0$ less than or equal to $\mathscr{E}[Q_k^0]$ for arbitrary $p_1, \ldots, p_{k+1}$?

40 A psychiatrist newly employed by a medical clinic remarked at a staff meeting that about 40 percent of all chronic headaches were of the psychosomatic variety. His disbelieving colleagues mixed some pills of plain flour and water, giving them to all such patients on the clinic's rolls with the story that they were a new headache remedy and asking for comments. When the comments were all in they could be fairly accurately classified as follows: (i) better than aspirin, 8, (ii) about the same as aspirin, 3, (iii) slower than aspirin, 1, and (iv) worthless, 29. While the doctors were somewhat surprised by these results, they nevertheless accused the psychiatrist of exaggeration. Did they have good grounds?

41 A die was cast 300 times with the following results:

Occurrence:	1	2	3	4	5	6
Frequency:	43	49	56	45	66	41

Are the data consistent at the .05 level with the hypothesis that the die is true?

42 Of 64 offspring of a certain cross between guinea pigs, 34 were red, 10 were black, and 20 were white. According to the genetic model, these numbers should be in the ratio 9/3/4. Are the data consistent with the model at the .05 level?

43 A prominent baseball player's batting average dropped from .313 in one year to .280 in the following year. He was at bat 374 times during the first year and 268 times during the second. Is the hypothesis tenable at the .05 level that his hitting ability was the same during the two years?

44 Using the data of Prob. 43, assume that one has a sample of 374 from one Bernoulli population and 268 from another. Derive the generalized likelihood-ratio test for testing whether the probability of a hit is the same for the two populations. How does this test compare with the ordinary test for a 2×2 contingency table?

45 The progeny of a certain mating were classified by a physical attribute into three groups, the numbers being 10, 53, and 46. According to a genetic model the frequencies should be in the ratios $p^2/2p(1-p)/(1-p)^2$. Are the data consistent with the model at the .05 level?

46 A thousand individuals were classified according to sex and according to whether or not they were color-blind as follows:

	Male	**Female**
Normal	442	514
Color-blind	38	6

According to the genetic model these numbers should have relative frequencies given by

$$\begin{array}{c|c} \dfrac{p}{2} & \dfrac{p^2}{2}+pq \\ \hline \dfrac{q}{2} & \dfrac{q^2}{2} \end{array}$$

where $q = 1 - p$ is the proportion of color-blind individuals in the population. Are the data consistent with the model?

47 Treating the table of Prob. 46 as a 2×2 contingency table, test the hypothesis that color blindness is independent of sex.

48 Gilby classified 1725 school children according to intelligence and apparent family economic level. A condensed classification follows:

	Dull	**Intelligent**	**Very capable**
Very well clothed	81	322	233
Well clothed	141	457	153
Poorly clothed	127	163	48

Test for independence at the .01 level.

49 A serum supposed to have some effect in preventing colds was tested on 500 individuals, and their records for 1 year were compared with the records of 500 untreated individuals as follows:

	No colds	**One cold**	**More than one cold**
Treated	252	145	103
Untreated	224	136	140

Test at the .05 level whether the two trinomial populations may be regarded as the same.

50 According to the genetic model the proportion of individuals having the four blood types should be given by:

$$\begin{aligned} O&: q^2 \\ A&: p^2 + 2pq \\ B&: r^2 + 2qr \\ AB&: 2pr \end{aligned}$$

where $p + q + r = 1$. Given the sample O, 374; A, 436; B, 132; AB, 58; how would you test the correctness of the model?

51 Galton investigated 78 families, classifying children according to whether or not they were light-eyed, whether or not they had a light-eyed parent, and whether or not they had a light-eyed grandparent. The following $2 \times 2 \times 2$ table resulted:

		Grandparent			
		Light		Not	
		Parent			
		Light	Not	Light	Not
Child	Light	1928	552	596	508
	Not	303	395	225	501

Test for complete independence at the .01 level. Test whether the child classification is independent of the other two classifications at the .01 level.

52 Compute the exact distribution of Λ for a 2×2 contingency table with marginal totals $N_{1.} = 4$, $N_{2.} = 7$, $N_{.1} = 6$, $N_{.2} = 5$. What is the exact probability that $-2 \log_e \Lambda$ exceeds 3.84, the .05 level of a chi-square distribution for one degree of freedom?

53 In testing independence in a 2×2 contingency table, find the exact distribution of the generalized likelihood-ratio for a sample of size 2. Do the same for samples of size 3 and 4. Discuss.

54 Let $X_1, \ldots, X_n$ be a random sample from $N(\mu, \sigma^2)$, where σ^2 is known. Let Λ denote the generalized likelihood-ratio for testing $\mathscr{H}_0: \mu = \mu_0$ versus $\mathscr{H}_1: \mu \neq \mu_0$. Find the exact distribution of $-2 \log_e \Lambda$, and compare it with the corresponding asymptotic distribution when $\mathscr{H}_0$ is true. HINT: $\sum (X_i - \bar{X})^2 = \sum (X_i - \mu)^2 - n(\bar{X} - \mu)^2$.

55 Here is an actual sequence of outcomes for independent Bernoulli trials. Do you think p (the probability of success) equals $\frac{1}{2}$?

s f f f s, s f s f f, s f f f f, f f s f s, s f f f f,
s f f s f, f f f f s, f f s f f, s f f f f, s s f f f.

If you do not think p is $\frac{1}{2}$, what do you think p is? Give a confidence-interval estimate of p. If the above data were generated by tossing two dice, then what would you think p is? If the data were generated by tossing two coins, then what would you think p is? (If the data were generated by tossing two dice, assume that the possible values of p are $j/36$, $j = 0, \ldots, 36$. If the data were generated by tossing two coins, assume that the possible values of p are $j/4$, $j = 0, \ldots, 4$.)

56 In sampling from a Bernoulli distribution, test the null hypothesis that $p = \frac{1}{4}$ against the alternative that $p = \frac{1}{3}$. Let p refer to the probability of two heads when tossing two coins, and carry through the test by tossing two coins, using $\alpha = \beta = .10$. (The alternative was obtained by reasoning that tossing two coins can result in the three outcomes: two heads, two tails, or one head and one tail, and then assuming each of the three outcomes equally likely.)

57 Show that the SPRT (sequential probability ratio test) of $\mu = \mu_0$ versus $\mu = \mu_1$ for the mean of the normal distribution with known variance may be performed by plotting the two lines

$$y = \frac{\sigma^2}{\mu_1 - \mu_0} \log_e k_0 + \frac{\mu_0 + \mu_1}{2} n$$

and

$$y = \frac{\sigma^2}{\mu_1 - \mu_0} \log_e k_1 + \frac{\mu_0 + \mu_1}{2} n$$

in the ny plane and then plotting $\sum_1^n X_i$ against n as the observations are made. The test ends when one of the lines is crossed.

58 Consider sampling from $f(x; \theta) = (1/\theta)I_{(0,\theta)}(x)$, $\theta > 0$. Discuss the sequential probability ratio test of $\theta = \theta_0$ versus $\theta = \theta_1$ with $\theta_0 < \theta_1$.

59 Let $X_1, X_2, \ldots, X_n, \ldots$ be independent random variables all having the same Bernoulli distribution given by $P[X_n = 1] = \theta = 1 - P[X_n = 0]$. To test $\mathscr{H}_0: \theta = \frac{1}{4}$ versus $\mathscr{H}_1: \theta = \frac{3}{4}$, the following sequential test was used: Continue sampling as long as $n/2 - 2 < \sum x_i < n/2 + 2$; if and when $\sum x_i$ is first less than or equal to $n/2 - 2$, accept $\mathscr{H}_0$; and if and when $\sum x_i$ is first greater than or equal to $n/2 + 2$, accept $\mathscr{H}_1$. Is this test a SPRT?

60 Assume that X has a Poisson distribution with mean θ. Consider testing $\mathscr{H}_0: \theta = 1$ versus $\mathscr{H}_1: \theta = 2$. Fix $\alpha = \beta = .05$.

(a) Find the fixed sample size necessary to achieve the prescribed error sizes.

(b) Derive the (approximate) sequential probability ratio test, and show that it can be based on the statistics $\sum_{i=1}^n X_i$, $n = 1, 2, \ldots$.

(c) Find the approximate expected sample sizes for the sequential probability ratio test.

X

LINEAR MODELS

1 INTRODUCTION AND SUMMARY

The purpose of this chapter is to discuss a special case of the linear statistical model. There is a large amount of material available on this subject, but we will only discuss the special case of the simple linear model. Some authors refer to this as the *theory of straight-line regression*. To study this model will require the use of some of the theory in previous chapters, such as distribution theory, point and interval estimation concepts, and some material from hypothesis testing. This chapter will demonstrate how these concepts can be utilized for a situation, the simple linear model, that is important in applied statistics.

In Sec. 2 two examples are given to illustrate how the simple linear model can be used to simulate real-world problems. In Sec. 3 the simple linear model will be rigorously defined and put into a framework that will allow us to study it by using statistical procedures from previous chapters. In the remaining sections discussion will be centered around point estimation, interval estimation, and testing hypotheses on the parameters in the model under two different assumptions about the distributions of the random variables in the model.

2 EXAMPLES OF THE LINEAR MODEL

In this section we will give two examples to illustrate how the linear model arises in applied problems.

EXAMPLE 1 The distance s that a particle travels in time t is given by the formula $s = \beta_0 + \beta_1 t$, where β_1 is the average speed and β_0 is the position at time $t = 0$. If β_0 and β_1 are unknown, then s can be observed for two distinct values of t and the resulting two equations solved for β_0 and β_1. For example, suppose that s is observed to be 2 when $t = 1$, and s is 11 when $t = 4$. This gives $2 = \beta_0 + \beta_1$ and $11 = \beta_0 + 4\beta_1$, and the solution is $\beta_0 = -1$, $\beta_1 = 3$; so $s = -1 + 3t$. Suppose that for some reason the distance cannot be observed accurately, but there is a measurement error which is of a random nature. Therefore s cannot be observed, but suppose that we can observe Y, where $Y = s + E$ and E is a random error whose mean is 0. Substituting for s gives us

$$Y = \beta_0 + \beta_1 t + E, \tag{1}$$

where Y is an observable random variable, t is an observable nonrandom variable, E is an unobservable random variable, and β_0 and β_1 are unknown parameters. We cannot solve for β_0 and β_1 by observing two sets of values of Y and t, as we did with s and t above, since there is no functional relationship between Y and t. The objective in this model is to find β_0 and β_1 and hence evaluate $s = \beta_0 + \beta_1 t$ for various values of t. Since s is subject to errors and cannot be observed, we cannot know β_0 and β_1, but by observing various sets of Y and t values statistical methods can be used to obtain estimates of β_0, β_1, and s. This type of model is a *functional-relationship model* with a measurement error. ////

EXAMPLE 2 For another example, consider the relationship between the height h and weight w of individuals in a certain city. Certainly there is no functional relationship between w and h, but there does seem to be some kind of relation. We shall consider them as random variables and shall postulate that (W, H) has a bivariate normal distribution. Then the expected value of H for a given value w of W is given by

$$\mathscr{E}[H \mid W = w] = \beta_0 + \beta_1 w, \tag{2}$$

where β_0 and β_1 are functions of the parameters in a bivariate normal density. Although there is no functional relationship between H and W, if they are assumed jointly normal, there is a linear functional relationship between the weights and the *average* value of the heights. Thus we can write the following: H and W are jointly normal, and

$$\mathscr{E}[H \mid W = w] = \beta_0 + \beta_1 w;$$

or we can write

$$H_w = \beta_0 + \beta_1 w + E,$$

where E is a normally distributed random variable denoting error. This is a *regression model*, and although it came from a somewhat different problem than the functional relationship in Example 1, they both are special cases of a *linear statistical model*, which will be discussed in this chapter. ////

3 DEFINITION OF LINEAR MODEL

Let $\mu(\cdot)$ be a linear function of a real variable x. This is defined by $\mu(x) = \beta_0 + \beta_1 x$, where x is in a domain D. Quite often D will be the entire real line, a half line, or a bounded interval on the real line. To model the situations referred to in Examples 1 and 2 above, we assume that there exists a family of c.d.f.'s (one c.d.f. for each x in D) such that the mean of the c.d.f. corresponding to a given x (say x_0) in D is $\beta_0 + \beta_1 x_0$. Thus the means of the c.d.f.'s are on the line defined by $\mu(x) = \beta_0 + \beta_1 x$. See Fig. 1. The objective is to sample some of the c.d.f.'s and on the basis of the sample to make statistical inferences about β_0, β_1, etc.

The sampling is accomplished as follows:

(i) A set of n x's in D is observed and denoted by $x_1, x_2, \ldots, x_n$. The x's are not random variables, but they may be selected either by some random procedure or by purposeful selection.

(ii) Each x_i determines a c.d.f. whose mean is $\beta_0 + \beta_1 x_i$ and whose variance is σ^2. From this c.d.f. a value is selected at random and denoted by Y_i. (Y_i is a shortened notation for Y_{x_i}.)

Thus we have a set of n pairs of observations, which we denote by (Y_1, x_1), $(Y_2, x_2), \ldots, (Y_n, x_n)$. We have assumed that

$$\mathscr{E}[Y_i] = \beta_0 + \beta_1 x_i$$

and

$$\text{var}\,[Y_i] = \sigma^2.$$

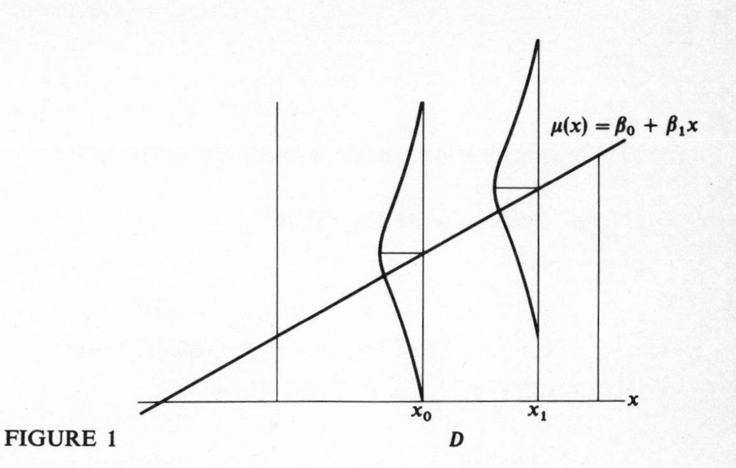

FIGURE 1

Therefore we can define random variables $E_1, E_2, \ldots, E_n$ by

$$E_i = Y_i - \beta_0 - \beta_1 x_i \qquad \text{for } i = 1, 2, \ldots, n,$$

and the E_i satisfy

$$\mathscr{E}[E_i] = 0$$

and

$$\text{var}\ [E_i] = \sigma^2.$$

So we can write

$$Y_i = \beta_0 + \beta_1 x_i + E_i \qquad \text{for } i = 1, 2, \ldots, n,$$

where

$$\mathscr{E}[E_i] = 0 \qquad \text{and} \qquad \text{var}\ [E_i] = \sigma^2,$$

and this defines a linear model. We summarize these ideas below.

Definition 1 Linear model Let the function $\mu(\cdot)$ be defined by $\mu(x) = \beta_0 + \beta_1 x$ for all x in a set D. For each x in D let $F_{Y_x}(\cdot)$ be a c.d.f. with a mean equal to $\mu(x)$, that is, $\beta_0 + \beta_1 x$, and variance σ^2. Let $x_1, x_2, \ldots, x_n$ be an observed set of nx's from D. For x_i let Y_i be a random sample of size 1 from the c.d.f. $F_{Yx_i}(\cdot)$ for $i = 1, 2, \ldots, n$. Then (Y_1, x_1), $(Y_2, x_2), \ldots, (Y_n, x_n)$ is a set of n observations related by

$$\mathscr{E}[Y_i] = \beta_0 + \beta_1 x_i$$

and (3)

$$\text{var}\,[Y_i] = \sigma^2, i = 1, 2, \ldots, n.$$

These specifications define a *linear statistical model.* ////

Note We can write Eq. (3) as

$$\begin{aligned} Y_i &= \beta_0 + \beta_1 x_i + E_i \\ \mathscr{E}[E_i] &= 0 \\ \text{var}\,[E_i] &= \sigma^2, \end{aligned} \tag{4}$$

where $i = 1, 2, \ldots, n$. ////

Note The word "linear" in "linear statistical model" refers to the fact that the function $\mu(\cdot)$ is linear in the unknown *parameters.* In the simple example we have referred to, $\mu(\cdot)$ is defined by $\mu(x) = \beta_0 + \beta_1 x$; x in D, and this is linear in x, but this is not an essential part of the definition of this linear model. For example, $Y = \mu(x) + E$, where $\mu(x) = \beta_0 + \beta_1 e^x$ is a linear statistical model. ////

Note In many situations some additional assumptions on the c.d.f. $F_{Y_x}(\cdot)$ will be made, such as normality. Also, generally the sampling procedure will be such that the Y_i will be either jointly independent or pairwise uncorrelated. In fact we shall discuss inference procedures for two sets of assumptions on the random variables defined in Cases A and B below. ////

Case A For this case we assume that the n random variables are jointly independent and each Y_i is a normal random variable. ////

Case B For this case we assume only that the Y_i are pairwise uncorrelated; that is, cov $[Y_i, Y_j] = 0$ for all $i \neq j = 1, 2, \ldots, n$. ////

For Case A we shall discuss the following:

(i) Point estimation of β_0, β_1, σ^2, and $\mu(x)$ for any x in D
(ii) Confidence interval for β_0, β_1, σ^2, and $\mu(x)$ for any x in D
(iii) Tests of hypotheses on β_0, β_1, and σ^2

For Case B we shall discuss the following:

(iv) Point estimation of β_0, β_1, σ^2, and $\mu(x)$ for any x in D

4 POINT ESTIMATION—CASE A

For this case $Y_1, Y_2, \ldots, Y_n$ are independent normal random variables with means $\beta_0 + \beta_1 x_1, \beta_0 + \beta_1 x_2, \ldots, \beta_0 + \beta_1 x_n$ and variances σ^2. To find point estimators, we shall use the method of maximum likelihood. The likelihood function is

$$\begin{aligned} L(\beta_0, \beta_1, \sigma^2) &= L(\beta_0, \beta_1, \sigma^2; y_1, y_2, \ldots, y_n) \\ &= \prod_{i=1}^{n} \left\{ \left(\frac{1}{2\pi\sigma^2} \right)^{1/2} \exp\left[-\tfrac{1}{2} \left(\frac{y_i - \beta_0 - \beta_1 x_i}{\sigma} \right)^2 \right] \right\}, \end{aligned} \tag{5}$$

and

$$\log L(\beta_0, \beta_1, \sigma^2) = -\frac{n}{2} \log 2\pi - \frac{n}{2} \log \sigma^2 - \frac{1}{2\sigma^2} \sum_{i=1}^{n} (y_i - \beta_0 - \beta_1 x_i)^2.$$

The partial derivatives of $\log L(\beta_0, \beta_1, \sigma^2)$ with respect to β_0, β_1, and σ^2 are obtained and set equal to 0. We let $\hat{\beta}_0$, $\hat{\beta}_1$, $\tilde{\sigma}^2$ denote the solutions of the resulting three equations. The three equations are given below (with some minor simplifications):

$$\begin{aligned} \sum_{i=1}^{n} (y_i - \hat{\beta}_0 - \hat{\beta}_1 x_i) &= 0 \\ \sum_{i=1}^{n} (y_i - \hat{\beta}_0 - \hat{\beta}_1 x_i) x_i &= 0 \\ \sum_{i=1}^{n} (y_i - \hat{\beta}_0 - \hat{\beta}_1 x_i)^2 &= n\tilde{\sigma}^2 \end{aligned} \tag{6}$$

The first two equations are called the *normal equations* for determining $\hat{\beta}_0$ and $\hat{\beta}_1$. They are linear in $\hat{\beta}_0$ and $\hat{\beta}_1$ and are readily solved. We obtain

$$\hat{\beta}_1 = \frac{\sum (y_i - \bar{y})(x_i - \bar{x})}{\sum (x_i - \bar{x})^2} \tag{7}$$

$$\hat{\beta}_0 = \bar{y} - \hat{\beta}_1 \bar{x} \tag{8}$$

$$\tilde{\sigma}^2 = \frac{1}{n} \sum_{i=1}^{n} (y_i - \hat{\beta}_0 - \hat{\beta}_1 x_i)^2. \tag{9}$$

These are maximum-likelihood estimates of β_1, β_0, and σ^2, respectively. We notice that the x_i's must be such that $\sum (x_i - \bar{x})^2 \neq 0$; that is, there must be at least two distinct values for the x_i.

Note that since

$$f_{Y_i}(y_i; \beta_0, \beta_1, \sigma) = (2\pi\sigma^2)^{-\frac{1}{2}} \exp\left[-\tfrac{1}{2}\left(\frac{y_i - \beta_0 - \beta_1 x_i}{\sigma}\right)^2\right]$$

$$= (2\pi\sigma^2)^{-\frac{1}{2}} \exp\left[-\frac{1}{2\sigma^2}(\beta_0 + \beta_1 x_i)^2\right]$$

$$\times \exp\left(-\frac{1}{2\sigma^2} y_i^2 + \frac{\beta_0}{\sigma^2} y_i + \frac{\beta_1}{\sigma^2} x_i y_i\right),$$

$f_{Y_i}(y_i; \beta_0, \beta_1, \sigma)$ is a member of a three-parameter exponential family; hence, by a generalization of Theorem 16 of Chap. VII

$$\sum_{i=1}^{n} Y_i^2, \qquad \sum_{i=1}^{n} Y_i, \qquad \sum_{i=1}^{n} x_i Y_i \tag{10}$$

is a set of minimal sufficient and jointly complete statistics. Furthermore, since the set of statistics given in Eq. (10) is a one-to-one transformation of the estimators (statistics) defined by Eqs. (7) to (9), the estimators are themselves minimal sufficient and jointly complete.

To further examine the properties that the estimators possess, we shall find the joint distribution of statistics corresponding to $\hat{\beta}_0$, $\hat{\beta}_1$, $\tilde{\sigma}^2$. To do this, we shall first find the moment generating function of $\hat{\Theta}_1$, $\hat{\Theta}_2$, and $\hat{\Theta}_3$, which are random quantities with values defined by

$$\hat{\theta}_1 = \frac{\hat{\beta}_0 - \beta_0}{\sigma}, \qquad \hat{\theta}_2 = \frac{\hat{\beta}_1 - \beta_1}{\sigma}, \qquad \hat{\theta}_3 = \frac{n\tilde{\sigma}^2}{\sigma^2}. \tag{11}$$

By Definition 25 of Chap. IV the joint moment generating function of $\hat{\Theta}_1$, $\hat{\Theta}_2$, $\hat{\Theta}_3$ is defined to be

$$m(t_1, t_2, t_3) = \mathscr{E}[e^{t_1\hat{\Theta}_1 + t_2\hat{\Theta}_2 + t_3\hat{\Theta}_3}]$$

if the expectation exists for $-h < t_i < h$ for some $h > 0$. We obtain

$$m(t_1, t_2, t_3) = \int_{-\infty}^{\infty} \cdots \int_{-\infty}^{\infty} e^{t_1\hat{\theta}_1 + t_2\hat{\theta}_2 + t_3\hat{\theta}_3}$$

$$\times \frac{\exp\left[-\frac{1}{2\sigma^2}\sum(y_i - \beta_0 - \beta_1 x_i)^2\right]}{(2\pi\sigma^2)^{n/2}} \, dy_1 \cdots dy_n,$$

where in the integral the quantities $\hat{\theta}_1$, $\hat{\theta}_2$, $\hat{\theta}_3$ will be written in terms of y_i and x_i. This integral is straightforward but tedious to evaluate, and the result is

$$m(t_1, t_2, t_3) = \left(\exp \tfrac{1}{2}\left\{t_1^2 \frac{\sum x_i^2/n}{\sum(x_i - \bar{x})^2} + 2t_1 t_2\left[-\frac{\bar{x}}{\sum(x_i - \bar{x})^2}\right]\right.\right.$$

$$\left.\left.+ t_2^2 \frac{1}{\sum(x_i - \bar{x})^2}\right\}\right) \times (1 - 2t_3)^{-(n-2)/2}, \text{ for } t_3 < \tfrac{1}{2}.$$

From this moment generating function we can learn a number of things:

(i) It factors into a function of t_1 and t_2 only times a function of t_3 only. We write this result as $m(t_1, t_2, t_3) = m_1(t_1, t_2)m_2(t_3)$. By using Theorem 10 of Chap. IV we know that the random variables associated with t_1 and t_2 are independent of the random variable associated with t_3; that is, $\hat{\Theta}_1$ and $\hat{\Theta}_2$ are independent of $\hat{\Theta}_3$, which implies that the maximum-likelihood estimators of β_0 and β_1 are jointly independent of the maximum-likelihood estimator of σ^2.

(ii) Since by a generalization of Theorem 7 in Chap. II a moment generating function uniquely determines the distribution of the random variables involved, we shall try to recognize the form of $m_1(t_1, t_2)$ and the form of $m_2(t_3)$. We note by Theorem 12 of Chap. IV that $m_1(t_1, t_2)$ is the moment generating function of a bivariate normal distribution, and, of course, we obtain the means, variances, and covariance. We see that the random variables, say $\hat{\mathbf{B}}_0$ and $\hat{\mathbf{B}}_1$, associated with $\hat{\beta}_0$ and $\hat{\beta}_1$ are bivariate normal random variables with means (β_0, β_1) and covariance matrix

$$\begin{bmatrix} \dfrac{\sigma^2 \sum x_i^2}{n \sum (x_i - \bar{x})^2} & \dfrac{-\sigma^2 \bar{x}}{\sum (x_i - \bar{x})^2} \\ \dfrac{-\sigma^2 \bar{x}}{\sum (x_i - \bar{x})^2} & \dfrac{\sigma^2}{\sum (x_i - \bar{x})^2} \end{bmatrix} \tag{12}$$

Another way to state this is the following: $(\hat{\mathbf{B}}_0, \hat{\mathbf{B}}_1)$ is a bivariate normal random variable with parameters

$$\mathscr{E}[\hat{\mathbf{B}}_0] = \beta_0,$$

$$\mathscr{E}[\hat{\mathbf{B}}_1] = \beta_1,$$

$$\operatorname{var}[\hat{\mathbf{B}}_0] = \frac{\sigma^2 \sum x_i^2}{n \sum (x_i - \bar{x})^2},$$

$$\operatorname{var}[\hat{\mathbf{B}}_1] = \frac{\sigma^2}{\sum (x_i - \bar{x})^2},$$

and

$$\operatorname{cov}[\hat{\mathbf{B}}_0, \hat{\mathbf{B}}_1] = \frac{-\sigma^2 \bar{x}}{\sum (x_i - \bar{x})^2}.$$

(iii) We recognize that $m_2(t_3)$ is the moment generating function of a chi-square random variable with $n - 2$ degrees of freedom. Hence we have

$$\frac{n\hat{\boldsymbol{\sigma}}^2}{\sigma^2} = \frac{1}{\sigma^2} \sum (Y_i - \hat{\mathbf{B}}_0 - \hat{\mathbf{B}}_1 x_i)^2,$$

which is distributed as a chi-square distribution with $n-2$ degrees of freedom. (Here, and in the rest of this chapter, $\hat{\boldsymbol{\sigma}}^2$ and $\tilde{\boldsymbol{\sigma}}^2$ are used to denote the random variables with values $\hat{\sigma}^2$ and $\tilde{\sigma}^2$ respectively.) By Eq. (22) of Chap. VI we get

$$\mathscr{E}\left[\frac{n\tilde{\boldsymbol{\sigma}}^2}{\sigma^2}\right] = n-2;$$

so we define $\hat{\sigma}^2$ by

$$\hat{\sigma}^2 = \frac{n}{n-2}\tilde{\sigma}^2 = \frac{1}{n-2}\sum_{i=1}^{n}(y_i - \hat{\beta}_0 - \hat{\beta}_1 x_i)^2.$$

We shall summarize these results in the following theorem.

Theorem 1 Consider Case A of the simple linear model given in Definition 1. The maximum-likelihood estimators of β_1, β_0, and σ^2 (corrected for bias) are given by

$$\hat{\mathbf{B}}_1 = \frac{\sum(Y_i - \bar{Y})(x_i - \bar{x})}{\sum(x_i - \bar{x})^2}, \qquad \hat{\mathbf{B}}_0 = \bar{Y} - \hat{\mathbf{B}}_1\bar{x},$$

$$\hat{\boldsymbol{\sigma}}^2 = \frac{1}{n-2}\sum(Y_i - \hat{\mathbf{B}}_0 - \hat{\mathbf{B}}_1 x_i)^2. \tag{13}$$

These estimators satisfy the following:

(i) They are jointly complete sufficient statistics.
(ii) They are unbiased estimators of their respective parameters.
(iii) $(\hat{\mathbf{B}}_0, \hat{\mathbf{B}}_1)$ is independent of $\hat{\boldsymbol{\sigma}}^2$.
(iv) $(\hat{\mathbf{B}}_0, \hat{\mathbf{B}}_1)$ has a bivariate normal distribution with mean (β_0, β_1) and covariance matrix given by Eq. (12).
(v) $(n-2)\hat{\boldsymbol{\sigma}}^2/\sigma^2$ is a chi-square random variable with $n-2$ degrees of freedom. ////

In Chap. VII, we noted that maximum-likelihood estimators possess a number of good properties, but they, in general, are not minimum-variance unbiased estimators. We now employ a minor generalization of Theorem 17 of Chap. VII along with the results of Theorem 1 above, to state a strong optimal property about the estimators $\hat{\mathbf{B}}_0$, $\hat{\mathbf{B}}_1$, $\hat{\boldsymbol{\sigma}}^2$ of β_0, β_1, σ^2.

Theorem 2 Consider the simple linear model given in Definition 1. Let $\tau(\beta_0, \beta_1, \sigma^2)$ be any known function of the parameters β_0, β_1, and σ^2 for which an unbiased estimator exists. Then there exists an unbiased estimator of $\tau(\beta_0, \beta_1, \sigma^2)$ that is a function of $\hat{\mathbf{B}}_0$, $\hat{\mathbf{B}}_1$, and $\hat{\boldsymbol{\sigma}}^2$. We denote this estimator by $\not{t}(\hat{\mathbf{B}}_0, \hat{\mathbf{B}}_1, \hat{\boldsymbol{\sigma}}^2)$, and it is the UMVUE of $\tau(\beta_0, \beta_1, \sigma^2)$.

PROOF This result follows from a generalization of Theorem 17 of Chap. VII, since $\hat{\mathbf{B}}_0$, $\hat{\mathbf{B}}_1$, $\hat{\boldsymbol{\sigma}}^2$ is a set of sufficient complete statistics. ////

Corollary The UMVUE of each of the parameters β_0, β_1, and σ^2 is given by $\hat{\mathbf{B}}_0$, $\hat{\mathbf{B}}_1$, and $\hat{\boldsymbol{\sigma}}^2$, respectively, in Theorem 1. ////

Corollary The UMVUE of $\mu(x) = \beta_0 + \beta_1 x$ for any x in the domain D is $\hat{\boldsymbol{\mu}}(x)$, where $\hat{\boldsymbol{\mu}}(x) = \hat{\mathbf{B}}_0 + \hat{\mathbf{B}}_1 x$. ($\hat{\boldsymbol{\mu}}(x)$ is the random variable with values $\hat{\mu}(x) = \hat{\beta}_0 + \hat{\beta}_1 x$.) ////

Corollary For any two known constants c_1 and c_2 the UMVUE of $c_1\beta_0 + c_2\beta_1$ is $c_1\hat{\mathbf{B}}_0 + c_2\hat{\mathbf{B}}_1$. ////

5 CONFIDENCE INTERVALS—CASE A

To obtain a γ-level confidence interval on σ^2, we note by Theorem 1 that

$$U = \frac{(n-2)\hat{\boldsymbol{\sigma}}^2}{\sigma^2}$$

is distributed as a chi-square random variable with $n-2$ d.f. (degrees of freedom). Hence U is a pivotal quantity, and we get

$$P[\chi^2_{(1-\gamma)/2}(n-2) \le U \le \chi^2_{(1+\gamma)/2}(n-2)] = \gamma.$$

If we substitute for U and simplify, we get

$$P\left[\frac{(n-2)\hat{\boldsymbol{\sigma}}^2}{\chi^2_{(1+\gamma)/2}(n-2)} \le \sigma^2 \le \frac{(n-2)\hat{\boldsymbol{\sigma}}^2}{\chi^2_{(1-\gamma)/2}(n-2)}\right] = \gamma, \tag{14}$$

and this is a 100γ percent confidence interval on σ^2.

To obtain a γ-level confidence interval on β_0, we note that by Theorem 1:

(i) $Z = (\hat{\mathbf{B}}_0 - \beta_0)\sqrt{\sum (x_i - \bar{x})^2 n/\sigma^2 \sum x_i^2}$ is distributed as a standard normal random variable.

(ii) $(n-2)\hat{\boldsymbol{\sigma}}^2/\sigma^2 = U$ is distributed as a chi-square random variable with $n-2$ d.f.

(iii) Z and U are independent.

Hence, by Theorem 10 of Chap. VI

$$T = \frac{\hat{\mathbf{B}}_0 - \beta_0}{\hat{\boldsymbol{\sigma}}} \sqrt{\frac{n \sum (x_i - \bar{x})^2}{\sum x_i^2}}$$

is distributed as Student's t distribution with $n-2$ d.f. Hence T is a pivotal quantity. We get

$$P[-t_{(1+\gamma)/2}(n-2) \le T \le t_{(1+\gamma)/2}(n-2)] = \gamma,$$

and if we substitute for T, we get

$$P\left[-t_{(1+\gamma)/2}(n-2) \le \frac{\hat{\mathbf{B}}_0 - \beta_0}{\hat{\boldsymbol{\sigma}}} \sqrt{\frac{n\sum(x_i-\bar{x})^2}{\sum x_i^2}} \le t_{(1+\gamma)/2}(n-2)\right] = \gamma.$$

After simplifying we get the following for a 100γ percent confidence interval on β_0:

$$P\left[\hat{\mathbf{B}}_0 - t_{(1+\gamma)/2}(n-2)\hat{\boldsymbol{\sigma}}\sqrt{\frac{\sum x_i^2}{n\sum(x_i-\bar{x})^2}}\right.$$

$$\left.\le \beta_0 \le \hat{\mathbf{B}}_0 + t_{(1+\gamma)/2}(n-2)\hat{\boldsymbol{\sigma}}\sqrt{\frac{\sum x_i^2}{n\sum(x_i-\bar{x})^2}}\right] = \gamma.$$

We note that by Theorem 1 we get

$$\operatorname{var}[\hat{\mathbf{B}}_0] = \sigma^2 \frac{\sum x_i^2}{n\sum(x_i-\bar{x})^2},$$

and the estimated variance of $\hat{\mathbf{B}}_0$, which we write as $\widehat{\operatorname{var}}[\hat{\mathbf{B}}_0]$, is given by

$$\widehat{\operatorname{var}}[\hat{\mathbf{B}}_0] = \hat{\boldsymbol{\sigma}}^2 \frac{\sum x_i^2}{n\sum(x_i-\bar{x})^2}.$$

Then the confidence statement can be written as

$$P[\hat{\mathbf{B}}_0 - t_{(1+\gamma)/2}(n-2)\sqrt{\widehat{\operatorname{var}}[\hat{\mathbf{B}}_0]} \le \beta_0 \le \hat{\mathbf{B}}_0 + t_{(1+\gamma)/2}(n-2)\sqrt{\widehat{\operatorname{var}}[\hat{\mathbf{B}}_0]}] = \gamma. \tag{15}$$

To obtain a γ-level confidence interval on β_1, we note that by Theorem 1:

(i) $Z = (\hat{\mathbf{B}}_1 - \beta_1)\sqrt{\sum(x_i-\bar{x})^2/\sigma^2}$ is distributed as a standard normal random variable.

(ii) $U = (n-2)\hat{\boldsymbol{\sigma}}^2/\sigma^2$ is distributed as a chi-square random variable with $n-2$ d.f.

(iii) Z and U are independent.

Hence by Theorem 10 of Chap. VI

$$T = (\hat{\mathbf{B}}_1 - \beta_1)\sqrt{\frac{\sum(x_i-\bar{x})^2}{\hat{\boldsymbol{\sigma}}^2}}$$

is distributed as Student's t distribution with $n-2$ d.f. Hence T is a pivotal quantity, and we get

$$P[-t_{(1+\gamma)/2}(n-2) \le T \le t_{(1+\gamma)/2}(n-2)] = \gamma.$$

If we substitute for T, we obtain

$$P\left[-t_{(1+\gamma)/2}(n-2) \le (\hat{\mathrm{B}}_1 - \beta_1)\sqrt{\frac{\sum (x_i - \bar{x})^2}{\hat{\boldsymbol{\sigma}}^2}} \le t_{(1+\gamma)/2}(n-2)\right] = \gamma.$$

After some simplification we get

$$P\left[\hat{\mathrm{B}}_1 - t_{(1+\gamma)/2}(n-2)\sqrt{\frac{\hat{\boldsymbol{\sigma}}^2}{\sum (x_i - \bar{x})^2}} \le \beta_1 \le \hat{\mathrm{B}}_1 + t_{(1+\gamma)/2}(n-2)\sqrt{\frac{\hat{\boldsymbol{\sigma}}^2}{\sum (x_i - \bar{x})^2}}\right] = \gamma,$$

and this is a 100γ percent confidence interval on β_1. We note from Theorem 1 that

$$\operatorname{var}[\hat{\mathrm{B}}_1] = \frac{\sigma^2}{\sum (x_i - \bar{x})^2}$$

and that the estimated variance of $\hat{\mathrm{B}}_1$, which is denoted by $\widehat{\operatorname{var}}[\hat{\mathrm{B}}_1]$, is given by

$$\widehat{\operatorname{var}}[\hat{\mathrm{B}}_1] = \frac{\hat{\boldsymbol{\sigma}}^2}{\sum (x_i - \bar{x})^2}.$$

If we substitute this into the confidence-interval statement, we obtain

$$P[\hat{\mathrm{B}}_1 - t_{(1+\gamma)/2}(n-2)\sqrt{\widehat{\operatorname{var}}[\hat{\mathrm{B}}_1]} \le \beta_1 \le \hat{\mathrm{B}}_1 + t_{(1+\gamma)/2}(n-2)\sqrt{\widehat{\operatorname{var}}[\hat{\mathrm{B}}_1]}] = \gamma. \tag{16}$$

To obtain a γ-level confidence interval on $\mu(x)$ for any x in the domain D, we note that

(i) $\mu(x) = \beta_0 + \beta_1 x.$

(ii) $\hat{\boldsymbol{\mu}}(x) = \hat{\mathrm{B}}_0 + \hat{\mathrm{B}}_1(x).$

(iii) $\mathscr{E}[\hat{\boldsymbol{\mu}}(x)] = \mu(x).$

(iv)
$$\begin{aligned}\operatorname{var}[\hat{\boldsymbol{\mu}}(x)] &= \operatorname{var}[\hat{\mathrm{B}}_0 + \hat{\mathrm{B}}_1 x] \\ &= \operatorname{var}[\hat{\mathrm{B}}_0] + 2x\operatorname{cov}[\hat{\mathrm{B}}_0, \hat{\mathrm{B}}_1] + x^2 \operatorname{var}[\hat{\mathrm{B}}_1] \\ &= \frac{\sigma^2}{\sum (x_i - \bar{x})^2}\left(\frac{\sum x_i^2}{n} - 2x\bar{x} + x^2\right) \\ &= \frac{\sigma^2}{\sum (x_i - \bar{x})^2}\left[(\bar{x} - x)^2 + \frac{1}{n}\sum (x_i - \bar{x})^2\right] \\ &= \sigma^2\left[\frac{1}{n} + \frac{(\bar{x} - x)^2}{\sum (x_i - \bar{x})^2}\right].\end{aligned}$$

(v) $Z = [\hat{\mu}(x) - \mu(x)]/\sqrt{\text{var}\ [\hat{\mu}(x)]}$ is distributed as a standard normal random variable.

(vi) $U = (n-2)\hat{\sigma}^2/\sigma^2$ is distributed as a chi-square random variable with $n-2$ d.f.

(vii) U and Z are independent.

(viii) $$T = \frac{\hat{\mu}(x) - \mu(x)}{\sqrt{\widehat{\text{var}}[\hat{\mu}(x)]}} = \frac{\hat{B}_0 + \hat{B}_1 x - \beta_0 - \beta_1 x}{\sqrt{\hat{\sigma}^2[1/n + (\bar{x} - x)^2/\sum (x_i - \bar{x})^2]}}$$

is distributed as Student's t distribution with $n-2$ d.f. Hence T is a pivotal quantity, and we obtain

$$P[-t_{(1+\gamma)/2}(n-2) \le T \le t_{(1+\gamma)/2}(n-2)] = \gamma.$$

If we substitute for T and simplify, we get

$$P\left[\hat{B}_0 + \hat{B}_1 x - t_{(1+\gamma)/2}(n-2)\sqrt{\hat{\sigma}^2\left[\frac{1}{n} + \frac{(\bar{x}-x)^2}{\sum (x_i - \bar{x})^2}\right]} \le \beta_0 + \beta_1 x \right.$$
$$\left. \le \hat{B}_0 + \hat{B}_1 x + t_{(1+\gamma)/2}(n+2)\sqrt{\hat{\sigma}^2\left[\frac{1}{n} + \frac{(\bar{x}-x)^2}{\sum (x_i - \bar{x})^2}\right]}\right] = \gamma$$

or

$$P[\hat{B}_0 + \hat{B}_1 x - t_{(1+\gamma)/2}(n-2)\sqrt{\widehat{\text{var}}\ [\hat{\mu}(x)]} \le \beta_0 + \beta_1 x$$
$$\le \hat{B}_0 + \hat{B}_1 x + t_{(1+\gamma)/2}(n-2)\sqrt{\widehat{\text{var}}\ [\hat{\mu}(x)]}] = \gamma,$$

and a 100γ percent confidence interval for $\beta_0 + \beta_1 x$ is obtained.

6 TESTS OF HYPOTHESES—CASE A

In the linear model there are many tests that could be of interest to an investigator. For example, he may want to test whether the line goes through the origin, i.e., to test if the intercept is equal to zero, or perhaps test whether the intercept is positive (or negative). These are indicated by

$$\mathscr{H}_0: \beta_0 = 0 \text{ versus } \mathscr{H}_1: \beta_0 \ne 0,$$
$$\mathscr{H}_0: \beta_0 \ge 0 \text{ versus } \mathscr{H}_1: \beta_0 < 0,$$
$$\mathscr{H}_0: \beta_0 \le 0 \text{ versus } \mathscr{H}_1: \beta_0 > 0.$$

These tests indicate that there is no interest in the slope β_1 or the variance σ^2

On the other hand the interest may be in the slope rather than the intercept, and an investigator could be interested in testing

$$\mathscr{H}_0: \beta_1 = 0 \text{ versus } \mathscr{H}_1: \beta_1 \neq 0,$$
$$\mathscr{H}_0: \beta_1 \leq 0 \text{ versus } \mathscr{H}_1: \beta_1 > 0,$$

etc. Rather than testing whether the intercept (or slope) is equal to 0 an investigator may be interested in testing whether it is equal to a given number. For example he may be interested in testing

$$\mathscr{H}_0: \beta_0 = 2 \text{ versus } \mathscr{H}_1: \beta_0 < 2,$$
$$\mathscr{H}_0: \beta_1 = 1 \text{ versus } \mathscr{H}_1: \beta_1 \neq 1,$$

etc. We shall derive a test of the hypothesis

$$\mathscr{H}_0: \beta_1 = 0 \text{ versus } \mathscr{H}_1: \beta_1 \neq 0.$$

We could just as well derive a test of the hypothesis

$$\mathscr{H}_0: \beta_1 \leq 0 \text{ versus } \mathscr{H}_1: \beta_1 > 0$$

or a test of the hypothesis

$$\mathscr{H}_0: \beta_1 \geq 0 \text{ versus } \mathscr{H}_1: \beta_1 < 0.$$

To test

$$\mathscr{H}_0: \beta_1 = 0 \text{ versus } \mathscr{H}_1: \beta_1 \neq 0,$$

one obvious choice for a test statistic is

$$T = \frac{\hat{\mathbf{B}}_1}{\sqrt{\widehat{\text{var}}\,[\hat{\mathbf{B}}_1]}}.$$

Under $\mathscr{H}_0$ the random variable T is distributed as Student's t distribution with $n - 2$ degrees of freedom. Thus a test procedure with size α is the following:

$$\text{Reject } \mathscr{H}_0 \text{ if and only if } |T| > t_{1-\alpha/2}(n-2).$$

By comparing this with Eq. (16) we notice that this test is equivalent to the procedure of setting a $1 - \alpha$ confidence interval on the parameter β_1 and rejecting the hypothesis if and only if the confidence interval does *not* contain 0.

We will now show that this test is a generalized likelihood-ratio test. Corresponding to the notation in Chap. IX we note that in testing

$$\mathscr{H}_0: \beta_1 = 0 \text{ versus } \mathscr{H}_1: \beta_1 \neq 0$$

the parameter spaces $\bar{\Theta}$, $\bar{\Theta}_0$, and $\bar{\Theta}_1$ are as given below, where $\theta = (\beta_0, \beta_1, \sigma^2)$:

$$\bar{\Theta} = \{(\beta_0, \beta_1, \sigma^2): -\infty < \beta_0 < \infty; -\infty < \beta_1 < \infty; \sigma^2 > 0\}$$
$$\bar{\Theta}_0 = \{(\beta_0, \beta_1, \sigma^2): -\infty < \beta_0 < \infty; \beta_1 = 0; \sigma^2 > 0\}$$
$$\bar{\Theta}_1 = \bar{\Theta} - \bar{\Theta}_0.$$

We must determine λ, where

$$\lambda = \frac{\sup_{\theta \in \bar{\Theta}_0} L(\theta; y_1, \ldots, y_n)}{\sup_{\theta \in \bar{\Theta}} L(\theta; y_1, \ldots, y_n)}. \tag{17}$$

$$\begin{aligned} L(\theta; y_1, \ldots, y_n) &= L(\beta_0, \beta_1, \sigma^2) \\ &= \frac{1}{(2\pi\sigma^2)^{n/2}} \exp\left[-\frac{1}{2\sigma^2} \sum (y_i - \beta_0 - \beta_1 x_i)^2\right], \end{aligned} \tag{18}$$

and the values of β_0, β_1, σ^2 that maximize this for $\theta \in \bar{\Theta}$ are the maximum-likelihood estimates given in Eqs. (7) to (9). Thus we get

$$\begin{aligned} \sup_{\theta \in \bar{\Theta}} L(\theta; y_1, \ldots, y_n) &= \frac{1}{(2\pi\tilde{\sigma}^2)^{n/2}} \exp\left[-\frac{\sum (y_i - \hat{\beta}_0 - \hat{\beta}_1 x_i)^2}{2\tilde{\sigma}^2}\right] \\ &= (2\pi\tilde{\sigma}^2)^{-n/2} e^{-n/2}, \end{aligned}$$

where $\tilde{\sigma}^2 = \frac{1}{n} \sum (y_i - \hat{\beta}_0 - \hat{\beta}_1 x_i)^2$. To find $\sup_{\theta \in \bar{\Theta}_0} L(\theta; y_1, \ldots, y_n)$, we substitute $\beta_1 = 0$ into Eq. (18) above and get

$$L(\beta_0, \sigma^2) = \frac{1}{(2\pi\sigma^2)^{n/2}} \exp\left[-\frac{1}{2\sigma^2} \sum (y_i - \beta_0)^2\right].$$

But this is the likelihood function for a random sample of size n from a normal distribution with mean β_0 and variance σ^2. The values of β_0 and σ^2 that maximize the likelihood function are the maximum-likelihood estimates

$$\beta_0^* = \bar{y}$$

and

$$\sigma^{*^2} = \frac{1}{n} \sum (y_i - \bar{y})^2.$$

Thus

$$\begin{aligned} \sup_{\theta \in \bar{\Theta}_0} L(\theta; y_1, \ldots, y_n) &= (\sigma^{*^2} 2\pi)^{-n/2} \exp\left[-\frac{1}{2\sigma^{*^2}} \sum (y_i - \beta_0^*)^2\right] \\ &= (\sigma^{*^2} 2\pi)^{-n/2} e^{-n/2}. \end{aligned}$$

We obtain

$$\lambda = \left(\frac{\tilde{\sigma}^2}{\sigma^{*2}}\right)^{n/2}$$

for the generalized likelihood-ratio. Instead of λ we will examine the quantity $(n-2)(\lambda^{-2/n}-1)$, which is a monotonic function of λ and hence will give an equivalent test function. We get

$$\lambda^{-2/n} - 1 = \frac{\sigma^{*2} - \tilde{\sigma}^2}{\tilde{\sigma}^2} = \frac{\sum (y_i - \bar{y})^2 - \sum (y_i - \hat{\beta}_0 - \hat{\beta}_1 x_i)^2}{\sum (y_i - \hat{\beta}_0 - \hat{\beta}_1 x_i)^2}.$$

Replace $\hat{\beta}_0$ with $\hat{\beta}_0 = \bar{y} - \hat{\beta}_1 \bar{x}$ in the numerator, and get

$$\lambda^{-2/n} - 1 = \frac{\sum (y_i - \bar{y})^2 - \sum [(y_i - \bar{y}) - \hat{\beta}_1(x_i - \bar{x})]^2}{\sum (y_i - \hat{\beta}_0 - \hat{\beta}_1 x_i)^2}.$$

Hence,

$$(n-2)(\lambda^{-2/n} - 1) = \frac{\hat{\beta}_1^2 \sum (x_i - \bar{x})^2}{\hat{\sigma}^2} = \frac{\hat{\beta}_1^2 \sum (x_i - \bar{x})^2/\sigma^2}{\hat{\sigma}^2/\sigma^2},$$

which is the ratio of the values of two independent chi-square random variables (under $\mathscr{H}_0\colon \beta_1 = 0$) divided by their respective degrees of freedom, which are 1 for the numerator and $n-2$ for the denominator. Thus $(n-2)(\Lambda^{-2/n}-1)$ has an F distribution with 1 and $n-2$ degrees of freedom under $\mathscr{H}_0$. The generalized likelihood-ratio test says to reject $\mathscr{H}_0$ if and only if $\lambda \le \lambda_0$, or if and only if

$$(n-2)(\lambda^{-2/n} - 1) \ge (n-2)(\lambda_0^{-2/n} - 1) = \lambda_0^* \text{ (say)},$$

or if and only if

$$[\hat{\beta}_1^2 \sum (x_i - \bar{x})^2]/\hat{\sigma}^2 \ge \lambda_0^*,$$

where λ_0^* is chosen for a desirable size of Type I error.

Note that $(n-2)(\Lambda^{-2/n}-1)$ is the square of

$$\frac{\hat{\mathrm{B}}_1}{\sqrt{\widehat{\text{var}}\,[\hat{\mathrm{B}}_1]}},$$

and recall that the square of a Student's t-distributed random variable with $n-2$ degrees of freedom has an F distribution with 1 and $n-2$ degrees of freedom. Thus we have verified that if the confidence-interval statement in Eq. (16) is used to test $\mathscr{H}_0\colon \beta_1 = 0$ versus $\mathscr{H}_1\colon \beta_1 \ne 0$, it is a generalized likelihood-ratio test.

We will generalize this result slightly in the following theorem.

Theorem 3 In the linear model given in Definition 1 the generalized likelihood-ratio test of size α of $\mathscr{H}_0$: $\beta_1 = b_1$ (b_1 is a given constant) versus $\mathscr{H}_1$: $\beta_1 \neq b_1$ is given by the following: Use Eq. (16) to set a $1 - \alpha$ confidence interval on β_1, and reject $\mathscr{H}_0$ if and only if the confidence interval does not include b_1. ////

We shall state a theorem concerning a test of hypothesis on β_0, and the proof will be asked for in Prob. 18.

Theorem 4 In the linear model given in Definition 1 the generalized likelihood-ratio test of size α of $\mathscr{H}_0$: $\beta_0 = b_0$ (b_0 is a given constant) versus $\mathscr{H}_1$: $\beta_0 \neq b_0$ is given by the following: Use Eq. (15) to set a $1 - \alpha$ confidence interval on β_0, and reject $\mathscr{H}_0$ if and only if the confidence interval does not include b_0. ////

There are many other tests that are of interest for the linear model and the interested reader can consult Refs. 17, 29, 31, and 32.

7 POINT ESTIMATION—CASE B

For this case $Y_1, Y_2, \ldots, Y_n$ are pairwise uncorrelated random variables with means $\beta_0 + \beta_1 x_1, \beta_0 + \beta_1 x_2, \ldots, \beta_0 + \beta_1 x_n$ and variances σ^2. Since the joint density of the Y_i is not specified, maximum-likelihood estimators of β_0, β_1, and σ^2 cannot be obtained. In models when the joint density of the observable random variables is not given, a method of estimation called *least-squares* can be utilized.

Definition 2 Least-squares Let (Y_i, x_i), $i = 1, 2, \ldots, n$, be n pairs of observations that satisfy the linear model given in Definition 1. The values of β_0 and β_1 that minimize the sum of squares

$$\sum_{i=1}^{n} (Y_i - \beta_0 - \beta_1 x_i)^2$$

are defined to be the *least-squares estimators* of β_0 and β_1. ////

To find the least-squares estimators of β_0 and β_1, we must find the values that minimize

$$L(\beta_0, \beta_1) = \sum_{i=1}^{n} (Y_i - \beta_0 - \beta_1 x_i)^2,$$

and clearly these are the same values that maximize the likelihood function in Eq. (5). Hence we have the following theorem.

Theorem 5 In Case B of the simple linear model given in Definition 1 the least-squares estimators of β_0 and β_1 are given by $\hat{\mathbf{B}}_0$ and $\hat{\mathbf{B}}_1$, where

$$\hat{\mathbf{B}}_1 = \frac{\sum (Y_i - \bar{Y})(x_i - \bar{x})}{\sum (x_i - \bar{x})^2}, \qquad \hat{\mathbf{B}}_0 = \bar{Y} - \hat{\mathbf{B}}_1 \bar{x}. \tag{19}$$

////

The least-squares method gives no estimator for σ^2, but an estimator of σ^2 *based* on the least-squares estimators of β_0 and β_1 is

$$\hat{\boldsymbol{\sigma}}^2 = \frac{1}{n-2}\left[\sum_{i=1}^{n} (Y_i - \hat{\mathbf{B}}_0 - \hat{\mathbf{B}}_1 x_i)^2\right].$$

For Case A the maximum-likelihood estimators of β_0, β_1, and σ^2 had some desirable optimum properties. The first corollary of Theorem 2 states that $\hat{\mathbf{B}}_0$ and $\hat{\mathbf{B}}_1$ are uniformly minimum-variance unbiased estimators. That is, in the class of all unbiased estimators of β_0 and β_1, the estimators $\hat{\mathbf{B}}_0$ and $\hat{\mathbf{B}}_1$ in Eq. (13) have uniformly minimum variance. No such desirable property as this is enjoyed by least-squares estimators for Case B. For Case A the assumptions are much stronger than for Case B, where the distribution of the random variables Y_i is assumed to be unknown; so we should not expect as strong an optimality in the estimators for Case B.

For Case B, we shall restrict our class of estimating functions and determine if the least-squares estimators have any optimal properties in the restricted class. Since $\mathscr{E}[Y_i] = \beta_0 + \beta_1 x_i$, we see that β_0 (and β_1) can be given by the expected value of linear functions of the Y_i. Within this class of linear functions we will define minimum-variance unbiased estimators.

Definition 3 Best linear unbiased estimators Let $Y_1, Y_2, \ldots, Y_n$ be observable random variables such that $\mathscr{E}[Y_i] = \tau_i(\theta)$, where $\tau_i(\cdot)$ are known functions that contain unknown parameters θ (θ may be vector-valued). To estimate any θ_j in θ, consider only the class of estimators that are linear functions of the random variables Y_i. In this class consider only the subclass of estimators that are unbiased for θ_j. If in this restricted class an estimator of θ_j exists which has smaller variance than any other estimator of θ_j in this restricted class, it is defined to be the *best linear unbiased estimator* of θ_j ("best" refers to minimum variance). ////

It should be noted that there are two restrictions on the estimating functions before the property of minimum variance is considered. First, the class of estimating functions is restricted to linear functions of the Y_i. Second, in the class of linear functions of the Y_i only unbiased estimators are considered. Finally, then, consideration is given to finding a minimum-variance estimator in the class of estimating functions that are linear and unbiased.

We will now prove an important theorem that gives optimum properties for the point estimators of β_0 and β_1 derived by the method of least squares for Case B. This theorem is often referred to as the *Gauss-Markov theorem*.

Theorem 6 Consider the linear model given in Definition 1, and let the assumptions for Case B hold. Then the least-squares estimators for β_1 and β_0 given in Eq. (19) are the respective best linear unbiased estimators for β_1 and β_0.

PROOF We shall demonstrate the proof for β_0; the proof for β_1 is similar. Since we are restricting the class of estimators to be linear, we have $\hat{B}_0 = \sum a_j Y_j$. We must determine the constant a_j such that:

(i) $\mathscr{E}[\hat{B}_0] = \beta_0$; that is, $\hat{B}_0$ is an unbiased estimator of β_0.
(ii) var $[\hat{B}_0]$ is a minimum among all estimators satisfying (i).

For (i) we must have

$$\beta_0 = \mathscr{E}[\hat{B}_0] = \sum a_j \mathscr{E}[Y_j] = \sum a_j(\beta_0 + \beta_1 x_j).$$

This gives the two equations which must be satisfied

$$\sum a_j = 1 \quad \text{and} \quad \sum a_j x_j = 0. \tag{20}$$

Now

$$\begin{aligned} \text{var}\,[\hat{B}_0] &= \mathscr{E}[(\hat{B}_0 - \beta_0)^2] = \mathscr{E}[(\sum a_j Y_j - \beta_0)^2] \\ &= \mathscr{E}[[\sum a_j(\beta_0 + \beta_1 x_j + E_j) - \beta_0]^2] \\ &= \mathscr{E}[(\beta_0 \sum a_j + \beta_1 \sum a_j x_j + \sum a_j E_j - \beta_0)^2]. \end{aligned}$$

By the restrictions of Eq. (20)

$$\text{var}[\hat{B}_0] = \mathscr{E}[(\sum a_j E_j)^2] = \mathscr{E}\left[\sum_j a_j^2 E_j^2 + \sum_{\substack{j \\ j \neq i}} \sum_i a_j a_i E_j E_i\right].$$

The quantity $\mathscr{E}[E_i E_j]$ is 0 if $i \neq j$ since, by assumption, the E_i are uncorrelated and have means 0. Hence

$$\text{var}[\hat{\text{B}}_0] = \sigma^2 \sum a_j^2.$$

Since σ^2 is a constant, to minimize var $[\hat{\text{B}}_0]$ we need to minimize $\sum a_j^2$. Thus constants a_j must be found which minimize $\sum a_j^2$ subject to the restrictions of Eq. (20). Using the *theory of Lagrange multipliers*, we must minimize

$$L = \sum a_j^2 - \lambda_1(\sum a_j - 1) - \lambda_2 \sum a_j x_j.$$

Taking derivatives, one finds

$$\frac{\partial L}{\partial a_t} = 2a_t - \lambda_1 - \lambda_2 x_t = 0, \qquad t = 1, 2, \ldots, n \tag{21}$$

$$\frac{\partial L}{\partial \lambda_1} = -\sum a_j + 1 = 0,$$

$$\frac{\partial L}{\partial \lambda_2} = -\sum a_j x_j = 0.$$

If we sum over the first n equations, we get $\left(\text{using } \sum a_t = 1\right)$

$$2 = n\lambda_1 + \lambda_2 \sum x_i. \tag{22}$$

If we multiply the jth equation in (21) by x_j and add, we get

$$2 \sum x_j a_j = \lambda_1 \sum x_j + \lambda_2 \sum x_j^2,$$

or since $\sum a_j x_j = 0$, this becomes

$$\lambda_1 = -\lambda_2 \frac{\sum x_i^2}{\sum x_i}. \tag{23}$$

If we substitute this into (22), we get

$$\lambda_2 = \frac{-2 \sum x_i/n}{\sum x_i^2 - n\bar{x}^2} = \frac{-2\bar{x}}{\sum (x_i - \bar{x})^2}$$

and

$$\lambda_1 = \frac{2 \sum x_i^2/n}{\sum (x_i - \bar{x})^2}.$$

Substituting λ_1 and λ_2 into the tth equation in (21) and solving for a_t gives

$$a_t = \frac{(\sum x_i^2/n) - \bar{x}x_t}{\sum (x_i - \bar{x})^2}.$$

The best linear unbiased estimator of β_0 is therefore

$$\hat{B}_0 = \sum a_t Y_t = \frac{\bar{Y}\sum x_i^2 - \bar{x}\sum Y_t x_t}{\sum (x_i - \bar{x})^2} = \bar{Y} - \hat{B}_1\bar{x},$$

which is the one given by least squares, and so the proof is complete. A similar proof holds for β_1. ////

PROBLEMS

1 Assume that the data below satisfy the simple linear model given in Definition 1 for Case A.

y:	−6.1	−0.5	7.2	6.9	−0.2	−2.1	−3.9	3.8
x:	−2.0	0.6	1.4	1.3	0.0	−1.6	−1.7	0.7

Find the maximum-likelihood estimates of β_0, β_1, and σ^2.

2 In Prob. 1 find the UMVUE of $\beta_0 + 3\beta_1$.

3 In Prob. 1 find a 95 percent confidence interval on β_0; on β_1; on σ^2.

4 In Prob. 1 find a 90 percent confidence interval on $\mu(x)$ for $x = -1.0$.

5 In the simple linear model for Case A find the maximum-likelihood estimator of θ, where $\theta = \beta_0 + 3\beta_1 + 2\sigma^2$.

6 In Prob. 5 find the UMVUE of θ.

7 In the simple linear model for Case A, show that p proportion of the distribution of Y at $x = x_0$ is below ξ_p, where $\xi_p = \beta_0 + \beta_1 x_0 + z_p\sigma$ and z_p is given by $\Phi(z_p) = p$.

8 In Prob. 7 find the UMVUE of ξ_p.

9 Use the data in Prob. 1 to evaluate the UMVUE of ξ_p in Prob. 7.

10 The hardness Y of the shells of eggs laid by a certain breed of chickens was assumed to be roughly linearly related to the amount x of a certain food supplement put into the diet of the chickens. The model was assumed to be a simple linear model for Case A. Data were collected and are given below:

y_i:	.70	.98	1.16	1.75	.76	.82	.95	1.24	1.75	1.95
x_i:	.12	.21	.34	.61	.13	.17	.21	.34	.62	.71

Test the hypothesis that $\beta_1 = 1.00$ versus the hypothesis $\beta_1 \neq 1.00$. Use a Type I error probability of 5 percent.

11 In Prob. 10 test the hypothesis $\beta_1 > 1$ versus the hypothesis $\beta_1 \leq 1$.

12 In Prob. 10 test the hypothesis $\mu(.50) > 1.5$ versus the hypothesis $\mu(.50) \leq 1.5$. Use a Type I error probability of 10 percent.

13 In Prob. 10 compute a 90 percent confidence interval on 2σ.

14 In the simple linear model for Case A find the UMVUE of β_1/σ^2.

15 Consider the simple linear model given in Definition 1 except var $[Y_i] = a_i^{-1}\sigma^2$, where a_i, $i = 1, 2, \ldots, n$, are known positive numbers. Find the maximum-likelihood estimators of β_0 and β_1.

16 What are the conditions on the x_i in the simple linear model for Case A so that $\hat{\mathbf{B}}_0$ and $\hat{\mathbf{B}}_1$ are independent?

17 In the simple linear model for Case A show that $\bar{Y}$ and $\hat{\mathbf{B}}_1$ are uncorrelated. Are they independent?

18 Prove Theorem 4.

19 In Theorem 6 give the proof for the best linear unbiased estimator of β_1.

20 For the simple linear model for Case B prove that the best (minimum-variance) linear unbiased estimator of $\beta_0 + \beta_1$ is $\hat{\mathbf{B}}_0 + \hat{\mathbf{B}}_1$, where $\hat{\mathbf{B}}_0$ and $\hat{\mathbf{B}}_1$ are the least-squares estimators of β_0 and β_1, respectively.

21 Extend Prob. 20 to $c_0\beta_0 + c_1\beta_1$, where c_0 and c_1 are given constants.

XI

NONPARAMETRIC METHODS

1 INTRODUCTION AND SUMMARY

The important place ascribed to the normal distribution in statistical theory is well justified on the basis of the central-limit theorem. However, often it is not known whether the basic distribution is such that the central-limit theorem applies or whether the approximation to the normal distribution is good enough that the resulting confidence intervals and tests of hypotheses based on normal theory are as accurate as desired. For example, if a random sample of size n is taken from a population with a normal density and a .95 confidence interval is set about the mean (see Sec. 3.1 of Chap. VIII) then the frequency interpretation is the following: If repeated random samples are taken from this population and if a 95 percent confidence interval is obtained for each random sample, in the long run 95 percent of these intervals will contain the mean of the density. If sampling is from a density that is not normal, then, instead of 95 percent of the intervals containing the mean, it may be 99 or 90 percent, or some other percentage. If it is close to 95 percent, say 93 to 97 percent, usually the experimenter will be satisfied. However, if it deviates a large amount from the desired percentage, then the experimenter will probably not be satisfied. In cases where it is known that the conventional methods based on the assumption of a normal density are not applicable, an alternative method is desired. If the basic distribution is known (but is not necessarily normal), one may be able to derive exact (or sufficiently accurate) tests of hypotheses and confidence intervals based on that distribution. *In many cases an experimenter does not know the form of the basic distribution* and needs statistical techniques which are applicable regardless of the form of the density. These techniques are called *nonparametric* or *distribution-free* methods.

The term "nonparametric" arises from considerations of testing hypotheses (Chap. IX). In forming the generalized likelihood-ratio, for example, one deals with a parameter space which defines a family of distributions as the parameters in the functional form of the distribution vary over the parameter space. The methods to be developed in this chapter make no use of functional forms or parameters of such forms. They apply to very wide families of distributions rather than only to families specified by a particular functional form. The term "distribution-free" is also often used to indicate similarly that the methods do not depend on the functional form of distribution functions.

The nonparametric methods that will be considered will, for the most part, be based on the order statistics. Also, although the methods to be presented are applicable to both continuous and discrete random variables, we shall direct our attention almost entirely to the continuous case.

Section 2 will be devoted to considerations of statistical inferences that concern the cumulative distribution function of the population to be sampled. The sample cumulative distribution function will be used in three types of inference, namely, point estimation, interval estimation, and testing. Population quantiles have been defined for any distribution function regardless of the form of that distribution. Section 3 deals with distribution-free statistical methods of making inferences regarding population quantiles. Section 4 studies an important concept, that of *tolerance limits*. The similarities and differences of tolerance limits and confidence limits are noted.

In Sec. 5 we return to an important problem in the application of the theory of statistics. It is the problem of testing the homogeneity of two populations. This problem was first mentioned in Subsec. 4.3 of Chap. IX when we tested the equality of the means of two normal populations. It was considered again in Subsec. 5.3 of Chap. IX when we tested the equality of two multinomial populations. We indicated there that the derived test using a chi-square-type statistic could be used to test the equality of two arbitrary populations, and so we had really anticipated this chapter inasmuch as we derived a *distribution-free* test. Other distribution-free tests of the homogeneity of two populations will be presented in Sec. 5. Included will be the sign test, the run test, the median test, and the rank-sum test.

In this chapter we present only a very brief introduction to nonparametric statistical methods. This chapter is similar to the last inasmuch as it includes use of the three basic kinds of inference that were the focus of our attention in Chaps. VII to IX. We shall see that much of the required distributional theory is elementary, seldom using anything more complicated than the basic principles of probability that were considered in Chap. I and the binomial distribution.

2 INFERENCES CONCERNING A CUMULATIVE DISTRIBUTION FUNCTION

2.1 Sample or Empirical Cumulative Distribution Function

In Subsec. 5.4 of Chap. VI, we defined the sample cumulative distribution function (c.d.f.). We indicated there that it could be used to estimate the cumulative distribution function from which we sampled. In this subsection some results about the sample c.d.f. will be reviewed and used to formulate point estimates. In the two following subsections the sample c.d.f. will be utilized to test a hypothesis (in Subsec. 2.2) and to set a confidence interval (in Subsec. 2.3).

Recall that (see Definition 13 in Chap. VI) the sample c.d.f. is defined by

$$F_n(x) = \frac{1}{n}(\text{number of } X_i \text{ less than or equal to } x)$$

$$= \frac{1}{n}\sum_{i=1}^{n} I_{(-\infty, x]}(X_i), \tag{1}$$

where $X_1, \ldots, X_n$ is a random sample from some c.d.f. $F(\cdot)$. According to Theorem 17 of Chap. VI,

$$P\left[F_n(x) = \frac{k}{n}\right] = \binom{n}{k}[F(x)]^k[1 - F(x)]^{n-k}, \qquad k = 0, 1, \ldots, n, \tag{2}$$

where $F_n(\cdot)$ is the sample c.d.f. corresponding to c.d.f. $F(\cdot)$. From Eq. (2), we see that

$$\mathscr{E}[F_n(x)] = \sum_{k=0}^{n} \frac{k}{n}\binom{n}{k}[F(x)]^k[1 - F(x)]^{n-k} = F(x) \tag{3}$$

and similarly

$$\text{var}\,[F_n(x)] = \frac{1}{n}F(x)[1 - F(x)]. \tag{4}$$

In fact, since $F_n(x)$ is the sample mean of random variables $I_{(-\infty, x]}(X_1)$, $\ldots$, $I_{(-\infty, x]}(X_n)$, we know by the central-limit theorem that $F_n(x)$ is asymptotically normally distributed with mean $F(x)$ and variance $(1/n)F(x)[1 - F(x)]$.

Equations (3) and (4) show that for fixed x, $F_n(x)$ is an unbiased and mean-squared-error consistent estimator of $F(x)$, regardless of the form of $F(\cdot)$. If one is interested in estimating $F(x)$ for every x (rather than for a fixed x), then one is interested in saying something about how close $F_n(x)$ is to $F(x)$ jointly over all values x; hence the following result is of interest:

$$P[\sup_{-\infty<x<\infty} |F_n(x) - F(x)| \xrightarrow[n\to\infty]{} 0] = 1. \tag{5}$$

Equation (5), known as the *Glivenko-Cantelli theorem*, states that with probability one the convergence of $F_n(x)$ to $F(x)$ is uniform in x. We can define

$$D_n = \sup_{-\infty<x<\infty} |F_n(x) - F(x)|. \tag{6}$$

D_n is a *random quantity* that measures how far $F_n(\cdot)$ deviates from $F(\cdot)$. Equation (5) states that $P[\lim_{n\to\infty} D_n = 0] = 1$; so, in particular, the c.d.f. of D_n, say $F_{D_n}(\cdot)$, converges to the discrete c.d.f. that has all its mass at 0. In the next subsection we will consider the limiting distribution of $\sqrt{n}\, D_n$. Equation (5) tells us that the estimating function $F_n(x)$ of the c.d.f. $F(x)$ converges to $F(x)$ uniformly for all x with probability one.

Instead of a point estimate of $F(x) = P[X \le x]$, one might be interested in a point estimate of $F(y) - F(x) = P[x < X \le y]$ for fixed $x < y$. The following remark is useful in showing that $F_n(y) - F_n(x)$ is an unbiased mean-squared-error consistent estimator of $F(y) - F(x)$.

Remark

$$\text{cov}\,[F_n(x), F_n(y)] = \frac{1}{n} F(x)[1 - F(y)] \qquad \text{for } y \ge x. \tag{7}$$

PROOF

$$\begin{aligned}
\text{cov}\,[F_n(x), F_n(y)] &= \text{cov}\left[\frac{1}{n}\sum_{i=1}^{n} I_{(-\infty, x]}(X_i), \frac{1}{n}\sum_{j=1}^{n} I_{(-\infty, y]}(X_j)\right] \\
&= \left(\frac{1}{n}\right)^2 \text{cov}\left[\sum_{i=1}^{n} I_{(-\infty, x]}(X_i), \sum_{j=1}^{n} I_{(-\infty, y]}(X_j)\right] \\
&= \left(\frac{1}{n}\right)^2 \sum_{i=1}^{n}\sum_{j=1}^{n} \text{cov}\,[I_{(-\infty, x]}(X_i), I_{(-\infty, y]}(X_j)] \\
&= \frac{1}{n} \text{cov}\,[I_{(-\infty, x]}(X_1), I_{(-\infty, y]}(X_1)]
\end{aligned}$$

$$= \frac{1}{n} \{\mathscr{E}[I_{(-\infty, x]}(X_1) I_{(-\infty, y]}(X_1)]$$
$$- \mathscr{E}[I_{(-\infty, x]}(X_1)] \mathscr{E}[I_{(-\infty, y]}(X_1)]\}$$
$$= \frac{1}{n} [F(x) - F(x)F(y)]$$
$$= \frac{1}{n} F(x)[1 - F(y)]. \qquad ////$$

Using Eq. (7), one sees immediately that

$$\operatorname{var}[F_n(y) - F_n(x)] = \operatorname{var}[F_n(y)] - 2\operatorname{cov}[F_n(x), F_n(y)] + \operatorname{var}[F_n(x)]$$
$$= \frac{1}{n} [F(y) - F(x)][1 - F(y) + F(x)];$$

mean-squared-error consistency of $F_n(y) - F_n(x)$ as an estimator of $F(y) - F(x)$ follows immediately.

Rather than estimating $P[x < X \leq y]$, i.e., the probability that X falls in some interval, one might consider estimating $P[X \in B]$, i.e., the probability that X falls in some set B. It can be shown that (see the Problems) $(1/n) \sum_{i=1}^{n} I_B(X_i)$ is an unbiased estimator of $P[X \in B]$, and

$$\operatorname{var}\left[\frac{1}{n} \sum_{i=1}^{n} I_B(X_i)\right] = \frac{1}{n} P[X \in B](1 - P[X \in B]),$$

hence, is mean-squared-error consistent.

2.2 Kolmogorov-Smirnov Goodness-of-fit Test

We noted above that $F_n(x)$ has an asymptotic normal distribution. Equivalently, $\sqrt{n}[F_n(x) - F(x)]$ has a limiting normal distribution with mean 0 and variance $F(x)[1 - F(x)]$. We now state (without proof) a result that gives the limiting distribution of

$$\sqrt{n} D_n = \sqrt{n} \sup_{-\infty < x < \infty} |F_n(x) - F(x)|.$$

Theorem 1 Let $X_1, \ldots, X_n, \ldots$ be independent identically distributed random variables having common continuous c.d.f. $F_X(\cdot) = F(\cdot)$. Define

$$D_n = d_n(X_1, \ldots, X_n) = \sup_{-\infty < x < \infty} |F_n(x) - F(x)|,$$

where $F_n(x)$ is the sample c.d.f. Then

$$\lim_{n\to\infty} F_{\sqrt{n}D_n}(x) = \lim_{n\to\infty} P[\sqrt{n}D_n \le x] \tag{8}$$

$$= \left[1 - 2\sum_{j=1}^{\infty}(-1)^{j-1}e^{-2j^2x^2}\right] I_{(0,\infty)}(x) = H(x), \qquad \text{say.} \qquad ////$$

The c.d.f. given in Eq. (8) does not depend on the c.d.f. from which the sample was drawn (other than that it be continuous); that is, the limiting distribution of $\sqrt{n}\, D_n$ is *distribution-free*. This fact allows D_n to be broadly used as a test statistic for goodness of fit. For instance, suppose one wishes to test that the distribution that is being sampled from is some specified continuous distribution; that is, test $\mathscr{H}_0$: $X_i \sim F_0(\cdot)$, where $F_0(\cdot)$ is some completely specified continuous c.d.f. If $\mathscr{H}_0$ is true,

$$K_n = k_n(X_1, \ldots, X_n) = \sqrt{n} \sup_{-\infty<x<\infty} |F_n(x) - F_0(x)| \tag{9}$$

is approximately distributed as $H(\cdot)$, the c.d.f. given in Eq. (8). If $\mathscr{H}_0$ is false, then $F_n(\cdot)$ will tend to be near the true c.d.f. $F(\cdot)$ and not near $F_0(\cdot)$, and consequently $\sup_{-\infty<x<\infty} |F_n(x) - F_0(x)|$ will tend to be large; hence a reasonable test criterion is to reject $\mathscr{H}_0$ if $\sup_{-\infty<x<\infty} |F_n(x) - F_0(x)|$ is large. Since $K_n = \sqrt{n} \sup_{-\infty<x<\infty} |F_n(x) - F_0(x)|$ is approximately distributed as $H(\cdot)$ when $\mathscr{H}_0$ is true and $H(\cdot)$ has been tabulated, $k_{1-\alpha}$ can be determined so that $1 - H(k_{1-\alpha}) = \alpha$, and hence $P[K_n > k_{1-\alpha}] \approx \alpha$. That is, the test defined by "Reject $\mathscr{H}_0$ if and only if $K_n > k_{1-\alpha}$" has approximate size α. Such a test is often labeled the *Kolmogorov-Smirnov goodness-of-fit test*. It tests how well a given set of observations fits some specified c.d.f. $F_0(\cdot)$. The fit is measured by the so-called *Kolmogorov statistic* $\sup_{-\infty<x<\infty} |F_n(x) - F_0(x)|$. Theorem 1 gives an asymptotic distribution for D_n. The exact distribution of D_n has been tabled for various n. See Ref. 44.

EXAMPLE 1 A question of at least curious interest is the following: Are the times of birth uniformly distributed over the hours of the day? For 37 consecutive births (actual data) the following times were observed: 7:02 P.M., 11:08 P.M., 3:56 A.M., 8:12 A.M., 8:40 A.M., 12:25 P.M., 1:24 A.M., 8:25 A.M., 2:02 P.M., 11:46 P.M., 10:07 A.M., 1:53 P.M., 6:45 P.M., 9:06 A.M., 3:57 P.M., 7:40 A.M., 3:02 A.M., 10:45 A.M., 3:06 P.M., 6:26 A.M., 4:44 P.M., 12:26 A.M., 2:17 P.M., 11:45 P.M., 5:08 A.M., 5:49 A.M., 6:32 A.M., 12:40 P.M., 1:30 P.M., 12:55 P.M., 3:22 P.M., 4:09 P.M., 7:46 P.M., 2:28 A.M., 10:06 A.M., 11:19 A.M., 4:31 P.M. Both the hypothesized uniform c.d.f. and the sample c.d.f. are sketched in Fig. 1.

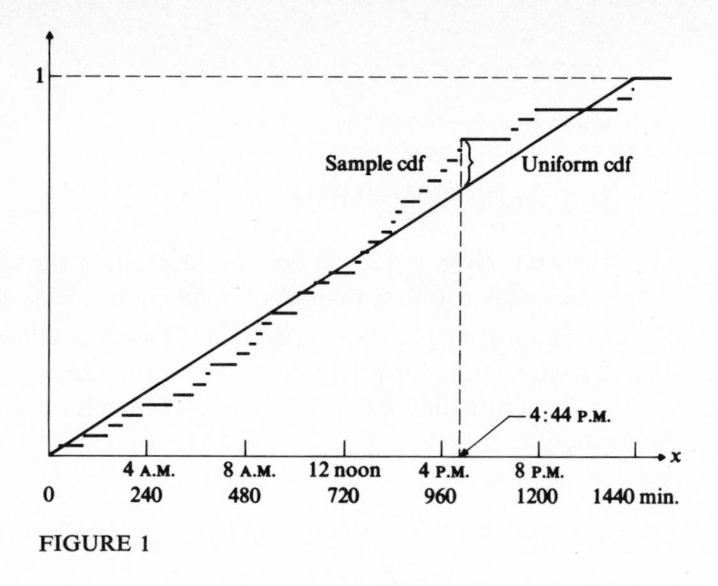

FIGURE 1

One can calculate $\sqrt{n} \sup_x |F_n(x) - F(x)| = \sqrt{37}|\frac{31}{37} - \frac{1004}{1440}| \approx .85$. The critical value for size $\alpha = .10$ is greater than 1.22; so, according to the Kolmogorov-Smirnov goodness-of-fit test, the data do not indicate that the hypothesis that times of birth are uniformly distributed throughout the hours of the day should be rejected. ////

The Kolmogorov-Smirnov goodness-of-fit test assumed that the null hypothesis was simple; that is, the null hypothesis completely specified (no unknown parameters) the distribution of the population. One might inquire as to whether such a goodness-of-fit testing procedure can be extended to a composite null hypothesis which states that the distribution of the population belongs to some parametric family of distributions, say $\{F(\cdot\,; \theta): \theta \in \underline{\overline{\Theta}}\}$. For such null hypotheses, $\sup_x |F_n(x) - F(x; \theta)|$ is no longer a statistic since it depends on an unknown parameter θ. An obvious way of removing the dependence on θ is to replace θ by an estimator, say $\hat{\Theta}$, similar to what was done in the classical chi-square goodness-of-fit test. The test statistic then becomes $\sup_x |F_n(x) - F(x; \hat{\Theta})|$. The distribution of such a test statistic is not known and, in general, depends on the hypothesized parametric family. Although some studies (often Monte Carlo) have been reported in the literature, much remains to be done before a Kolmogorov-Smirnov goodness-of-fit test for composite hypotheses becomes a practical testing tool.

2.3 Confidence Bands for Cumulative Distribution Function

Theorem 1 can also be used to set confidence bands on the c.d.f. $F(\cdot)$ sampled from. Let k_γ be defined by $H(k_\gamma) = \gamma$, where $H(\cdot)$ is the c.d.f. in Eq. (8). A brief table of γ and k_γ is

γ	.99	.95	.90	.85	.80
k_γ	1.63	1.36	1.22	1.14	1.07

It follows that

$$P[\sqrt{n} \sup_x |F_n(x) - F(x)| \leq k_\gamma] \approx \gamma,$$

but

$$P[\sqrt{n} \sup_x |F_n(x) - F(x)| \leq k_\gamma] = P[\sup_x |F_n(x) - F(x)| \leq k_\gamma/\sqrt{n}]$$

$$= P\left[F_n(x) - \frac{k_\gamma}{\sqrt{n}} \leq F(x) \leq F_n(x) + \frac{k_\gamma}{\sqrt{n}} \text{ for all } x\right],$$

noting that

$$\sup_x |F_n(x) - F(x)| \leq \frac{k_\gamma}{\sqrt{n}}$$

if and only if

$$F_n(x) - \frac{k_\gamma}{\sqrt{n}} \leq F(x) \leq F_n(x) + \frac{k_\gamma}{\sqrt{n}}$$

for all x. Using the fact that $0 \leq F(x) \leq 1$, we have

$$P\left[\max\left[0, F_n(x) - \frac{k_\gamma}{\sqrt{n}}\right] \leq F(x) \leq \min\left[F_n(x) + \frac{k_\gamma}{\sqrt{n}}, 1\right] \text{ for all } x\right] \approx \gamma; \tag{10}$$

that is, the band with lower boundary defined by $L(x) = \max [0, F_n(x) - k_\gamma/\sqrt{n}]$ and upper boundary defined by $U(x) = \min [F_n(x) + k_\gamma/\sqrt{n}, 1]$ is an approximate 100γ percent confidence band for the c.d.f. $F(\cdot)$, where the meaning of the confidence band is given in Eq. (10).

3 INFERENCES CONCERNING QUANTILES

3.1 Point and Interval Estimates of a Quantile

Throughout this section, we will assume that we are sampling from a continuous c.d.f., say $F(\cdot)$. Recall (see Definition 17 of Chap. II) that the qth quantile of c.d.f. $F(\cdot)$, denoted by ξ_q, is defined by $F(\xi_q) = q$ for fixed q, $0 < q < 1$. In particular, for $q = \frac{1}{2}$, $\xi_{\frac{1}{2}}$ is called the *median*. We saw in Subsec. 4.6 of Chap. II that quantiles can be used to measure location and dispersion of a c.d.f. For instance, $\xi_{\frac{1}{2}}$, $(\xi_q + \xi_{1-q})/2$, etc., are measures of location, and $\xi_{.9} - \xi_{.1}$, $\xi_{.75} - \xi_{.25}$, etc., are measures of dispersion.

In Subsec. 2.1, we considered estimating $F(x)$ for fixed x; now, we consider estimating ξ_q such that $F(\xi_q) = q$ for fixed q. We know that if X is a continuous random variable with c.d.f. $F(\cdot)$, then the random variable $F(X)$ has a uniform distribution (see Theorem 12 in Chap. V) over the interval (0, 1). Hence $F(Y_j)$ has the same distribution as the jth order statistic from a uniform distribution, and we know that $\mathscr{E}[F(Y_j)] = j/(n+1)$. (As usual, $Y_1, \ldots, Y_n$ are the order statistics corresponding to the random sample $X_1, \ldots, X_n$.) Consequently, we might estimate ξ_q with Y_j if $q \approx j/(n+1)$. [If $j/(n+1) < q < (j+1)/(n+1)$, one could estimate ξ_q by interpolating between the order statistics Y_j and Y_{j+1}.]

A confidence-interval estimate of ξ_q can be obtained by using two order statistics, the interval between them constituting the confidence interval. We are interested in computing the confidence coefficient for a pair of order statistics.

$$
\begin{aligned}
P[Y_j \le \xi_q \le Y_k] &= P[F(Y_j) \le F(\xi_q) = q \le F(Y_k)] \\
&= 1 - P[F(Y_j) > q] - P[F(Y_k) < q] \\
&= P[F(Y_j) \le q] - P[F(Y_k) < q].
\end{aligned}
$$

Recall that

$$
f_{Y_j}(y) = \frac{n!}{(j-1)!\,(n-j)!}[F(y)]^{j-1}[1 - F(y)]^{n-j}f(y);
$$

hence for $Z = F(Y_j)$

$$
\frac{dz}{dy} = f(y),
$$

and so,

$$
f_Z(z) = \left|\frac{1}{dz/dy}\right| f_{Y_j}(y) = \frac{n!}{(j-1)!\,(n-j)!} z^{j-1}(1-z)^{n-j} \qquad \text{for } 0 < z < 1.
$$

Thus,

$$
\begin{aligned}
P[F(Y_j) \le u] &= \int_0^u f_Z(z)\, dz \\
&= \frac{1}{\mathrm{B}(j, n-j+1)} \int_0^u z^{j-1}(1-z)^{n-j+1-1}\, dz \\
&= \mathrm{IB}_u(j, n-j+1),
\end{aligned}
$$

called the *incomplete beta function*, which is extensively tabulated. Hence,

$$P[Y_j \le \xi_q \le Y_k] = \mathrm{IB}_q(j, n-j+1) - \mathrm{IB}_q(k, n-k+1),$$

which is the confidence coefficient of the interval (Y_j, Y_k). In practice, of course, we are interested in going in the other direction; that is, for fixed γ pick j and k (and consequently order statistics Y_j and Y_k) such that

$$\mathrm{IB}_q(j, n-j+1) - \mathrm{IB}_q(k, n-k+1) = \gamma,$$

and then (Y_j, Y_k) is a 100γ percent confidence interval for ξ_q. Of course, for arbitrary γ there will not exist a j and k so that the confidence coefficient is exactly γ.

The confidence coefficient can be obtained another way.

$$P[Y_j \le \xi_q \le Y_k] = P[Y_j \le \xi_q] - P[Y_k < \xi_q].$$

But

$$
\begin{aligned}
P[Y_j \le \xi_q] &= P[j\text{th order statistic} \le \xi_q] \\
&= P[j \text{ or more observations} \le \xi_q] \\
&= \sum_{i=j}^{n} P[\text{exactly } i \text{ observations} \le \xi_q] \\
&= \sum_{i=j}^{n} \binom{n}{i} [F(\xi_q)]^i [1 - F(\xi_q)]^{n-i} \\
&= \sum_{i=j}^{n} \binom{n}{i} q^i (1-q)^{n-i};
\end{aligned}
$$

hence,

$$
\begin{aligned}
P[Y_j \le \xi_q \le Y_k] &= \sum_{i=j}^{n} \binom{n}{i} q^i (1-q)^{n-i} - \sum_{i=k}^{n} \binom{n}{i} q^i (1-q)^{n-i} \\
&= \sum_{i=j}^{k-1} \binom{n}{i} q^i (1-q)^{n-i}.
\end{aligned}
$$

Note that a table of the binomial distribution can now be used to evaluate the confidence coefficient.

EXAMPLE 2 For a sample of size 10, what is the confidence coefficient of the interval (Y_2, Y_9), which is a confidence-interval estimator of the population median? We have

$$P[Y_2 \le \xi_{\frac{1}{2}} \le Y_9] = \sum_{2}^{8} \binom{n}{i} \left(\frac{1}{2}\right)^n = \sum_{2}^{8} \binom{10}{i} \left(\frac{1}{2}\right)^{10} = .9784. \qquad ////$$

We have presented one way, using order statistics, of obtaining point estimates or confidence-interval estimates for a quantile.

Besides being extremely general in that the method requires few assumptions about the form of the distribution function, the method is extraordinarily simple. No complex analysis or distribution theory was needed; the simple binomial distribution provided the necessary equipment to determine the confidence coefficient. The only inconvenience was the paucity of confidence levels that could be attained.

3.2 Tests of Hypotheses Concerning Quantiles

Let $X_1, \ldots, X_n$ denote a random sample from a probability density function, say $f(\cdot)$. Suppose that it is desired to test that the qth quantile of the population sampled from is a specified value, say ξ. That is, it is desired to test

$$\mathscr{H}_0: \xi_q = \xi \text{ versus } \mathscr{H}_1: \xi_q \neq \xi,$$

where $f(\cdot)$ is unspecified (other than being a probability density function). The confidence-interval method of deriving a test (see Subsec. 3.4 of Chap. IX) can be used; for instance, obtain a 100γ percent confidence interval for ξ_q, and accept $\mathscr{H}_0$ if and only if the derived confidence interval contains ξ. Such a test has size $1 - \gamma$.

An alternative test is the so-called *one-sample sign test*. It is a very simple test based on the value of a statistic that represents the number of the n transformed observations that have a positive sign. To illustrate the principle involved in the sign test, consider testing $\mathscr{H}_0: \xi_q = \xi$ versus $\mathscr{H}_1: \xi_q \neq \xi$ for a random sample $X_1, \ldots, X_n$ from some unspecified probability density function $f(\cdot)$. Let Z denote the number of X_i's that exceed ξ. Equivalently, Z is the number of $X_1 - \xi, \ldots, X_n - \xi$ that have a positive sign. If $\mathscr{H}_0$ is true, Z has a binomial distribution with parameters n and $p = 1 - q = \int_{\xi}^{\infty} f(x)\,dx$. So if $\mathscr{H}_0$ is true, one would expect Z to be near np, and hence an intuitively appealing test is to accept $\mathscr{H}_0$ if and only if Z is near np. Since the distribution of Z is known, one can determine what is meant by "near" by fixing the size of the test.

For example, suppose $q = \frac{1}{2}$ so that $\xi_q = \xi_{\frac{1}{2}}$ = median; then a possible test of $\mathscr{H}_0: \xi_{\frac{1}{2}} = \xi$ versus $\mathscr{H}_1: \xi_{\frac{1}{2}} \neq \xi$ is to accept $\mathscr{H}_0$ if and only if $|Z - np| = |Z - n/2| \leq c$, where c is a constant determined by

$$P[|Z - n/2| \leq c] = 1 - \alpha,$$

where α is the desired size of the test. Now

$$P[|Z - n/2| \leq c] = \sum_{n/2-c}^{n/2+c} \binom{n}{j} \left(\frac{1}{2}\right)^j \left(\frac{1}{2}\right)^{n-j};$$

so c can be determined from a binomial table. (For small sample sizes, not many α's are possible, unless randomized tests are used.) The power function of such a test can be readily obtained since the distribution of Z is still binomial even when the null hypothesis is false; Z has the binomial distribution with parameters n and $p = P[X > \xi]$. Such a power function could be sketched as a function of p.

Note also that the sign test can be used to test one-sided hypotheses. For instance, in testing $\mathscr{H}_0: \xi_q \leq \xi$ versus $\mathscr{H}_1: \xi_q > \xi$, the sign test says to reject $\mathscr{H}_0$ if and only if Z, defined as above, is large. Again the power function can be easily obtained.

4 TOLERANCE LIMITS

An automatic machine in a ball-bearing factory is supposed to manufacture bearings .25 inch in diameter. The bearings are regarded as acceptable from an engineering standpoint if the diameter falls between the limits .249 and .251 inch. Production is regularly checked each day by measuring the diameter of a random sample of bearings and computing statistical tolerance limits L_1 and L_2 from their samples. If L_1 is above .249 and L_2 is below .251, the production is accepted. How large should the sample be so that one can be assured with 90 percent probability that the statistical tolerance limits will contain at least 80 percent of the population of bearing diameters? There is a simple nonparametric solution to problems of this kind.

In more general terms, let $f(\cdot)$ be a probability density function, and on the basis of a sample of n values it is desirable to determine two numbers, say L_1 and L_2, such that at least .80, say, of the area under $f(\cdot)$ is between L_1 and L_2. On the basis of a sample we cannot be certain that .80 of the area under $f(\cdot)$ is between L_1 and L_2, but we can specify a probability that it is so.

In other words, we want to find two functions $L_1 = l_1(X_1, \ldots, X_n)$ and $L_2 = l_2(X_1, \ldots, X_n)$ of the random sample $X_1, \ldots, X_n$ such that the probability that

$$\int_{L_1}^{L_2} f(x)\, dx \geq \beta \tag{11}$$

is equal to γ, for specified γ and β. We summarize with a definition.

Definition 1 Tolerance limits Let $X_1, \ldots, X_n$ be a random sample from continuous c.d.f. $F(\cdot)$ having a density function $f(\cdot)$. Let $L_1 = l_1(X_1, \ldots, X_n) < L_2 = l_2(X_1, \ldots, X_n)$ be two statistics which satisfy:

(i) The distribution of $F(L_2) - F(L_1)$ does not depend on $F(\cdot)$.
(ii) $P[F(L_2) - F(L_1) \geq \beta] = \gamma$.

Then L_1 and L_2 will be defined to be 100β percent *distribution-free tolerance limits* at probability level γ. ////

Remark Note that the random quantity $F(L_2) - F(L_1)$ represents the area under $f(\cdot)$ between L_1 and L_2. ////

For continuous random variables, order statistics Y_j and Y_k $(j < k)$ form tolerance limits. To obtain the coefficients β and γ in the definition of tolerance limits, we need the distribution of $F(L_2) - F(L_1)$. Recall that

$$f_{Y_j, Y_k}(y_j, y_k) = \frac{n!}{(j-1)!\,(k-1-j)!\,(n-k)!}$$
$$\times [F(y_j)]^{j-1}[F(y_k) - F(y_j)]^{k-1-j}[1 - F(y_k)]^{n-k} f(y_j) f(y_k).$$

Make the transformation $Z = F(Y_k) - F(Y_j)$ and $Y = F(Y_j)$, find the joint distribution of Y and Z, and then integrate out y to get the marginal distribution of Z. The following obtains:

$$f_Z(z) = \frac{n!}{(k-1-j)!\,(n-k+j)!} z^{k-1-j}(1-z)^{n-k+j} I_{(0,1)}(z), \tag{12}$$

which is a beta distribution with parameters $k - j$ and $n - k + j + 1$. Now

$$P[Z < \beta] = \int_0^\beta f_Z(z)\, dz = \text{IB}_\beta(k - j, n - k + j + 1),$$

the incomplete beta function, which is tabled. Also, recall that

$$\text{IB}_\beta(k - j, n - k + j + 1) = \sum_{k-j}^{n} \binom{n}{i} \beta^i (1-\beta)^{n-i}.$$

Thus for any β, the probability level γ can be computed.

EXAMPLE 3 For a random sample of size 5, use (Y_1, Y_5) as a tolerance interval for 75 percent of the population; that is, $\beta = .75$. What is the corresponding probability level γ? We seek

$$\begin{aligned}\gamma &= P[F(Y_5) - F(Y_1) \geq .75] \\ &= 1 - P[F(Y_5) - F(Y_1) < .75] \\ &= 1 - \sum_{i=4}^{5} \binom{5}{i} (.75)^i (.25)^{5-i} = .3672. \end{aligned}$$ ////

We might note that, in general,

$$\mathscr{E}[Z] = \mathscr{E}[F(Y_k)] - \mathscr{E}[F(Y_j)] = \frac{k}{n+1} - \frac{j}{n+1} = \frac{k-j}{n+1}. \tag{13}$$

For Example 3 $\mathscr{E}[Z] = \frac{4}{6} = \frac{2}{3}$.

EXAMPLE 4 Suppose that it is desired to determine how large a sample must be taken so that the probability is .90 that at least 99 percent of a future day's output of bearings will have diameters between the largest and smallest observations in the sample. The quantities are $\gamma = .90$ and $\beta = .99$, and we want to determine n such that

$$P[F(Y_n) - F(Y_1) \geq \beta] = \gamma,$$

where the density of $Z = F(Y_n) - F(Y_1)$ is given by Eq. (12) for $j = 1$ and $k = n$. We get

$$\gamma = P[Z \geq \beta] = \int_\beta^1 n(n-1)z^{n-2}(1-z)\, dz = 1 - n\beta^{n-1} + (n-1)\beta^n.$$

If we substitute for γ and β, we get the equation

$$.90 = 1 - n(.99)^{n-1} + (n-1)(.99)^n,$$

which can be solved to determine n. The solution is $n \approx 388$. ////

There are similarities and differences between tolerance limits and confidence limits. Tolerance limits, like confidence limits, are two statistics, one less than the other, that together form an interval with random end points. The user of either interval is reasonably confident (the degree of confidence being measured by the corresponding confidence level) that the interval obtained contains what it is claimed to. This is where the similarity ends. A confidence interval is an interval thought to contain a fixed unknown parameter value. On the other hand, a tolerance interval is an interval thought to contain a prescribed proportion of the values of the random variable under consideration. In other words, a confidence interval is an interval thought to contain an unknown fixed parameter value that characterizes the distribution of population values, whereas a tolerance interval is an interval thought to contain actual population values, and not some characteristic of them.

5 EQUALITY OF TWO DISTRIBUTIONS

5.1 Introduction

In this section various tests of the equality of two populations will be studied. As we mentioned in Sec. 1 above, we first studied the equality of two populations when we tested that the means from two normal populations were equal in Subsec. 4.3 of Chap. IX. Then again in Subsec. 5.3 of Chap. IX, we gave a test of homogeneity of two populations. A great many nonparametric methods have been developed for testing whether two populations have the same distribution. We shall consider only four of them; a fifth will be briefly mentioned at the end of this subsection.

The problem that we propose to consider is the following: Let $X_1, \ldots, X_m$ denote a random sample of size m from c.d.f. $F_X(\cdot)$ with a corresponding density function $f_X(\cdot)$, and let $Y_1, \ldots, Y_n$ denote a random sample of size n from c.d.f. $F_Y(\cdot)$ with a corresponding density function $f_Y(\cdot)$. (Note that we are departing from our usual convention of using Y's to represent the order statistics corresponding to the X's.) Further, assume that the observations from $F_X(\cdot)$ are independent of the observations from $F_Y(\cdot)$. Test $\mathscr{H}_0$: $F_X(z) = F_Y(z)$ for all z versus $\mathscr{H}_1$: $F_X(z) \neq F_Y(z)$ for at least one value of z. In Sec. 2 above we pointed out that the sample c.d.f. can be used to estimate the population c.d.f. In the case that $\mathscr{H}_0$ is true, that is, $F_X(z) = F_Y(z)$, we have two independent estimators of the common population c.d.f., one using the sample c.d.f. of the X's and the other using the sample c.d.f. of the Y's. Intuitively, then, one might consider using the closeness of the two sample c.d.f.'s to each other as a test criterion. Although we will not study it, a test, called the *two-sample Kolmogorov-Smirnov test*, has been devised that uses such a criterion.

We will assume throughout that the random variables under consideration are continuous and merely point out at this time that the methods to be presented can be extended to include discrete random variables as well. In our presentation, we will consider testing two-sided hypotheses and will not consider one-sided hypotheses, although the theory works equally well for one-sided hypotheses.

5.2 Two-sample Sign Test

The first test that we will consider is the two-sample sign test. We shall see that, in a certain sense, this test is a nonparametric analog of the paired t test (see Prob. 17 in Chap. IX). For this test we assume that the sampling situation is such that the X and Y observations are paired; that is, we observe (X_1, Y_1), $\ldots$, (X_n, Y_n). One could think that the X observation is "untreated" (a control) and the corresponding Y observation is "treated," the object of the test being to determine if there is a "*treatment*" *effect*. We wish to test $\mathscr{H}_0$: $F_X(\cdot) = F_Y(\cdot)$. Assume that $(X_1, Y_1), \ldots, (X_n, Y_n)$ is a random sample from some joint distribution $F_{X,Y}(\cdot, \cdot)$. Further assume that $F_{X,Y}(\cdot, \cdot)$ is such that $P[X > Y] = P[X < Y] = \frac{1}{2}$ when $\mathscr{H}_0$ is true. (Recall that we are assuming continuous random variables, and then such an assumption is satisfied if X and Y are independent.) Consider a test based on the signs of the differences $X_i - Y_i$, $i = 1, \ldots, n$. For instance, define

$$Z_i = I_{(0, \infty)}(X_i - Y_i);$$

then Z_i has a Bernoulli distribution, and consequently $S_n = \sum_{i=1}^{n} Z_i$ has a binomial distribution with parameters n and $p = P[X_i > Y_i]$. If $\mathscr{H}_0$ is true, $p = \frac{1}{2}$, and $\mathscr{E}[S_n] = n/2$. If the alternative hypothesis is two-sided so that $p = P[X_i > Y_i]$ can be either larger or smaller than $\frac{1}{2}$, then a possible test criterion is to accept $\mathscr{H}_0$ if S_n is close to $n/2$, that is, accept $\mathscr{H}_0$ if $|S_n - n/2| \leq k$, where k is determined by fixing the size of the test. k is easily determined from a binomial table, and we have a very simple test of the equality of the two populations.

One can see that avoidance of the assumption that X_i and Y_i are independent is desirable. For example, X_i might represent an observation on the ith entity before some "treatment" and Y_i the observation on the *same* entity after "treatment." In such a case one is not likely to have independence of X_i and Y_i since they are observations taken on the same entity, yet one can sometimes test that there is no "treatment" effect by testing that the "before" and "after" populations are the same.

5.3 Run Test

As before let $X_1, \ldots, X_m$ denote a random sample from $F_X(\cdot)$ and $Y_1, \ldots, Y_n$ a random sample from $F_Y(\cdot)$. A rather simple test of $\mathscr{H}_0$: $F_X(z) = F_Y(z)$ for all z is based on *runs* of values of X and values of Y. To understand the meaning of runs, combine the m x observations with the n y observations and then order (in ascending order of magnitude) the combined sample. For example, if $m = 4$ and $n = 5$, one might obtain

$$y \quad x \quad x \quad y \quad x \quad y \quad y \quad y \quad x. \tag{14}$$

A *run* is a sequence of letters of the same kind bounded by letters of another kind except for the first and last position. Thus, in Eq. (14) the ordering starts with a run of one y value, then follows a run of two x values, then a run of one y value, and so on; six runs are exhibited in Eq. (14). It is apparent that if the two samples are from the same population, the x's and y's will ordinarily be well mixed, and the total number of runs will be large. If the two populations are widely separated so that their range of values does not overlap, then the number of runs will be only two, and, in general, differences between the two populations will tend to reduce the number of runs. Thus the two populations may have the same mean or median, but if the x population is concentrated while the y population is dispersed, there will be a tendency to have a long y run on each end of the combined sample, and there will thus be a tendency to reduce the number of runs. A test then is performed by observing the total number of runs, say Z, in the combined sample and rejecting $\mathscr{H}_0$ if Z is less than or equal to some specified number z_0. Our task now is to determine the distribution of Z under $\mathscr{H}_0$ in order that for a given test size we may specify z_0.

If $\mathscr{H}_0$ is true, it can be argued that the possible arrangements of the m x values and n y values are equally likely. It is clear that there are exactly $\binom{m+n}{m}$ such arrangements. To find $P[Z = z]$, it is necessary now to count all arrangements with exactly z runs. Suppose z is even, say $2k$; then there must be k runs of x values and k runs of y values. To get k runs of x values, the m x's must be divided into k groups. We can form these k groups, or runs, by inserting $k - 1$ *dividers* into the $m - 1$ *spaces* between the m x values with no more than one divider per space. We can place the $k - 1$ dividers into the $m - 1$ spaces in $\binom{m-1}{k-1}$ ways. Similarly, we can construct the k runs of y values in $\binom{n-1}{k-1}$ ways. Any particular arrangement of the k runs of x values can be combined with any arrangement of the k runs of y values; furthermore, the first run in the combined arrangement can be either a run of x values or a run of y values; hence there are a total of $2\binom{m-1}{k-1}\binom{n-1}{k-1}$ arrangements having exactly $z = 2k$ runs. Hence

$$P[Z = z] = P[Z = 2k] = \frac{2\binom{m-1}{k-1}\binom{n-1}{k-1}}{\binom{m+n}{m}} \tag{15}$$

Similarly, for z odd

$$P[Z = z] = P[Z = 2k + 1] = \frac{\binom{m-1}{k}\binom{n-1}{k-1} + \binom{m-1}{k-1}\binom{n-1}{k}}{\binom{m+n}{m}}. \tag{16}$$

To test $\mathscr{H}_0$ with size of Type I error equal to α, one finds the integer z_0 so that (as nearly as possible)

$$\sum_{z=2}^{z_0} P[Z = z] = \alpha \tag{17}$$

and rejects $\mathscr{H}_0$ if the observed value of Z does not exceed z_0.

The computation involved in Eq. (17) can become quite tedious unless both m and n are small. Fortunately, the distribution of Z is approximately normal for large samples, and in fact the approximation is usually good enough for practical purposes when both m and n exceed 10. If $\mathscr{H}_0$ is true, the mean and variance of Z are

$$\mathscr{E}[Z] = \frac{2mn}{m+n} + 1 \tag{18}$$

and

$$\text{var}\,[Z] = \frac{2mn(2mn - m - n)}{(m+n)^2(m+n-1)}. \tag{19}$$

The asymptotic normal distribution of Z under $\mathscr{H}_0$ has mean and variance given in Eqs. (18) and (19). This asymptotic normal distribution can be used to determine the critical value z_0 for large samples.

The run test is sensitive to both differences in shape and differences in location between two distributions.

5.4 Median Test

Let $X_1, \ldots, X_m$ be a random sample from $F_X(\cdot)$ and $Y_1, \ldots, Y_n$ be a random sample from $F_Y(\cdot)$. As in the previous subsection, combine the two samples, and order them. Let $Z_1 < Z_2 < \ldots < Z_{m+n}$ be the combined ordered sample. The *median* test of $\mathscr{H}_0$: $F_X(u) = F_Y(u)$ for all u consists of finding the median, say $\tilde{z}$, of the z values and then counting the number of x values, say m_1, which exceed $\tilde{z}$ and the number of y values, say n_1, which exceed $\tilde{z}$. If $\mathscr{H}_0$ is true, m_1 should be approximately $m/2$ and n_1 approximately $n/2$. We can use either the statistic M_1 or the statistic N_1 to construct the test. Let us use M_1 = number of X's which exceed $\tilde{Z}$, the median of the combined sample. If $m + n$ is even, there are exactly $(m + n)/2$ of the observations (combined x's and y's) greater than the median of the combined sample. (Since we have an even number of continuous random variables, no two are equal, and the median is midway between the middle two.) It can be easily argued that

$$P[M_1 = m_1] = \frac{\binom{m}{m_1}\binom{n}{(m+n)/2 - m_1}}{\binom{m+n}{(m+n)/2}}$$

for $m + n$ even and $\mathscr{H}_0$ true. A similar expression obtains for $m + n$ odd. Such a distribution can be used to find a constant k such that

$$P\left[\left|M_1 - \frac{m}{2}\right| \geq k\right] \approx \alpha,$$

and our test is given by the following:

$$\text{Reject } \mathscr{H}_0 \text{ if and only if } \left| M_1 - \frac{m}{2} \right| \geq k.$$

Just as in the run test, an asymptotic normal distribution of M_1 can be derived, but we will not study it.

5.5 Rank-sum Test

A very interesting nonparametric test for two samples was described by Wilcoxon and studied by Mann and Whitney. Given two random samples $X_1, X_2, \ldots, X_m$ and $Y_1, \ldots, Y_n$ from populations with absolutely continuous c.d.f.'s $F_X(\cdot)$ and $F_Y(\cdot)$, respectively, one arranges the $m + n$ observations in ascending order and then replaces the smallest observation by 1, the next by 2, and so on, the largest being replaced by $m + n$. These integers are called the *ranks* of the observations. Let T_x denote the sum of the ranks of the m x values and T_y the sum of the ranks of the n y values. Note that $T_x + T_y = \sum_{j=1}^{m+n} j = (m + n + 1)(m + n)/2$; so T_y is a linear function of T_x. We could base a test on either statistic T_x or T_y. Let us use T_x. T_x is linearly related to another statistic, which we denote by U. Set

$$U = \sum_{j=1}^{n} \sum_{i=1}^{m} I_{[Y_j, \infty)}(X_i), \tag{20}$$

the number of times an X exceeds a Y. For a given set of observations, let $r_1, r_2, \ldots, r_m$ denote the ranks of the x values, and let $x'_1, \ldots, x'_m$ denote the ordered x values. Clearly x'_1 exceeds $(r_1 - 1)$ y-values, x'_2 exceeds $(r_2 - 2)$ y-values, and so on, and x_m exceeds $(r_m - m)$ y-values. Hence

$$u = \sum_{i=1}^{m} (r_i - i) = \sum r_i - \sum i = t_x - \frac{m(m+1)}{2},$$

or

$$U = T_x - \frac{m(m+1)}{2}. \tag{21}$$

To find the first two moments of T_x, we find the first two moments of U.

$$\begin{aligned} \mathscr{E}[U] &= \mathscr{E}\left[\sum\sum I_{[Y_j, \infty)}(X_i)\right] = \sum\sum \mathscr{E}[I_{[Y_j, \infty)}(X_i)] \\ &= \sum\sum P[X_i \geq Y_j] = \sum\sum p = mnp, \end{aligned}$$

where

$$p = P[X_i \geq Y_j] = \int P[Y \leq x \mid X = x] f_X(x)\, dx = \int F_Y(x) f_X(x)\, dx.$$

If $\mathscr{H}_0$ is true,

$$p = \int F_X(x) f_X(x)\, dx = \int_0^1 u\, du = \tfrac{1}{2}.$$

Similarly, the variance of U can be found. The derivation is somewhat more complicated since one needs the expected value of U^2. From the mean and variance of U, the mean and variance of T_x can be obtained. If $\mathscr{H}_0$ is true, they are given by

$$\mathscr{E}[T_x] = \frac{m(m+n+1)}{2} \tag{22}$$

and

$$\text{var}\,[T_x] = \frac{mn(m+n+1)}{12}. \tag{23}$$

The exact distribution of T_x turns out to be a very troublesome problem for large m and n. However, Mann and Whitney have calculated the distribution for small m and n, have shown that T_x is approximately normally distributed for large m and n, and have demonstrated that the normal approximation is quite accurate when m and n are larger than 7. Thus for samples of reasonable size one can use the normal approximation with mean and variance given by Eqs. (22) and (23) to find a critical region for testing $\mathscr{H}_0$: $F_X(z) = F_Y(z)$ for all z versus $\mathscr{H}_1$: $F_X(z) \neq F_Y(z)$. The test would be the following:

$$\text{Reject } \mathscr{H}_0 \text{ if } |T_x - \mathscr{E}[T_x]| \text{ is large;}$$

that is,

$$\text{Reject } \mathscr{H}_0 \text{ if and only if } |T_x - \mathscr{E}[T_x]| \geq k,$$

where k is determined by fixing the size of the test and using the asymptotic normal distribution of T_x.

EXAMPLE 5 Find the exact distribution of T_x under $\mathscr{H}_0$ for $m = 3$ and $n = 2$. Each of the following arrangements is equally likely if $\mathscr{H}_0$ is true:

$$x\,x\,x\,y\,y,\ x\,x\,y\,x\,y,\ x\,x\,y\,y\,x,\ x\,y\,x\,x\,y,\ x\,y\,x\,y\,x,$$
$$x\,y\,y\,x\,x,\ y\,x\,x\,x\,y,\ y\,x\,x\,y\,x,\ y\,x\,y\,x\,x,\ y\,y\,x\,x\,x.$$

The corresponding T_x values are, respectively, 6, 7, 8, 8, 9, 10, 9, 10, 11, 12; so

$$P[T_x = 6] = P[T_x = 7] = \tfrac{1}{10},$$
$$P[T_x = 8] = P[T_x = 9] = P[T_x = 10] = \tfrac{2}{10},$$

and

$$P[T_x = 11] = P[T_x = 12] = \tfrac{1}{10}.$$ ////

PROBLEMS

1 Show that $T = \frac{1}{n} \sum_{i=1}^{n} I_B(X_i)$ is an unbiased estimator of $P[X \in B]$. Find var $[T]$, and show that T is a mean-squared-error consistent estimator of $P[X \in B]$.

2 Define $F_n(B_j) = \frac{1}{n} \sum_{i=1}^{n} I_{B_j}(X_i)$ for $j = 1, 2$. Find cov $[F_n(B_1), F_n(B_2)]$.

3 Let $Y_1, \ldots, Y_n$ be the order statistics corresponding to a random sample of size n from a continuous c.d.f. $F(\cdot)$.
(*a*) Find the density of $F(Y_j)$.
(*b*) Find the joint density of $F(Y_i)$ and $F(Y_j)$.
(*c*) Find the density of $[F(Y_n) - F(Y_2)]/[F(Y_n) - F(Y_1)]$.

4 Let $X_1, \ldots, X_n$ be independent and identically distributed random variables having common continuous c.d.f. $F(\cdot)$. Let $Y_1 < \cdots < Y_n$ be the corresponding order statistics, and define $F_n(\cdot)$ to be the sample c.d.f. Set $D_n = \sup_{-\infty < x < \infty} |F_n(x) - F(x)|$.
(*a*) Find the exact distribution of D_n for $n = 1$.
(*b*) Do the same for $n = 2$. HINT: Does $D_n = \max [F(Y_1), \frac{1}{2} - F(Y_1), F(Y_2) - \frac{1}{2}, 1 - F(Y_2)]$?
(*c*) Argue that the exact distribution of D_n will not depend on $F(\cdot)$.

5 Show that the expected value of the larger of a random sample of two observations from a normal population with mean 0 and unit variance is $1/\sqrt{\pi}$ and hence that for the general normal population the expected value is $\mu + \sigma/\sqrt{\pi}$.

6 If (X, Y) is an observation from a bivariate normal population with means 0, unit variances, and correlation ρ, show that the expected value of the larger of X and Y is $\sqrt{(1-\rho)/\pi}$.

7 We have seen that the sample mean for a distribution with infinite variance (such as the Cauchy distribution) is not necessarily a consistent estimator of the population mean. Is the sample median a consistent estimator of the population median?

8 Construct a (approximate) 90 percent confidence band for the data of Example 1. Does your band include the appropriate uniform distribution?

9 Let $Y_1 < \cdots < Y_5$ be the order statistics corresponding to a random sample from some continuous c.d.f. Compute $P[Y_1 < \xi_{.50} < Y_5]$ and $P[Y_2 < \xi_{.50} < Y_4]$. Compute $P[Y_1 < \xi_{.20} < Y_2]$. Compute $P[Y_3 < \xi_{.75} < Y_5]$.

10 Let Y_1 and Y_n be the first and last order statistics of a random sample of size n from some continuous c.d.f. $F(\cdot)$. Find the smallest value of n such that $P[F(Y_n) - F(Y_1) \geq .75] \geq .90$.

11 Test as many ways as you know how at the 5 percent level that the following two samples came from the same population:

x	1.3	1.4	1.4	1.5	1.7	1.9	1.9
y	1.6	1.8	2.0	2.1	2.1	2.2	2.3

12 Let $X_1, \ldots, X_5$ denote a random sample of size 5 from the density $f(x; \theta) = I_{(\theta - \frac{1}{2}, \theta + \frac{1}{2})}(x)$. Consider estimating θ.

(a) Determine the confidence coefficient of the confidence interval (Y_1, Y_5).

(b) Find a confidence interval for θ that has the same confidence coefficient as in part (a) using the pivotal quantity $(Y_1 + Y_5)/2 - \theta$.

(c) Compare the expected lengths of the confidence intervals of parts (a) and (b).

13 Find var $[U]$ when $F_X(\cdot) \equiv F_Y(\cdot)$. See Eq. (20).

14 Equation (21) shows that U and T_x are linearly related. Find the exact distribution of U or T_x when $\mathscr{H}_0$ is true for small sample sizes. For example, take $m = 1$, $n = 2$; $m = 1$, $n = 3$; $m = 2$, $n = 1$; $m = 3$, $n = 1$; and $m = n = 2$.

15 We saw that $\mathscr{E}[U] = mnp$. Is U/mn an unbiased estimator of $p = P[X_i \geq Y_j]$ whether or not $\mathscr{H}_0$ is true? Is U a consistent estimator of p?

16 A common measure of association for random variables X and Y is the *rank correlation*, or *Spearman's correlation*. The X values are ranked, and the observations are replaced by their ranks; similarly the Y observations are replaced by their ranks. For example, for a sample of size 5 the observations

x	20.4	19.7	21.8	20.1	20.7
y	9.2	8.9	11.4	9.4	10.3

are replaced by

$r(x)$	3	1	5	2	4
$r(y)$	2	1	5	3	4

Let $r(X_i)$ denote the rank of X_i and $r(Y_i)$ the rank of Y_i. Using these paired ranks, the ordinary sample correlation is computed:

$$\text{Spearman's correlation} = S = \frac{\sum [r(X_i) - \bar{r}(X)][r(Y_i) - \bar{r}(Y)]}{\sqrt{\sum [r(X_i) - \bar{r}(X)]^2 \sum [r(Y_i) - \bar{r}(Y)]^2}},$$

where $\bar{r}(X) = \sum r(X_i)/n$ and $\bar{r}(Y) = \sum r(Y_i)/n$.

(*a*) Show that $S = 1 - 6 \sum D_i^2/(n^3 - n)$, where $D_i = r(X_i) - r(Y_i)$.

(*b*) Compute the ordinary correlation and Spearman's correlation for the above data.

17 Argue that the distribution of S in Prob. 16 is independent of the form of the distributions of X and Y provided that X and Y are continuous and independently distributed random variables. Hence S can be used as a test statistic in a nonparametric test of the null hypothesis of independence.

18 Show that the mean and variance of S (in Prob. 17) under the hypothesis of independence are 0 and $1/(n - 1)$, respectively.

APPENDIX A
MATHEMATICAL ADDENDUM

1 INTRODUCTION

The purpose of this appendix is to provide the reader with a ready reference to some mathematical results that are used in the book. This appendix is divided into two main sections: The first, Sec. 2 below, gives results that are, for the most part, combinatorial in nature, and the last gives results from calculus. No attempt is made to prove these results, although sometimes a method of proof is indicated.

2 NONCALCULUS

2.1 Summation and Product Notation

A sum of terms such as $n_3 + n_4 + n_5 + n_6 + n_7$ is often designated by the symbol $\sum_{i=3}^{7} n_i$. $\sum$ is the capital Greek letter sigma, and in this connection it is often called the *summation sign*. The letter i is called the *summation index*. The term following $\sum$ is called the *summand*. The "$i = 3$" below $\sum$ indicates that the first term of the sum is obtained by putting $i = 3$ in the summand. The "7" above the $\sum$ indicates that the final term of the sum is obtained by putting $i = 7$ in the summand. The other terms of the sum are obtained by giving i the integral values between the limits 3 and 7. Thus

$$\sum_{j=2}^{5} (-1)^{j-2} j x^{2j} = 2x^4 - 3x^6 + 4x^8 - 5x^{10}.$$

An analogous notation for a product is obtained by substituting the capital Greek letter $\prod$ for $\sum$. In this case the terms resulting from substituting the integers for the index are multiplied instead of added. Thus

$$\prod_{i=1}^{5} \left[a + (-1)^i \frac{i}{b}\right] = \left(a - \frac{1}{b}\right)\left(a + \frac{2}{b}\right)\left(a - \frac{3}{b}\right)\left(a + \frac{4}{b}\right)\left(a - \frac{5}{b}\right).$$

EXAMPLE 1 Some useful formulas involving summations are listed below. They can be proved using mathematical induction.

$$\sum_{i=1}^{n} i = \frac{n(n+1)}{2}. \tag{1}$$

$$\sum_{i=1}^{n} i^2 = \frac{n(n+1)(2n+1)}{6}. \tag{2}$$

$$\sum_{i=1}^{n} i^3 = \left[\frac{n(n+1)}{2}\right]^2. \tag{3}$$

$$\sum_{i=1}^{n} i^4 = \frac{n(n+1)(2n+1)(3n^2+3n-1)}{30}. \tag{4}$$

Equation (1) can be used to derive the following formula for an *arithmetic series* or *progression:*

$$\sum_{j=1}^{n} [a + (j-1)d] = na + \frac{d}{2} n(n-1). \tag{5}$$

A companion series, the finite *geometric series,* or *progression,* is given by

$$\sum_{j=0}^{n-1} ar^j = a \frac{1-r^n}{1-r}. \tag{6}$$

////

2.2 Factorial and Combinatorial Symbols and Conventions

A product of a positive integer n by all the positive integers smaller than it is usually denoted by $n!$ (read "*n factorial*"). Thus

$$n! = n(n-1)(n-2) \cdot \cdots \cdot 1 = \prod_{j=0}^{n-1} (n-j). \tag{7}$$

$0!$ is defined to be 1.

A product of a positive integer n by the next $k-1$ smaller positive integers is usually denoted by $(n)_k$. Thus

$$(n)_k = n(n-1)\cdot \cdots \cdot (n-k+1)$$

$$= \prod_{j=1}^{k} (n-j+1). \tag{8}$$

Note that there are k terms in the product in Eq. (8).

Remark $(n)_k = n!/(n-k)!$, and $(n)_n = n!/0! = n!$. The *combinatorial* symbol $\binom{n}{k}$ is defined as follows:

$$\binom{n}{k} = \frac{(n)_k}{k!} = \frac{n!}{(n-k)!k!}. \tag{9}$$

$\binom{n}{k}$ is read "combination of n things taking k at a time" or more briefly as "n pick k"; it is also called a *binomial coefficient*. Define

$$\binom{n}{k} = 0 \quad \text{if } k < 0 \quad \text{or} \quad k > n. \tag{10}$$

////

Remark

$$\binom{n}{0} = \binom{n}{n} = 1.$$

$$\binom{n}{k} = \binom{n}{n-k}.$$

$$\binom{n+1}{k} = \binom{n}{k} + \binom{n}{k-1} \qquad \text{for } n = 1, 2, \ldots \text{ and } k = 0, \pm 1, \pm 2, \ldots. \tag{11}$$

Equation (11) is a useful recurrent formula that is easily proved. ////

Both $(n)_k$ and the combinatorial symbol $\binom{n}{k}$ can be generalized from a positive integer n to *any real number* t by defining

$$(t)_k = t(t-1)\cdot \cdots \cdot (t-k+1), \quad \binom{t}{k} = \frac{t(t-1)\cdot \cdots \cdot (t-k+1)}{k!} \qquad \text{for } k = 1, 2, \ldots, \tag{12}$$

and $\binom{t}{k} = 1$ for $k = 0$.

Remark

$$\binom{-n}{k} = \frac{(-n)(-n-1)\cdot\cdots\cdot(-n-k+1)}{k!}$$

$$= (-1)^k \frac{n(n+1)\cdot\cdots\cdot(n+k-1)}{k!}$$

$$= (-1)^k \binom{n+k-1}{k}. \qquad ////$$

2.3 Stirling's Formula

In finding numerical values of probabilities, one is often confronted with the evaluation of long factorial expressions which can be troublesome to compute by direct multiplication. Much labor may be saved by using *Stirling's formula,* which gives an approximate value of $n!$. Stirling's formula is

$$n! \approx (2\pi)^{\frac{1}{2}} e^{-n} n^{n+\frac{1}{2}} \tag{13}$$

or

$$n! = (2\pi)^{\frac{1}{2}} e^{-n} n^{n+\frac{1}{2}} e^{r(n)/12n}, \tag{14}$$

where $1 - 1/(12n+1) < r(n) < 1$. To indicate the accuracy of Stirling's formula, 10! was evaluated using five-place logarithms and Eq. (13), and 3,599,000 was obtained. The actual value of 10! is 3,628,800. The percent error is less than 1 percent, and the percent error will decrease as n increases.

2.4 The Binomial and Multinomial Theorems

The *binomial theorem* is often given as

$$(a+b)^n = \sum_{j=0}^{n} \binom{n}{j} a^j b^{n-j} \tag{15}$$

for n, a positive integer. The binomial theorem explains why the $\binom{n}{j}$ are sometimes called binomial coefficients. Four special cases are noted in the following remark.

Remark

$$(1+t)^n = \sum_{j=0}^{n} \binom{n}{j} t^j, \tag{16}$$

$$(1-t)^n = \sum_{j=0}^{n} \binom{n}{j} (-1)^j t^j, \tag{17}$$

$$2^n = \sum_{j=0}^{n} \binom{n}{j}, \tag{18}$$

and

$$0 = \sum_{j=0}^{n} (-1)^j \binom{n}{j}. \tag{19}$$

////

Expanding both sides of

$$(1+x)^a(1+x)^b = (1+x)^{a+b}$$

and then equating coefficients of x to the nth power gives

$$\sum_{j=0}^{n} \binom{a}{j}\binom{b}{n-j} = \binom{a+b}{n}, \tag{20}$$

a formula that is particularly useful in considerations of the hypergeometric distribution.

A generalization of the binomial theorem is the *multinomial theorem*, which is

$$\left(\sum_{j=1}^{k} a_j\right)^n = \sum \frac{n!}{\prod_{i=1}^{k} n_i!} \prod_{i=1}^{k} a_i^{n_i}, \tag{21}$$

where the summation is over all nonnegative integers $n_1, n_2, \ldots, n_k$ which sum to n. A special case is

$$\left(\sum_{j=1}^{k} a_j\right)^2 = \left(\sum_{i=1}^{k} a_i\right)\left(\sum_{j=1}^{k} a_j\right) = \sum_{i=1}^{k}\sum_{j=1}^{k} a_i a_j. \tag{22}$$

Also note that

$$\left(\sum_{i=1}^{m} a_i\right)\left(\sum_{j=1}^{n} b_j\right) = \sum_{i=1}^{m}\sum_{j=1}^{n} a_i b_j. \tag{23}$$

3 CALCULUS

3.1 Preliminaries

It is assumed that the reader is familiar with the concepts of limits, continuity, differentiation, integration, and infinite series. A particular limit that is referred to several times in the book is the limit expression for the number e; that is,

$$\lim_{x \to 0} (1+x)^{1/x} = e. \tag{24}$$

Equation (24) can be derived by taking logarithms and utilizing l'Hospital's rule, which is reviewed below. There are a number of variations of Eq. (24), for instance,

$$\lim_{x \to \infty} (1 + x^{-1})^x = e \tag{25}$$

and

$$\lim_{x \to 0} (1 + \lambda x)^{1/x} = e^{\lambda} \qquad \text{for constant } \lambda. \tag{26}$$

A rule that is often useful in finding limits is the following so-called *l'Hospital's rule:* If $f(\cdot)$ and $g(\cdot)$ are functions for which $\lim_{x \to a} f(x) = \lim_{x \to a} g(x) = 0$ and if

$$\lim_{x \to a} \frac{f'(x)}{g'(x)}$$

exists, then so does

$$\lim_{x \to a} \frac{f(x)}{g(x)},$$

and

$$\lim_{x \to a} \frac{f(x)}{g(x)} = \lim_{x \to a} \frac{f'(x)}{g'(x)}.$$

EXAMPLE 2 Find $\lim_{x \to 0} [(1/x) \log_e (1 + x)]$. Let $f(x) = \log_e (1 + x)$ and $g(x) = x$; then

$$\lim_{x \to 0} \frac{f'(x)}{g'(x)} = \lim_{x \to 0} \frac{1}{1 + x} = 1 = \lim_{x \to 0} \left[\frac{1}{x} \log_e (1 + x)\right]. \qquad ////$$

Another rule that we use in the book is *Leibniz' rule for differentiating an integral:* Let

$$I(t) = \int_{g(t)}^{h(t)} f(x; t)\, dx,$$

where $f(\cdot\,;\cdot)$, $g(\cdot)$, and $h(\cdot)$ are assumed differentiable. Then

$$\frac{dI}{dt} = \int_{g(t)}^{h(t)} \frac{\partial f}{\partial t}\, dx + f(h(t); t) \frac{dh}{dt} - f(g(t); t) \frac{dg}{dt}. \tag{27}$$

Several important special cases derive from Leibniz' rule; for example, if the integrand $f(x; t)$ does not depend on t, then

$$\frac{d}{dt}\left[\int_{g(t)}^{h(t)} f(x)\, dx\right] = f(h(t)) \frac{dh}{dt} - f(g(t)) \frac{dg}{dt}; \tag{28}$$

in particular, if $g(t)$ is constant and $h(t) = t$, Eq. (28) simplifies to

$$\frac{d}{dt}\left[\int_{c}^{t} f(x)\, dx\right] = f(t). \tag{29}$$

3.2 Taylor Series

The *Taylor series* for $f(x)$ about $x = a$ is defined as

$$f(x) = f(a) + f^{(1)}(a)(x-a) + \frac{f^{(2)}(a)(x-a)^2}{2!} + \cdots + \frac{f^{(n)}(a)(x-a)^n}{n!} + R_n, \qquad (30)$$

where

$$f^{(i)}(a) = \left.\frac{d^i f(x)}{dx^i}\right|_{x=a}; \quad R_n = \frac{f^{(n+1)}(c)(x-a)^{n+1}}{(n+1)!} \qquad \text{and} \qquad a \le c \le x.$$

R_n is called the *remainder*. $f(x)$ is assumed to have derivatives of at least order $n+1$. If the remainder is not too large, Eq. (30) gives a polynomial (of degree n) approximation, when R_n is dropped, of the function $f(\cdot)$. The infinite series corresponding to Eq. (30) will converge in some interval if $\lim_{n\to\infty} R_n = 0$ in this interval. Several important infinite Taylor series, along with their intervals of convergence, are given in the following examples.

EXAMPLE 3 Suppose $f(x) = e^x$ and $a = 0$. Then

$$e^x = 1 + x + \frac{x^2}{2!} + \frac{x^3}{3!} + \cdots$$

$$= \sum_{j=0}^{\infty} \frac{x^j}{j!} \qquad \text{for } -\infty < x < \infty. \qquad (31)$$

////

EXAMPLE 4 Suppose $f(x) = (1-x)^t$ and $a = 0$; then $f^{(1)}(x) = -t(1-x)^{t-1}$, $f^{(2)}(x) = t(t-1)(1-x)^{t-2}, \ldots, f^{(j)}(x) = (-1)^j t(t-1) \cdot \cdots \cdot (t-j+1)(1-x)^{t-j}$, and hence

$$f(x) = (1-x)^t = \sum_{j=0}^{\infty} (-1)^j (t)_j \frac{x^j}{j!}$$

$$= \sum_{j=0}^{\infty} \binom{t}{j} (-x)^j \qquad \text{for } -1 < x < 1. \qquad (32)$$

////

There are several interesting special cases of Eq. (32). $t = -n$ gives

$$(1-x)^{-n} = \sum_{j=0}^{\infty} \binom{-n}{j} (-x)^j = \sum_{j=0}^{\infty} \binom{n+j-1}{j} x^j \qquad \text{for } -1 < x < 1; \qquad (33)$$

$t = -1$ gives the geometric series

$$(1-x)^{-1} = \sum_{j=0}^{\infty} x^j; \qquad (34)$$

$t = -2$ gives

$$(1 - x)^{-2} = \sum_{j=0}^{\infty} (j + 1)x^j. \tag{35}$$

EXAMPLE 5 Suppose $f(x) = \log_e (1 + x)$ and $a = 0$; then

$$\log_e (1 + x) = x - \frac{x^2}{2} + \frac{x^3}{3} - \frac{x^4}{4} + \cdots \qquad \text{for } -1 < x \leq 1. \tag{36}$$

////

The Taylor series for functions of one variable given in Eq. (30) can be generalized to the Taylor series for functions of several variables. For example, the Taylor series for $f(x, y)$ about $x = a$ and $y = b$ can be written as

$$f(x, y) = f(a, b) + f_x(a, b)(x - a) + f_y(a, b)(y - b) +$$
$$\frac{1}{2!} [f_{xx}(a, b)(x - a)^2 + 2f_{xy}(a, b)(x - a)(y - b) + f_{yy}(a, b)(y - b)^2] + \cdots,$$

where

$$f_x(a, b) = \left. \frac{\partial f}{\partial x} \right|_{x=a, y=b},$$

$$f_{xy}(a, b) = \left. \frac{\partial^2 f}{\partial y \, \partial x} \right|_{x=a, y=b},$$

and similarly for the others.

3.3 The Gamma and Beta Functions

The *gamma function*, denoted by $\Gamma(\cdot)$, is defined by

$$\Gamma(t) = \int_0^{\infty} x^{t-1} e^{-x} \, dx \qquad \text{for } t > 0. \tag{37}$$

$\Gamma(t)$ is nothing more than a notation for the definite integral that appears on the right-hand side of Eq. (37). Integration by parts yields

$$\Gamma(t + 1) = t\Gamma(t), \tag{38}$$

and, hence, if $t = n$ (an integer),

$$\Gamma(n + 1) = n!. \tag{39}$$

If n is an integer,

$$\Gamma(n + \tfrac{1}{2}) = \frac{1 \cdot 3 \cdot 5 \cdot \cdots \cdot (2n - 1)}{2^n} \sqrt{\pi}, \tag{40}$$

and, in particular,

$$\Gamma(\tfrac{1}{2}) = 2\Gamma(\tfrac{3}{2}) = \sqrt{\pi}. \tag{41}$$

The *beta function*, denoted by $B(\cdot, \cdot)$, is defined by

$$B(a, b) = \int_0^1 x^{a-1}(1-x)^{b-1}\,dx \qquad \text{for } a > 0,\, b > 0. \tag{42}$$

Again, $B(a, b)$ is just a notation for the definite integral that appears on the right-hand side of Eq. (42). A simple variable substitution gives $B(a, b) = B(b, a)$. The beta function is related to the gamma function according to the following formula:

$$B(a, b) = \frac{\Gamma(a)\Gamma(b)}{\Gamma(a+b)}. \tag{43}$$

APPENDIX B

TABULAR SUMMARY OF PARAMETRIC FAMILIES OF DISTRIBUTIONS

1 INTRODUCTION

The purpose of this appendix is to provide the reader with a convenient reference to the parametric families of distributions that were introduced in Chap. III. Given are two tables, one for discrete distributions and the other for continuous distributions.

Table 1 DISCRETE DISTRIBUTIONS

Name of parametric family of distributions	Discrete density functions $f(\cdot)$	Parameter space	Mean $\mu = \mathscr{E}[X]$
Discrete uniform	$f(x) = \frac{1}{N} I_{\{1, \ldots, N\}}(x)$	$N = 1, 2, \ldots$	$\frac{N+1}{2}$
Bernoulli	$f(x) = p^x q^{1-x} I_{\{0, 1\}}(x)$	$0 \le p \le 1$ $(q = 1 - p)$	p
Binomial	$f(x) = \binom{n}{x} p^x q^{n-x} I_{\{0, 1, \ldots, n\}}(x)$	$0 \le p \le 1$ $n = 1, 2, 3, \ldots$ $(q = 1 - p)$	np
Hypergeometric	$f(x) = \frac{\binom{K}{x}\binom{M-K}{n-x}}{\binom{M}{n}} I_{\{0, 1, \ldots, n\}}(x)$	$M = 1, 2, \ldots$ $K = 0, 1, \ldots, M$ $n = 1, 2, \ldots, M$	$n\frac{K}{M}$
Poisson	$f(x) = \frac{e^{-\lambda}\lambda^x}{x!} I_{\{0, 1, \ldots\}}(x)$	$\lambda > 0$	λ
Geometric	$f(x) = pq^x I_{\{0, 1, \ldots\}}(x)$	$0 < p \le 1$ $(q = 1 - p)$	$\frac{q}{p}$
Negative binomial	$f(x) = \binom{r+x-1}{x} p^r q^x I_{\{0, 1, \ldots\}}(x)$	$0 < p \le 1$ $r > 0$ $(q = 1 - p)$	$\frac{rq}{p}$

Variance $\sigma^2 = \mathscr{E}[(X-\mu)^2]$	Moments $\mu'_r = \mathscr{E}[X^r]$ or $\mu_r = \mathscr{E}[(X-\mu)^r]$ and/or cumulants κ_r	Moment generating function $\mathscr{E}[e^{tX}]$
$\frac{N^2-1}{12}$	$\mu'_3 = \frac{N(N+1)^2}{4}$ $\mu'_4 = \frac{(N+1)(2N+1)(3N^2+3N-1)}{30}$	$\sum_{j=1}^{N} \frac{1}{N} e^{jt}$
pq	$\mu'_r = p$ for all r	$q + pe^t$
npq	$\mu_3 = npq(q-p)$ $\mu_4 = 3n^2p^2q^2 + npq(1-6pq)$	$(q+pe^t)^n$
$n\frac{K}{M}\frac{M-K}{M}\frac{M-n}{M-1}$	$\mathscr{E}[X(X-1)\cdots(X-r+1)] = r!\frac{\binom{K}{r}\binom{n}{r}}{\binom{M}{r}}$	not useful
λ	$\kappa_r = \lambda$ for $r = 1, 2, \ldots$ $\mu_3 = \lambda$ $\mu_4 = \lambda + 3\lambda^2$	$\exp[\lambda(e^t - 1)]$
$\frac{q}{p^2}$	$\mu_3 = \frac{q+q^2}{p^2}$ $\mu_4 = \frac{q+7q^2+q^3}{p^4}$	$\frac{p}{1-qe^t}$
$\frac{rq}{p^2}$	$\mu_3 = \frac{r(q+q^2)}{p^3}$ $\mu_4 = \frac{r[q+(3r+4)q^2+q^3]}{p^4}$	$\left(\frac{p}{1-qe^t}\right)^r$

Table 2 CONTINUOUS DISTRIBUTIONS

Name of parametric family of distributions	Cumulative distribution function $F(\cdot)$ or probability density function $f(\cdot)$	Parameter space	Mean $\mu = \mathscr{E}[X]$
Uniform or rectangular	$f(x) = \frac{1}{b-a} I_{[a,b]}(x)$	$-\infty < a < b < \infty$	$\frac{a+b}{2}$
Normal	$f(x) = \frac{1}{\sqrt{2\pi}\sigma} \exp[-(x-\mu)^2/2\sigma^2]$	$-\infty < \mu < \infty$ $\sigma > 0$	μ
Exponential	$f(x) = \lambda e^{-\lambda x} I_{(0,\infty)}(x)$	$\lambda > 0$	$\frac{1}{\lambda}$
Gamma	$f(x) = \frac{\lambda^r}{\Gamma(r)} x^{r-1} e^{-\lambda x} I_{(0,\infty)}(x)$	$\lambda > 0$ $r > 0$	$\frac{r}{\lambda}$
Beta	$f(x) = \frac{1}{\mathrm{B}(a,b)} x^{a-1}(1-x)^{b-1} I_{(0,1)}(x)$	$a > 0$ $b > 0$	$\frac{a}{a+b}$
Cauchy	$f(x) = \frac{1}{\pi\beta\{1 + [(x-\alpha)/\beta]^2\}}$	$-\infty < \alpha < \infty$ $\beta > 0$	Does not exist
Lognormal	$f(x) = \frac{1}{x\sqrt{2\pi}\sigma} \exp[-(\log_e x - \mu)^2/2\sigma^2] I_{(0,\infty)}(x)$	$-\infty < \mu < \infty$ $\sigma > 0$	$\exp[\mu + \frac{1}{2}\sigma^2]$
Double exponential	$f(x) = \frac{1}{2\beta} \exp\left(-\frac{\lvert x-\alpha \rvert}{\beta}\right)$	$-\infty < \alpha < \infty$ $\beta > 0$	α

Variance $\sigma^2 = \mathscr{E}[(X-\mu)^2]$	Moments $\mu_r' = \mathscr{E}[X^r]$ or $\mu_r = \mathscr{E}[(X-\mu)^r]$ and/or cumulants κ_r	Moment generating function $\mathscr{E}[e^{tX}]$
$\frac{(b-a)^2}{12}$	$\mu_r = 0$ for r odd $\mu_r = \frac{(b-a)^r}{2^r(r+1)}$ for r even	$\frac{e^{bt}-e^{at}}{(b-a)t}$
σ^2	$\mu_r = 0$, r odd; $\mu_r = \frac{r!}{(r/2)!}\frac{\sigma^r}{2^{r/2}}$, r even; $\kappa_r = 0$, $r > 2$	$\exp[\mu t + \frac{1}{2}\sigma^2 t^2]$
$\frac{1}{\lambda^2}$	$\mu_r' = \frac{\Gamma(r+1)}{\lambda^r}$	$\frac{\lambda}{\lambda - t}$ for $t < \lambda$
$\frac{r}{\lambda^2}$	$\mu_j' = \frac{\Gamma(r+j)}{\lambda^j \Gamma(r)}$	$\left(\frac{\lambda}{\lambda - t}\right)^r$ for $t < \lambda$
$\frac{ab}{(a+b+1)(a+b)^2}$	$\mu_r = \frac{B(r+a, b)}{B(a, b)}$	not useful
Does not exist	Do not exist	Characteristic function is $e^{i\alpha t - \beta\|t\|}$
$\exp[2\mu + 2\sigma^2]$ $-\exp[2\mu + 2\sigma^2]$	$\mu_r' = \exp[r\mu + \frac{1}{2}r^2\sigma^2]$	not useful
$2\beta^2$	$\mu_r = 0$ for r odd; $\mu_r = r!\,\beta^r$ for r even	$\frac{e^{\alpha t}}{1-(\beta t)^2}$

(continued)

Table 2 CONTINUOUS DISTRIBUTIONS (*continued*)

Name of parametric family of distributions	Cumulative distribution function $F(\cdot)$ or probability density function $f(\cdot)$	Parameter space	Mean $\mu = \mathscr{E}[X]$
Weibull	$f(x) = abx^{b-1}\exp[-ax^b]I_{(0,\infty)}(x)$	$a > 0$ $b > 0$	$a^{-1/b}\Gamma(1 + b$
Logistic	$F(x) = [1 + e^{-(x-\alpha)/\beta}]^{-1}$	$-\infty < \alpha < \infty$ $\beta > 0$	α
Pareto	$f(x) = \dfrac{\theta x_0^\theta}{x^{\theta+1}} I_{(x_0,\infty)}(x)$	$x_0 > 0$ $\theta > 0$	$\dfrac{\theta x_0}{\theta - 1}$ for $\theta > 1$
Gumbel or extreme value	$F(x) = \exp(-e^{-(x-\alpha)/\beta})$	$-\infty < \alpha < \infty$ $\beta > 0$	$\alpha + \beta\gamma$, $\gamma \approx .577216$
t distribution	$f(x) = \dfrac{\Gamma[(k+1)/2]}{\Gamma(k/2)} \dfrac{1}{\sqrt{k\pi}} \dfrac{1}{(1 + x^2/k)^{(k+1)/2}}$	$k > 0$	$\mu = 0$ for $k > 1$
F distribution	$f(x) = \dfrac{\Gamma[(m+n)/2]}{\Gamma(m/2)\Gamma(n/2)} \left(\dfrac{m}{n}\right)^{m/2} \times \dfrac{x^{(m-2)/2}}{[1 + (m/n)x]^{(m+n)/2}} I_{(0,\infty)}(x)$	$m, n = 1, 2, \ldots$	$\dfrac{n}{n-2}$ for $n > 2$
Chi-square distribution	$f(x) = \dfrac{1}{\Gamma(k/2)} \left(\dfrac{1}{2}\right)^{k/2} x^{k/2-1} e^{-(1/2)x} I_{(0,\infty)}(x)$	$k = 1, 2, \ldots$	k

Variance $\sigma^2 = \mathscr{E}[(X-\mu)^2]$	Moments $\mu'_r = \mathscr{E}[X^r]$ or $\mu_r = \mathscr{E}[(X-\mu)^r]$ and/or cumulants κ_r	Moment generating function $\mathscr{E}[e^{tX}]$
$a^{-2/b}[\Gamma(1+2b^{-1}) - \Gamma^2(1+b^{-1})]$	$\mu'_r = a^{-r/b}\Gamma\left(1+\frac{r}{b}\right)$	$\mathscr{E}[X^t] = a^{-t/b}\Gamma\left(1+\frac{t}{b}\right)$
$\frac{\beta^2\pi^2}{3}$		$e^{\alpha t}\pi\beta t \csc(\pi\beta t)$
$\frac{\theta x_0^2}{(\theta-1)^2(\theta-2)}$ for $\theta > 2$	$\mu'_r = \frac{\theta x_0^r}{\theta - r}$ for $\theta > r$	does not exist
$\frac{\pi^2\beta^2}{6}$	$\kappa_r = (-\beta)^r\psi^{(r-1)}(1)$ for $r \geq 2$, where $\psi(\cdot)$ is digamma function	$e^{\alpha t}\Gamma(1-\beta t)$ for $t < 1/\beta$
$\frac{k}{k-2}$ for $k > 2$	$\mu_r = 0$ for $k > r$ and r odd; $\mu_r = \frac{k^{r/2}\mathrm{B}((r+1)/2, (k-r)/2)}{\mathrm{B}(\frac{1}{2}, k/2)}$ for $k > r$ and r even	does not exist
$\frac{2n^2(m+n-2)}{m(n-2)^2(n-4)}$ for $n > 4$	$\mu'_r = \left(\frac{n}{m}\right)^r \frac{\Gamma(m/2+r)\Gamma(n/2-r)}{\Gamma(m/2)\Gamma(n/2)}$ for $r < \frac{n}{2}$	does not exist
$2k$	$\mu'_j = \frac{2^j\Gamma(k/2+j)}{\Gamma(k/2)}$	$\left(\frac{1}{1-2t}\right)^{k/2}$ for $t < 1/2$

APPENDIX C
REFERENCES AND RELATED READING

MATHEMATICS BOOKS

1. PROTTER and MORREY: "Calculus with Analytic Geometry: A Second Course," Addison-Wesley Publishing Company, Inc., Reading, Mass., 1971.
2. THOMAS: "Calculus and Analytic Geometry," alternate ed., Addison-Wesley Publishing Company, Inc., Reading, Mass., 1972.
3. WIDDER: "Advanced Calculus," 2d ed., Prentice-Hall, Inc., Englewood Cliffs, N.J., 1961.
4. WYLIE: "Advanced Engineering Mathematics," 3d ed., McGraw-Hill Book Company, New York, 1966.

PROBABILITY BOOKS

5. ASH: "Basic Probability Theory," John Wiley & Sons, Inc., New York, 1970.
6. DRAKE: "Fundamentals of Applied Probability Theory," McGraw-Hill Book Company, New York, 1967.
7. DWASS: "Probability: Theory and Applications," W. A. Benjamin, Inc., New York, 1970.

8. FELLER: "An Introduction to Probability Theory and Its Applications," Vol. 1, 3d ed., John Wiley & Sons, Inc., New York, 1968.
9. FELLER: "An Introduction to Probability Theory and Its Applications," Vol. 2, John Wiley & Sons, Inc., New York, 1966.
10. PARZEN: "Modern Probability Theory and Its Applications," John Wiley & Sons, Inc., New York, 1960.

PROBABILITY AND STATISTICS BOOKS

Advanced (more advanced than MGB)

11. CRAMÉR: "Mathematical Methods of Statistics," Princeton University Press, Princeton, N.J., 1946.
12. FERGUSON: "Mathematical Statistics, A Decision Theoretic Approach," Academic Press, Inc., New York, 1967.
13. KENDALL and STUART: "The Advanced Theory of Statistics, Vol. 1, Distribution Theory," 2d ed., Hafner Publishing Company, Inc., New York, 1963.
14. KENDALL and STUART: "The Advanced Theory of Statistics, Vol. 2, Inference and Relationship," Hafner Publishing Company, Inc., New York, 1961.
15. KENDALL and STUART: "The Advanced Theory of Statistics, Vol. 3, Design and Analysis, and Time Series," Hafner Publishing Company, Inc., New York, 1966.
16. LEHMANN: "Testing Statistical Hypotheses," John Wiley & Sons, Inc., New York, 1959.
17. RAO: "Linear Statistical Inference and Its Applications," John Wiley & Sons, Inc., New York, 1965.
18. WILKS: "Mathematical Statistics," John Wiley & Sons, Inc., New York, 1962.
19. ZACKS: "The Theory of Statistical Inference," John Wiley & Sons, Inc., New York, 1971.

Intermediate (about the same level as MGB)

20. BRUNK: "An Introduction to Mathematical Statistics," 2d ed., Blaisdell Publishing Company, a division of Ginn and Company, Waltham, Mass., 1965.
21. DWASS: "Probability and Statistics," W. A. Benjamin, Inc., New York, 1970.
22. HOGG and CRAIG: "Introduction to Mathematical Statistics," 3d ed., The Macmillan Co. of Canada, Limited, Toronto, 1970.
23. LINDGREN: "Statistical Theory," 2d ed., The Macmillan Company, New York, 1968.
24. TUCKER: "An Introduction to Probability and Mathematical Statistics," Academic Press, Inc., New York, 1962.

Elementary (less advanced than MGB, but calculus prerequisite)

25. FREUND: "Mathematical Statistics," 2d ed., Prentice-Hall, Inc., Englewood Cliffs, N.J., 1971.
26. GUTTMAN, WILKS, and HUNTER: "Introductory Engineering Statistics," 2d ed., John Wiley & Sons, Inc., New York, 1971.
27. HOEL: "Introduction to Mathematical Statistics," 3d ed., John Wiley & Sons, Inc., New York, 1971.
28. HOEL, PORT, and STONE: "Introduction to Statistical Theory," Houghton Mifflin Company, Boston, 1971.

SPECIAL BOOKS

29. DANIEL and WOOD: "Fitting Equations to Data; Computer Analysis of Multifactor Data for Scientists and Engineers," Interscience Publishers, a division of John Wiley & Sons, Inc., New York, 1971.
30. DAVID: "Order Statistics," John Wiley & Sons, Inc., New York, 1970.
31. DRAPER and SMITH: "Applied Regression Analysis," John Wiley & Sons, Inc., New York, 1966.
32. GIBBONS: "Nonparametric Statistical Inference," McGraw-Hill Book Company, New York, 1971.
33. GRAYBILL: "An Introduction to Linear Statistical Models," Vol. 1, McGraw-Hill Book Company, New York, 1961.
34. JOHNSON and KOTZ: "Discrete Distributions," Houghton Mifflin Company, Boston, 1969.
35. JOHNSON and KOTZ: "Continuous Univariate Distributions—1," Houghton Mifflin Company, Boston, 1970.
36. JOHNSON and KOTZ: "Continuous Univariate Distributions—2," Houghton Mifflin Company, Boston, 1970.
37. KEMPTHORNE and FOLKS: "Probability, Statistics, and Data Analysis," The Iowa State University Press, Ames, 1971.
38. MORRISON: "Multivariate Statistical Methods," McGraw-Hill Book Company, New York, 1967.
39. RAJ: "Sampling Theory," McGraw-Hill Book Company, New York, 1968.

PAPERS

40. JOINER and ROSENBLATT: "Some Properties of the Range in Samples from Tukey's Symmetric Lambda Distributions," *Journal of the American Statistical Association*, Vol. 66 (1971), 394–399.

41. PITMAN: "The Estimation of the Location and Scale Parameters of a Continuous Population of Any Given Form," *Biometrika*, Vol. 30 (1939), pp. 391–421.
42. WOLFOWITZ: "The Minimum Distance Method," *Annals of Mathematical Statistics*, Vol. 28(1) (1957), pp. 75–88.
43. ZEHNA: "Invariance of Maximum Likelihood Estimation," *Annals of Mathematical Statistics*, Vol. 37 (1966), p. 744.

BOOKS OF TABLES

44. "Handbook of Tables for Probability and Statistics," Chemical Rubber Company, Cleveland, 1966.
45. MOLINA: "Poisson's Exponential Binomial Limit," D. Van Nostrand Company, Inc., Princeton, N.J., 1942.
46. OWEN: "Handbook of Statistical Tables," Addison-Wesley Publishing Company, Inc., Reading, Mass., 1962.

APPENDIX D

TABLES

1 DESCRIPTION OF TABLES

Table 1 Ordinates of the Normal Density Function

This table gives values of

$$\phi(x) = \frac{1}{\sqrt{2\pi}} e^{-x^2/2}$$

for values of x between 0 and 4 at intervals of .01. For negative values of x one uses the fact that $\phi(-x) = \phi(x)$.

Table 2 Cumulative Normal Distribution

This table gives values of

$$\Phi(x) = \int_{-\infty}^{x} \frac{1}{\sqrt{2\pi}} e^{-t^2/2}\, dt = \int_{-\infty}^{x} \phi(t)\, dt$$

for values of x beteenn 0 and 3.5 at intervals of .01. For negative values of x, one uses the relation $\Phi(-x) = 1 - \Phi(x)$. Values of x corresponding to a few special values of Φ are given separately beneath the main table.

Table 3 Cumulative Chi-Square Distribution

This table gives values of u corresponding to a few selected values of $F(u)$, where

$$F(u) = \int_0^u \frac{x^{(n-2)/2} e^{-x/2}\, dx}{2^{n/2}\Gamma(n/2)}$$

for n, the number of degrees of freedom, equal to 1, 2, ..., 30. For larger values of n, a normal approximation is quite accurate. The quantity $\sqrt{2u} - \sqrt{2n-1}$ is nearly normally distributed with mean 0 and unit variance. Thus u_α, the αth quantile point of the distribution, may be computed by

$$u_\alpha = \tfrac{1}{2}(z_\alpha + \sqrt{2n-1})^2,$$

where z_α is the αth quantile point of the cumulative normal distribution. As an illustration, we may compute the .95 value of u for $n = 30$ degrees of freedom:

$$u_{.95} = \tfrac{1}{2}(1.645 + \sqrt{59})^2$$
$$= 43.5,$$

which is in error by less than 1 percent.

Table 4 Cumulative F Distribution

This table gives values of F corresponding to five values of

$$G(F) = \int_0^F \frac{\Gamma\left(\frac{m+n}{2}\right) m^{m/2} n^{n/2} x^{(m-2)/2} (n + mx)^{-(m+n)/2}\, dx}{\Gamma\left(\frac{m}{2}\right)\Gamma\left(\frac{n}{2}\right)}$$

for selected values of m and n; m is the number of degrees of freedom in the numerator of F, and n is the number of degrees of freedom in the denominator of F. The table also provides values corresponding to $G = .10$, .05, .025, .01, and .005 because $F_{1-\alpha}$ for m and n degrees of freedom is the reciprocal of F_α for n and m degrees of freedom. Thus for $G = .05$ with three and six degrees of freedom, one finds

$$F_{.05}(3, 6) = \frac{1}{F_{.95}(6, 3)} = \frac{1}{8.94} = .112$$

One should interpolate on the reciprocals of m and n as in Table **5** for good accuracy.

Table 5 Cumulative Students t Distribution

This table gives values of t corresponding to a few selected values of

$$F(t) = \int_{-\infty}^{t} \frac{\Gamma\left(\frac{n+1}{2}\right)}{\Gamma\left(\frac{n}{2}\right)\sqrt{\pi n}\left(1+\frac{x^2}{n}\right)^{(n+1)/2}}\,dx$$

with $n = 1, 2, \ldots, 30, 40, 60, 120, \infty$. Since the density is symmetrical in t, it follows that $F(-t) = 1 - F(t)$. One should not interpolate linearly between degrees of freedom but on the reciprocal of the degrees of freedom, if good accuracy in the last digit is desired. As an illustration, we shall compute the .975th quantile point for 40 degrees of freedom. The values for 30 and 60 are 2.042 and 2.000. Using the reciprocals of n, the interpolated value is

$$2.042 - \frac{\frac{1}{30} - \frac{1}{40}}{\frac{1}{30} - \frac{1}{60}}(2.042 - 2.000) = 2.021,$$

which is the correct value. Interpolating linearly, one would have obtained 2.028.

Table 1 ORDINATES OF THE NORMAL DENSITY FUNCTION

$$\phi(x) = \frac{1}{\sqrt{2\pi}} e^{-x^2/2}$$

x	.00	.01	.02	.03	.04	.05	.06	.07	.08	.09
.0	.3989	.3989	.3989	.3988	.3986	.3984	.3982	.3980	.3977	.3973
.1	.3970	.3965	.3961	.3956	.3951	.3945	.3939	.3932	.3925	.3918
.2	.3910	.3902	.3894	.3885	.3876	.3867	.3857	.3847	.3836	.3825
.3	.3814	.3802	.3790	.3778	.3765	.3752	.3739	.3725	.3712	.3697
.4	.3683	.3668	.3653	.3637	.3621	.3605	.3589	.3572	.3555	.3538
.5	.3521	.3503	.3485	.3467	.3448	.3429	.3410	.3391	.3372	.3352
.6	.3332	.3312	.3292	.3271	.3251	.3230	.3209	.3187	.3166	.3144
.7	.3123	.3101	.3079	.3056	.3034	.3011	.2989	.2966	.2943	.2920
.8	.2897	.2874	.2850	.2827	.2803	.2780	.2756	.2732	.2709	.2685
.9	.2661	.2637	.2613	.2589	.2565	.2541	.2516	.2492	.2468	.2444
1.0	.2420	.2396	.2371	.2347	.2323	.2299	.2275	.2251	.2227	.2203
1.1	.2179	.2155	.2131	.2107	.2083	.2059	.2036	.2012	.1989	.1965
1.2	.1942	.1919	.1895	.1872	.1849	.1826	.1804	.1781	.1758	.1736
1.3	.1714	.1691	.1669	.1647	.1626	.1604	.1582	.1561	.1539	.1518
1.4	.1497	.1476	.1456	.1435	.1415	.1394	.1374	.1354	.1334	.1315
1.5	.1295	.1276	.1257	.1238	.1219	.1200	.1182	.1163	.1145	.1127
1.6	.1109	.1092	.1074	.1057	.1040	.1023	.1006	.0989	.0973	.0957
1.7	.0940	.0925	.0909	.0893	.0878	.0863	.0848	.0833	.0818	.0804
1.8	.0790	.0775	.0761	.0748	.0734	.0721	.0707	.0694	.0681	.0669
1.9	.0656	.0644	.0632	.0620	.0608	.0596	.0584	.0573	.0562	.0551
2.0	.0540	.0529	.0519	.0508	.0498	.0488	.0478	.0468	.0459	.0449
2.1	.0440	.0431	.0422	.0413	.0404	.0396	.0387	.0379	.0371	.0363
2.2	.0355	.0347	.0339	.0332	.0325	.0317	.0310	.0303	.0297	.0290
2.3	.0283	.0277	.0270	.0264	.0258	.0252	.0246	.0241	.0235	.0229
2.4	.0224	.0219	.0213	.0208	.0203	.0198	.0194	.0189	.0184	.0180
2.5	.0175	.0171	.0167	.0163	.0158	.0154	.0151	.0147	.0143	.0139
2.6	.0136	.0132	.0129	.0126	.0122	.0119	.0116	.0113	.0110	.0107
2.7	.0104	.0101	.0099	.0096	.0093	.0091	.0088	.0086	.0084	.0081
2.8	.0079	.0077	.0075	.0073	.0071	.0069	.0067	.0065	.0063	.0061
2.9	.0060	.0058	.0056	.0055	.0053	.0051	.0050	.0048	.0047	.0046
3.0	.0044	.0043	.0042	.0040	.0039	.0038	.0037	.0036	.0035	.0034
3.1	.0033	.0032	.0031	.0030	.0029	.0028	.0027	.0026	.0025	.0025
3.2	.0024	.0023	.0022	.0022	.0021	.0020	.0020	.0019	.0018	.0018
3.3	.0017	.0017	.0016	.0016	.0015	.0015	.0014	.0014	.0013	.0013
3.4	.0012	.0012	.0012	.0011	.0011	.0010	.0010	.0010	.0009	.0009
3.5	.0009	.0008	.0008	.0008	.0008	.0007	.0007	.0007	.0007	.0006
3.6	.0006	.0006	.0006	.0005	.0005	.0005	.0005	.0005	.0005	.0004
3.7	.0004	.0004	.0004	.0004	.0004	.0004	.0003	.0003	.0003	.0003
3.8	.0003	.0003	.0003	.0003	.0003	.0002	.0002	.0002	.0002	.0002
3.9	.0002	.0002	.0002	.0002	.0002	.0002	.0002	.0002	.0001	.0001

Table 2 CUMULATIVE NORMAL DISTRIBUTION

$$\Phi(x) = \int_{-\infty}^{x} \frac{1}{\sqrt{2\pi}} e^{-t^2/2}\, dt$$

x	.00	.01	.02	.03	.04	.05	.06	.07	.08	.09
.0	.5000	.5040	.5080	.5120	.5160	.5199	.5239	.5279	.5319	.5359
.1	.5398	.5438	.5478	.5517	.5557	.5596	.5636	.5675	.5714	.5753
.2	.5793	.5832	.5871	.5910	.5948	.5987	.6026	.6064	.6103	.6141
.3	.6179	.6217	.6255	.6293	.6331	.6368	.6406	.6443	.6480	.6517
.4	.6554	.6591	.6628	.6664	.6700	.6736	.6772	.6808	.6844	.6879
.5	.6915	.6950	.6985	.7019	.7054	.7088	.7123	.7157	.7190	.7224
.6	.7257	.7291	.7324	.7357	.7389	.7422	.7454	.7486	.7517	.7549
.7	.7580	.7611	.7642	.7673	.7704	.7734	.7764	.7794	.7823	.7852
.8	.7881	.7910	.7939	.7967	.7995	.8023	.8051	.8078	.8106	.8133
.9	.8159	.8186	.8212	.8238	.8264	.8289	.8315	.8340	.8365	.8389
1.0	.8413	.8438	.8461	.8485	.8508	.8531	.8554	.8577	.8599	.8621
1.1	.8643	.8665	.8686	.8708	.8729	.8749	.8770	.8790	.8810	.8830
1.2	.8849	.8869	.8888	.8907	.8925	.8944	.8962	.8980	.8997	.9015
1.3	.9032	.9049	.9066	.9082	.9099	.9115	.9131	.9147	.9162	.9177
1.4	.9192	.9207	.9222	.9236	.9251	.9265	.9279	.9292	.9306	.9319
1.5	.9332	.9345	.9357	.9370	.9382	.9394	.9406	.9418	.9429	.9441
1.6	.9452	.9463	.9474	.9484	.9495	.9505	.9515	.9525	.9535	.9545
1.7	.9554	.9564	.9573	.9582	.9591	.9599	.9608	.9616	.9625	.9633
1.8	.9641	.9649	.9656	.9664	.9671	.9678	.9686	.9693	.9699	.9706
1.9	.9713	.9719	.9726	.9732	.9738	.9744	.9750	.9756	.9761	.9767
2.0	.9772	.9778	.9783	.9788	.9793	.9798	.9803	.9808	.9812	.9817
2.1	.9821	.9826	.9830	.9834	.9838	.9842	.9846	.9850	.9854	.9857
2.2	.9861	.9864	.9868	.9871	.9875	.9878	.9881	.9884	.9887	.9890
2.3	.9893	.9896	.9898	.9901	.9904	.9906	.9909	.9911	.9913	.9916
2.4	.9918	.9920	.9922	.9925	.9927	.9929	.9931	.9932	.9934	.9936
2.5	.9938	.9940	.9941	.9943	.9945	.9946	.9948	.9949	.9951	.9952
2.6	.9953	.9955	.9956	.9957	.9959	.9960	.9961	.9962	.9963	.9964
2.7	.9965	.9966	.9967	.9968	.9969	.9970	.9971	.9972	.9973	.9974
2.8	.9974	.9975	.9976	.9977	.9977	.9978	.9979	.9979	.9980	.9981
2.9	.9981	.9982	.9982	.9983	.9984	.9984	.9985	.9985	.9986	.9986
3.0	.9987	.9987	.9987	.9988	.9988	.9989	.9989	.9989	.9990	.9990
3.1	.9990	.9991	.9991	.9991	.9992	.9992	.9992	.9992	.9993	.9993
3.2	.9993	.9993	.9994	.9994	.9994	.9994	.9994	.9995	.9995	.9995
3.3	.9995	.9995	.9995	.9996	.9996	.9996	.9996	.9996	.9996	.9997
3.4	.9997	.9997	.9997	.9997	.9997	.9997	.9997	.9997	.9997	.9998

x	1.282	1.645	1.960	2.326	2.576	3.090	3.291	3.891	4.417
$\Phi(x)$	.90	.95	.975	.99	.995	.999	.9995	.99995	.999995
$2[1-\Phi(x)]$	.20	.10	.05	.02	.01	.002	.001	.0001	.00001

Table 3 CUMULATIVE CHI-SQUARE DISTRIBUTION*

$$F(u) = \int_0^u \frac{x^{(n-2)/2} e^{-x/2} \, dx}{2^{n/2}\Gamma(n/2)}$$

n \ F	.005	.010	.025	.050	.100	.250	.500	.750	.900	.950	.975	.990	.995
1	$.0^4393$	$.0^3157$	$.0^3982$	$.0^2393$	.0158	.102	.455	1.32	2.71	3.84	5.02	6.63	7.88
2	.0100	.0201	.0506	.103	.211	.575	1.39	2.77	4.61	5.99	7.38	9.21	10.6
3	.0717	.115	.216	.352	.584	1.21	2.37	4.11	6.25	7.81	9.35	11.3	12.8
4	.207	.297	.484	.711	1.06	1.92	3.36	5.39	7.78	9.49	11.1	13.3	14.9
5	.412	.554	.831	1.15	1.61	2.67	4.35	6.63	9.24	11.1	12.8	15.1	16.7
6	.676	.872	1.24	1.64	2.20	3.45	5.35	7.84	10.6	12.6	14.4	16.8	18.5
7	.989	1.24	1.69	2.17	2.83	4.25	6.35	9.04	12.0	14.1	16.0	18.5	20.3
8	1.34	1.65	2.18	2.73	3.49	5.07	7.34	10.2	13.4	15.5	17.5	20.1	22.0
9	1.73	2.09	2.70	3.33	4.17	5.90	8.34	11.4	14.7	16.9	19.0	21.7	23.6
10	2.16	2.56	3.25	3.94	4.87	6.74	9.34	12.5	16.0	18.3	20.5	23.2	25.2
11	2.60	3.05	3.82	4.57	5.58	7.58	10.3	13.7	17.3	19.7	21.9	24.7	26.8
12	3.07	3.57	4.40	5.23	6.30	8.44	11.3	14.8	18.5	21.0	23.3	26.2	28.3
13	3.57	4.11	5.01	5.89	7.04	9.30	12.3	16.0	19.8	22.4	24.7	27.7	29.8
14	4.07	4.66	5.63	6.57	7.79	10.2	13.3	17.1	21.1	23.7	26.1	29.1	31.3
15	4.60	5.23	6.26	7.26	8.55	11.0	14.3	18.2	22.3	25.0	27.5	30.6	32.8
16	5.14	5.81	6.91	7.96	9.31	11.9	15.3	19.4	23.5	26.3	28.8	32.0	34.3
17	5.70	6.41	7.56	8.67	10.1	12.8	16.3	20.5	24.8	27.6	30.2	33.4	35.7
18	6.26	7.01	8.23	9.39	10.9	13.7	17.3	21.6	26.0	28.9	31.5	34.8	37.2
19	6.84	7.63	8.91	10.1	11.7	14.6	18.3	22.7	27.2	30.1	32.9	36.2	38.6
20	7.43	8.26	9.59	10.9	12.4	15.5	19.3	23.8	28.4	31.4	34.2	37.6	40.0
21	8.03	8.90	10.3	11.6	13.2	16.3	20.3	24.9	29.6	32.7	35.5	38.9	41.4
22	8.64	9.54	11.0	12.3	14.0	17.2	21.3	26.0	30.8	33.9	36.8	40.3	42.8
23	9.26	10.2	11.7	13.1	14.8	18.1	22.3	27.1	32.0	35.2	38.1	41.6	44.2
24	9.89	10.9	12.4	13.8	15.7	19.0	23.3	28.2	33.2	36.4	39.4	43.0	45.6
25	10.5	11.5	13.1	14.6	16.5	19.9	24.3	29.3	34.4	37.7	40.6	44.3	46.9
26	11.2	12.2	13.8	15.4	17.3	20.8	25.3	30.4	35.6	38.9	41.9	45.6	48.3
27	11.8	12.9	14.6	16.2	18.1	21.7	26.3	31.5	36.7	40.1	43.2	47.0	49.6
28	12.5	13.6	15.3	16.9	18.9	22.7	27.3	32.6	37.9	41.3	44.5	48.3	51.0
29	13.1	14.3	16.0	17.7	19.8	23.6	28.3	33.7	39.1	42.6	45.7	49.6	52.3
30	13.8	15.0	16.8	18.5	20.6	24.5	29.3	34.8	40.3	43.8	47.0	50.9	53.7

* This table is abridged from "Tables of percentage points of the incomplete beta function and of the chi-square distribution," *Biometrika*, Vol. 32 (1941). It is here published with the kind permission of its author, Catherine M. Thompson, and the editor of *Biometrika*.

Table 4 CUMULATIVE F DISTRIBUTION* (m degrees of freedom in numerator; n in denominator)

$$G(F) = \int_0^F \frac{\Gamma\left(\frac{m+n}{2}\right) m^{m/2} n^{n/2} x^{(m-2)/2} (n+mx)^{-(m+n)/2}\,dx}{\Gamma\left(\frac{m}{2}\right)\Gamma\left(\frac{n}{2}\right)}$$

G	n \ m	1	2	3	4	5	6	7	8	9	10	12	15	20	30	60	120	∞
.90		39.9	49.5	53.6	55.8	57.2	58.2	58.9	59.4	59.9	60.2	60.7	61.2	61.7	62.3	62.8	63.1	63.3
.95		161	200	216	225	230	234	237	239	241	242	244	246	248	250	252	253	254
.975	1	648	800	864	900	922	937	948	957	963	969	977	985	993	1000	1010	1010	1020
.99		4,050	5,000	5,400	5,620	5,760	5,860	5,930	5,980	6,020	6,060	6,110	6,160	6,210	6,260	6,310	6,340	6,370
.995		16,200	20,000	21,600	22,500	23,100	23,400	23,700	23,900	24,100	24,200	24,400	24,600	24,800	25,000	25,200	25,400	25,500
.90		8.53	9.00	9.16	9.24	9.29	9.33	9.35	9.37	9.38	9.39	9.41	9.42	9.44	9.46	9.47	9.48	9.49
.95		18.5	19.0	19.2	19.2	19.3	19.3	19.4	19.4	19.4	19.4	19.4	19.4	19.5	19.5	19.5	19.5	19.5
.975	2	38.5	39.0	39.2	39.2	39.3	39.3	39.4	39.4	39.4	39.4	39.4	39.4	39.4	39.5	39.5	39.5	39.5
.99		98.5	99.0	99.2	99.2	99.3	99.3	99.4	99.4	99.4	99.4	99.4	99.4	99.4	99.5	99.5	99.5	99.5
.995		199	199	199	199	199	199	199	199	199	199	199	199	199	199	199	199	199
.90		5.54	5.46	5.39	5.34	5.31	5.28	5.27	5.25	5.24	5.23	5.22	5.20	5.18	5.17	5.15	5.14	5.13
.95		10.1	9.55	9.28	9.12	9.01	8.94	8.89	8.85	8.81	8.79	8.74	8.70	8.66	8.62	8.57	8.55	8.53
.975	3	17.4	16.0	15.4	15.1	14.9	14.7	14.6	14.5	14.5	14.4	14.3	14.3	14.2	14.1	14.0	13.9	13.9
.99		34.1	30.8	29.5	28.7	28.2	27.9	27.7	27.5	27.3	27.2	27.1	26.9	26.7	26.5	26.3	26.2	26.1
.995		55.6	49.8	47.5	46.2	45.4	44.8	44.4	44.1	43.9	43.7	43.4	43.1	42.8	42.5	42.1	42.0	41.8
.90		4.54	4.32	4.19	4.11	4.05	4.01	3.98	3.95	3.93	3.92	3.90	3.87	3.84	3.82	3.79	3.78	3.76
.95		7.71	6.94	6.59	6.39	6.26	6.16	6.09	6.04	6.00	5.96	5.91	5.86	5.80	5.75	5.69	5.66	5.63
.975	4	12.2	10.6	9.98	9.60	9.36	9.20	9.07	8.98	8.90	8.84	8.75	8.66	8.56	8.46	8.36	8.31	8.26
.99		21.2	18.0	16.7	16.0	15.5	15.2	15.0	14.8	14.7	14.5	14.4	14.2	14.0	13.8	13.7	13.6	13.5
.995		31.3	26.3	24.3	23.2	22.5	22.0	21.6	21.4	21.1	21.0	20.7	20.4	20.2	19.9	19.6	19.5	19.3
.90		4.06	3.78	3.62	3.52	3.45	3.40	3.37	3.34	3.32	3.30	3.27	3.24	3.21	3.17	3.14	3.12	3.11
.95		6.61	5.79	5.41	5.19	5.05	4.95	4.88	4.82	4.77	4.74	4.68	4.62	4.56	4.50	4.43	4.40	4.37
.975	5	10.0	8.43	7.76	7.39	7.15	6.98	6.85	6.76	6.68	6.62	6.52	6.43	6.33	6.23	6.12	6.07	6.02
.99		16.3	13.3	12.1	11.4	11.0	10.7	10.5	10.3	10.2	10.1	9.89	9.72	9.55	9.38	9.20	9.11	9.02
.995		22.8	18.3	16.5	15.6	14.9	14.5	14.2	14.0	13.8	13.6	13.4	13.1	12.9	12.7	12.4	12.3	12.1
.90		3.78	3.46	3.29	3.18	3.11	3.05	3.01	2.98	2.96	2.94	2.90	2.87	2.84	2.80	2.76	2.74	2.72
.95		5.99	5.14	4.76	4.53	4.39	4.28	4.21	4.15	4.10	4.06	4.00	3.94	3.87	3.81	3.74	3.70	3.67
.975	6	8.81	7.26	6.60	6.23	5.99	5.82	5.70	5.60	5.52	5.46	5.37	5.27	5.17	5.07	4.96	4.90	4.85
.99		13.7	10.9	9.78	9.15	8.75	8.47	8.26	8.10	7.98	7.87	7.72	7.56	7.40	7.23	7.06	6.97	6.88
.995		18.6	14.5	12.9	12.0	11.5	11.1	10.8	10.6	10.4	10.2	10.0	9.81	9.59	9.36	9.12	9.00	8.88
.90		3.59	3.26	3.07	2.96	2.88	2.83	2.78	2.75	2.72	2.70	2.67	2.63	2.59	2.56	2.51	2.49	2.47
.95		5.59	4.74	4.35	4.12	3.97	3.87	3.79	3.73	3.68	3.64	3.57	3.51	3.44	3.38	3.30	3.27	3.23
.975	7	8.07	6.54	5.89	5.52	5.29	5.12	4.99	4.90	4.82	4.76	4.67	4.57	4.47	4.36	4.25	4.20	4.14
.99		12.2	9.55	8.45	7.85	7.46	7.19	6.99	6.84	6.72	6.62	6.47	6.31	6.16	5.99	5.82	5.74	5.65
.995		16.2	12.4	10.9	10.1	9.52	9.16	8.89	8.68	8.51	8.38	8.18	7.97	7.75	7.53	7.31	7.19	7.08
.90		3.46	3.11	2.92	2.81	2.73	2.67	2.62	2.59	2.56	2.54	2.50	2.46	2.42	2.38	2.34	2.31	2.29
.95		5.32	4.46	4.07	3.84	3.69	3.58	3.50	3.44	3.39	3.35	3.28	3.22	3.15	3.08	3.01	2.97	2.93
.975	8	7.57	6.06	5.42	5.05	4.82	4.65	4.53	4.43	4.36	4.30	4.20	4.10	4.00	3.89	3.78	3.73	3.67
.99		11.3	8.65	7.59	7.01	6.63	6.37	6.18	6.03	5.91	5.81	5.67	5.52	5.36	5.20	5.03	4.95	4.86
.995		14.7	11.0	9.60	8.81	8.30	7.95	7.69	7.50	7.34	7.21	7.01	6.81	6.61	6.40	6.18	6.06	5.95

.90		3.36	3.01	2.81	2.69	2.61	2.55	2.51	2.47	2.44	2.42	2.38	2.34	2.30	2.25	2.21	2.18	2.16
.95		5.12	4.26	3.86	3.63	3.48	3.37	3.29	3.23	3.18	3.14	3.07	3.01	2.94	2.86	2.79	2.75	2.71
.975	9	7.21	5.71	5.08	4.72	4.48	4.32	4.20	4.10	4.03	3.96	3.87	3.77	3.67	3.56	3.45	3.39	3.33
.99		10.6	8.02	6.99	6.42	6.06	5.80	5.61	5.47	5.35	5.26	5.11	4.96	4.81	4.65	4.48	4.40	4.31
.995		13.6	10.1	8.72	7.96	7.47	7.13	6.88	6.69	6.54	6.42	6.23	6.03	5.83	5.62	5.41	5.30	5.19
.90		3.29	2.92	2.73	2.61	2.52	2.46	2.41	2.38	2.35	2.32	2.28	2.24	2.20	2.15	2.11	2.08	2.06
.95		4.96	4.10	3.71	3.48	3.33	3.22	3.14	3.07	3.02	2.98	2.91	2.84	2.77	2.70	2.62	2.58	2.54
.975	10	6.94	5.46	4.83	4.47	4.24	4.07	3.95	3.85	3.78	3.72	3.62	3.52	3.42	3.31	3.20	3.14	3.08
.99		10.0	7.56	6.55	5.99	5.64	5.39	5.20	5.06	4.94	4.85	4.71	4.56	4.41	4.25	4.08	4.00	3.91
.995		12.8	9.43	8.08	7.34	6.87	6.54	6.30	6.12	5.97	5.85	5.66	5.47	5.27	5.07	4.86	4.75	4.64
.90		3.18	2.81	2.61	2.48	2.39	2.33	2.28	2.24	2.21	2.19	2.15	2.10	2.06	2.01	1.96	1.93	1.90
.95		4.75	3.89	3.49	3.26	3.11	3.00	2.91	2.85	2.80	2.75	2.69	2.62	2.54	2.47	2.38	2.34	2.30
.975	12	6.55	5.10	4.47	4.12	3.89	3.73	3.61	3.51	3.44	3.37	3.28	3.18	3.07	2.96	2.85	2.79	2.72
.99		9.33	6.93	5.95	5.41	5.06	4.82	4.64	4.50	4.39	4.30	4.16	4.01	3.86	3.70	3.54	3.45	3.36
.995		11.8	8.51	7.23	6.52	6.07	5.76	5.52	5.35	5.20	5.09	4.91	4.72	4.53	4.33	4.12	4.01	3.90
.90		3.07	2.70	2.49	2.36	2.27	2.21	2.16	2.12	2.09	2.06	2.02	1.97	1.92	1.87	1.82	1.79	1.76
.95		4.54	3.68	3.29	3.06	2.90	2.79	2.71	2.64	2.59	2.54	2.48	2.40	2.33	2.25	2.16	2.11	2.07
.975	15	6.20	4.77	4.15	3.80	3.58	3.41	3.29	3.20	3.12	3.06	2.96	2.86	2.76	2.64	2.52	2.46	2.40
.99		8.68	6.36	5.42	4.89	4.56	4.32	4.14	4.00	3.89	3.80	3.67	3.52	3.37	3.21	3.05	2.96	2.87
.995		10.8	7.70	6.48	5.80	5.37	5.07	4.85	4.67	4.54	4.42	4.25	4.07	3.88	3.69	3.48	3.37	3.26
.90		2.97	2.59	2.38	2.25	2.16	2.09	2.04	2.00	1.96	1.94	1.89	1.84	1.79	1.74	1.68	1.64	1.61
.95		4.35	3.49	3.10	2.87	2.71	2.60	2.51	2.45	2.39	2.35	2.28	2.20	2.12	2.04	1.95	1.90	1.84
.975	20	5.87	4.46	3.86	3.51	3.29	3.13	3.01	2.91	2.84	2.77	2.68	2.57	2.46	2.35	2.22	2.16	2.09
.99		8.10	5.85	4.94	4.43	4.10	3.87	3.70	3.56	3.46	3.37	3.23	3.09	2.94	2.78	2.61	2.52	2.42
.995		9.94	6.99	5.82	5.17	4.76	4.47	4.26	4.09	3.96	3.85	3.68	3.50	3.32	3.12	2.92	2.81	2.69
.90		2.88	2.49	2.28	2.14	2.05	1.98	1.93	1.88	1.85	1.82	1.77	1.72	1.67	1.61	1.54	1.50	1.46
.95		4.17	3.32	2.92	2.69	2.53	2.42	2.33	2.27	2.21	2.16	2.09	2.01	1.93	1.84	1.74	1.68	1.62
.975	30	5.57	4.18	3.59	3.25	3.03	2.87	2.75	2.65	2.57	2.51	2.41	2.31	2.20	2.07	1.94	1.87	1.79
.99		7.56	5.39	4.51	4.02	3.70	3.47	3.30	3.17	3.07	2.98	2.84	2.70	2.55	2.39	2.21	2.11	2.01
.995		9.18	6.35	5.24	4.62	4.23	3.95	3.74	3.58	3.45	3.34	3.18	3.01	2.82	2.63	2.42	2.30	2.18
.90		2.79	2.39	2.18	2.04	1.95	1.87	1.82	1.77	1.74	1.71	1.66	1.60	1.54	1.48	1.40	1.35	1.29
.95		4.00	3.15	2.76	2.53	2.37	2.25	2.17	2.10	2.04	1.99	1.92	1.84	1.75	1.65	1.53	1.47	1.39
.975	60	5.29	3.93	3.34	3.01	2.79	2.63	2.51	2.41	2.33	2.27	2.17	2.06	1.94	1.82	1.67	1.58	1.48
.99		7.08	4.98	4.13	3.65	3.34	3.12	2.95	2.82	2.72	2.63	2.50	2.35	2.20	2.03	1.84	1.73	1.60
.995		8.49	5.80	4.73	4.14	3.76	3.49	3.29	3.13	3.01	2.90	2.74	2.57	2.39	2.19	1.96	1.83	1.69
.90		2.75	2.35	2.13	1.99	1.90	1.82	1.77	1.72	1.68	1.65	1.60	1.54	1.48	1.41	1.32	1.26	1.19
.95		3.92	3.07	2.68	2.45	2.29	2.18	2.09	2.02	1.96	1.91	1.83	1.75	1.66	1.55	1.43	1.35	1.25
.975	120	5.15	3.80	3.23	2.89	2.67	2.52	2.39	2.30	2.22	2.16	2.05	1.94	1.82	1.69	1.53	1.43	1.31
.99		6.85	4.79	3.95	3.48	3.17	2.96	2.79	2.66	2.56	2.47	2.34	2.19	2.03	1.86	1.66	1.53	1.38
.995		8.18	5.54	4.50	3.92	3.55	3.28	3.09	2.93	2.81	2.71	2.54	2.37	2.19	1.98	1.75	1.61	1.43
.90		2.71	2.30	2.08	1.94	1.85	1.77	1.72	1.67	1.63	1.60	1.55	1.49	1.42	1.34	1.24	1.17	1.00
.95		3.84	3.00	2.60	2.37	2.21	2.10	2.01	1.94	1.88	1.83	1.75	1.67	1.57	1.46	1.32	1.22	1.00
.975	∞	5.02	3.69	3.12	2.79	2.57	2.41	2.29	2.19	2.11	2.05	1.94	1.83	1.71	1.57	1.39	1.27	1.00
.99		6.63	4.61	3.78	3.32	3.02	2.80	2.64	2.51	2.41	2.32	2.18	2.04	1.88	1.70	1.47	1.32	1.00
.995		7.88	5.30	4.28	3.72	3.35	3.09	2.90	2.74	2.62	2.52	2.36	2.19	2.00	1.79	1.53	1.36	1.00

* This table is abridged from "Tables of percentage points of the inverted beta distribution," *Biometrika*, Vol. 33 (1943). It is here published with the kind permission of its authors, Maxine Merrington and Catherine M. Thompson, and the editor of *Biometrika*.

Table 5 CUMULATIVE STUDENT'S t DISTRIBUTION*

$$F(t) = \int_{-\infty}^{t} \frac{\Gamma\left(\frac{n+1}{2}\right)}{\Gamma(n/2)\sqrt{\pi n}\left(1+\frac{x^2}{n}\right)^{(n+1)/2}}\, dx$$

n \ F	.75	.90	.95	.975	.99	.995	.9995
1	1.000	3.078	6.314	12.706	31.821	63.657	636.619
2	.816	1.886	2.920	4.303	6.965	9.925	31.598
3	.765	1.638	2.353	3.182	4.541	5.841	12.941
4	.741	1.533	2.132	2.776	3.747	4.604	8.610
5	.727	1.476	2.015	2.571	3.365	4.032	6.859
6	.718	1.440	1.943	2.447	3.143	3.707	5.959
7	.711	1.415	1.895	2.365	2.998	3.499	5.405
8	.706	1.397	1.860	2.306	2.896	3.355	5.041
9	.703	1.383	1.833	2.262	2.821	3.250	4.781
10	.700	1.372	1.812	2.228	2.764	3.169	4.587
11	.697	1.363	1.796	2.201	2.718	3.106	4.437
12	.695	1.356	1.782	2.179	2.681	3.055	4.318
13	.694	1.350	1.771	2.160	2.650	3.012	4.221
14	.692	1.345	1.761	2.145	2.624	2.977	4.140
15	.691	1.341	1.753	2.131	2.602	2.947	4.073
16	.690	1.337	1.746	2.120	2.583	2.921	4.015
17	.689	1.333	1.740	2.110	2.567	2.898	3.965
18	.688	1.330	1.734	2.101	2.552	.2878	3.922
19	.688	1.328	1.729	2.093	2.539	2.861	3.883
20	.687	1.325	1.725	2.086	2.528	2.845	3.850
21	.686	1.323	1.721	2.080	2.518	2.831	3.819
22	.686	1.321	1.717	2.074	2.508	2.819	3.792
23	.685	1.319	1.714	2.069	2.500	2.807	3.767
24	.685	1.318	1.711	2.064	2.492	2.797	3.745
25	.684	1.316	1.708	2.060	2.485	2.787	3.725
26	.684	1.315	1.706	2.056	2.479	2.779	3.707
27	.684	1.314	1.703	2.052	2.473	2.771	3.690
28	.683	1.313	1.701	2.048	2.467	2.763	3.674
29	.683	1.311	1.699	2.045	2.462	2.756	3.659
30	.683	1.310	1.697	2.042	2.457	2.750	3.646
40	.681	1.303	1.684	2.021	2.423	2.704	3.551
60	.679	1.296	1.671	2.000	2.390	2.660	3.460
120	.677	1.289	1.658	1.980	2.358	2.617	3.373
∞	.674	1.282	1.645	1.960	2.326	2.576	3.291

* This table is abridged from the "Statistical Tables" of R. A. Fisher and Frank Yates published by Oliver & Boyd, Ltd., Edinburgh and London, 1938. It is here published with the kind permission of the authors and their publishers.

INDEX